APPLICATIONS INDEX (cont'd)

(Applications index is continued on inside back cover.)

ELEMENTARY STATISTICS

In a World of Applications
SECOND EDITION

ELEMENTARY STATISTICS

In a World of Applications
SECOND EDITION

RAMAKANT KHAZANIE
Humboldt State University

SCOTT, FORESMAN AND COMPANY
Glenview, Illinois   London, England

To the memory of Dada, Aai, and my brother, Suresh

AVAILABLE SUPPLEMENTS

Study Guide: includes chapter summaries and listing of key terms, additional examples and exercises with corresponding solutions, a selection of answers or solutions to chapter tests in the book, and assorted MINITAB applications.

Statistical Toolkit (by Tony Patricelli of Northeastern Illinois University): two-sided 48K Apple II or compatible diskette (color monochrome option) which provides both computational experience and graphical displays for key introductory concepts.

Library of Congress Cataloging-in-Publication Data

Khazanie, Ramakant.
 Elementary statistics in a world of applications.

  1. Statistics. I. Title. II. Title: Elementary statistics.
QA276.12.K5 1986   519.5   85-27654
ISBN 0-673-16618-X

23456-RRC-91 90 89888786

PREFACE

This book is a substantial revision of the first edition. Considerable portions of the text have been entirely rewritten with greater attention to clarity of exposition. Some new material has been added, such as stem-and-leaf diagrams, Chebyshev's Theorem, significance probability, paired t-tests, and prediction intervals. Certain topics have been reorganized to streamline the material and make it more teachable. A large number of drill exercises have been added in appropriate places to prepare the student to tackle word problems. Also, there are many new applied exercises, and the chapter tests have been expanded. An additional feature of this edition is the use of color to highlight important statements and results.

More and more academic disciplines are requiring a course in introductory statistics. Statistical methodology has become an important component of scientific reasoning and the areas of its application are numerous. Statistical techniques are now employed in many fields such as engineering, education, agriculture, business, biology, medicine, fisheries, natural resources, geology, communications, psychology, and ecology. For this reason, the treatment in this book is not slanted toward any particular discipline but touches on diverse areas of interest in all walks of life. The student's acquaintance with the fundamentals of statistics developed here should provide a basis for studying specialized methods within his or her field.

In preparing the second edition, the basic format of the book is left untouched, and the order of coverage is much the same as in the earlier edition. The following changes have shaped this revised edition:

Chapter 1 is substantially reorganized, having fewer sections now. Also, stem-and-leaf diagrams have been added in the graphing section.

Chapter 2 on measures of central tendency and dispersion is a result of considerable rearrangement of topics in Chapters 2 and 3 of the first edition. In particular, mean and variance for grouped data are now discussed in a separate section after having been introduced for ungrouped data in the earlier sections. Also, the definition of sample variance is now in conformity with the more traditional usage where the divisor $n - 1$ is used instead of n. (In the first edition, understandably, the divisor n was an irritant for some users.) Finally, a small discussion on Chebyshev's Theorem is added.

Section 3–2 has been almost entirely rewritten in order to give the student a better feeling for the concepts of sample space and events. Also, more illustrative examples with detailed explanations have been added in the section on counting techniques. In the final analysis, it is my considered opinion that counting techniques *per se* are extraneous to understanding the basic concepts of probability. An instructor who wishes to stay clear of Section 3-4 can do so without impairing a systematic development of later topics. The only place where these techniques will come up in the text is in the treatment of the binomial distribution and its approximation to the normal distribution in Chapter 6. The parts that might be skipped without losing continuity are clearly indicated at the appropriate places.

Chapter 7 on estimation (single population) and Chapter 9 on testing of hypotheses have undergone major changes. One change is that, whereas in the first edition the t distribution, the chi-square distribution, and the F distribution were discussed in one section of the chapter on sampling distributions, in the current edition they are discussed as and when they are needed. Also added, under testing of hypotheses, are significance probability (P-value) and the paired t-test.

Substantial changes have been effected in Section 11–2 on linear regression and linear correlation. The prediction interval for y is an addition to this section.

Chapter 12 on analysis of variance and Chapter 13 on nonparametric tests have also been revised somewhat, though not in a major way.

This textbook covers topics in elementary statistics that, in recent years, have been offered at the freshman-sophomore level in various colleges and universities. The treatment presented here employs mathematics at the most elementary level, with the express purpose of stimulating interest among students who have no special interest in or taste for mathematical formalism. Instead of addressing such formalism, constant appeal is made to the reader's intuition. A background in high school algebra would suffice, although a course in finite mathematics could help the student attain a certain degree of maturity and self-confidence.

In developing the ideas, the overriding concern has been to present the material in as clear a way as possible so that any student who wishes to read it on his or her own can do so with a minimum of difficulty. Each new concept is introduced through an example often drawn from a real-life situation to which the student can easily relate. This introduction is followed by numerous detailed examples. Many of these examples are based on hypothetical data. The book is replete with exercises at the end of each section giving students ample opportunities to test their skills and to fortify their understanding of the subject matter. These exercises serve an additional purpose—to show how pervasively statistical applications are used in everyday life. As a final measure of the competency of the student, a test is provided at the end of each chapter. To derive maximum benefit, the chapter test should be treated as such—a test—and should be attempted only after acquiring a certain degree of proficiency in the topics covered in the chapter.

The contents of this book could be divided into three parts. The first part, consisting of Chapters 1–2, gives an account of descriptive statistics and is intended to familiarize the student with field data summary. The second part, Chapters 3–6, involves elementary probability. In this part the goal is to build the tools essential to understanding the concepts of statistical inference. In the third part, which deals with the inferential aspect of statistics, Chapters 7–9 should be considered by far the most important. Of the remaining chapters, I feel the order of importance is as follows: Chapters 11, 10, 13, 12. However, this will vary from instructor to instructor.

The topics treated in the book should prove adequate for an introductory one-semester course meeting three hours a week or a one-quarter course meeting four hours a week. In my teaching of a one-semester course meeting three hours a week or during a one-quarter course meeting four hours a week, I have invariably covered the first nine chapters. There is enough optional material from which to fashion a two-quarter course meeting three hours a week. The chapters on linear regression and linear correlation, goodness of fit, analysis of variance, and nonparametric methods are included to enable flexibility in structuring such an offering.

In a first course such as this, one can hardly be expected to attain mastery of the vast subject of statistics. This book sets for itself the very modest goal of introducing the student to some aspects of statistical methodology. It is hoped that a student who has acquired a sound understanding of the topics developed here will be in a position to realize the importance of statistical reasoning in work and in life, to discern the basic statistical assumptions underlying a given situation, and to pick an appropriate test if the problem falls within the framework of his or her introduction to the subject. But more than anything, it is hoped that students will be able to interpret the results of a statistical investigation and decipher the barrage of statistical information aimed at them constantly.

In the course of developing this edition, I have availed myself of the many valuable suggestions made by my colleagues and users of the first edition. I have also benefited from the many helpful comments of the reviewers. I express my thanks and appreciation to the following individuals who contributed in many ways to the first edition and this one: Jerald Ball, Chabot College; Leonard Deaton, California State University, Los Angeles; Robert Goosey, Eastern Michigan University; John Skillings, Miami University; Frank C. Hammons, Sinclair Community College; Wilbur Waggoner, Central Michigan University; Shu-ping Hodgson, Central Michigan University; Ian W. McKeague, Florida State University; Jan F. Bjørnstad, University of Texas at Austin; Donald L. Meyer, University of Pittsburgh; Robert J. Lacher, South Dakota State University; Arthur Dull, Diablo Valley College; Donald Fridshal, California State University, Chico; August Zarcone, College of DuPage; Andre Lehre and Victor Tang, Humboldt State University. I extend my special appreciation to my colleague, Professor Charles Biles, who took a keen interest in the project, read the entire manuscript, and made many helpful suggestions. My thanks are also due to Professor Virgil Anderson of Purdue University who inculcated in me a taste and liking for statistical applications. I also express my appreciation to Kathleen McCutcheon, who typed parts of the manuscript for this edition. Finally, I thank my family for enduring patiently during the course of this project.

I am grateful to the Literary Executor of the late Sir Ronald A. Fisher, F.R.S., to Dr. Frank Yates, F.R.S. and to Longman Group, Ltd., London, for permission to reprint Table III from their book, *Statistical Tables for Biological, Agricultural and Medical Research* (6th edition, 1974).

Ramakant Khazanie

CONTENTS

1

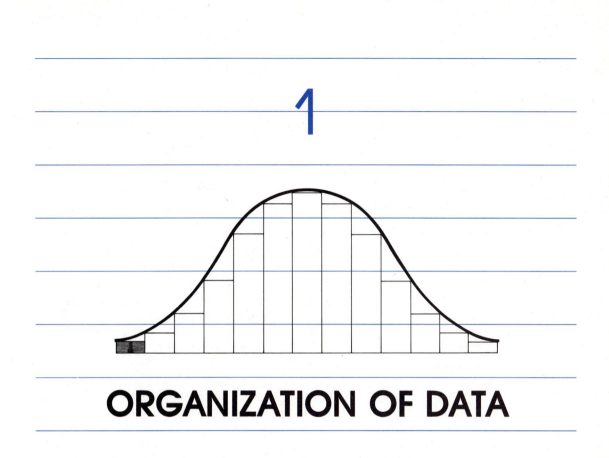

ORGANIZATION OF DATA

INTRODUCTION

What is statistics? Allusions to statistics are abundant and a penchant to quote statistics is almost universal. Strangely enough, however, skepticism about statistics is widespread. Much of this reaction stems from the fact that misleading conclusions are often drawn either from poor planning of an experiment or because some pertinent information is withheld. In this context the following observation is appropriate:

> The secret language of statistics, so appealing in a fact-minded culture, is employed to sensationalize, inflate, confuse and oversimplify. Statistical methods and statistical terms are necessary in reporting the mass data of social and economic trends, business conditions, "opinion" polls, and census. But without writers who use the words with honesty and understanding and readers who know what they mean, the result can only be semantic nonsense.*

Statistics is a science that deals with the methods of collecting, organizing, and summarizing data in such a way that valid conclusions can be drawn from them. Statistical investigations and analyses of data fall into two broad categories—*descriptive statistics* and *inductive statistics.*

Descriptive statistics deals with processing data without attempting to draw any inferences from it. It refers to the presentation of data in the form of tables as in Table 1-1 and graphs as in Figure 1-1 and to the description of some of its features, such as averages, which we will consider in Chapter 2. To most people it is this notion that is conveyed by the word *statistics:* a compilation of a huge mass of data and charts dealing with taxes, accidents, epidemics, farm outputs, incomes, university enrollments, brands of televisions in use, sports statistics, and so on.

TABLE 1-1										
Statistical information presented as a table **Newspaper Advertising—Expenditures for 64 Cities: (in millions of dollars)**										
Type of Advertising	*1957*	*1970*	*1974*	*1975*	*1976*	*1977*	*1978*	*1979*	*1980*	*1981*
Total	1,611	3,120	3,845	4,117	5,352	5,696	6,666	7,641	8,186	9,575
Automotive	63	93	109	93	127	145	151	196	182	226
Classified	313	724	967	982	1,342	1,522	1,892	2,179	2,196	2,515
Financial	42	117	135	131	148	147	203	244	297	387
General	326	426	514	547	731	752	827	982	1,122	1,380
Retail	866	1,759	2,120	2,364	3,005	3,129	3,593	4,040	4,389	5,068
Source: Compiled by Media Records, Inc. Current data in U.S. Bureau of Economic Analysis, *Survey of Current Business*, monthly.										

*From Darrell Huff, *How to Lie With Statistics* (New York: W. W. Norton & Co.), p. 8.

As distinguished from descriptive statistics we have **inductive statistics,** also called **inferential statistics,** which is a scientific discipline concerned with developing and using mathematical tools to make forecasts and inferences. In this role of devising procedures for carrying out analyses of data, statistics caters to the needs of research scientists in such diverse disciplines as industrial engineering, biological sciences, social sciences, marketing research, and economics, to mention just a few. Basic to the development and understanding of inductive statistics are the concepts of probability theory.

In this and the next chapter we shall concentrate specifically on some aspects of descriptive statistics. In subsequent chapters we shall be concerned mainly with inductive statistics, developing concepts in Chapters 3 through 6 and then, in the remainder of the book, applying these to decision-making procedures.

FIGURE 1-1

Statistical information presented through charts.

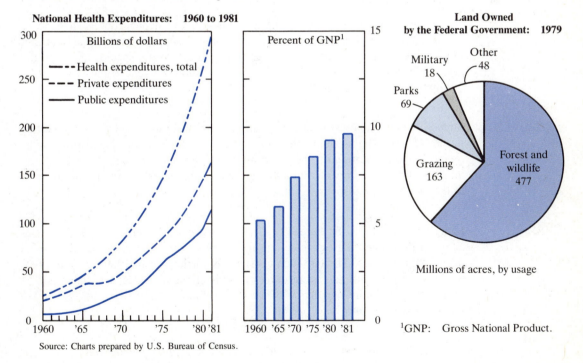

Source: Charts prepared by U.S. Bureau of Census.

1-1 POPULATION, SAMPLE, AND VARIABLES

One of the goals of a statistical investigation is to explore the characteristics of a large group of items on the basis of a few. Sometimes it is physically, economically, or for some other reason almost impossible to examine each item in a group under study. In such a situation the only recourse is to examine a subcollection of items from this group.

For example, if in a primary election we are interested in finding the disposition of the voters toward a certain candidate, it would not be feasible in terms of time and money to interview every potential voter. Instead we might interview only a few of the voters, the selection being carried out in such a way as to provide a good cross section of the voters. Since we are interviewing only a few and drawing conclusions about the entire set of voters, our inferences are subject to an element of uncertainty and so the statements should be couched in the language of probability.

All the conceivable members of a group under study constitute a **population,** or a **universe.** The words *population* and *universe* do not carry the usual dictionary meanings but refer simply to the totality of observations relevant to a given discussion.

What group of items make up a population will depend upon what the researcher is interested in investigating. For example, suppose a retailer is interested in knowing whether the bolts in a large shipment are within specifications. In this case the shipment of bolts constitutes the population relevant to the retailer's investigation.

EXAMPLE 1   Suppose an ornithologist is interested in investigating migration patterns of birds in the Northern Hemisphere. Then all the birds in the Northern Hemisphere will represent the population of interest to him. His choice of the universe restricts him, for it does not include birds that are native to Australia and do not migrate to the Northern Hemisphere. ▬▬

EXAMPLE 2   Every ten years the U.S. Bureau of Census conducts a census of the entire population of the United States accounting for every person regarding sex, age, and other characteristics. The last such census was carried out in 1980. Only the federal government with its vast resources can undertake a task of this magnitude. In this case the entire population of the United States is the population in the statistical sense. ▬▬

A population can be finite or infinite. For a *finite population,* if we start counting the members, the counting process will ultimately come to an end. This is not the case for an *infinite population.* All the examples given above are examples of finite populations. The following are examples of infinite populations: the possible weights of people; the possible temperature readings on a day (before the day starts one can conceive of any possible reading over a range); the possible heights that a balloon can attain.

A **sample** is a subcollection of items drawn from the population under study.

EXAMPLE 3 A fisheries researcher is interested in the behavior pattern of Dungeness crab *(Cancer Magister Dana)* along the Pacific Coast of North America. It would be inconceivable and impossible to investigate every crab individually. The only way to make any kind of educated guess about their behavior would be by examining a small subcollection, that is, a sample. ■

EXAMPLE 4 Suppose a machine has produced 10,000 electric bulbs and we are interested in getting some idea about how long the bulbs will last. It would not be practical to test all the bulbs because the bulbs that are tested for their life will never reach the market. So we might pick 50 of these bulbs to test. Our interest is in learning about the 10,000 bulbs and we study 50. The 10,000 bulbs constitute the population, and the 50 bulbs the sample. ■

Any quantitative measure that describes a characteristic of a population is called a **parameter.** Parameters are the constants that are peculiar to a given population. A quantitative measure that describes a characteristic of a sample is called a **statistic.**

For example, the proportion of Democrats among all the voters in the United States is a parameter. If we interview 1000 people among all the voters in the United States, then the proportion of Democrats in this sample is a statistic.

As another example, suppose we are interested in the heights of the people in the United States at a given time. Then the tallest person at that instant in the United States is a parameter. The height of the tallest person in a sample of 500 people, for example, is a statistic.

Usually access to the entire population is limited by such factors as the time and costs involved, the physical size of the population, and, quite often, the very nature of the investigation, as in Example 4. An understanding of the population is acquired via a sample. When we draw a sample generally we are interested not in studying the sample per se but in extrapolating the nature of the population from which the sample is drawn. On the basis of the findings from the sample, if it is representative of the population, we acquire a better understanding of the population.

Professional pollsters such as Gallup, Lou Harris, and Roper, various advertising agencies interested in consumer preferences, and research scientists invariably use samples to make projections regarding population characteristics, that is, parameters based on their findings from the sample. Thus the aspect of statistics they are dealing with is inferential.

Statistical data or information that we gather is obtained by interviewing people, by inspecting items, or by many other different ways. The characteristic that is being studied is called a *variable*. Heights of people, grades on a high school test, the time it takes for the bus to arrive at the bus stop, and hair color are examples of variables. There are two kinds of variables—*qualitative* and *quantitative*.

A **qualitative variable** can be identified simply by noting its presence. The color of an object is an example of a qualitative variable. In the same way, the outcome on tossing a coin is an example of a qualitative variable. The outcome is either heads or tails and has no numerical value. Although a qualitative variable has no numerical value, it is possible to assign numerical values to a qualitative variable by giving values to each quality. For example, we could assign the value of 1 to heads and the value of 0 to tails and in this way quantify the qualitative information.

A **quantitative variable** consists of numerical values. For example, the height of an individual when expressed in feet or inches is a quantitative variable. Diameters of bolts (in inches) produced by a machine, waiting time at a bus stop (in minutes), the price of a stock (in dollars), the annual income of a family (in dollars), the number of bacteria in a culture, the volume of sales in a day (in dollars) are other examples of quantitative variables.

A familiar example of a variable which is given qualitatively as well as quantitatively is a score on an examination. If the score is given numerically as 80 points, 65 points, and so on, then the variable is quantitative; but if the score is given in letter grades A, B, C, D, F, then the same variable becomes qualitative.

As another example, the height of an individual measured in feet is a quantitative variable; the same variable recorded as tall, medium, or short is a qualitative variable.

We can classify quantitative variables further as continuous or discrete. If the variable can assume any numerical value over an interval or intervals, it is a **continuous variable.** Weight, height, and time are examples of continuous variables. The observations can be measured to any degree of accuracy on a numerical scale. For example, if we choose, we can report the time of travel as 2 hours, 2.5 hours, or 2.5378 hours, although, as a rule, we are not interested in such precision. Loosely speaking, if a variable is continuous, then there are no ''breaks'' in the possible values the variable may assume.

In contrast with a continuous variable, a **discrete variable** is one whose possible values consist of breaks between successive values. For our purpose this definition will suffice. The number of Democrats is an example of a discrete variable, since the only values that the variable can take are integral values such as 0, 1, 2, and so forth. For instance, we would not report that there were 2.5 Democrats in a group. Other examples of discrete variables are the number of bacteria in a culture, the number of defective bulbs, the number of telephone calls received during a one-hour period, and the number of unfilled orders at the end of a day.

A classification of variables can be displayed schematically as follows:

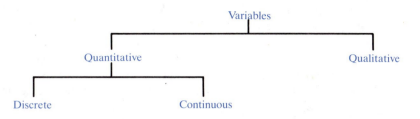

EXAMPLE 5   The following is an excerpt from an article that appeared in Time magazine in the Economy and Business section (Dec. 12, 1983).

> The "little guy" is back in the stock market in a big way and, as it turns out, is not so likely to be a guy. Women have rushed to buy stocks during the past two years, more than men. They constitute 57% of all new shareholders.

The information in the statement "Women have rushed to buy stocks during the past two years, more than men" is qualitative in nature where the variable of interest is the sex of the shareholder—male or female.

Presumably the study is conducted to describe the new shareholders during the period mentioned. Naturally, then, "all new shareholders during the two-year period" would constitute the population.

The statement that women "constitute 57% of all new shareholders" gives a quantitative measure of the percent of women shareholders in the population. Hence 57 percent is the value of the parameter. ▪

EXAMPLE 6   The following table based on data from U.S. International Trade Commission, *Synthetic Organic Chemicals,* annual, is reproduced from the Statistical Abstract of the United States 1982–83.

There are several variables involved in describing the data in Table 1-2.

TABLE 1-2

Synthetic Organic Pesticides—Production and Sales

Item	Unit	1960	1965	1970	1975	1976	1977	1978	1979	1980
Production, total	Mil. lb	648	877	1,034	1,603	1,364	1,388	1,416	1,429	1,468
Herbicides	Mil. lb	102	263	404	788	656	674	664	657	806
Insecticides	Mil. lb	366	490	490	660	566	570	605	617	506
Fungicides	Mil. lb	179	124	140	155	142	143	147	155	156
Production, value[1]	Mil. dol	307	577	1,058	2,900	2,880	3,116	3,342	3,685	4,269
Sales, total	Mil. lb	570	764	881	1,317	1,193	1,263	1,300	1,369	1,406
Sales value[1]	Mil. dol	262	497	870	2,359	2,410	2,808	3,041	3,631	4,078

[1]Manufacturers unit value multipled by production.

Source: U.S. Dept. of Agriculture, Agriculture Stabilization and Conservation Service, *The Pesticide Review,* 1979. Based on data from U.S. International Trade Commission, *Synthetic Organic Chemicals,* annual

One of the variables is the type of pesticide. It is a qualitative variable with three qualities, namely, herbicides, insecticides, and fungicides. Another variable is the number of pounds of each type of pesticide, which is a quantitative variable. In theory we can record weight to any degree of accuracy, so this is a continuous variable. In practice, however, the weights must have been recorded to the nearest pound. Two other variables are production value (in dollars) and sales value (in dollars). They are both quantitative variables and are discrete.

Data collected in an investigation and not organized systematically is called **raw data.** The arrangement of this data in an ascending or descending order of magnitude is called an **array.** (Without the aid of electronic sorting facilities, arranging such data could be very tedious, especially with a large number of observations.) The difference between the largest and the smallest value is called the **range.** For example, suppose the prices of a stock during a five-day trading period were $10, $8, $7, $9, and $12. Since $7 was the lowest price and $12 was the highest price, the range of prices is $12 − $7, or $5.

Table 1-3 records the heights (in inches) of eight students. Column 1 presents the raw data and column 2 illustrates the arrangement in an array. The largest value is 73 and the smallest value is 65. Hence, the range is 73 − 65, or 8 inches.

TABLE 1-3	
Heights presented as raw data and as an array	
Raw data	*Array*
66	65
68	66
72	66
65	68
66	68
73	69
68	72
69	73

A range computed from the sample data gives the *sample range;* it is a statistic. A range computed from the entire population gives the *population range;* consequently it is a parameter.

SECTION 1-1 EXERCISES

1. Which of the following variables are qualitative and which are quantitative?

   (a) The life of a light bulb (in hours)
   (b) The scent of a flower
   (c) A person's nationality
   (d) Monthly telephone bill (in dollars)
   (e) The salary of an individual (in dollars)
   (f) A person's sex
   (g) The tensile strength of an alloy
   (h) The verdict of a jury
   (i) Dividend (in dollars) paid by a company
   (j) Type of security (whether common stock, preferred stock, or bond) owned by an individual

2. Give three other examples each of qualitative and quantitative variables.

3. Indicate which of the following quantitative variables are continuous and which are discrete:

   (a) The life of a light bulb
   (b) The number of children born in a family
   (c) The height of a person
   (d) The number of registered Republicans in some given year
   (e) The salary of an individual
   (f) The speed of a car
   (g) The breaking strength of a certain type of cable
   (h) The length of time one has to wait for a bus to arrive
   (i) The amount of water in a reservoir
   (j) The number of redwood trees on an acre of land

4. Give three other examples each of discrete and continuous variables.

5. An operator at a plant that manufactures bicycle tires is interested in the durability of these tires. Describe what constitutes the population in this case. Is it finite or infinite? Describe a suitable sample that the operator might draw.

6. An opinion survey is conducted to find out the attitude of adult Americans toward decriminalizing marijuana.

   (a) Describe the population. Is it finite or infinite?
   (b) Describe a suitable sample that you might draw.

7. When the final election returns were in, a candidate received 624,725 votes in the state of California. If this figure represents the response in the entire population, for which office did the candidate run? (Two answers possible.) If it represents the response in a sample, what is the office?

8. When a fair coin was tossed 20 times, it showed heads 12 times.

   (a) Describe the population. Is it finite or infinite?
   (b) Describe the sample.

9. The following figures refer to the amount (in dollars) spent on groceries during 20 weeks:

41.10	36.20	32.78	27.63	13.41
29.80	12.85	8.31	34.81	22.67
28.50	52.68	37.38	18.69	64.85
18.60	10.87	23.76	17.68	15.87

Find

(a) the highest amount spent during a week
(b) the lowest amount spent during a week
(c) the range of spending
(d) the number of weeks during which less than $30 were spent.

10. The data below gives the ages of 20 volunteers working in a hospital:

30	28	32	43	62	39	48	41	70	62
28	54	46	49	56	31	52	71	58	53

Find

(a) the range of age
(b) how many volunteers are over the age of 45.

1-2 FREQUENCY DISTRIBUTIONS

In this section we shall consider the organization of data in tabular form to give what are called *frequency distributions*. A statistical table is useful because it summarizes and often readily identifies some of the striking features of the data. Data in frequency distributions may be ungrouped or grouped.

UNGROUPED DATA

As we saw in Section 1-1, an array is an arrangement of data in an ascending or descending order of magnitude. In forming an array, any value is repeated as many times as it appears. The number of times a value appears in the listing is referred to as its **frequency.** Thus if the value 20.5 appears three times, we will list it that many times and call 3 the frequency of the value 20.5. In giving the frequency of a value, we answer the question, "How frequently does the value occur in the listing?"

When the data is arranged in tabular form by giving the frequencies, the table is called a *frequency table*. The arrangement itself is called a **frequency distribution.**

The **relative frequency** of any observation is obtained by dividing the actual frequency of the observation by the total frequency (that is, the sum of all the frequencies). If the relative frequencies are multiplied by 100 and thus expressed as percents, we get what are called **percentage frequencies.**

For example, suppose a value occurs 9 times out of a total of 60 observations. Then the relative frequency of the value is 9/60, or 0.15, and its percentage frequency is 0.15 × 100, or 15 percent.

An advantage of expressing frequencies as relative frequencies is that the frequency distributions of two sets of data can be compared.

EXAMPLE 1

The following data were obtained when a die was tossed 30 times:

1	2	4	2	2	6	3	5	6	3
3	1	3	1	3	4	5	3	5	3
5	1	6	3	1	2	4	2	4	4

Construct a frequency table.

SOLUTION

There are six observed values, namely, 1, 2, 3, 4, 5, 6, and so we get the frequency distribution given in Table 1-4.

TABLE 1-4				
Frequency table giving the number of times each face on the die appears in thirty throws				
Number on the die	*Tally*	*Frequency*	*Relative frequency*	*Percentage frequency*
1	⫻⫻	5	5/30 = 0.1667	16.67
2	⫻⫻	5	5/30 = 0.1667	16.67
3	⫻⫻ ///	8	8/30 = 0.2666	26.66
4	⫻⫻	5	5/30 = 0.1667	16.67
5	////	4	4/30 = 0.1333	13.33
6	///	3	3/30 = 0.1000	10.00
Sum		30	1.0	100.00

The face that shows up is recorded in column 2 by a tally mark (/). The total number of tallies, when counted, gives the frequency. To facilitate counting, the tallies are grouped in blocks of five.

The relative frequency column is obtained by dividing each frequency by 30, the total frequency. The percentage frequency column is obtained by multiplying the corresponding relative frequencies by 100. ■■■

GROUPED DATA, CLASSES OF EQUAL LENGTH

When there is a huge mass of data and when too many of the observed values are *distinct* values, it is inconvenient and almost impossible to give a frequency table based on individual values as we did in Example 1. In such a situation, it is better to divide the entire range of values and group the data into classes. For example, if we are interested in the distribution of ages of people, we could form the classes 0–19, 20–39, 40–59, 60–79, and 80–99. A class such as 40–59 represents all the people with ages between 40 and 59 years, inclusive. When the data are arranged in this way, they are called *grouped data*. The number of individuals in a class is called the *class frequency*.

As an example, consider the figures recorded in Table 1-5, which represent the examination scores of 80 students. (Note that the data are quantitative and discrete.)

TABLE 1-5
Scores earned by 80 students on an examination

62	73	85	42	68	54	38	27	32	63
68	69	75	59	52	58	36	85	88	72
52	52	63	68	29	73	29	76	29	57
46	43	28	32	9	66	72	68	42	76
38	38	39	28	19	12	78	72	92	82
72	33	92	69	28	39	85	59	68	52
85	59	76	80	72	74	54	48	29	36
10	82	58	88	68	58	46	37	29	35

The instructor wants to evaluate the performance of his students. Are the students performing poorly? Are there many good students? From a look at the mass of data given above not much can be said regarding the performance of the students. The problem would be even more unwieldy if there were a thousand students. Therefore, we condense the data.

Our first step is to form classes. We should be careful that pertinent information is not lost by taking too few classes; nor should there be so many classes that the data are not condensed enough to make the arrangement into classes worthwhile. There are different rules to determine the number of classes, but depending upon the bulk of data, we use our own judgment and, as a general guide, choose anywhere from six to twenty classes.

Let us agree to divide the data above into six classes. If we scan the data, we find that the range of scores is from a low of 9 points to a high of 92 points. The difference between the highest and the lowest score is 83 points which is the range that we wish to cover. Since we have agreed to form six classes, we have 83/6 or about 13.8 points. Thus, the length of each class is about 14 points. We shall round this number to 15 points in order to provide convenient interval sizes. With a class length of 15 points, we can form the following classes:

5–19, 20–34, 35–49, 50–64, 65–79, and 80–94.

The first class is set arbitrarily as 5–19. We could have formed the following classes as well:

7–21, 22–36, 37–51, 52–66, 67–81, and 82–96.

We must only be sure that the classes completely cover the entire range of values.

Once the classes are formed, we scan the data systematically and match the values with the classes in which they fall. There is no overlap between classes and, consequently, any observation belongs to only one class. We identify the values by tally marks. The number of values that fall in a class represents the frequency of the class. Table 1-6 gives the classes, the corresponding frequencies, the relative frequencies (obtained by dividing the class frequencies by 80), and the percentage frequencies. It is a good practice to include these four columns to describe a standard frequency distribution.

TABLE 1-6				
Distribution of test scores given in Table 1-5				
Class	*Tally*	*Frequency*	*Relative frequency*	*Percentage frequency*
5–19	////	4	0.0500	5.00
20–34	₸₦ ₸₦ //	12	0.1500	15.00
35–49	₸₦ ₸₦ ₸₦	15	0.1875	18.75
50–64	₸₦ ₸₦ ₸₦ /	16	0.2000	20.00
65–79	₸₦ ₸₦ ₸₦ ₸₦ //	22	0.2750	27.50
80–94	₸₦ ₸₦ /	11	0.1375	13.75
Sum		80	1.0000	100.00

An overall picture emerges from the frequency distribution shown in Table 1-6. For example, we see that the class 65–79 had the maximum number of students, namely, 22; and that 22 + 11 (or 33) students scored over 64 points.

In summary, the following set of steps are suggested for forming a frequency distribution from raw data:

1. *Range.* Scan the raw data and find the smallest and the largest values in the data. The difference between the largest value and the smallest value gives the range.
2. *Number of classes.* Decide on a suitable number of classes, depending upon what information the table is supposed to present.
3. *Class size.* Divide the range by the number of classes. Round this figure to a convenient value to obtain the class size and form the classes. Be sure that the classes completely cover the entire range of values.
4. *Frequency.* Find the number of observations in each class.

EXAMPLE 2   The following data give the amounts (in dollars) spent on groceries by a young couple during 40 weeks.

32	22	19	18	43	42	40	43	18	21
31	26	22	25	47	40	26	32	22	34
28	35	47	26	35	38	35	28	19	38
35	38	36	25	22	45	48	26	34	41

Construct a frequency distribution using seven classes.

SOLUTION   The smallest value is 18 and the largest value is 48. Therefore, the range is 48 − 18, or 30.

Since we are asked to form seven classes, the approximate size of a class interval is 30/7, or 4.29. Rounding to 5, we take the size of a class interval as 5.

A convenient value to start the first class is at 15. Thus the first class would be 15–19. Other classes, each 5 units long, are 20–24, 25–29, . . . , 45–49. The resulting frequency distribution is given in Table 1-7.

TABLE 1-7

Frequency distribution of the amount spent on groceries during 40 weeks

Class	Tally	Frequency	Relative frequency	Percentage frequency
15–19	////	4	0.100	10.0
20–24	////	5	0.125	12.5
25–29	//// ///	8	0.200	20.0
30–34	////	5	0.125	12.5
35–39	//// ///	8	0.200	20.0
40–44	//// /	6	0.150	15.0
45–49	////	4	0.100	10.0
Sum		40	1.000	100.0

GROUPED DATA, CLASSES OF UNEQUAL LENGTHS

It is recommended that as far as possible, all the classes be of the same length. However, this may not always be possible. The class lengths will often be dictated by the nature of the data and by the particular aspect of the distribution that we wish to stress.

Look at the data given in Table 1-8, which lists the prices of 40 stocks on a given day. The smallest price is 10.0 and the highest price is 600. However, a quick glance shows that a majority of the prices are in the range of 10 to 70. If we try to cover the entire range in equal intervals of 10 units each, then we will need about sixty class intervals. Too many! But if we decide to make ten equal classes of 60 dollars each, we will put most of the data in one class interval, namely, 1–60. Such concentration of data in one class will condense the data to such an extent that we will sacrifice relevant information. A reasonable set of classes would be 10.0–19.9, 20.0–29.9, 30.0–39.9, 40.0–49.9, 50.0–59.9, 60.0–69.9, and 70.0–600.0.

TABLE 1-8							
Prices of 40 stocks on a given day							
12.5	36.0	30.6	30.9	30.0	55.8	39.6	48.6
18.5	42.0	26.5	28.5	150.0	200.0	45.0	10.0
28.0	48.0	22.0	22.0	260.0	28.0	120.5	68.2
39.0	51.3	17.6	272.0	85.0	62.0	185.0	70.0
28.7	55.7	600.0	556.0	20.6	538.0	48.0	70.0

CLASS INTERVALS, CLASS MARKS, AND CLASS BOUNDARIES

In Table 1-6 we presented a frequency distribution of the scores earned by eighty students. The blocks such as 5–19, 20–34, and so on, are called **class intervals.** The lower ends of the class intervals are called *lower limits* and their upper ends are called *upper limits*. Thus, 5, 20, 35, 50, 65, and 80 constitute the lower limits and 19, 34, 49, 64, 79, and 94 the upper limits. A class that does not have either an upper limit or a lower limit is called an open-ended class. The **class size,** also called class length or class width, is equal to the difference between two consecutive upper limits around that class. For example, the two consecutive upper limits around the class 20–34 are 34 and 19. Hence the size of the class 20–34 is $34 - 19$, or 15.

The **class mark** is defined as the midpoint of a class interval. It is computed by adding the lower and upper class limits of a class interval and then dividing the sum by 2. Thus,

$$\text{Class mark} = \frac{\left(\begin{array}{c}\text{lower limit} \\ \text{of the class}\end{array} + \begin{array}{c}\text{upper limit} \\ \text{of the class}\end{array}\right)}{2}.$$

The midpoints of the class intervals 5–19 and 20–34 are, respectively,

$$\frac{5 + 19}{2} = 12 \quad \text{and} \quad \frac{20 + 34}{2} = 27.$$

Because the class mark is the midpoint of the class, it serves as a representative value of the class. This is especially realistic if the observations in the class are evenly distributed over the class.

The halfway point between the two class intervals 5–19 and 20–34 is 19.5. Note that 19.5 is equal to $(19 + 20)/2$. For the class intervals 20–34 and 35–49, the halfway point is 34.5.

A point that represents the halfway, or dividing, point between successive classes is called a **class boundary.** So a class boundary is a point where the lower class leaves off and the higher class takes over. In general it is obtained by adding the upper limit of the lower class and the lower limit of the upper class and then dividing by 2, and is reported to one more decimal place than the original data.

TABLE 1-9

Class limits, class boundaries, and class marks for frequency distribution presented in Table 1-6

Class	Lower limit	Upper limit	Lower boundary	Upper boundary	Class mark
5–19	5	19	4.5	19.5	12
20–34	20	34	19.5	34.5	27
35–49	35	49	34.5	49.5	42
50–64	50	64	49.5	64.5	57
65–79	65	79	64.5	79.5	72
80–94	80	94	·79.5	94.5	87

In Table 1-9 we give a summary with reference to the frequency distribution given in Table 1-6. It is worth noting that the upper boundary of one class is the lower boundary of the next.

EXAMPLE 3   The following frequency distribution gives the amount of time (in minutes) that a doctor spent with 28 of his patients:

Class	Frequency
2–6	5
7–11	7
12–16	10
17–27	6

Find

(a) the upper and lower limits of all the classes
(b) the lengths of all the class intervals
(c) the class marks of the classes
(d) the boundaries of the classes.

SOLUTION   (a) For example, the lower end of the class interval 7–11 is 7, and so 7 is a lower limit. The upper end of 7–11 is 11, and therefore 11 is an upper limit. Other lower and upper limits are obtained similarly and are as given in columns 2 and 3 of Table 1-10.

(b) The length of the class interval 7–11 is 11 − 6, or 5, the difference between two consecutive upper limits. The lengths of other class intervals are given in column 4 of Table 1-10.

(c) The class mark of the class 7–11 is obtained by dividing the sum of the lower and upper class limits by 2. Thus the class mark of 7–11 is (7 + 11)/2, or 9. Other class marks are given in column 5 of Table 1-10.

(d) The lower boundary of the class 7–11 is the halfway point between the classes 2–6 and 7–11. Therefore, the lower class boundary of 7–11 is $(6 + 7)/2$, or 6.5. Notice also that 6.5 is the upper boundary of 2–6. We complete columns 6 and 7 following this procedure.

TABLE 1-10

Class limits, class marks, and class boundaries

	Class limits		Length of class interval	Class mark	Class boundaries	
Class	Lower	Upper			Lower	Upper
2–6	2	6	5	4	1.5	6.5
7–11	7	11	5	9	6.5	11.5
12–16	12	16	5	14	11.5	16.5
17–27	17	27	11	22	16.5	27.5

SECTION 1-2 EXERCISES

1. A quality control engineer is interested in determining whether a machine is properly adjusted to dispense one pound (16 ounces) of sugar. The data given below refer to the net weight (in ounces) packed in 30 one-pound bags after the machine was adjusted.

15.6	15.9	16.2	16.0	15.6	16.2	16.2	16.0
15.9	16.0	15.6	15.6	16.0	16.2	15.9	16.2
15.6	15.9	16.2	15.6	16.2	15.8	16.2	16.0
16.0	15.8	15.9	16.2	15.8	15.8		

Arrange the data in a frequency table.
[*Hint:* Notice that there are only a few distinct values.]

2. Prepare a frequency table having the classes 2.2–2.3, 2.4–2.5, 2.6–2.7, 2.8–2.9, 3.0–3.1, 3.2–3.3, 3.4–3.5, and 3.6–3.7 for the following set of data:

3.0	3.2	2.9	3.4	3.3	3.1	3.3	3.5	3.3	2.9
3.2	3.4	3.4	3.0	3.2	3.6	3.4	3.6	3.5	2.2
3.1	3.2	2.9	3.6	3.1	3.0	2.3	2.6	2.7	3.7
2.5	3.3	3.5	2.4	2.6	3.1	2.7	3.6	2.7	3.0

3. Arrange the data given below into a frequency table using the classes 5.00–9.99, 10.00–14.99, 15.00–19.99, 20.00–24.99, 25.00–29.99, and 30.00–34.99:

21.10	16.20	12.78	7.63	13.41	16.48	6.89
9.80	12.85	8.31	14.81	22.67	10.45	8.88
28.50	32.68	17.38	18.69	14.85	11.78	9.99
18.60	10.87	23.76	17.68	15.87	29.99	8.76

4. The weekly earnings of workers in a factory varied from a low of $278.35 to a high of $431.50. If it is agreed to have eight classes with equal class intervals, give suitable classes for grouping the data into a frequency table.

5. The following are the hourly wages (in dollars) of 30 factory workers:

9.60	11.50	8.85	12.20	8.75	9.30	10.10	9.90
9.25	9.10	11.35	11.20	8.90	9.60	9.75	10.25
9.80	10.65	10.15	9.75	11.10	10.15	10.85	9.70
9.35	10.60	10.15	10.60	10.45	11.20		

Construct a frequency distribution by arranging the data into a suitable number of classes.

6. The following figures give the telephone bills (in dollars) of 40 residents of Metroville:

15.80	23.05	17.72	44.18	33.38	23.20	68.50	43.47
34.05	16.10	18.10	29.65	52.25	27.28	68.90	57.12
46.04	27.00	36.07	19.16	17.78	48.19	16.89	26.55
25.28	37.13	18.50	38.59	50.25	40.51	37.51	47.97
33.26	16.81	31.72	65.49	58.64	21.30	34.21	20.40

(a) What percent of the residents paid over $24?
(b) What percent of the residents paid less than $36 but more than $24?
(c) Arrange the data into a frequency distribution.

7. Fifty students were interviewed regarding the number of hours per week that they spend on their studies. The response was grouped in a frequency table and is given below:

Number of hours	Number of students
0–9	3
10–19	7
20–29	17
30–39	12
40–49	10
50 and over	1

Answer the following questions, where possible. Find how many students studied

(a) less than 20 hours   (b) less than 22 hours
(c) less than 29 hours   (d) more than 20 hours
(e) more than 22 hours   (f) more than 29 hours

8. The frequency table below gives the distribution of the amounts (in dollars) of charitable contributions made during a year by 80 families:

Amount contributed	Number of families
0.00–49.99	5
50.00–99.99	14
100.00–149.99	19
150.00–199.99	17
200.00–249.99	12
250.00–299.99	10
300 and over	3

Where possible, find how many families contributed

(a) less than $200   (b) more than $200
(c) $150 or more   (d) less than $250
(e) $50 or more but less than $200.

9. The number of quarts of milk consumed during one week by a certain number of
   families is presented in a frequency table. If the classes are given as 6–10, 11–15,
   16–20, 21–25, 26–30, and 31–35, find

   (a) the class marks   (b) the class boundaries
   (c) the sizes of the class intervals.

10. A civil engineer analyzed the breaking strength (in pounds) of 100 cables. The fol-
    lowing table gives the distribution of her findings:

Breaking strength	Number of cables
1400–1499	8
1500–1599	20
1600–1699	12
1700–1799	35
1800–1899	18
1900–1999	7

    Find

    (a) the class marks   (b) the class boundaries
    (c) the class boundary below which there are 75 percent of the cables
    (d) the class boundary above which there are 60 percent of the cables.

11. Mac, the meat cutter, recorded the amount of ground chuck that he sold on a particu-
    lar day. He sold a total of 90 packages and the following is the distribution of the
    weight (in ounces) of the packages:

Weight	Number of packages
12.1–16.0	8
16.1–20.0	14
20.1–28.0	20
28.1–32.0	23
32.1–40.0	18
40.1–48.0	7

    Find

    (a) the class marks   (b) the class boundaries.

12. The figures below pertain to the distribution of the voltage of 100 batteries:

Voltage	Number of batteries
8.75–8.80	12
8.81–8.88	20
8.89–9.00	28
9.01–9.06	26
9.07–9.15	10
9.16–9.30	4

Find

(a) the width of each class   (b) the class marks

(c) the class boundaries.

13. The distribution of the earnings of a free-lance photographer during 60 months is presented below:

Earnings (in dollars)	Number of months
900–1099	5
1100–1299	9
1300–1499	27
1500–1899	13
1900–2299	2
2300–2699	2
2700–3300	2

(a) Why is it desirable to arrange the frequency distribution with unequal classes?

(b) Find the class marks.

(c) Find the class boundaries.

(d) Find the length of the longest class.

(e) Find the class with the largest frequency.

14. The class marks in a frequency distribution of the tips (in dollars) earned by a waiter during a week are given as

84.5, 104.5, 124.5, 144.5, 164.5.

Assuming equal class sizes, find

(a) the size of each class interval   (b) the class boundaries

(c) each class.

15. The class marks in a frequency distribution of the lives of a number of light bulbs (in hours) are

400, 425, 450, 475, 500, 525.

Assuming that the class intervals are equal, find

(a) the class boundaries   (b) the size of each class interval

(c) each class.

1-3 GRAPHS AND CHARTS

As previously discussed, data presented in tabular form can reveal some features which we would not be able to discern simply by looking at raw data. It is often even more helpful to present this information in a *graphical* form so that it can make a stronger visual impact. Pictorial presentation has the advantage that even a novice without any technical expertise can assimilate the information by looking at a chart or graph.

FIGURE 1-2

Number of students enrolled in the Tippegawa School District during 1980–85

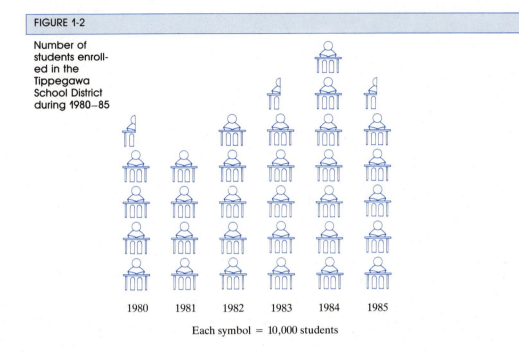

1980   1981   1982   1983   1984   1985

Each symbol = 10,000 students

FIGURE 1-3

Annual budget of Mondavia during the fiscal year 1984–85

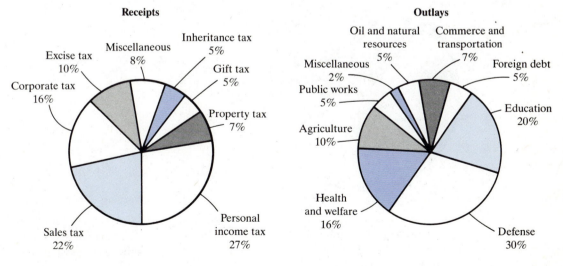

FIGURE 1-4

The annual
budget of
Tippegawa
School District
during 1980–
1985

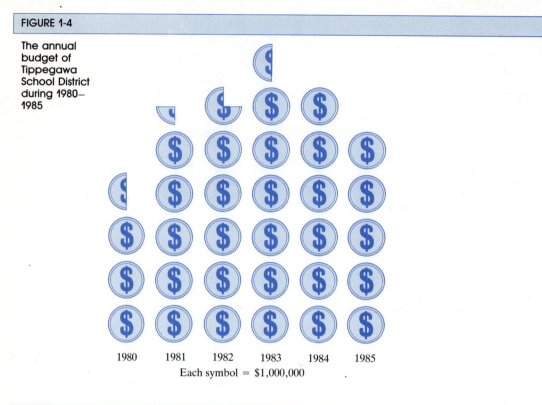

1980   1981   1982   1983   1984   1985
Each symbol = $1,000,000

FIGURE 1-5

Pie chart for
data in
Table 1-11.

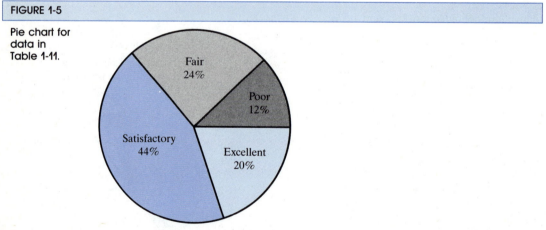

We can make a point pictorially in several ways if we are imaginative enough.
Charts such as Figures 1-2, 1-3, and 1-4 are probably familiar to most people
through their reading of newspapers and magazines.

The diagram in Figure 1-3 is called a **pie chart** for the obvious reason that it resembles a pie cut into wedge-shaped slices. Pie charts are useful for presenting information regarding qualitative variables when we are interested in showing what percentage of the total is accounted for by each category.

Suppose a category has a relative frequency 0.05 or, as a percentage frequency, 5 percent. Since the total number of degrees in a circle is 360 degrees and 5 percent of 360 is 360(0.05), that is, 18 degrees, we allot a central angle of 18 degrees to this category. The pie chart in Figure 1-5 is constructed for the data in Table 1-11, which gives the distribution of responses of 150 patients regarding the relief provided when a pain-killing drug was administered to them.

TABLE 1-11			
Response regarding the relief provided by a pain-killing drug			
Response	*Frequency*	*Relative frequency*	*Central angle (in degrees)*
Excellent	30	0.20	(0.20)(360) = 72
Satisfactory	66	0.44	(0.44)(360) = 158.4
Fair	36	0.24	(0.24)(360) = 86.4
Poor	18	0.12	(0.12)(360) = 43.2
Sum	150	1.00	360

For frequency distributions describing quantitative data, graphical arrangements include the *histogram* and the *ogive*.

THE HISTOGRAM

A **histogram** is a graphic presentation of a frequency distribution, where vertical rectangles are erected on the horizontal axis with the centers of the bases located at the class marks.

A histogram is called a *frequency histogram* when frequencies are plotted along the vertical axis. Instead of frequencies we might plot relative frequencies along the vertical axis to obtain a *relative frequency histogram*. The two histograms can be drawn to look identical by choosing scales along the vertical axis appropriately.

In the definition above a tacit assumption is that all classes are of the same length. If this is not the case, then we should first divide the frequency of each class by the corresponding class size to obtain the **frequency densities** of the classes. Thus,

$$\left(\begin{array}{c} \text{frequency density} \\ \text{of a class} \end{array} \right) = \frac{\text{frequency of the class}}{\text{the class size}}.$$

We now plot frequency densities along the vertical axis to complete the histogram.

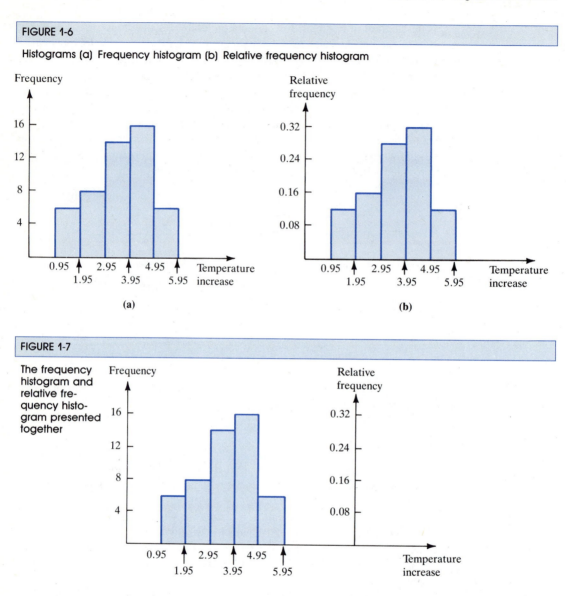

FIGURE 1-6

Histograms (a) Frequency histogram (b) Relative frequency histogram

(a)

(b)

FIGURE 1-7

The frequency histogram and relative frequency histogram presented together

EXAMPLE 1   The first two columns of Table 1-12 present the distribution of temperature increase (in degrees Celsius) of a coolant used in a compressor chamber recorded on 50 occasions.

Draw a frequency histogram.

SOLUTION   Since the class sizes are equal, there is no need to find the frequency densities. We find the class boundaries in column 4 of Table 1-12 and, using a suitable scale, mark them along the horizontal axis. We then use a suitable scale and plot the frequencies along the vertical axis. Finally, in Figure 1-6(a) we erect a rec-

TABLE 1-12

Distribution of temperature increase			
Temperature increase	Frequency f	Relative frequency f/50	Class boundary
			0.95
1.0–1.9	6	0.12	
			1.95
2.0–2.9	8	0.16	
			2.95
3.0–3.9	14	0.28	
			3.95
4.0–4.9	16	0.32	
			4.95
5.0–5.9	6	0.12	
			5.95

tangle of height 6 over the interval 0.95–1.95, a rectangle of height 8 over the interval 1.95–2.95, and so on to obtain the histogram.

In Figure 1-6(b) we obtain exactly the same graph by plotting relative frequencies along the vertical axis. Of course, along the vertical axis we have used a scale which is 50 times that in the first graph.

Both the graphs can be presented together as shown in Figure 1-7. ▬▬

EXAMPLE 2 The first two columns of Table 1-13 below show the distribution of test scores of 77 students. Plot the histogram.

TABLE 1-13

Distribution of test scores			
Class	Frequency f	Boundary point	Frequency density (f/class size)
		4.5	
5–20	8		8/16 = 0.50
		20.5	
21–40	12		12/20 = 0.60
		40.5	
41–55	15		15/15 = 1.00
		55.5	
56–87	40		40/32 = 1.25
		87.5	
88–95	2		2/8 = 0.25
		95.5	

SOLUTION Since the class sizes are unequal, column 4 gives the frequency densities. Column 3 gives the boundary points.

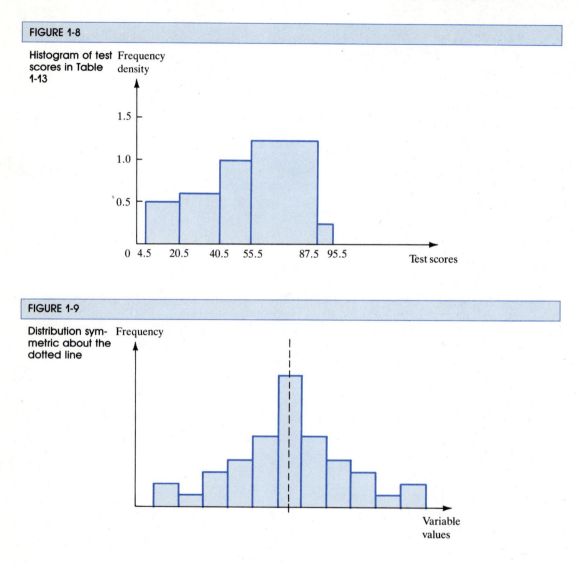

FIGURE 1-8

Histogram of test scores in Table 1-13

FIGURE 1-9

Distribution symmetric about the dotted line

We now use a suitable scale to locate the class boundaries on the horizontal axis. Next, with the intervals between successive class boundaries representing the base, we use a suitable scale and erect rectangles with their heights to correspond to the frequency densities of the respective classes.

Thus, we obtain the histogram in Figure 1-8. ▪

Graphical presentation of data in the form of a histogram will be helpful later in our study of a probability distribution, which is the theoretical counterpart of a frequency distribution.

If the histogram of a distribution is such that when the graph is folded along a vertical axis the two halves coincide (as in Figure 1-9 where the axis is indicated by the dotted line), then the distribution is a **symmetric distribution.** Otherwise it is an **asymmetric distribution.**

For an asymmetric distribution, if there is a long tail on the right-hand side as in Figure 1-10(a), the distribution is said to be *skewed to the right,* and if there is a long tail to the left as in Figure 1-10(b), it is said to be *skewed to the left.*

As we conclude our discussion of the histogram, we mention a rather interesting method of displaying data called the **stem-and-leaf diagram** developed by Professor John Tukey. It gives the actual data in a display that resembles a histogram. First we must write each number in two parts. The first part of a number serves as a *stem;* the part that follows it is a *leaf.* We illustrate the basic approach in the following two examples.

FIGURE 1-10

Asymmetric distributions

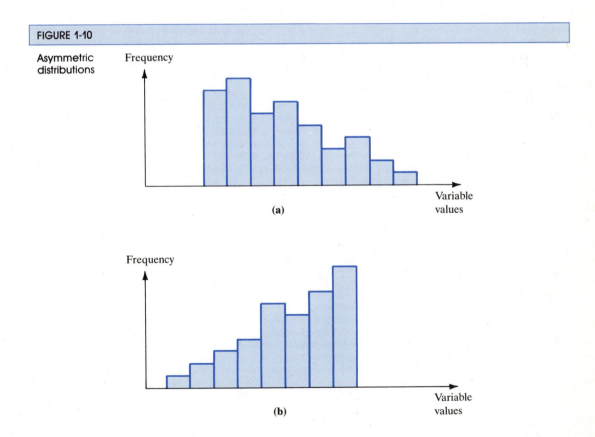

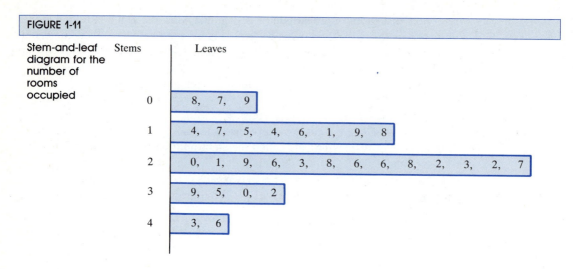

FIGURE 1-11

Stem-and-leaf diagram for the number of rooms occupied

Stems	Leaves
0	8, 7, 9
1	4, 7, 5, 4, 6, 1, 9, 8
2	0, 1, 9, 6, 3, 8, 6, 6, 8, 2, 3, 2, 7
3	9, 5, 0, 2
4	3, 6

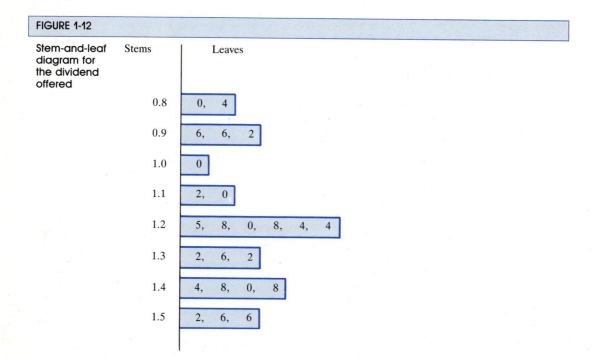

FIGURE 1-12

Stem-and-leaf diagram for the dividend offered

Stems	Leaves
0.8	0, 4
0.9	6, 6, 2
1.0	0
1.1	2, 0
1.2	5, 8, 0, 8, 4, 4
1.3	2, 6, 2
1.4	4, 8, 0, 8
1.5	2, 6, 6

EXAMPLE 3   The following figures represent the number of rooms occupied in a motel on 30 days. Draw a stem-and-leaf diagram.

20	14	21	29	43	17	15	26	8	14
39	23	16	46	28	11	26	35	26	28
30	22	23	7	32	19	22	18	27	9

SOLUTION   We consider each number as a two-digit number. The first digit in each case constitutes a stem; the second, a leaf. For instance, 2 is the stem and 0 is the leaf for the first number 20; 1 is the stem and 4 is the leaf for the second number 14; and so on. Altogether, there are five stems, namely 0, 1, 2, 3, and 4, since our data range from 07 to 46. These stems are written to the left of a vertical line and the corresponding leaves to the right as in Figure 1-11. From the diagram, reading horizontally along stem 3, for instance, we see that 39, 35, 30, and 32 rooms were occupied on four days.   ▬

EXAMPLE 4   Consider the following information on per-share annual dividends offered by 24 corporations listed on the New York Stock Exchange.

1.25	0.80	1.52	1.44	1.56	1.48
1.40	1.48	0.96	1.56	1.28	1.32
1.36	1.32	1.20	1.28	0.96	1.12
1.24	0.84	0.92	1.00	1.10	1.24

Plot a stem-and-leaf diagram.

SOLUTION   Because of the nature of the data, we shall take the stem for each number to be the first part of the number which includes the first decimal place. For example, for the first number 1.25, the part 1.2 will constitute a stem and the part that follows, namely 5, will form a leaf. Scanning the data systematically we obtain the diagram in Figure 1-12.   ▬

THE OGIVE

Before we can sketch an *ogive* (pronounced ō-jīv), we should explain what is meant by a *cumulative frequency distribution*. As the name suggests, it represents accumulated frequencies. For any class boundary a **cumulative frequency distribution** gives the total number of observations which have a value "less than" that boundary.

EXAMPLE 5   The data given in Table 1-14 refer to the net weight of sugar (in ounces) packed in 185 one-pound bags. Obtain the cumulative frequency distribution.

TABLE 1-14

Weight class	Number of bags
15.1–15.3	5
15.4–15.6	10
15.7–15.9	45
16.0–16.2	65
16.3–16.5	46
16.6–16.8	14

TABLE 1-15

The "less than" cumulative frequency distribution

Weight	Cumulative frequency (number of bags)			
less than 15.05	0			
less than 15.35	5			
less than 15.65	15			
less than 15.95	60			
less than 16.25	125			
less than 16.55	171	=	125	+ 46
less than 16.85	185		↑	↑
			Previous cumulative frequency	frequency of the class.

SOLUTION   First, we find the class boundaries and form the "less than" column (column 1) in Table 1-15. To obtain the cumulative frequency distribution we have to determine how many observations are less than any given boundary point. For example, how many bags weigh less than 15.05 ounces? There are none. Therefore, the cumulative frequency of bags weighing less than 15.05 ounces is 0.

The cumulative frequency up to and including the class interval 15.7–15.9, that is, the cumulative frequency of bags weighing less than 15.95 ounces, is 5 + 10 + 45, or 60. Continuing, we get Table 1-15, which gives the cumulative frequency distribution.   ▬

An **ogive** is a line graph obtained by representing the upper class boundaries along the horizontal axis and the corresponding cumulative frequencies along the vertical axis. It is also referred to as a *cumulative frequency polygon*.

For the distribution of sugar weights in Example 5 we obtained the cumulative frequency distribution in Table 1-15. The ogive for this distribution is now plotted in Figure 1-13.

SECTION 1-3   EXERCISES

1. Complete the following table for the categorical data:

Category	Frequency	Relative frequency	Central angle (in degrees)
Category I	85	0.274	(0.274)(360) = 98.7
Category II	156	_____	_____
Category III	69	_____	_____
Sum	310	1.000	360

2. In a genetic experiment, when crosses were made involving a variety of pink-flower plants, it was found that among 800 flowers collected there were 224 red, 380 pink, and 196 white. Construct a pie chart showing these results.

3. The following figures give the distribution of land in a certain county:

Forest land	Farm land	Urban	Other land
30%	40%	10%	20%

Describe the distribution with a pie chart.

4. Of the 200 patients in a hospital, 60 were of blood type O, 75 of type A, 45 of type B, and 20 of type AB. Construct a pie chart to depict the distribution.

5. At the end of a training program, 400 workers in a plant are judged by a panel of examiners as excellent, very good, good, or fair. The following data give the distribution:

Excellent	Very good	Good	Fair
80	140	120	60

Represent the figures by a pie chart.

6. The following are the results of a poll when the question was asked, ''To what degree do you approve of the budget cuts proposed by President Reagan in the fiscal 1986 budget?''

A great deal	21%
Somewhat	45%
Not at all	30%
No opinion	4%

Draw a pie chart to show the response.

FIGURE 1-13

Ogive of the distribution of net weight of sugar

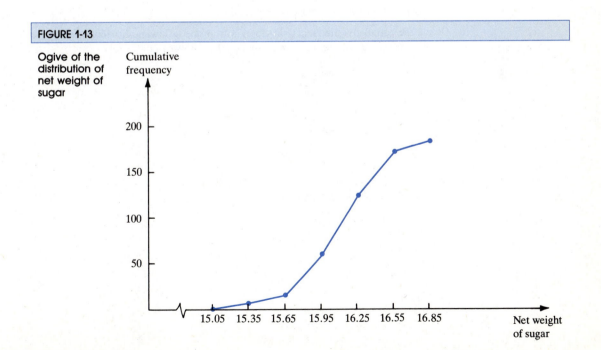

7. Draw the histogram for the following frequency distribution of the lives of 400 light bulbs:

Life of bulb (in hours)	Number of bulbs
600–699	85
700–799	77
800–899	124
900–999	78
1000–1099	36

8. Draw the histogram for the following distribution of the heights of 50 students:

Height (in inches)	Number of students
60–62	4
63–65	12
66–68	14
69–71	9
72–74	11

9. The following measurements were obtained for the leaf lengths (in centimeters) of 30 rhododendron leaves:

10.5	12.5	8.5	7.0	11.5	7.0	11.0	11.0
13.5	15.5	7.2	8.3	12.5	12.0	8.2	6.5
13.5	9.5	12.8	13.2	8.7	9.5	11.7	12.2
7.9	10.3	7.6	14.7	9.1	11.7		

(a) Choose a suitable number of classes and obtain a frequency distribution.

(b) Construct a histogram representing the data.

10. Obtain the cumulative frequency distribution from the following distribution of the number of days absent during a year among 50 employees in a factory:

Number of days absent	Number of employees
0–5	4
6–10	12
11–15	14
16–20	9
21–25	11

11. The following data give the cumulative frequency distribution of the weights (in milligrams) of the adrenal glands of 120 mice:

Weight (in milligrams)	Cumulative frequency
less than 2.45	0
less than 2.95	6
less than 3.45	26
less than 3.95	49
less than 4.45	74
less than 4.95	102
less than 5.45	120

Find how many glands weighed

(a) less than 3.95 milligrams

(b) not less than 4.45 milligrams

(c) 3.45 milligrams or more

(d) less than 4.45 milligrams but 3.45 milligrams or more

12. Draw the ogive of the distribution of the lives of 400 light bulbs given below:

Life of bulb (in hours)	Number of bulbs
600–699	85
700–799	77
800–899	124
900–999	78
1000–1099	36

13. Draw the ogive of the distribution of incomes presented below of 1000 families in an economically depressed area.

Income (in dollars)	Number of families
less than 8000	58
8000–10,999	120
11,000–13,999	278
14,000–16,999	384
17,000–19,999	121
20,000–22,999	39

14. Sketch the ogive of the distribution of the weights of 90 packages of ground chuck given in the following table.

Weight (in ounces)	Number of packages
12.1–16.0	8
16.1–20.0	14
20.1–24.0	20
24.1–28.0	23
28.1–32.0	18
32.1–36.0	7

15. The highway police department conducted a survey and clocked the speeds of 200 cars on a highway. The following distribution was obtained.

Speed (in miles per hour)	Number of cars
48–50	12
51–53	32
54–56	50
57–59	85
60–62	15
63–65	6

(a) Obtain the cumulative frequency distribution.

(b) Draw the histogram of the distribution.

(c) Draw the ogive of the frequency distribution.

16. Plot a stem-and-leaf diagram for the following cholesterol readings (mg/deciliter of blood) of 20 patients:

| 185 | 230 | 195 | 186 | 240 | 190 | 238 | 254 | 225 | 237 |
| 210 | 224 | 214 | 197 | 203 | 233 | 198 | 215 | 216 | 205 |

KEY TERMS AND EXPRESSIONS

descriptive statistics

inductive, or inferential statistics

population, or universe

sample

parameter

statistic

qualitative variable

quantitative variable

continuous variable

discrete variable

raw data

array

range

frequency

frequency distribution

relative frequency

percentage frequency

class intervals

class size

class mark

class boundary

pie chart

histogram

frequency density

symmetric distribution

asymmetric distribution

stem-and-leaf diagram

ogive

cumulative frequency distribution

CHAPTER 1   TEST

1. Define the following terms:

   (a) population; sample

   (b) parameter; statistic

   (c) continuous variable; discrete variable

   (d) qualitative variable; quantitative variable

   (e) frequency; relative frequency; percentage frequency

   (f) class mark; class boundary

   (g) cumulative frequency

2. Classify the following variables as to whether they are qualitative or quantitative. If the variable is quantitative state whether it is continuous or discrete.

   (a) Number of tickets issued by a policeman

   (b) Size of a shirt specified large, medium, small

   (c) Number of letters on a page

   (d) Height of a redwood tree (in feet)

   (e) Depth of an ocean basin (in feet)

   (f) Number of eggs in a bird nest

(g) Number of units a student carries in a semester

(h) Time it takes to solve a certain problem (in minutes)

(i) Number of organizations in the United Nations

(j) Price of a house (in dollars)

(k) Score on the Scholastic Aptitude Test

(l) Color of the eye of a drosophila fly

(m) Attitude toward an election proposition (Possible response: favor, do not favor, indifferent)

3. In a poll conducted by a polling organization for a news magazine, the following question was asked: In your opinion, how likely is it that the United States and the Soviet Union will be able to reach an agreement in the near future to reduce the size of their nuclear arsenal? The response was:

Very likely	7%
Fairly likely	26%
Not too likely	36%
Not at all likely	27%
Don't know	4%

Draw a pie chart to show the response.

4. A laboratory experiment to measure the tensile strength (in tons per square inch) of 30 specimens of an alloy yielded the following readings:

2.58	2.65	2.40	2.46	2.44	2.31	2.30	2.71
2.38	2.46	2.73	2.43	2.39	2.59	2.68	2.73
2.64	2.52	2.80	2.64	2.71	2.73	2.43	2.59
2.48	2.46	2.54	2.68	2.66	2.60		

(a) Arrange the data in a frequency distribution having the classes

2.30–2.38, 2.39–2.47, 2.48–2.56, 2.57–2.65, 2.66–2.74, 2.75–2.83.

(b) Obtain the class marks and class boundaries.

(c) Draw the histogram of the distribution.

(d) Convert the distribution obtained in Part (a) into a cumulative frequency distribution.

(e) Draw the ogive of the frequency distribution.

5. Plot a stem-and-leaf diagram for the data in Exercise 4.

6. The following figures give the distribution of sales (in dollars) made by a mail-order service during the pre-Christmas period:

Sales (in dollars)	Number of orders
0–199	120
200–399	85
400–599	74
600–799	69
800–999	6

(a) Obtain the cumulative frequency distribution.

(b) Draw the histogram of the distribution.

(c) Draw the ogive of the frequency distribution.

7. The data below gives the concentration of a poison (parts per million) found in the brains of 25 birds that were found dead near an industrial zone.

8.2	7.3	14.2	7.6	9.8
8.9	9.6	14.5	10.6	7.5
14.4	12.3	11.5	8.8	11.6
11.9	8.8	14.8	12.2	10.8
12.0	12.5	13.0	14.9	8.4

Plot a stem-and-leaf diagram.

8. The following table records the distribution of the amount of magnesium (mg per liter) found in 60 samples of drinking water.

Amount of magnesium (mg per liter)	Frequency
6.1– 7.0	6
7.1– 8.0	8
8.1– 9.0	14
9.1–10.0	16
10.1–11.0	10
11.1–12.0	6

(a) Obtain the cumulative frequency distribution.

(b) Draw the histogram of the distribution.

(c) Draw the ogive.

2

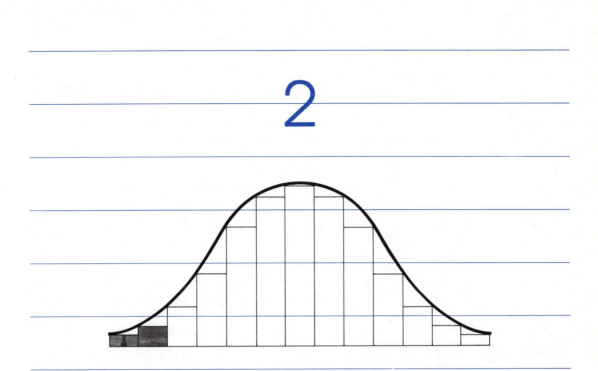

MEASURES OF CENTRAL TENDENCY AND DISPERSION

2-1 Summation Notation

2-2 Measures of Central Tendency

2-3 Comparison of Mean, Median, and Mode

2-4 Measures of Dispersion

2-5 Mean and Variance for Grouped Data

INTRODUCTION

In the preceding chapter we considered useful methods of summarizing and presenting data to bring out an overall picture of the data. We shall now discuss methods of summarizing data by computing values from the observed data. These computations come under two broad headings—*averages,* also called *measures of central tendency,* and *measures of dispersion.*

To the "average" reader the word *average* is familiar from its frequent everyday use. For example, we talk about a batting average in baseball, an average family, the average income of a person, the average annual rainfall, a grade point average, and so on. But most people's definition of an "average family" may only be a vague notion of a "typical" or "representative" family, whatever that means.

In statistical terminology an **average** is a value that typically and effectively represents a given set of data. Since an extreme value, such as the smallest or the largest, is not the most typical observation, a better choice for a representative value should be a number on a numerical scale somewhere at the center of the distribution of the data. For this reason an average is referred to as a **measure of central tendency** or a *measure of location* of the distribution.

The commonly used measures of central tendency are (1) the arithmetic mean, (2) the median, and (3) the mode.

A measure of central tendency describes only one of the important characteristics of a distribution. To understand the distribution well, we must also know the extent of **variability.** The variability of the data is also referred to as the *spread, dispersion,* or *scatter* of the data. Incomes of people with a certain level of education will exhibit some spread; as will heights of people, life lengths of electric bulbs, scores of students on a test, the amount of coffee packed in jars, the number of accidents on a day, and so on.

The commonly used measures of variability are (1) the range, (2) the mean deviation, and (3) the variance (and a related quantity, the standard deviation).

We shall find a notation called *sigma notation,* or *summation notation,* useful in working with averages and dispersion. ■■■

2-1 SUMMATION NOTATION

Suppose we are recording values of a quantitative variable, for instance, the weights of a group of people. Instead of writing, in a cumbersome way, that the weight of the first person is 150 pounds, the weight of the second person is 175 pounds, and so on, we can symbolically write $x_1 = 150$ lb, $x_2 = 175$ lb, and so on, where the letter x stands for the weight variable and the subscript for the person number. With this understanding, x_{50}, for instance, will represent the weight of the fiftieth person.

If we have six people in the group, then we can represent their weights as x_1, x_2, x_3, x_4, x_5, x_6. If we wish to add these weight values, we can write their sum as

$$x_1 + x_2 + x_3 + x_4 + x_5 + x_6.$$

A mathematical shorthand symbol that abbreviates the sum above is

$$\sum_{i=1}^{6} x_i.$$

This mathematical shorthand is called sigma notation, or **summation notation.** The symbol Σ, uppercase Greek sigma, tells us to carry out the summation; $i = 1$ at the bottom of Σ indicates that the subscript of the first term in the summation is 1, and the 6 at the top indicates that the subscript of the last term in the summation is 6.

In general, if there are n terms, or observations, $x_1, x_2, \ldots, x_n$, then $\sum_{i=1}^{n} x_i$ is a shorthand expression for their sum. Thus,

$$\sum_{i=1}^{n} x_i = x_1 + x_2 + \cdots + x_n.$$

(Here the three dots indicate the observations that are not listed.)

EXAMPLE 1

Suppose $x_1 = 10$, $x_2 = 5$, $x_3 = 7$, $x_4 = 9$, $x_5 = 16$, $x_6 = 10$, and $x_7 = 11$. Find

(a) $\sum_{i=1}^{7} x_i$ (b) $\sum_{i=1}^{5} (x_i + 1)$.

SOLUTION

(a) To find $\sum_{i=1}^{7} x_i$ we add every term from the first to the seventh. Hence

$$\sum_{i=1}^{7} x_i = x_1 + x_2 + x_3 + x_4 + x_5 + x_6 + x_7$$
$$= 10 + 5 + 7 + 9 + 16 + 10 + 11 = 68.$$

(b) Here we obtain

$$\sum_{i=1}^{5} (x_i + 1) = (x_1 + 1) + (x_2 + 1) + (x_3 + 1) + (x_4 + 1) + (x_5 + 1)$$
$$= (10 + 1) + (5 + 1) + (7 + 1) + (9 + 1) + (16 + 1)$$
$$= 52. \quad \blacksquare$$

EXAMPLE 2

Suppose $x_1 = 13$, $x_2 = 9$, $x_3 = 10$, $x_4 = 20$, $f_1 = 3$, $f_2 = 5$, $f_3 = 2$, and $f_4 = 7$. Find

(a) $\sum_{i=1}^{4} f_i x_i$ (b) $\sum_{i=1}^{4} f_i x_i^2$ (c) $\sum_{i=1}^{4} f_i (x_i - 10)^2$.

SOLUTION

(a) The sigma notation tells us that we are adding four terms starting with $i = 1$ and ending with $i = 4$. The general term is given as $f_i x_i$, which is $f_1 x_1$ when $i = 1$, $f_2 x_2$ when $i = 2$, $f_3 x_3$ when $i = 3$, and $f_4 x_4$ when $i = 4$. Hence we have

$$\sum_{i=1}^{4} f_i x_i = f_1 x_1 + f_2 x_2 + f_3 x_3 + f_4 x_4$$
$$= 3(13) + 5(9) + 2(10) + 7(20)$$
$$= 244.$$

(b) Arguing as in (a), we have

$$\sum_{i=1}^{4} f_i x_i^2 = f_1 x_1^2 + f_2 x_2^2 + f_3 x_3^2 + f_4 x_4^2$$
$$= 3(13)^2 + 5(9)^2 + 2(10)^2 + 7(20)^2$$
$$= 3912.$$

(c) Here we have

$$\sum_{i=1}^{4} f_i (x_i - 10)^2 = f_1 (x_1 - 10)^2 + f_2 (x_2 - 10)^2 + f_3 (x_3 - 10)^2$$
$$+ f_4 (x_4 - 10)^2$$
$$= 3(13 - 10)^2 + 5(9 - 10)^2 + 2(10 - 10)^2$$
$$+ 7(20 - 10)^2$$
$$= 732. \quad \blacksquare$$

SECTION 2-1   EXERCISES

1. Write out the following sums explicitly. Do not compute the sums.

   (a) $\displaystyle\sum_{i=1}^{6} 3^i$ (b) $\displaystyle\sum_{k=1}^{4} \frac{2}{k + 5}$ (c) $\displaystyle\sum_{j=1}^{4} (2^j + j^2)$

   (d) $\displaystyle\sum_{j=1}^{5} j(j - 1)$ (e) $\displaystyle\sum_{k=1}^{4} (k^2 + 2k - 1)$ (f) $\displaystyle\sum_{j=1}^{5} \frac{j}{j + 3}$

2. Compute $\displaystyle\sum_{i=1}^{5} x_i$ if $x_1 = 2.2$, $x_2 = 3.8$, $x_3 = 4.9$, $x_4 = 2.5$, and $x_5 = 4.9$.

3. Compute $\displaystyle\sum_{i=1}^{6} (x_i - 4)$ if $x_1 = 5$, $x_2 = 7$, $x_3 = -2$, $x_4 = 3$, $x_5 = 0$, $x_6 = 4$.

4. Suppose $x_1 = 3$, $x_2 = 5$, $x_3 = -2$, $x_4 = 5$, $x_5 = -8$, and $x_6 = 4$. Compute

   (a) $\displaystyle\sum_{i=1}^{4} x_i$ (b) $\displaystyle\sum_{i=1}^{6} x_i + 1$ (c) $\displaystyle\sum_{i=1}^{6} (1 + x_i)$

(d) $\displaystyle\sum_{i=1}^{4} (x_i + 1)^2$ (e) $\displaystyle\sum_{i=1}^{4} x_i^2 + 1$ (f) $\displaystyle\sum_{i=1}^{6} x_i(x_i - 1)$.

5. Suppose $x_1, x_2, \ldots, x_n$ and $y_1, y_2, \ldots, y_n$ are given values. Write the following expressions using summation notation:

(a) $3x_1 + 3x_2 + \cdots + 3x_n + 2y_1 + 2y_2 + \cdots + 2y_n$

(b) $2x_1^2 + 2x_2^2 + \cdots + 2x_n^2 + 2y_1^2 + 2y_2^2 + \cdots + 2y_n^2$

(c) $2(x_1 + y_1)^2 + 2(x_2 + y_2)^2 + \cdots + 2(x_n + y_n)^2$

(d) $x_1^2 y_1 + x_2^2 y_2 + \cdots + x_n^2 y_n$.

6. (a) Suppose $x_1, x_2, x_3, x_4,$ and x_5 are given numbers. Write the sum $\displaystyle\sum_{i=1}^{5} (3x_i - 2)^2$ explicitly without sigma notation.

(b) Compute $\displaystyle\sum_{i=1}^{5} (3x_i - 2)^2$ if $x_1 = 2$, $x_2 = 3$, $x_3 = -1$, $x_4 = 0$, and $x_5 = 6$.

7. If $x_1 = 3$, $x_2 = -4$, $x_3 = 5$, $x_4 = -2$, $f_1 = 2$, $f_2 = 3$, $f_3 = 10$, and $f_4 = 1$, compute

(a) $\displaystyle\sum_{i=1}^{4} f_i x_i$ (b) $\displaystyle\sum_{i=1}^{4} f_i x_i^2$ (c) $\left(\displaystyle\sum_{i=1}^{4} f_i x_i\right)^2$ (d) $\displaystyle\sum_{i=1}^{4} (f_i x_i)^2$.

8. If $x_1 = 3$, $x_2 = -4$, $x_3 = 5$, $x_4 = -2$, $y_1 = -2$, $y_2 = -3$, $y_3 = 2$, and $y_4 = 4$, compute

(a) $\displaystyle\sum_{i=1}^{4} x_i y_i$ (b) $\displaystyle\sum_{i=1}^{4} (x_i - 1)(y_i - 3)$

(c) $\left(\displaystyle\sum_{i=1}^{4} x_i\right)\left(\displaystyle\sum_{i=1}^{4} y_i\right)$ (d) $\displaystyle\sum_{i=1}^{4} x_i^2 y_i^2$.

9. If $\displaystyle\sum_{i=1}^{10} x_i = 238.6$, $x_1 = 12.6$, and $x_{10} = -28.0$, find

(a) $\displaystyle\sum_{i=2}^{10} x_i$ (b) $\displaystyle\sum_{i=2}^{9} x_i$.

10. If $\dfrac{\displaystyle\sum_{i=1}^{10} x_i}{10} = 22.6$ and $\displaystyle\sum_{i=1}^{9} x_i = 188.2$, find x_{10}.

11. If $\displaystyle\sum_{i=1}^{10} x_i = 150$ and $\displaystyle\sum_{i=1}^{10} x_i^2 = 2350$, compute

$$\sum_{i=1}^{10} x_i^2 - \frac{\left(\displaystyle\sum_{i=1}^{10} x_i\right)^2}{10}.$$

12. If $x_1 = 1$, $x_2 = 2$, $x_3 = 4$, and $y_1 = -1$, $y_2 = 0$, $y_3 = 2$, compute

$$\sum_{i=1}^{3} x_i y_i - \frac{\left(\displaystyle\sum_{i=1}^{3} x_i\right)\left(\displaystyle\sum_{i=1}^{3} y_i\right)}{3}.$$

2-2 MEASURES OF CENTRAL TENDENCY

THE ARITHMETIC MEAN

The *arithmetic mean* for any set of measurements is computed by adding all the values in the data and then dividing this total by the number of values in the sum.

Thus, if the amounts spent on groceries during four weeks are given as $14, $17, $24, and $9, then the mean amount (in dollars) spent per week is equal to

$$\frac{14 + 17 + 24 + 9}{4} = 16.$$

If the data represent the entire population, then, naturally, the mean of the values is referred to as the **population mean** and is customarily denoted by the Greek letter μ. (A technicality involved in this connection will be clarified later in Example 8 in Section 4-2.) Since this mean is a numerical measure describing a population characteristic, it is, of course, a parameter.

On the other hand, if the data constitute a sample drawn from a larger population, then the mean is referred to as the **sample mean** which is, therefore, a statistic. If the variable values are denoted by x, the symbol customarily used for the sample mean is $\bar{x}$ (read *x bar*). Usually the population data will not be available in practice and the data at hand will be measurements from a sample. It is for this reason that the symbol $\bar{x}$ is commonly encountered in reference to the mean.

THE ARITHMETIC MEAN

Suppose there are n observations with numerical values which we denote as $x_1, x_2, \ldots, x_n$. Then, in sumation notation,

$$\text{mean} = \frac{\text{sum of values}}{\text{number of values}} = \frac{\sum_{i=1}^{n} x_i}{n}.$$

The statistical function $\bar{x}$ is built into many scientific pocket calculators and can be obtained by pressing the appropriate keys.

EXAMPLE 1   Six motorists stopping at a gas station bought the following amounts of gasoline (in gallons):

6.2,   10.8,   16.5,   8.0,   18.4,   4.5.

Calculate the mean amount of gasoline bought.

SOLUTION   The mean amount of gasoline (in gallons) bought is

$$\frac{6.2 + 10.8 + 16.5 + 8.0 + 18.4 + 4.5}{6} = \frac{64.4}{6} = 10.73.$$

A word about rounding: As a general rule, in any set of computations, rounding should never be carried out in the intermediate steps but only when presenting the final answer. Other than this, the number of significant figures that should appear in the final answer depends on the problem involved. In presenting the mean we shall agree to have one significant place more than in the original data. If the mean is to be used in subsequent formulas, then as many significant figures as possible should be retained.

Properties of the Mean

The mean has the following important properties:

1. Suppose there are two sets of data and the mean for the first set is $\bar{x}$ and that for the second set is $\bar{y}$. *If the two sets have the same number of observations,* then

$$\text{the mean of the combined data} = \frac{\bar{x} + \bar{y}}{2}.$$

This result is not true if the number of observations are different.

2. If a fixed value d is added to each of the observations in the data, then

$$\text{the mean of the new data} = d + \text{mean of the old data}.$$

For example, suppose the mean monthly salary of the employees working for a company is 1600 dollars. If each employee gets a monthly raise of 100 dollars, then the mean monthly salary after the raise will be $100 + 1600$, or 1700 dollars.

3. If each observation in the data is multiplied by a fixed constant c, then

$$\text{the mean of the new data} = c \ times \text{ the mean of the old data}.$$

For example, suppose the height of students in a classroom is measured in feet and the mean height is found to be 4.8 feet. If it is desired to express the mean height in inches, then the mean height is 12 (4.8), or 57.6 inches. (This saves us the trouble of converting the height of each student into inches and then finding the mean.)

THE MEDIAN

The *median* is another commonly used measure of central tendency. The reader might have heard the median mentioned in many contexts such as median annual salary, median age of college graduates, median family income, and so on. Median conveys the notion of the middlemost value, dividing the distribution into two halves.

When there is an *odd* number of observations in the data the **median** of a set of values is defined as the middle value when the observations are arranged in an array in order of magnitude from the smallest to the largest (or vice versa).

For example, suppose we want to find the median amount spent on groceries when seven customers spent the following amounts (in dollars):

15,   18,   10,   4,   27,   5,   32.

Arranging the data in an ascending order of magnitude, we get

Customer number	1	2	3	4	5	6	7
Amount spent	4	5	10	15	18	27	32

bottom 3 values   median   top 3 values

Three of the customers spent less than 15 dollars and three spent more than 15 dollars. Therefore, the median amount spent on groceries is 15 dollars.

If the number of observations is *even,* then the **median** is defined as the mean of the two middle values.

As an example, suppose we want to find the median amount when eight customers spent the following amounts (in dollars):

15,   18,   10,   4,   27,   5,   32,   42.

Arranging these data in an ascending order of magnitude, we get

Customer number	1	2	3	4	5	6	7	8
Amount spent	4	5	10	15	18	27	32	42

$$\text{median} = \frac{15 + 18}{2} = 16.5$$

Since there is an even number of observations, the median is the mean of the two middle observations, 15 and 18. Therefore, the median is $(15 + 18)/2$, or 16.5 dollars.

With proper interpretation one can define the median of n observations as the $\frac{n + 1}{2}$ th item in the array (ascending or descending). The proper interpretation is called for when n is even. For instance, when $n = 8$, then $(n + 1)/2 = (8 + 1)/2 = 4.5$. In this case, we will interpret the median as the mean of the 4th and 5th observations, the two ordinal numbers flanking 4.5. [*Caution:* The median is *not* $(n + 1)/2$. It is the $\frac{n + 1}{2}$ th largest observation.]

EXAMPLE 2   Find the median of 14, 15, 14, 14, and 23.

SOLUTION   The data can be arranged in an ascending order of magnitude as 14, 14, 14, 15, and 23. There are 5 observations—an odd number. The median is the third value, since $(5 + 1)/2 = 3$. The median is 14. ■

EXAMPLE 3 The numbers of school years completed by ten adults were given as:

6,   5,   15,   6,   14,   0,   18,   6,   11,   10.

Find the median school years completed.

SOLUTION We first arrange the data in an ascending order of magnitude as follows:

Adult number	1	2	3	4	5	6	7	8	9	10
Years completed	0	5	6	6	6	10	11	14	15	18

$$\text{median} = \frac{6 + 10}{2} = 8 \text{ years}$$

There are ten adults—an even number. Since $(10 + 1)/2 = 5.5$, the median is equal to the mean of the 5th and 6th values, which are 6 and 10, respectively. Their mean gives the median as $(6 + 10)/2$, or 8 years. ■

While we are discussing the median, it is appropriate to introduce some other measures of location as well. These are *percentiles* and *quartiles*.

A **percentile** is a value below which a given percent of the observations in a distribution lie.

For example, the 35th percentile of a distribution is a value such that at least 35 percent of the observations are less than or equal to it and at least 65 percent of the observations are larger than or equal to it.

The 25th percentile is the first quartile (or lower quartile) and the 75th percentile is the third quartile (or upper quartile). The median, which is the 50th percentile, is also the second quartile.

THE MODE

The word *mode* in French means "fashionable," and in the context of a frequency distribution, mode means the most common value. The **mode** is therefore defined as the value that occurs most frequently in the data. Of all the measures of central tendency, the mode is the easiest to obtain once the data are arranged in a frequency table. We find it simply by inspection.

It may happen that there are some values in the data that occur equally often, but more frequently than the rest of the values. Each of these values is a mode. For example, suppose the data consist of

1.0,   1.1,   1.2,   1.2,   1.2,   1.4,   1.4,   1.4.

The values 1.2 and 1.4 occur equally often, namely, three times, which is more often than the other observations 1.0 and 1.1. Our data have two modes, 1.2 and

1.4. Therefore, the data are called *bimodal*. In general, if the data have more than one mode, they are said to be *multimodal*.

On the other hand, the data might consist of values each of which occurs the same number of times. In this case there exists no mode. For instance, the set of values 1.2, 1.2, 1.4, 1.4, 2.0, and 2.0 has no mode. Thus, it is possible for a given set of data to have more than one mode or no mode at all.

EXAMPLE 4   Discuss the nature of the mode for the following sets of data:

(a) 2, 3, 7, 7, 7, 8, 9, 9

(b) 2, 3, 7, 7, 7, 7, 8, 9, 9, 9, 9

(c) 22, 23, 32, 40, 49

SOLUTION   (a) Since 7 is the value that occurs most often (three times), 7 is the mode. Also, 7 is the only mode.

(b) This set of values has two modes, 7 and 9. They each occur four times. Thus, the data are bimodal.

(c) This set of values has no mode because there is no one particular value that occurs more often than any other.   ▆▆▆

When the data are grouped in a frequency distribution, the class with the highest frequency is called the *modal class*. The class mark of the modal class is then called the mode. Of course in this context we must require that all the classes be of the same length.

SECTION 2-2   EXERCISES

1. The scores of five students were 52, 68, 76, 55, and 84. Find the arithmetic mean of the scores.

2. The mean speed of six race drivers was found to be 120 miles per hour. It was realized, however, that the speed of one of the drivers was not recorded and that the speeds of the other five drivers were 140, 110, 120, 130, and 110 miles per hour. Find the missing speed.

3. Among forty people, fifteen weighed 150 pounds each, twenty weighed 170 pounds each, and five weighed 200 pounds each. Find the mean weight.

4. A company employs 80 junior executives, 26 senior executives, and 8 vice-presidents. During one year the mean salary of the junior executives was $48,000, that of the senior executives was $58,000, and that of the vice-presidents was $88,000. Find the mean salary of the 114 employees combined.

5. In a class of ten students, the lowest score on a test was 40 points and the highest score was 80 points. State which of the following statements are true:

(a)   The mean score is $(80 + 40)/2$, or 60.

(b) The mean score is at least 40.

(c) The mean score is at most 80.

6. During the first ten weeks of 1984, the mean weekly receipts of a city transit company were $840,000. In the next ten weeks, the mean weekly receipts were $860,000. Find the mean weekly receipts for the first twenty weeks.

7. If the mean weekly receipts of a transit company for the first ten weeks were $840,000 and for the next twenty weeks were $860,000, find the mean weekly receipts during the first thirty weeks.

8. The mean weight of ten items in one box was 30 pounds, of fifteen items in a second box was 60 pounds, and of twenty items in a third box was 20 pounds. Find the mean weight of all the items in the three boxes.

9. A warehouse must store 18,000 gallons of a lubricant. If there are 310 containers available with a mean capacity of 59 gallons per container, are there enough containers?

10. The mean score of a student on the first three tests is 81 points. How many points must she score on the next two tests if her overall mean score has to be 85 points?

11. Tom traveled a distance of 240 miles at a mean speed of 40 miles per hour. He then returned the same distance at a mean speed of 60 miles per hour. Find his mean hourly speed for the entire trip.

12. (a) Suppose the mean of the first twenty observations is 29.8. What would the next observation have to be for the mean of the first twenty-one observations to be 30?

    (b) Suppose the mean of the first 1000 observations is 29.8. What would the next observation have to be for the mean of the first 1001 observations to be 30?

    If you compare your answers in Parts (a) and (b) above, what conclusion can you draw?

13. Using the formula for the mean, find the mean of the data in Part (a). *Use this answer* to find the means of the data in (b) and (c).

    (a) 24, 32, 28, 26, 25   (b) 74, 82, 78, 76, 75

    (c) 10,124, 10,132, 10,128, 10,126, 10,125

    [*Hint:* The data in Part (b) are obtained by adding 50 to each observation in Part (a); and the data in Part (c) are obtained by adding 10,100 to each observation in Part (a).]

14. Using the formula for the mean, find the mean of the data in Part (a). *Use this answer* to find the means of the data in Parts (b) and (c).

    (a) 8, 6, 9, 5

    (b) 24, 18, 27, 15

    (c) 56, 42, 63, 35

    [*Hint:* The data in Part (b) are obtained by multiplying by 3 each observation in Part (a); and the data in Part (c) are obtained by multiplying by 7 each observation in Part (a).]

15. A worker made ten flags. The mean length of cloth per flag was found to be 2.5 yards. Express the mean length in inches.

16. During the month of February, the mean daily temperature of a ski resort was found

to be 20 degrees Fahrenheit. Express the mean in degrees Celsius.

> [*Hint:* A Fahrenheit reading is converted to a Celsius reading by first sub-
> tracting 32 and then multiplying by 5/9.]

17. A quality control engineer was interested in whether a machine was properly adjusted to dispense one pound of sugar in each bag. Thirty bags filled on the machine were weighed. The data given below refer to the net weight of sugar (in ounces) packed in each bag.

15.6	15.9	16.2	16.0	15.6	16.2	16.2	16.0
15.9	16.0	15.6	15.6	16.0	16.2	15.9	16.2
15.6	15.9	16.2	15.6	16.2	15.8	16.2	16.0
16.0	15.8	15.9	16.2	15.8	15.8		

Find the mean weight per bag.

18. The number of guests in a hotel during nine weeks were

$$455, \quad 150, \quad 327, \quad 169, \quad 321, \quad 2568, \quad 398, \quad 1352, \quad 387.$$

Find the median number of guests.

19. The following are the speeds (in miles per hour) attained by eight race drivers:

$$128, \quad 140, \quad 126, \quad 137, \quad 148, \quad 130, \quad 142, \quad 133.$$

Find the median speed.

20. The following figures refer to the diameters (in inches) of six bolts produced by a machine:

$$2.0, \quad 2.3, \quad 2.1, \quad 2.0, \quad 2.1, \quad 2.1.$$

Find the median diameter.

21. Give two different sets of data, each with five observations and the same median.

22. Suppose we have two sets of data with no elements in common. If each set has an odd number of observations, is it possible for the two sets to have the same median? Explain. What if there is an even number of observations?

23. A Girl Scout who visited 30 households obtained the following contributions (in dollars):

0.50	0.75	1.00	0.50	0.25	0.75	1.50	0.50	0.25
0.75	1.50	0.50	0.25	0.75	0.25	0.75	0.50	0.50
0.50	1.50	0.50	0.25	0.75	0.25	0.25	1.50	0.75
1.50	0.25	0.75						

Find the median amount contributed.

24. The median height of ten fifth-grade students is 57 inches and the median height of seven sixth-grade students is 59 inches. What information would you need in order to find the median height of the seventeen students in the two grades?

25. The following figures give the diameters (in feet) at the base of the trunks of 20 lumber trees in a forest:

2.8	3.3	2.4	3.9	4.5	1.8	2.7	4.8	3.7	1.6
2.3	3.8	3.3	4.1	3.7	4.3	1.9	3.8	2.2	4.9

Find

(a) the median diameter   (b) the mean diameter.

26. During the three-hour period from 9 A.M. to 12 noon on a certain day, a bank teller observed that customers arrived at the following hours:

    9:05,   9:35,   9:50,   10:15,   10:30,   10:40,   10:55,   11:15,
    11:28,   11:40,   11:58.

    If the time between any two successive arrivals is called *interarrival time,* find

    (a) the mean interarrival time   (b) the median interarrival time.

    Find the mode(s), if they exist, for the sets of data in Exercises 27-33.

27. 14, 15, 16, 14, 14, 15, 16, 10

28. 14, 14, 13, 12, 15, 15

29. 14, 17, 18, 27, 26

30. 4.2, 3.5, 3.9, 3.9, 3.8, 3.9, 4.2, 4.2

31. $-3.1, -4.0, -5.3, -6.8, -4.0, -6.8$

32. 2.8, 2.8, 3.1, 4.5, 3.1, 2.8, 3.1

33. $-4, -5, -7, -7, -5, -7, -4, -4, -5$

34. On a fishing trip Harry caught ten trout. The following figures refer to their sizes (in inches):

    10.2,   11.5,   10.6,   10.6,   12.2,   11.5,   12.1,   11.5,
    10.2,   10.2

    Find the mode if it exists.

35. A hardware store has in stock nails of the following different sizes (in inches):

    0.5,   1.0,   1.5,   2.0,   2.5,   3.0,   3.5,   4.0,   4.5,
    5.0

    If 10 percent of the nails are of size 2.0, 15 percent are of size 3.0, and 53 percent are of size 3.5, can you find the mode? Justify your answer.

2-3 COMPARISON OF THE MEAN, MEDIAN, AND MODE

If the distribution of the data is symmetric with a single mode, then the mean, median, and mode will coincide, as in Figure 2-1(a). However, if the distribution is skewed, the three measures will differ from each other, as in Figure 2-1(b).

Of the three measures of central tendency that we have discussed, the arithmetic mean is by far the most commonly used measure and the mode, the least used. The mode is not even well-defined in many data sets since it may not be unique or, for some sets of data, may not even exist. We shall list below the advantages and disadvantages of these measures:

1. The mean can always be computed when all the measurements are available, and every observation is involved in its computation. Computation of the median or the mode does not take into account individual values in the data. If the mode exists, it is the easiest measure to find.

2. From the standpoint of inductive statistics the mean is more tractable mathematically than either the median or the mode and has many desirable properties for sound tests. This aspect of the mean will be discussed in greater detail in Chapter 7.

3. The mean is sensitive to extreme values, which are not representative of the rest of the data; that is, it is strongly affected by an extremely large value or an extremely small value. This disadvantage does not occur with either the median or the mode, which are positional averages. For example, the median of 100, 200, and 300 is 200, and the median of 100, 200, and 30,000 is also 200. Similarly, the mode of the set of values 10, 15, 15, 18, and 16 is 15, and that of -2, 15, 15, 2000, and 30,000 is also 15.

4. If we are given means for several sets of data, together with the number in each set, we can find the mean for the combined data without having to compute it from the original data. This is not possible for the median and mode.

Clearly the main disadvantage of the mean is that it is affected strongly by extreme values. It gives a distorted picture if the data contain just a few observations that are disproportionately higher (or lower) than the rest of the data. The median or the mode are to be preferred in such situations. For example, if the annual incomes of five families are

$8000, $10,800, $19,720, $12,760, $185,000,

then the mean, equal to $47,256, gives a distorted picture of the incomes, whereas the median, which is $12,760, gives a more realistic picture.

EXAMPLE 1   Two food chains in California, Food Baron and Food Market, gave the following sales figures for the amount of meat sold (in pounds) on seven days:

Food Baron (pounds of meat sold)	Food Market (pounds of meat sold)
1700	1000
1600	1100
1400	1600
1300	4200
1600	1000
1300	1300
1600	1000

Compare the performances of the two chains by considering the measures of central tendency.

FIGURE 2-1

(a) Symmetric distribution; mean, median, mode coincide. (b) Distribution skewed to the right.

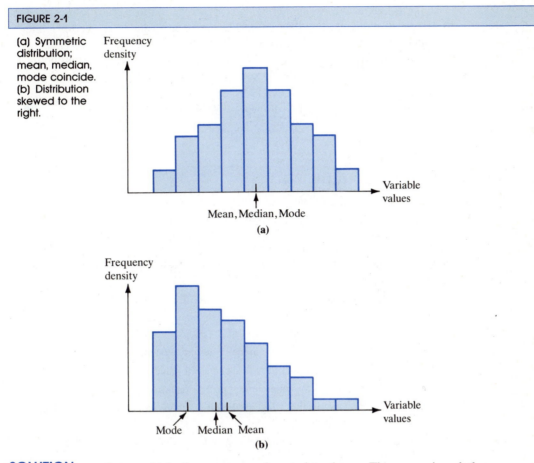

(a)

(b)

SOLUTION

Let us obtain the measures of central tendency. These are given below.

	Food Baron	*Food Market*
mean	1500	1600
median	1600	1100
mode	1600	1000

If we used the mean as the criterion for deciding which chain fared better, we would certainly have to conclude that Food Market is ahead in its performance. However, a closer look at the sales figures conveys an entirely different picture. The performance of Food Baron is very consistent, having consistently sold around 1500 pounds. On the other hand, the performance of Food Market is erratic, and the one figure that has inflated the mean of the Food Market data is 4200 pounds.

If we look at the median amounts sold, we get a more realistic picture. On 50 percent of the days, Food Baron sold at least 1600 pounds. The corresponding figure for Food Market is only 1100 pounds.

Finally, considering the mode, we find that Food Baron sold 1600 pounds most often, namely, three times. In contrast, the amount sold most often by Food Market is 1000 pounds.

In terms of consistent performance, the Food Baron chain is better.   ■■■

SECTION 2-3   EXERCISES

1. We know that if the distribution is symmetric, then the mean, the median, and the mode coincide. Is it true that if the mean, the median, and the mode coincide, then the distribution must be symmetric?

2. What measure of central tendency would you use if you were seeking the most common, or typical, value?

3. What measure of central tendency would you use if an exceptionally high or low value was affecting the arithmetic mean?

4. Which of the following statements are true?

   (a) The mean, the median, and the mode represent frequencies of values.

   (b) The mean, the median, and the mode represent values of the variable.

   (c) Only the median and the mode represent frequencies.

5. Find the mean and the median when all the observations in the data are identical, say, equal to 20.

6. The following figures give the medical expenses (in dollars) incurred during a year by ten families:

820	430	390	10,760	550
1950	476	340	25,570	330

   Find (a) the mean medical expenses and (b) the median medical expenses. Which measure do you think is more appropriate as a measure of central tendency?

7. The following figures represent the closing prices of a stock (in dollars) on eight trading days: 12, 11, 15, 12, 13, 18, 48, and 91. Find (a) the mean closing price, (b) the median closing price, and (c) the modal closing price. Which measure of central tendency do you think is least satisfactory?

8. In a frog race with ten frogs, three frogs strayed away. For the remaining frogs that completed the race, the following times (in seconds) were recorded: 185, 200, 140, 145, 245, 580, and 600. Choose an appropriate measure of central tendency and find its value.

9. A psychologist assigned an identical task to each of eight students in special education classes. The following figures give the time (in minutes) that each child took to complete the task: 28, 36, 38, 50, 52, 60, 140, and 168. What measure of central tendency do you think is most appropriate? What is its value?

10. Suppose you are a retailer of sportswear and have data available on sizes of jackets sold during this fiscal year. If you are interested in replenishing your stocks for the next year, what measure of central tendency would be of greatest use to you?

2-4 MEASURES OF DISPERSION

As useful as the measures of central location are in providing some understanding about the data in a distribution, total reliance on the information conveyed by the mean, the median, and the mode can be misleading, as we shall now illustrate.

Suppose a test is given to two sections of a class, Section A and Section B. Table 2-1 gives the scores (in points) recorded in the two sections.

TABLE 2-1	
Test scores in Sections A and B	
Section A scores	*Section B scores*
56	30
58	35
60	60
60	60
60	60
62	85
64	90

We see that both sections have the same mean score (60), the same median score (60), and the same mode (60). Solely on the basis of this information, however, it would be wrong to infer that the two sets of data are identical. In Section A, the group of students is homogeneous, all of them scoring in the vicinity of 60 points. In Section B, on the other hand, the performance is very erratic, from a low of 30 points to a high of 90 points. If we choose any student from Section A, we can assume that the student's score will be close to 60 points. We cannot say the same about a student from Section B.

Often a revealing picture emerges by considering histograms of the distributions as in Figure 2-2. The two distributions give the tensile strength of cables manufactured by two processes. In both cases we have the same mean, median, and mode. However, there is smaller variability in the data for the histogram in Figure 2-2(a) than in the one for the histogram in Figure 2-2(b).

Variability of values in data collected is a very common phenomenon, and its importance should be acknowledged. For instance, a company manufacturing electric bulbs will be interested not only in the average life of the bulbs but also in how consistent the performance of the bulbs is. The manufacturer interested in marketing

FIGURE 2-2

Two distributions having the same mean, median and mode, but the spread of data in (b) is more than in (a)

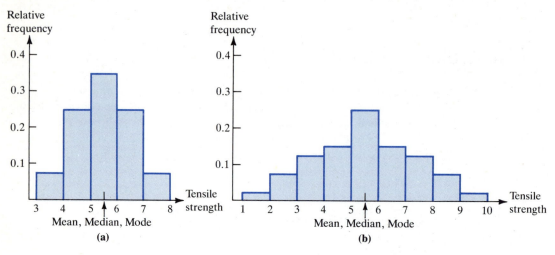

(a)   (b)

bulbs with a mean life of 1000 hours will not be satisfied even if this goal is realized if, in fact, there is a very high ratio of bulbs that only last for up to 200 hours. In other words, the success of the bulbs will depend on the variability of life; the smaller the variability the better.

THE RANGE

The simplest measure of dispersion is the **range.** As we saw in Section 1-1, the range is defined as the difference between the largest and smallest value in the data. Thus

range = largest value − smallest value.

A familiar example where range is used as the measure of variability is a weatherman's daily temperature report given as, for example, a low of 40 degrees to a high of 70 degrees. Here the range is $70 - 40$, or 30, degrees.

If we take the highest and the lowest observations, the range is easy to compute. However, it ignores the intermediate values and provides no information at all about the set of observations between the highest and lowest values. Consider, for instance, the following two sets of data:

$$3, \quad 7, \quad 9, \quad 11, \quad 16 \quad \text{and} \quad 3, \quad 12, \quad 12, \quad 12, \quad 16.$$

Each set has the same range, but the two distributions are markedly different. Also, by describing data solely in terms of extreme values, the presence of an unusually extreme value will inflate the range out of proportion. The range, therefore, is the least satisfactory of all the measures of dispersion that we shall consider in this chapter.

MEAN DEVIATION

The variability of the data depends upon the extent to which individual observations are spread about a measure of central tendency. If the values are widely scattered, the variability will be large. On the other hand, if the values are compactly distributed about this measure, the variability will be small. Because the arithmetic mean is by far the most important measure of central tendency, the deviations of the individual observations are usually taken as deviations from the mean $\bar{x}$.

Thus, if $x_1, x_2, \ldots, x_n$ represent n observations, then $x_1 - \bar{x}, x_2 - \bar{x}, \ldots, x_n - \bar{x}$ will be the n deviations from the mean. Some of these deviations will be positive, others negative. Their sum, however, is *always* zero, irrespective of the scatter of the data. That is, for every set of data,

$$\sum_{i=1}^{n} (x_i - \bar{x}) = 0.$$

Since the sum of the deviations is always zero, $\sum_{i=1}^{n} (x_i - \bar{x})$ is of no value to us as a measure of variability. But notice that if the mean $\bar{x}$ is 50, then the value 42 is as much removed from the mean as is the value 58, and the two will contribute to the same extent toward the variability. Thus, we can overcome the difficulty, giving weight only to the magnitude of the deviation, by considering all the deviations positive. These values are called *absolute deviations* and are denoted as $|x_1 - \bar{x}|, |x_2 - \bar{x}|, \ldots, |x_n - \bar{x}|$. The arithmetic mean of the absolute deviations is called the **mean deviation** and is found by dividing the sum of the absolute deviations by the number of observations.

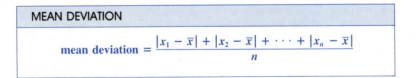

MEAN DEVIATION

$$\text{mean deviation} = \frac{|x_1 - \bar{x}| + |x_2 - \bar{x}| + \cdots + |x_n - \bar{x}|}{n}$$

In contrast with the range, the mean deviation takes into account the magnitude of all the observations in the data.

EXAMPLE 1   A secretary typed six letters. The times (in minutes) spent on these letters were 6, 8, 13, 10, 8, and 9. Find the mean deviation.

SOLUTION   We see that the mean is

$$\bar{x} = \frac{6 + 8 + 13 + 10 + 8 + 9}{6} = 9.$$

Therefore,

$$\binom{\text{mean}}{\text{deviation}} = \frac{|6 - 9| + |8 - 9| + |13 - 9| + |10 - 9| + |8 - 9| + |9 - 9|}{6}$$

$$= \frac{3 + 1 + 4 + 1 + 1 + 0}{6} = \frac{5}{3}. \quad \blacksquare$$

VARIANCE AND THE STANDARD DEVIATION

The use of absolute values causes considerable problems in treating mean deviation mathematically. For this reason, mean deviation is not suitable, and we shall seldom refer to it again in the remainder of the book.

An alternate measure which is more commonly used is provided by squaring the individual deviations, thereby producing positive numbers for both positive and negative deviations. This measure is called the *variance*. If the data represent a population, then the variance is called the **population variance** and is denoted by σ^2 where σ represents a positive number (σ is the lowercase Greek letter sigma, and σ^2 is read *sigma squared*). On the other hand, if the data constitute a sample, then the variance is called the **sample variance** and is denoted by s^2 where s is a positive number. The definitions follow.

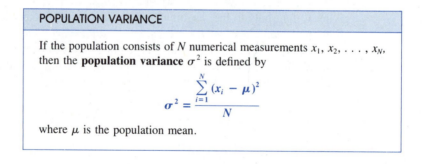

POPULATION VARIANCE

If the population consists of N numerical measurements $x_1, x_2, \ldots, x_N$, then the **population variance** σ^2 is defined by

$$\sigma^2 = \frac{\sum\limits_{i=1}^{N} (x_i - \mu)^2}{N}$$

where μ is the population mean.

(More will be said about this definition in Example 8 in Section 4-2.) A close scrutiny of the formula reveals that population variance is simply the arithmetic mean of the squared deviations from the population mean μ. The positive square root of the population variance, namely σ, is the **population standard deviation.**

We might expect to define the sample variance in a manner analogous to the population variance, as the arithmetic mean of squared deviations from the sample mean. However, sample variance is defined slightly different, as follows.

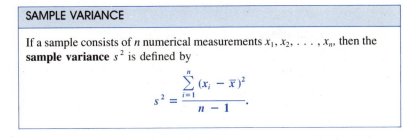

SAMPLE VARIANCE

If a sample consists of n numerical measurements $x_1, x_2, \ldots, x_n$, then the **sample variance** s^2 is defined by

$$s^2 = \frac{\sum_{i=1}^{n} (x_i - \bar{x})^2}{n - 1}.$$

Rationale for dividing by $n - 1$

For finding s^2, the sum of squares of deviations from $\bar{x}$ is divided by $n - 1$ instead of by n. The rationale is as follows. Suppose we are told to pick *any* three numbers with the condition that their sum is to be 18. Are we really free to pick three numbers? Of course not. We have freedom to pick only two (that is, $3 - 1$) numbers since the third number is determined automatically by the constraint. About the same sort of thing happens when finding the sample variance. We are told that we have n observations from which to compute the sample variance but first we must compute the sample mean using these observations since it is involved in the formula. It turns out that we have, in effect, $n - 1$ observations with which to compute the sample variance. The number $n - 1$ is called the degrees of freedom (abbreviated df) associated with the sample variance. (Also, later in Chapter 7 we will meet a criterion called unbiasedness which makes it desirable to divide by $n - 1$).

The positive square root of the sample variance, denoted s, is the **sample standard deviation.** (The statistical function s is built into many scientific pocket calculators and can be obtained simultaneously with the mean $\bar{x}$.)

As we have mentioned earlier, in practice the data at hand will be sample data. Hence published results will usually give s values in quoting the standard deviation.

EXAMPLE 2   The price (in dollars) of a certain commodity on eight trading sessions was

38,   11,   8,   60,   52,   68,   32,   19.

Find the variance and standard deviation.

SOLUTION   The computations are presented in Table 2-2 on page 58.

In column 1 we find the mean $\bar{x} = 288/8$, or 36. The second column contains deviations from the mean; that is, $x_i - 36$. As a check, notice that the sum in this column is 0. The third column presents squares of deviations from the mean. The sum in this column yields the sum of squares of deviations. Thus,

$$\sum_{i=1}^{8} (x_i - \bar{x})^2 = 3{,}574.$$

Since $n = 8$, the variance given in (dollars)^2 is

TABLE 2-2		
Computation of the variance of the price of a commodity.		
Price x_i	Deviation $x_i - 36$	Square of deviation $(x_i - 36)^2$
38	2	4
11	-25	625
8	-28	784
60	24	576
52	16	256
68	32	1,024
32	-4	16
19	-17	289
$\sum\limits_{i=1}^{8} x_i = 288$ mean $\bar{x} = 36$	$\sum\limits_{i=1}^{8} (x_i - \bar{x}) = 0$	$\sum\limits_{i=1}^{8} (x_i - \bar{x})^2 = 3{,}574$

$$s^2 = \frac{\sum\limits_{i=1}^{8}(x_i - \bar{x})^2}{n - 1} = \frac{3574}{8 - 1} = 510.57.$$

The positive square root of the variance is the standard deviation given in dollars, or

$$s = \sqrt{510.5714} = 22.6. \quad \blacksquare$$

A word about rounding: We shall retain two decimal places more than in the original data in presenting the sample variance and one decimal place more than in the original data in presenting the standard deviation. Be careful not to find the standard deviation from the rounded variance.

Through some algebraic simplifications, the above formula for s^2 leads to the following shortcut formula which is convenient for computational purposes.

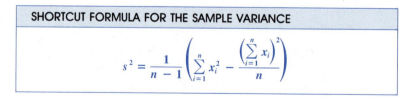

SHORTCUT FORMULA FOR THE SAMPLE VARIANCE

$$s^2 = \frac{1}{n - 1}\left(\sum_{i=1}^{n} x_i^2 - \frac{\left(\sum\limits_{i=1}^{n} x_i\right)^2}{n}\right)$$

This version is preferred especially if $\bar{x}$ contains a large number of significant places.

Let us compute s^2 for the data in Example 2 using the shortcut formula. We have seen that $n = 8$ and $\sum\limits_{i=1}^{8} x_i = 288$. Also,

$$\sum_{i=1}^{8} x_i^2 = (38)^2 + (11)^2 + (8)^2 + (60)^2 + (52)^2$$
$$+ (68)^2 + (32)^2 + (19)^2$$
$$= 1{,}444 + 121 + 64 + 3{,}600 + 2{,}704$$
$$+ 4{,}624 + 1{,}024 + 361$$
$$= 13{,}942.$$

Substituting in the shortcut formula, we then get

$$s^2 = \frac{1}{(8-1)}\left(13{,}942 - \frac{(288)^2}{8}\right)$$
$$= 510.57.$$

This is the same answer we obtained in Example 2.

A word about the units to be employed in expressing the variance and the standard deviation: Since the variance is based on the sum of squared differences, it is expressed in squared units. Since the standard deviation is the square root of the variance, it is expressed in terms of the units that were employed in measuring the data.

For example, if the data are given in feet, then the units used for expressing the variance are feet squared, abbreviated $(ft)^2$, and those for the standard deviation are feet. In the same way, if the data are expressed in minutes, then the units for expressing the variance and the standard deviation are, respectively, minutes squared, or $(min)^2$, and minutes.

PROPERTIES OF THE VARIANCE

The variance of data has the following two important properties:

1. If a fixed value d is added to each of the observations in the data, then the variance is unchanged, that is,

$$\left(\begin{matrix}\text{the variance of}\\ \text{the new data}\end{matrix}\right) = \left(\begin{matrix}\text{the variance of the}\\ \text{original data}\end{matrix}\right).$$

For example, if the salary of each employee is raised by 100 dollars, then the variance of salaries before the raise is the same as the variance after the raise.

2. If each observation in the data is multiplied by a fixed constant c, then

$$\left(\begin{matrix}\text{the variance of}\\ \text{the new data}\end{matrix}\right) = c^2 \; times \; \left(\begin{matrix}\text{the variance of}\\ \text{the original data}\end{matrix}\right).$$

For example, suppose the height of students in a classroom is measured in feet and the variance is 0.8 $(ft)^2$. If the height is measured in inches, then the variance will be $(12^2)(0.8)$, or 115.2 $(in.)^2$.

FIGURE 2-3

Fraction in the interval which covers k standard deviations from $\bar{x}$ is at least $1 - 1/k^2$.

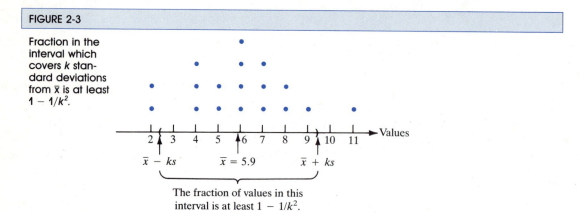

The fraction of values in this interval is at least $1 - 1/k^2$.

Finally, consider the significance of s in explaining the spread of the data about the mean $\bar{x}$. Intuitively, we suspect that a very small value of s will indicate that the measurements are concentrated around the mean and conversely. The following celebrated theorem of the great Russian mathematician P. L. Chebyshev (1821–1894) supports our intuition.

CHEBYSHEV'S THEOREM

For any set of data, the fraction of the measurements falling within k standard deviations of the mean is at least $1 - 1/k^2$.

The situation is shown graphically by means of the dot diagram in Figure 2-3 and by Figure 2-4. In Figure 2-3 the horizontal scale covers the range of values and individual values are plotted as dots. Computations will show that $\bar{x} = 5.9$ and $s = 2.3$ in the diagram. We have chosen $k = 1.5$ so that $\bar{x} - ks = 2.45$ and $\bar{x} + ks = 9.35$. The theorem claims that there should be at least a fraction $1 - 1/1.5^2$ or $5/9$ in the interval $(\bar{x} - 1.5s, \bar{x} + 1.5s)$. Actual count shows that the fraction is $15/18$, in agreement with the claim.

Thus, regardless of the source of data, the situation in Figure 2-4 is true:

EXAMPLE 3 In a survey conducted in a metropolitan area it was found that the mean income of 650 families was \$18,000 with a standard deviation of \$3,000.

(a) Use Chebyshev's theorem to find at least how many families in the survey have incomes between \$10,500 and \$25,500.

(b) Justify or refute the following statement: When the original data were scrutinized, 420 families were found to have incomes between \$12,000 and \$24,000.

SOLUTION   (a) The income range \$10,500 to \$25,500 can be written as $18{,}000 \pm (2.5)(3{,}000)$. Thus, clearly, $k = 2.5$. From Chebyshev's theorem it follows that the fraction of families having incomes in the given range must be at least

$$1 - \frac{1}{k^2} = 1 - \frac{1}{(2.5)^2} = 0.84.$$

FIGURE 2-4

Interpretation
of Chebyshev's
theorem for
$k = 1, 2, 3$.

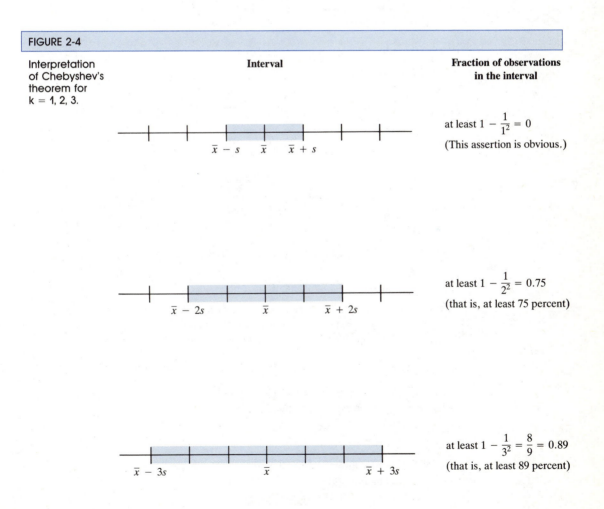

	Interval	Fraction of observations in the interval

$\bar{x} - s$ $\bar{x}$ $\bar{x} + s$

at least $1 - \dfrac{1}{1^2} = 0$

(This assertion is obvious.)

$\bar{x} - 2s$ $\bar{x}$ $\bar{x} + 2s$

at least $1 - \dfrac{1}{2^2} = 0.75$

(that is, at least 75 percent)

$\bar{x} - 3s$ $\bar{x}$ $\bar{x} + 3s$

at least $1 - \dfrac{1}{3^2} = \dfrac{8}{9} = 0.89$

(that is, at least 89 percent)

Since there are 650 families in the survey, at least 650(0.84) = 546 families must have incomes between $10,500 and $25,500.

(b) We can write the income range $12,000 to $24,000 as 18,000 ± 2(3000). Hence $k = 2$.

Chebyshev's theorem asserts that the fraction of families in the income range above is at least $1 - 1/2^2 = 0.75$. Therefore, *at least* 650(0.75), or 487.5, rounded to 487, families in the survey must have incomes between $12,000 and $24,000. The statement that there were 420 families is wrong. There is either a computational error or a wrong count.   ▄▄▄

SECTION 2-4   EXERCISES

1. The scores of six students on a test are

      75,   82,   56,   69,   87,   43.

   Arrange the data in an ascending order of magnitude and find the range of scores.

2. The amount of time spent on each of ten telephone calls was recorded (in minutes) as follows:

      25,   15,   8,   2,   5,   12,   34,   3,   12,   3.

   Find the range.

3. The altitudes (in feet) of six locations (above or below sea level) are

      1200 above,   140 above,   100 below,   10,500 above,   400 below,
      800 above.

   Find the range of heights.

4. The minimum daily temperature readings (°F) of a ski resort on eight days were

      22,   − 18,   0,   − 8,   16,   30,   5,   − 20.

   Find the temperature range for these days.

5. If the range of the price of a stock during a week was $15.75, and if the lowest price was $82.50, find the highest price during the week.

6. For the set of values 8, 6, 12, 10, and 14, find

   (a) $\sum_{i=1}^{5} (x_i - \bar{x})$ (b) $\sum_{i=1}^{5} |x_i - \bar{x}|$ (c) the mean deviation.

7. The following figures refer to the times (in minutes) that eight customers spent from the time they entered a store to the time they left it:

      6,   3,   2,   10,   2,   5,   4,   6.

   Find

   (a) the arithmetic mean $\bar{x}$ of the time the customers spent in the store

   (b) the absolute deviations

(c) the mean deviation.

In Exercises 8–11, compute mean deviations for the given sets of observations:

8. 14, −4, 10, 6, 10, −15, and 7

9. 1.8, 2.8, 2.3, 4.5, and 1.1

10. 6, −7, −5, −4, and 5

11. −8, −7, −6, −5, −10, and −12

12. Is it possible for the mean deviation to be negative? Justify your answer.

13. What would it mean if the mean deviation of a set of observations was equal to zero? Justify your answer.

14. Seven buses are expected to arrive at the bus depot at 8 A.M., 9 A.M., 10 A.M., 11 A.M., 12 noon, 1 P.M., and 2 P.M. They arrive at 8:04 A.M., 9:10 A.M., 9:52 A.M., 10:56 A.M., 11:55 A.M., 1:10 P.M., and 2:00 P.M., respectively. Find the mean deviation of the amount of time by which a bus is late. (If a bus arrives ahead of schedule, for example, by 4 minutes, then consider the late arrival time is negative, or −4.)

15. The following figures give the price of ground chuck (in dollars) at a supermarket during six weeks:

     1.98,   1.89,   1.80,   1.88,   1.84,   1.95.

    Find the number of weeks when the price is within one mean deviation of the mean price.

16. For the set of values 18, 19, 16, 12, 7, 10, 23, find
    (a) the arithmetic mean $\bar{x}$ (b) the deviations from the mean
    (c) the squared deviations   (d) the sample variance
    (e) the sample standard deviation.

    In each of Exercises 17–19, compute the variance and the standard deviation for the given sets of observations.

17. 14, −4, 10, 6, 10, −15, and 7

18. 1.8, 2.8, 2.3, 4.5, and 1.1

19. 6, −7, −5, −4, and 5

20. What is the variance of a set of scores when all the observations are identical?

21. What does it mean when the variance of a set of observed values is zero?

22. Can the standard deviation ever be negative? Justify your answer.

23. The following readings were obtained for the tensile strength (in tons per square inch) of six specimens of an alloy:

     2.58,   2.65,   2.40,   2.46,   2.44,   2.41.

    Find the mean tensile strength and the standard deviation of the tensile strengths.

24. The heights of four fifth-grade students are 58, 56, 60, and 62 inches, and those of five sixth-grade students are 66, 59, 63, 61, and 66 inches. Which grade has the higher variance?

25. Five samples of coal, each of which weighs 2.00 grams after air drying, loses the following amounts of moisture (in grams) on drying in an oven:

    0.158,   0.160,   0.156,   0.153,   0.163.

    Find the variance of the amount of moisture lost.

26. In a chemistry experiment seven students obtained the following percent of chlorine in a sample of pure sodium chloride:

    60.65,   60.68,   60.70,   60.60,   60.64,   60.65,   60.70.

    Find the standard deviation of the percent of chlorine.

27. A club has ten members whose weights are $x_1, x_2, \ldots, x_{10}$. Find the mean weight and the variance of the weights if $\sum_{i=1}^{10} x_i = 1500$ and $\sum_{i=1}^{10} x_i^2 = 325,000$.

28. Using the formula for the variance, find the variance of the data in Part (a). From this answer find the variance of the data in Parts (b) and (c).

    (a) 7, 8, 6, 8, and 6

    (b) 27, 28, 26, 28, and 26

    (c) 107, 108, 106, 108, and 106

    [*Hint:* the observations in Part (b) are obtained by adding 20 to each observation in Part (a); and those in Part (c) are obtained by adding 100 to each observation in Part (a).]

29. After grading a test, Professor Adams announced to the class that the mean score was 68 points with a standard deviation of 12 points. However, a student brought to his attention that one of the questions on the test was incorrect and impossible to solve. As a result, Professor Adams agreed to add 10 points to the score of each student. Find the mean, the variance, and the standard deviation of the new set of scores.

30. The mean salary of the employees in a certain company is $18,000 with a standard deviation of $2500. The company announces a flat raise of $1000 for each employee during the following year. Find the mean and the standard deviation of the new salaries.

31. The data given below are the chlorophyll content (in milligrams per gram) of 30 wheat leaves:

2.6	2.9	3.2	3.0	2.6	3.2	3.2	3.0	2.9	3.0
2.6	2.6	3.0	3.2	2.9	3.2	2.6	2.9	3.2	2.6
3.2	2.8	3.2	3.0	3.0	2.8	2.9	3.2	2.8	2.8

    Find the variance and the standard deviation of the chlorophyll content.

32. The following data are the number of rooms occupied in a motel on Interstate 101 over a period of 40 days:

20	14	21	29	43	17	15	26	10	14
39	23	14	30	28	11	26	35	19	24
29	10	18	15	22	8	34	13	30	24
11	14	7	27	20	14	31	18	14	23

    Compute the mean and the standard deviation of the number of rooms occupied.

33. Linda has recorded the price (in dollars) of a certain item at two different stores over a period of fifteen weeks. She has compiled the following information regarding the mean prices and the standard deviations:

	Mean price	Standard deviation
Store A	32.75	8.37
Store B	31.56	2.20

If Linda is a bargain hunter, in which store should she plan to shop?

34. The following figures give the cost (in cents) for a quart of a particular brand of oil at eight automotive supply centers:

    60,   65,   75,   68,   72,   65,   72,   75

Find the number of places where the price is within one standard deviation of the mean.

35. When ten 5-gram samples of an alloy were analyzed for silver content by the method of electrodeposition, the following amounts of silver (in grams) were found:

    1.273   1.276   1.265   1.276   1.274
    1.277   1.273   1.275   1.285   1.278

    (a) Find the mean and the standard deviation of the amount of silver.

    (b) Determine the interval which represents two standard deviations from the mean.

    (c) How many samples have amounts of silver in the interval obtained in Part (b)?

    (d) According to Chebyshev's theorem at least how many samples must have amounts of silver in the interval in Part (b)?

    (e) Are the answers in Parts (c) and (d) in agreement?

36. A quality control engineer for an electrical firm tested four hundred 60-watt bulbs and recorded the time (in hours) a bulb burns before it burns out. It was found that the mean was 650 hours with a standard deviation of 30 hours. At least how many bulbs burned between 575 and 725 hours before burning out?

37. When the maximal grip strength (in kilograms) of 60 athletes was measured it was found that their mean maximal grip strength was 41.2 kg, with a standard deviation of 4.3 kg.

    (a) Determine the interval $(\bar{x} - 2.4s, \bar{x} + 2.4s)$.

    (b) According to Chebyshev's theorem, at least what fraction of the athletes had maximal grip strength in the interval in Part (a)?

    (c) At least how many of the 60 athletes had maximal grip strength in the interval in Part (a)?

2-5 MEAN AND VARIANCE FOR GROUPED DATA

The basic formulas for computing the sample mean and the sample variance have been given in preceding sections. Now, if the data are given in a frequency table, then the formulas can be expressed in the following form:

SAMPLE MEAN AND VARIANCE WHEN FREQUENCY DISTRIBUTION IS GIVEN

If the data consisting of n observations is given in a frequency table having k distinct values $x_1, x_2, \ldots, x_k$ with respective frequencies $f_1, f_2, \ldots, f_k$, then the sample mean $\bar{x}$ and variance s^2 are given by

$$\bar{x} = \frac{\sum_{i=1}^{k} f_i x_i}{n}$$

and

$$s^2 = \frac{\sum_{i=1}^{k} f_i (x_i - \bar{x})^2}{n - 1}$$

where $\sum_{i=1}^{k} f_i = n$.

A more convenient version of the formula for computing the sample variance is the following:

$$s^2 = \frac{1}{(n-1)} \left[\sum_{i=1}^{k} f_i x_i^2 - \frac{\left(\sum_{i=1}^{k} f_i x_i \right)^2}{n} \right]$$

EXAMPLE 1   The data in Table 2-3 give the distribution of the diameters (in inches) of 20 tubes manufactured by a machine.

TABLE 2-3

Diameters of tubes manufactured by a machine

Diameter (in inches)	Frequency
2.0	2
2.2	4
2.3	6
2.8	3
3.0	5

Find

(a) the mean diameter   (b) the variance.

SOLUTION

The necessary computations are presented in Table 2-4.

TABLE 2-4				
Computations for finding the mean and the variance				
Diameter x_i	Frequency f_i	$f_i x_i$	x_i^2	$f_i x_i^2$
2.0	2	4.0	4.00	8.00
2.2	4	8.8	4.84	19.36
2.3	6	13.8	5.29	31.74
2.8	3	8.4	7.84	23.52
3.0	5	15.0	9.00	45.00
Sum	20	50.0		127.62

(a) The sums in columns 2 and 3 yield

$$n = \sum_{i=1}^{k} f_i = 20 \quad \text{and} \quad \sum_{i=1}^{k} f_i x_i = 50.0$$

respectively. Therefore, the mean (in inches) is

$$\bar{x} = \frac{\sum_{i=1}^{k} f_i x_i}{n}$$

$$= \frac{50.0}{20} = 2.5.$$

(b) From column 5 of Table 2-4 we have $\sum_{i=1}^{k} f_i x_i^2 = 127.62$. Substituting in the formula

$$s^2 = \frac{1}{(n-1)} \left[\sum_{i=1}^{k} f_i x_i^2 - \frac{\left(\sum_{i=1}^{k} f_i x_i \right)^2}{n} \right],$$

we get the sample variance (in (inches)^2) as

$$s^2 = \frac{1}{(20-1)} \left[127.62 - \frac{(50)^2}{20} \right]$$

$$= 0.138 \quad \blacksquare$$

In the next example in which we compute the mean and the variance for grouped data, the secret here is to find the class marks first and then treat them as though they are the observed values.

EXAMPLE 2   **(Grouped data.)** In columns 1 and 2 of Table 2-5 scores of 80 students are grouped into a frequency distribution.

TABLE 2-5				
Computations to find the mean and the variance of student scores				
Class	Frequency f_i	Class mark x_i	$f_i x_i$	$f_i x_i^2$
5-19	4	12	48	576
20-34	12	27	324	8,748
35-49	15	42	630	26,460
50-64	16	57	912	51,984
65-79	22	72	1,584	114,048
80-94	11	87	957	83,259
sum	80		4,455	285,075

Compute

(a) the mean   (b) the variance.

SOLUTION   The class marks are computed in column 3 of Table 2-5. Other computations are presented in columns 4 and 5 of the table.

(a) From columns 2 and 4, $n = 80$ and $\sum_{i=1}^{k} f_i x_i = 4{,}455$. Therefore, the mean (in points) is

$$\bar{x} = \frac{4{,}455}{80} = 55.7.$$

(b) From column 5 we have $\sum_{i=1}^{k} f_i x_i^2 = 285{,}075$. Hence, the sample variance (in (points)^2) is

$$s^2 = \frac{1}{(80 - 1)} \left[285{,}075 - \frac{(4{,}455)^2}{80} \right]$$

$$= 468.19. \quad \blacksquare$$

SECTION 2-5   EXERCISES

1. The distribution below gives the top speeds (in miles per hour) at which 30 racers were clocked in an auto race:

Top speed	Number of racers
145	9
150	8
160	11
170	2

Find the mean and the variance of the distribution of the top speeds.

2. Find the variance and the standard deviation of the following distribution of earnings during a year of 100 graduate students:

Earnings (in dollars)	Number of graduate students
5550	13
7550	26
9550	36
11,550	25

3. The data given below are the net weights of sugar (in ounces) packed in 30 one-pound bags:

15.6	15.9	16.2	16.0	15.6	16.2	16.2	16.0
15.9	16.0	15.6	15.6	16.0	16.2	15.9	16.2
15.6	15.9	16.2	15.6	16.2	15.8	16.2	16.0
16.0	15.8	15.9	16.2	15.8	15.8		

Find

(a) the mean amount per bag
(b) the variance.

[*Hint:* Arrange in a frequency table first.]

4. Complete the following table and compute the variance of the data given:

Class	Class mark x_i	Frequency f_i	$f_i x_i$	$f_i x_i^2$
2–6	—	23	—	—
7–11	—	14	—	—
12–16	14	34	476	6664
17–21	—	15	—	—
22–26	—	14	—	—
Sum				

5. For the distribution of life (in hours) of 400 cathode tubes, given in the table below, find the variance and the standard deviation of the life of the bulbs.

Life of tube (in hours)	Number of tubes
600–699	85
700–799	77
800–899	124
900–999	78
1000–1099	36

6. The table below gives the distribution of the breaking strength (in pounds) of 60 cables:

Breaking strength	Number of cables
140–149	8
150–159	10
160–169	18
170–179	12
180–189	8
190–199	4

Find the mean and the standard deviation of the breaking strength of the cables.

7. Fifty candidates entering an astronaut training program were given a psychological profile test. The following table gives the distribution of their scores.

Score (in points)	Number of candidates
60–79	8
80–99	16
100–119	12
120–139	8
140–159	6

Find (a) the mean score   (b) variance of the score.

8. Gary has 30 books on a shelf. When he grouped them according to the number of pages each has, he obtained the following distribution:

Number of pages	Number of books
100–199	3
200–299	6
300–399	8
400–499	8
500–599	5

Find the variance of the number of pages.

9. The table below gives the distribution of the amount of rainfall in December in Tippegawa County for a period of 30 years:

Rainfall in December (in inches)	Number of years
2.8–4.7	3
4.8–6.7	6
6.8–8.7	8
8.8–10.7	8
10.8–12.7	5

Find

(a) the mean amount of rainfall in December

(b) the standard deviation of the amount of rainfall.

10. The office of the Dean of Graduate Studies has compiled the following information regarding the ages of 26 students who will be awarded their Ph.D.s:

Age (in years)	Number of students
20–24	1
25–29	9
30–34	8
35–39	5
40–44	2
45–49	1

Find

(a) the mean age of the students awarded Ph.D.s

(b) the standard deviation of the age.

KEY TERMS AND EXPRESSIONS

average
measures of central tendency
variability
summation notation
arithmetic mean
population mean
sample mean
median

percentile
mode
range
mean deviation
population variance
sample variance
population standard deviation
sample standard deviation
Chebyshev's theorem

KEY FORMULAS

mean (ungrouped) (called the population mean μ or the sample mean $\bar{x}$ depending on whether the n observations constitute the entire population or the sample)

$$\text{mean} = \frac{\sum_{i=1}^{n} x_i}{n}$$

$$\text{mean} = \frac{\sum_{i=1}^{k} f_i x_i}{n}, \text{ where } f_i \text{ is the frequency of } x_i \text{ and } \sum_{i=1}^{k} f_i = n.$$

mean (grouped)

$$\text{mean} = \frac{\sum_{i=1}^{k} f_i x_i}{n}, \text{ where } x_i \text{ is the class mark and } f_i \text{ is its frequency}$$

range

$$\text{range} = \text{largest value} - \text{smallest value}$$

mean deviation

$$\text{mean deviation} = \frac{\sum_{i=1}^{n} |x_i - \bar{x}|}{n}$$

population variance

$$\sigma^2 = \frac{\sum_{i=1}^{N} (x_i - \mu)^2}{N}$$

sample variance (ungrouped)

$$s^2 = \frac{\sum_{i=1}^{n} (x_i - \bar{x})^2}{n-1} = \frac{1}{n-1}\left[\sum_{i=1}^{n} x_i^2 - \frac{\left(\sum_{i=1}^{n} x_i\right)^2}{n}\right]$$

$$s^2 = \frac{\sum_{i=1}^{k} f_i (x_i - \bar{x})^2}{n-1}$$

$$= \frac{1}{n-1}\left[\sum_{i=1}^{k} f_i x_i^2 - \frac{\left(\sum_{i=1}^{k} f_i x_i\right)^2}{n}\right], \text{ where } f_i \text{ is the frequency of } x_i \text{ and } \sum_{i=1}^{k} f_i = n$$

sample variance (grouped)

$$s^2 = \frac{1}{n-1}\left[\sum_{i=1}^{k} f_i x_i^2 - \frac{\left(\sum_{i=1}^{k} f_i x_i\right)^2}{n}\right], \text{ where } x_i \text{ is the class mark and } f_i \text{ is its frequency}$$

population standard deviation $= \sqrt{\text{population variance}}$

sample standard deviation $= \sqrt{\text{sample variance}}$

CHAPTER 2   TEST

1. Describe the three measures of central tendency and compare their advantages and disadvantages.

2. Suppose just one observation in the data is changed. Which measure of central tendency is bound to be affected?

3. Give a set of data for which the mean, the median, and the mode coincide.

4. The following figures were reported for annual suspended sediment (in cubic yards per square mile) from the Casper Creek watershed in Northern California during the period 1963–1967. (Raymond Rice et al of the Pacific Southwest Forest and Range Experiment Station)

Hydrologic year	Suspended sediment
1963	101.15
1964	94.00
1965	612.54
1966	744.04
1967	138.24

   Find

   (a) the mean annual suspended sediment during the period
   (b) the median annual sediment during the period.

5. A secretary typed twelve letters. The times (in minutes) spent on these letters were

   6,   8,   13,   10,   8,   9,   10,   6,   8,   8,   6,   10.

   Find the median amount of time spent.

6. A shipment of twenty boxes containing a certain type of electrode is received. Upon inspection it was found that three boxes contained no defectives, ten boxes contained 2 defectives each, four boxes contained 3 defectives each, and the remaining boxes contained 4 defectives each. Find

   (a) the mean number of defectives in a box
   (b) the median number of defectives in a box
   (c) the modal number of defectives in a box.

7. The table below gives the distribution of the amount of gasoline (in gallons) sold at a gas station on a certain day:

Amount sold	Number of customers
0.0–3.9	17
4.0–7.9	70
8.0–11.9	92
12.0–15.9	63
16.0–19.9	58

Find

(a) the mean amount of gasoline sold per customer

(b) the modal class and the mode.

8. Over a 40-day period, Tom has obtained a frequency distribution of the number of minutes that he had to wait for the bus to arrive, which is shown in the table below.

Extent of wait (in minutes)	Frequency f
0–4	11
5–9	14
10–14	8
15–19	4
20–24	3
Sum	40

Compute

(a) the mean waiting time

(b) the variance of the waiting time.

9. Cholestyramine is a powerful drug that lowers cholesterol. Suppose six patients, who had abnormally high levels of cholesterol, were kept on a low fat diet and were treated for a certain period with the drug. The following figures give, for each individual, the cholesterol level (in mg/deciliter of blood) before and after the treatment.

Patient	Cholesterol before (mg/dl)	Cholesterol after
Bob	280	224
Judith	290	240
Raphael	250	194
George	260	216
Laura	285	242
Roy	310	262

(a) Compute percent reduction of cholesterol for each individual.

[*Hint:* For example, in Bob's case the percent reduction is $100(280 - 224)/280$.]

(b) Compute the mean percent reduction.

10. Define the concept of variability in experimental data and compare the different measures of it.

11. Suppose the data consist of resistance of resistors measured in ohms. Give the units in which s^2 and s should be expressed.

12. The U.S. Agriculture Department reported the following figures (in billions of dollars) for net cash farm income during the period 1977–1981.

Year	Net cash farm income
1977	17.4
1978	25.9
1979	27.4
1980	21.9
1981	19.0

Calculate the mean deviation of the annual net cash farm income during the given period, and the standard deviation.

13. Of twenty dental patients, three had 1 cavity, six had 3 cavities, five had 4 cavities, three had 5 cavities, and three had 8 cavities. Find the variance and the standard deviation of the number of cavities a patient has.

14. The distribution of the earnings of a free-lance photographer during 40 months is presented below:

Earnings (in dollars)	Number of months
900–1299	5
1300–1699	9
1700–2099	17
2100–2499	3
2500–2899	6

Compute the variance and the standard deviation of the earnings.

15. For assigning grades to his students, Professor Bradley adopts a procedure which is shown below schematically:

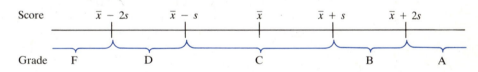

For instance, any student getting a score greater than $\bar{x} + s$ but less than or equal to $\bar{x} + 2s$ will earn a letter grade of B. In a statistics course that Professor Bradley taught, the following scores were recorded for twenty students.

75	71	88	99	80
77	76	70	74	68
49	61	87	75	88
72	74	69	77	59

Assign grades to each of the students.

16. In Exercise 15, according to Chebyshev's theorem at least how many students should get a grade of D, C, or B? Is your assigment of grades in agreement with this?

16. In Exercise 15, according to Chebyshev's theorem at least how many students should get a grade of D, C, or B? Is your assigment of grades in agreement with this?

17. Rainwater was collected in water collectors at thirty different sites near an industrial basin and the amount of acidity (pH level) was measured. The stem-and-leaf diagram presented below shows the figures (ranging from 2.6 to 6.3).

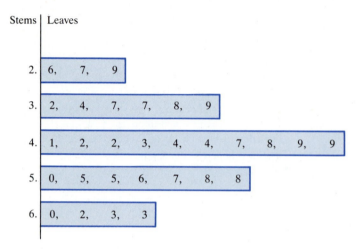

Stems	Leaves
2.	6, 7, 9
3.	2, 4, 7, 7, 8, 9
4.	1, 2, 2, 3, 4, 4, 7, 8, 9, 9
5.	0, 5, 5, 6, 7, 8, 8
6.	0, 2, 3, 3

Find

(a) the mean acidity   (b) the median acidity
(c) the standard deviation of acidity.

18. The following number of issues were traded on the New York Stock Exchange during seven trading sessions:

    1994,   2008,   2005,   1998,   1992,   2003,   2007.

Find the mean, the variance, and the standard deviation of the number of issues traded adopting the following procedure: subtract 2000 from each number, use the new data obtained, and apply the properties of the mean and the variance.

3

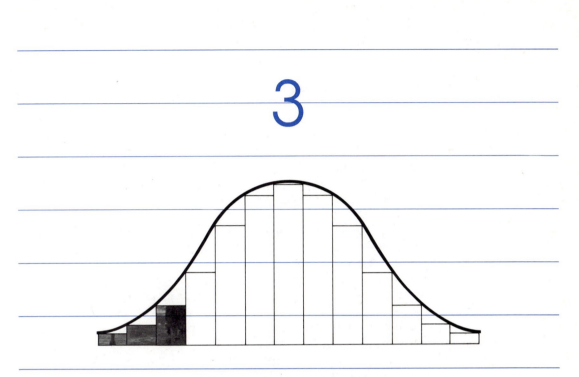

PROBABILITY

3-1 The Empirical Concept of Probability

3-2 Sample Space, Events, and the Meaning of Probability

3-3 Some Results of Elementary Probability

3-4 Counting Techniques and Applications to Probability

3-5 Conditional Probability and Independent Events

INTRODUCTION

In every walk of life we make statements that are probabilistic in nature and that carry overtones of chance. For example, we might talk about the probability that a bus will arrive on time, or that a child-to-be-born will be a son, or that the stock market will go up, and so on. What is the characteristic feature in all the phenomena above? It is that they all lack a deterministic nature. Past information, no matter how voluminous, will not allow us to formulate a rule to determine precisely what will happen when the experiment is repeated. Phenomena of this type are called *random phenomena*. The theory of probability involves their study.

Probability theory has a central place in the theory and applications of statistics since we are analyzing and interpreting data that involve an element of chance or uncertainty. The reliability of our conclusions is supported by accompanying probability statements.

The origin of probability theory goes back to the middle of the seventeenth century. The early study most likely started with Blaise Pascal (1623–1662). His interest and that of his contemporaries, Pierre Fermat among them, was occasioned by problems in gambling initially brought to their attention by a French nobleman, Antoine Gombauld, known as the Chevalier de Méré.

As interest in the natural sciences proliferated, so did demands for developing new laws of probability. Among prominent early contributors in this development were the nineteenth-century mathematicians Laplace, De Moivre, Gauss, and Poisson. Today, probability theory finds applications in diverse disciplines such as biology, economics, operations research, and astronomy. A biologist might be interested in the distribution of bacteria in a culture, an economist in economic forecasts, a production engineer in the inventory of a particular commodity, an astronomer in the distribution of stars in different galaxies, and so on. Considerable interest is also shown in this field as a mathematical discipline in its own right. A. Kolmogorov (1903–   ) should be mentioned as a pioneer in this area.   ■

3-1 THE EMPIRICAL CONCEPT OF PROBABILITY

A statement such as ''There is a 90 percent chance that the bus will arrive on time'' is not uncommon. When we consider a particular bus, either it arrives on time or it does not. Why does ''90 percent chance'' figure in our statement? This statement is based presumably on a person's experience over a long period of time. If we were to keep track of all the buses arriving at the bus stop to find the ratio of the buses that arrived on time to the total number of buses that stopped at the stop, then that ratio would be approximately 90 percent. The empirical, or experimental, idea behind this statement is the kind of rationale that is, in general, at the root of any probabilistic statement.

As another illustration of the *empirical concept of probability,* let us consider an experiment of tossing a normal coin. A person tossing the coin one thousand times might obtain 462 heads and 538 tails, thus getting heads 46.2 percent of the times. In general, if we were to toss the coin a large number of times, invariably we would find the ratio of the number of heads to the total number of tosses to be *approximately* 50 percent. The larger the number of tosses, the closer the approximation will be to 50 percent. This approximation (close to 1/2) that is inherent in the situation lends support to the widely accepted statement that the chance of getting heads on a coin is 1 out of 2.

In our discussion regarding the coin, the emphasis should not be on the fact that the ratio approaches 1/2 but that it stabilizes. In practice, this stable value is the measure that we take of probability. For instance, suppose we toss a coin repeatedly and, as in the illustration above, compute the ratio of the number of heads to the total number of tosses. If, after tossing the coin a large number of times, the ratio stabilizes around 1/3, then we would say, with a high degree of confidence, that the coin is a biased coin, or not fair, and the chance of getting heads on this particular coin is 1 out of 3.

> The **empirical concept of probability** is that of ''relative frequency,'' the ratio of the total number of occurrences of a situation to the total number of times the experiment is repeated.
>
> When the number of trials is large, the relative frequency provides a satisfactory measure of the probability associated with a situation of interest. This is one of the so-called laws of large numbers of probability theory.

In interpreting a probability statement, we should think in terms of repeated performance of the experiment, as in the following illustrations:

(a) The probability of drawing a spade from a deck of cards is 0.25. Empirically this means that if we were to pick a card from the deck repeatedly, every time returning the card to the deck and shuffling the deck properly, approximately 25 percent of the times we will get a spade. If the experiment is repeated a large number of times, the approximation will be very close to 0.25.

(b) The probability of precipitation today is 0.80. The weatherman has calculated from past records that when the weather conditions were such as they are today, it has rained approximately 80 percent of the time.

(c) A patient is told by his doctor that the probability of his recovery is 95 percent. Should the patient derive comfort from the doctor's statement? What the doctor's statement means presumably is that with similar patients, about 95 percent have been cured. If the doctor has been in practice for many years and has had many such patients, the figure 95 percent would be reliable as a measure of the doctor's success. Of course, the particular patient can take it for what it is worth as far as his own recovery is concerned.

SECTION 3-1   EXERCISES

Give empirical interpretations of the following probability statements. Philosophize if you have to.

1. For a person who enters a certain contest, the probability of his winning a prize is 0.12.
2. The probability that a child born in the ghetto will be successfully employed when grown up is 16 percent.
3. The probability that the stock market will go up on Monday is 0.62.
4. The probability of striking oil when a well is drilled in a certain region is 0.28.
5. In the United States, the probability that a person will catch cold during winter is 3/4.
6. The probability that the accused will be acquitted is 80 percent.
7. The probability that Senator Thornapple will vote for the farm bill is very slim.
8. The probability that transistors of a certain brand will last beyond 500 hours is 0.7.
9. The probability that the airline pilots will go on strike is 0.42.
10. The probability that candidate X will win in the mayoral primary election is 0.58.

3-2 SAMPLE SPACE, EVENTS, AND THE MEANING OF PROBABILITY

Because notions of set theory in mathematics are at the heart of probability theory, we begin by introducing some related terms.

A **set** is a gathering together of objects that we choose to isolate because they share some common characteristic. The intuitively familiar notion of a set is a collection of objects, requiring only that we can determine unambiguously whether or not any given object is a member of the collection.

When a complete list of the members, or elements, of a set is given, it is customary to write the list within braces, separated by commas. For example,

{apple, peach, pear, banana}

is a set listing four kinds of fruit.

Since we are interested only in what objects are in the set, there is no reason why the members should be written in any particular order. Thus the sets

FIGURE 3-1

Venn diagrams displaying relationships between two sets A and B.

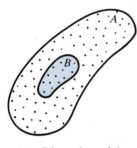   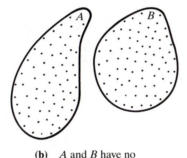   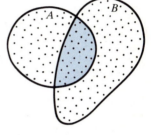

(a) B is a subset of A **(b)** A and B have no   **(c)**   The shaded portion shows
                                                 members in common   members common to A and B

$$\{\text{apple, peach, pear, banana}\} \quad \text{and} \quad \{\text{banana, peach, pear, apple}\}$$

represent the same collection and consequently the same set. Sets are commonly denoted using uppercase letters. For instance, we might use the letter F (rather suggestively for fruit) to denote the set above. Thus

$$F = \{\text{apple, peach, pear, banana}\}.$$

Any subcollection of members from a set A is called a **subset** of A.

Thus if $A = \{\text{Tom, Dick, Harry, Jane}\}$ and $B = \{\text{Dick, Jane}\}$, then B is a subset of A.

Sets and relations between them may be illustrated conveniently by geometric configurations. In such diagrams, called **Venn diagrams,** named for the English mathematician John Venn (1834–1923), sets are identified as regions and the members of sets as points in those regions. Venn diagrams are especially useful for visualizing the relationship between two or more sets.

The Venn diagrams in Figure 3-1 display relationships between two sets A and B. Part (a) shows that B is a subset of A, Part B shows that A and B have no members in common, and Part (c) shows that A and B have some members in common indicated by the shaded portion.

SAMPLE SPACE AND EVENTS

Let us now turn our attention back to probability. If asked the chance of getting a head when a coin is tossed, how many of us would not say that it is one chance in two? Put mathematically, the chance is $1/2$. The reasoning would be that there are two possibilities on a toss, namely, a head or a tail, and there is only one head! If we rolled a die, the chance of getting an even number would be $1/2$. Once again, our argument would be that there are six possible outcomes, namely, 1, 2, 3, 4, 5, 6, of which 2, 4, and 6 are even—three outcomes out of six!

FIGURE 3-2

The Venn diagram for the sample space.

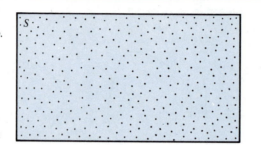

Our intuitive notion lays the groundwork for our formal definitions. As can be seen, the knowledge of all the possible outcomes when a coin is tossed, or a die is rolled, is important. The tossing of a coin or the rolling of a die constitutes an *experiment*. In probability and statistics the term **experiment** is used in a very wide sense and refers to any procedure that yields a collection of outcomes. In this sense, "a baby being born" is an experiment whose possible outcomes are: boy, girl.

The set whose elements are all the possible outcomes of an experiment is called the **sample space.** The elements of the sample space are called **sample points.**

Every conceivable outcome of the experiment is listed in the sample space. When the experiment is actually performed, it will result in exactly one outcome, that is, one sample point, of this set. The sample space, which will be denoted by the letter S, provides the basis for our discussion and we work within the scope of this set. Usually the Venn diagram for the sample space is shown as a rectangle as in Figure 3-2.

A sample space is called a **finite sample space** if it has a finite number of outcomes in it. If there are N outcomes in S, we can identify them as $e_1, e_2, \ldots, e_N$. This would permit us to write the sample space S as

$$S = \{e_1, e_2, \ldots, e_N\}.$$

As an example, consider a conventional die in the form of a cube with one to six dots on its faces. We can think of tossing such a die as an experiment. There are six possible outcomes:

so that

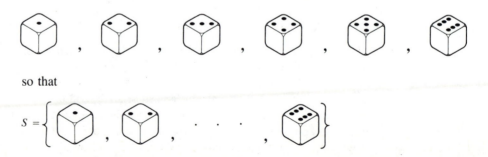

FIGURE 3-3

Venn diagram
displaying
event *E* as a
subset of *S*.

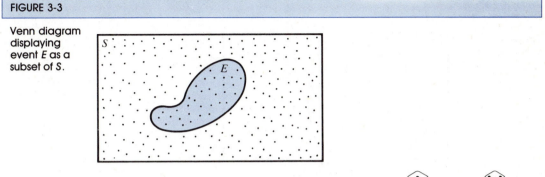

Here $N = 6$, and e_1 would then stand for the outcome ⚀ , e_2 for ⚁ , and so on.

If the number of outcomes in the sample space is not finite, then it is said to be an **infinite sample space.** For an example of an infinite sample space, suppose when Jill throws a dart at a circular dart board, we are interested in the point where she hits the board. The sample space is infinite because there is an infinite number of points on the board where the dart could conceivably hit. In what follows, we shall restrict our discussion only to finite sample spaces.

A particular situation of interest, such as ''an even number of dots appear when a die is tossed,'' is called an *event. We say that an event has occurred if one of the outcomes that make up the event takes place.* Thus the event ''an even number of dots appear when a die is tossed'' will occur if either 2, 4, or 6 dots show up. (We certainly do not expect 2, 4, and 6 dots to show up simultaneously on one toss; they cannot.) In other words, this particular event will be realized if any one of the outcomes in the set $\left\{ ⚁ , ⚃ , ⚅ \right\}$ takes place. Thus any verbal description of a situation can be characterized by a subset of outcomes of the sample space. We, therefore, define an event as follows:

An **event** is a subset of the sample space. A **simple event,** or *elementary event,* is one that contains only one outcome; that is, it contains one sample point.

An event E with reference to the sample space S can be expressed by means of a Venn diagram as in Figure 3-3.

Given any sample space with N outcomes $e_1, e_2, \ldots, e_N$, there are exactly N simple events $\{e_1\}, \{e_2\}, \ldots, \{e_N\}$. Any event E can be represented as comprised of simple events of outcomes in E. For example, if $E = \{e_2, e_5, e_8\}$, then E is made up of the three simple events $\{e_2\}, \{e_5\}, \{e_8\}$.

An **impossible event** is one that has no outcomes in it and, consequently, cannot occur. On the other hand, a **sure event** is one that has all the outcomes of the sample space in it and will, therefore, definitely occur when the experiment is performed. Thus the sample space constitutes a sure event.

For example, suppose we make a list of all the presidents of the United States before 1986, and choose one name from the list. The event ''the president is a female'' is an impossible event; whereas the events ''the president is a male'' and ''the president was born after 1400 A.D.'' are sure events.

FIGURE 3-4

Venn diagram
displaying the
sample space S
and the events
C and O.

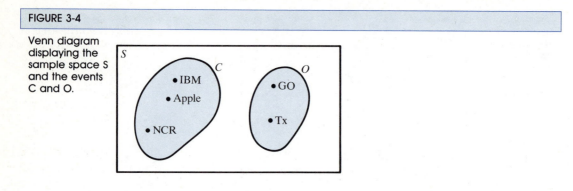

EXAMPLE 1   Mrs. Shapiro owns three computer stocks, namely, IBM, NCR, and Apple, and two oil stocks, GO and Tx. She calls her broker and asks him to sell one stock of his choosing.

"Which stock might the broker sell?" is the question that Mrs. Shapiro has in mind. Just one possibility is that the broker will sell NCR. All the possibilities that she contemplates comprise the sample space S, which can be written as

$$S = \{\text{IBM, NCR, Apple, GO, Tx}\}.$$

There are five simple events, namely,

$$\{\text{IBM}\}, \{\text{NCR}\}, \{\text{Apple}\}, \{\text{GO}\}, \{\text{Tx}\}.$$

We define below some events, describing them verbally and also listing them as sets. These events are also shown in Figure 3-4.

C: The broker sells a computer stock = {IBM, NCR, Apple}
O: The broker sells an oil stock = {GO, Tx}   ▬▬

EXAMPLE 2   As a slight alteration of Example 1, suppose Mrs. Shapiro asks the broker to sell two stocks of his choice.

(a) Give the appropriate sample space.
(b) List as sets the outcomes in the following events:

   (i) C: Both stocks are computer stocks.
   (ii) O: No computer stock is sold; that is, both stocks are oil stocks.
   (iii) B: The broker sells one computer stock and one oil stock.

SOLUTION   (a) It is conceivable that the broker sells IBM and NCR. We shall represent this outcome as [IBM, NCR]. There are nine other ways. The entire set S of ten outcomes is given by

$$S = \{[\text{IBM, NCR}], [\text{IBM, Apple}], [\text{IBM, GO}], [\text{IBM, Tx}],$$
$$[\text{NCR, Apple}], [\text{NCR, GO}], [\text{NCR, Tx}], [\text{Apple, GO}],$$
$$[\text{Apple, Tx}], [\text{Tx, GO}]\}.$$

Each pair is a sample point.

(b) (i) "Both stocks are computer stocks" describes those outcomes of S where both entries are computer stocks. Hence

$$C = \{[\text{IBM, NCR}], [\text{IBM, Apple}], [\text{NCR, Apple}]\}.$$

(ii) Here we consider those outcomes of S in which there is no computer entry. There is only one such outcome, namely, [Tx, GO]. Hence

$$O = \{[\text{Tx, GO}]\}.$$

Incidentally, O is a simple event.

(iii) The outcomes comprising this event will clearly have one computer entry and one oil entry. So

$$B = \{[\text{IBM, GO}], [\text{IBM, Tx}], [\text{NCR, Tx}], [\text{NCR, GO}],$$
$$[\text{Apple, Tx}], [\text{Apple, GO}]\}.$$

We observe in passing that the three events C, O, and B together account for the entire sample space. ■

EXAMPLE 3   Vincent and Joel play a game where they simultaneously exhibit their right hands with one, two, three, or four fingers extended.

(a) Write all the outcomes in the sample space.
(b) List the outcomes in the following events:

(i) M: Vincent and Joel extend the same number of fingers.
(ii) F: Vincent and Joel together extend four fingers.
(iii) E: Vincent shows an even number of fingers.

SOLUTION   (a) We shall give the outcomes as *ordered pairs* where the first component represents the number of fingers on Vincent's hand and the second component, those on Joel's hand. The sample space can be given as

$$S = \{(1, 1), (1, 2), (1, 3), (1, 4), (2, 1), (2, 2), (2, 3), (2, 4),$$
$$(3, 1), (3, 2), (3, 3), (3, 4), (4, 1), (4, 2), (4, 3), (4, 4)\}$$

and has 16 outcomes.

(b) The events are now easily given as subsets of S as follows:

(i) M: Vincent and Joel extend the same number of fingers $= \{(1, 1),$ $(2, 2), (3, 3), (4, 4)\}$
(ii) F: Vincent and Joel together extend four fingers $= \{(1, 3), (2, 2),$ $(3, 1)\}$
    The sum of the components of each sample point is 4.
(iii) E: Vincent shows an even number of fingers $= \{(2, 1), (2, 2), (2, 3),$ $(2, 4), (4, 1), (4, 2), (4, 3), (4, 4)\}$.
    Notice that the first component of each of the outcomes in this event is an even number, either a 2 or a 4. ■

THE THEORETICAL DEFINITION OF PROBABILITY

Now that we have discussed the important concepts of sample space and events, we can define the probability of an event. Let us assume that the sample space S has N outcomes $e_1, e_2, \ldots, e_N$, so that there are N simple events $\{e_1\}, \{e_2\}, \ldots, \{e_N\}$.

DEFINITION OF PROBABILITY

The **probability of a simple event** $\{e\}$ consisting of outcome e is a number which is assigned to $\{e\}$. It is denoted by $P(\{e\})$ and satisfies the following conditions:

1. $P(\{e\})$ is always between zero and one; that is, $0 \leqslant P(\{e\}) \leqslant 1$.

2. The sum of the probabilities of all the simple events is 1; that is,

$$P(\{e_1\}) + P(\{e_2\}) + \cdots + P(\{e_N\}) = 1.$$

The **probability of an event** A, denoted by $P(A)$, is defined as the sum of the probabilities assigned to the simple events that comprise the event A. As you might expect, the impossible event has probability 0 and the sure event S has probability 1.

Thus, if $A = \{e_2, e_4, e_7\}$, then $P(A) = P(\{e_2\}) + P(\{e_4\}) + P(\{e_7\})$.

EXAMPLE 4   A loaded die is rolled.

(a) Show that the following assignment of probabilities would be acceptable:

$$P(\{1\}) = \frac{1}{3} \qquad P(\{4\}) = \frac{1}{12}$$

$$P(\{2\}) = \frac{1}{4} \qquad P(\{5\}) = \frac{1}{6}$$

$$P(\{3\}) = \frac{1}{12} \qquad P(\{6\}) = \frac{1}{12}$$

(b) Find the probabilities of the following events:

(i)   The number is a multiple of 3.
(ii) The number is even.
(iii) The number is even or a multiple of 3.

SOLUTION   (a) We see that the probabilities of all the simple events are between 0 and 1. Also,

$$P(\{1\}) + P(\{2\}) + P(\{3\}) + P(\{4\}) + P(\{5\}) + P(\{6\})$$

$$= \frac{1}{3} + \frac{1}{4} + \frac{1}{12} + \frac{1}{12} + \frac{1}{6} + \frac{1}{12}$$

$$= 1.$$

Therefore, the given assignment of probabilities would be acceptable.

(b) (i) The probability that the number is a multiple of 3 is

$$P(\{3, 6\}) = P(\{3\}) + P(\{6\}) = \frac{1}{12} + \frac{1}{12} = \frac{1}{6}.$$

(ii) The probability that the number is even is

$$P(\{2, 4, 6\}) = P(\{2\}) + P(\{4\}) + P(\{6\})$$
$$= \frac{1}{4} + \frac{1}{12} + \frac{1}{12} = \frac{5}{12}.$$

(iii) The probability that the number is even or a multiple of 3 is

$$P(\{2, 3, 4, 6\}) = P(\{2\}) + P(\{3\}) + P(\{4\}) + P(\{6\})$$
$$= \frac{1}{4} + \frac{1}{12} + \frac{1}{12} + \frac{1}{12} = \frac{1}{2}. \quad \blacksquare$$

EXAMPLE 5 A certain number of playing cards—some spades, some hearts, some diamonds, and some clubs—are laid on a table with their faces down. Suppose when a card is picked, the probability of picking a spade is twice that of picking a heart; the probability of picking a heart is three times that of picking a diamond; and the probability of picking a diamond is four times that of picking a club. Find the probabilities of the following events:

(a) A: The card is red.
(b) B: The card is a spade or a diamond.

SOLUTION For our interest, the relevant sample space is $S = \{$spade, heart, diamond, club$\}$, with four outcomes. First, we shall find probabilities assigned to each of the simple events. If we assume that $P(\{$club$\}) = x$, then

$$P(\{\text{diamond}\}) = 4x$$
$$P(\{\text{heart}\}) = 3(4x) = 12x$$
$$P(\{\text{spade}\}) = 2(12x) = 24x.$$

Now, we know that the sum of the probabilities of the simple events should be 1. Therefore,

$$x + 4x + 12x + 24x = 1.$$

That is,

$$41x = 1$$

and, consequently,

$$x = \frac{1}{41}.$$

Therefore, the probabilities assigned to simple events are given by

$$P(\{\text{club}\}) = x = \frac{1}{41} \qquad P(\{\text{diamond}\}) = 4x = \frac{4}{41}$$

$$P(\{\text{heart}\}) = 12x = \frac{12}{41} \qquad P(\{\text{spade}\}) = 24x = \frac{24}{41}$$

(a) Since $A = \{\text{heart, diamond}\}$,

$$P(A) = P(\{\text{heart}\}) + P(\{\text{diamond}\}) = \frac{12}{41} + \frac{4}{41} = \frac{16}{41}.$$

(b) We have $B = \{\text{spade, diamond}\}$. Therefore,

$$P(B) = P(\{\text{spade}\}) + P(\{\text{diamond}\}) = \frac{24}{41} + \frac{4}{41} = \frac{28}{41}.$$ ■

We now consider an important special case, which arises when each outcome is assigned the same probability. In this case the outcomes are said to be **equally likely outcomes.** Since the sum of all the probabilities assigned to the simple events should be 1, it is obvious that each outcome will have a probability equal to $1/N$, where N is the number of outcomes in the sample space. Consequently, if an event A contains r outcomes, then on the basis of the definition of probability of an event, we get $P(A)$ by adding r terms, each equal to $1/N$. That is,

$$P(A) = \frac{1}{N} + \frac{1}{N} + \cdots + \frac{1}{N} = \frac{r}{N}.$$

This leads to the classical definition of probability.

> **CLASSICAL DEFINITION OF $P(A)$**
>
> If the sample space contains a finite number of outcomes, all equally likely, then
>
> $$P(A) = \frac{\text{number of outcomes in } A}{\text{number of outcomes in } S}.$$

The outcomes in A are often referred to as the outcomes favorable to A. Letting $n(E)$ be the number of outcomes in an event E, we can write

$$P(A) = \frac{n(A)}{n(S)},$$

for any event A.

In dealing with problems on probability, we will often use statements such as "An unbiased coin is tossed," "A fair die is rolled," "An object is picked at random," and so on. Such expressions are all meant to suggest that the outcomes of the experiment are equally likely.

EXAMPLE 6   Suppose a fair die is rolled. Find the probabilities of the following events:

(a) E: An even number of dots show up.
(b) T: The number of dots showing up is a multiple of 3.
(c) B: The number of dots showing up is even or a multiple of 3.

SOLUTION   The sets describing these events are given by

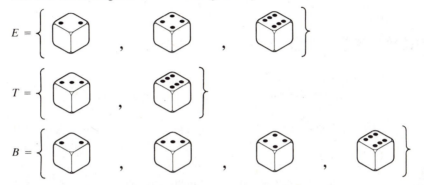

Because the die is a fair die, all six outcomes in the sample space are equally likely. Since there are three outcomes in E, two in T, and four in B, we get

(a) $P(E) = \dfrac{3}{6} = \dfrac{1}{2}$ (b) $P(T) = \dfrac{2}{6} = \dfrac{1}{3}$ (c) $P(B) = \dfrac{4}{6} = \dfrac{2}{3}.$ ▬

EXAMPLE 7   In Example 2, suppose the broker sells two stocks chosen at random. Find the probabilities of the following events:

(a) C: Both stocks are computer stocks.
(b) O: No computer stock is sold.
(c) B: One computer stock and one oil stock are sold.

SOLUTION   We have already discussed the sample space S and the events C, O, and B in Example 2. If we count the outcomes in each of these events we find

$$n(S) = 10 \qquad n(C) = 3 \qquad n(O) = 1 \qquad n(B) = 6.$$

Since the two stocks are chosen at random, which implies that the outcomes are equally likely, we obtain

$$P(C) = \frac{3}{10} \qquad P(O) = \frac{1}{10} \qquad P(B) = \frac{6}{10}. \quad ▬$$

EXAMPLE 8   With reference to Example 3, find the probabilities of the following events:

(a) M: Vincent and Joel extend the same number of fingers.
(b) F: Vincent and Joel together extend four fingers.
(c) E: Vincent shows an even number of fingers.

SOLUTION   As we have seen, there are four outcomes in the event M, three in the event F, and eight in the event E. Also, there are sixteen outcomes in S, which we shall assume to be equally likely. Hence,

(a) $P(M) = \dfrac{4}{16} = \dfrac{1}{4}$ (b) $P(F) = \dfrac{3}{16}$ (c) $P(E) = \dfrac{8}{16} = \dfrac{1}{2}$.   ▬

SECTION 3-2   EXERCISES

1. Three children are born in a family. On any birth the child could be a son or a daughter. Using s to represent a son and d to represent a daughter, give the sample space S.

2. A committee of 2 is selected from a group consisting of 5 people, Juan, Dick, Mary, Paul, and Jane. Write the following events as sets, listing the outcomes they contain:

   (a) The sample space.

   (b) Both members on the committee are males.

   (c) Exactly one member is a male.

3. A box contains 7 tubes, 3 of which are defective. Two tubes are selected at random.

   (a) List six of the outcomes in the sample space.

   (b) List six of the outcomes in the event specifying 1 defective and 1 nondefective tube

   > [*Hint:* In this exercise, identify the 7 tubes $G_1, G_2, G_3, G_4, D_1, D_2, D_3$, where G_1, G_2, G_3, G_4 represent the good tubes and D_1, D_2, D_3 represent the defective tubes. Now list the outcomes in a suitable way.]

4. Let a penny, a nickel, and a die be tossed. If you are observing the outcome on the penny, the nickel, and the die in that order, list all the outcomes in the following sets:

   (a) The sample space.

   (b) A: Three dots appear on the die.

   (c) B: An odd number of dots appear on the die.

   (d) C: Heads appear on the two coins.

   (e) D: A head appears on the penny and a tail on the nickel.

5. Suppose the sample space $S = \{e_1, e_2, e_3\}$. Indicate the cases where we have an acceptable assignment of probabilities to the simple events:

   (a) $P(\{e_1\}) = \dfrac{1}{3}$, $P(\{e_2\}) = \dfrac{1}{2}$, $P(\{e_3\}) = \dfrac{1}{3}$

   (b) $P(\{e_1\}) = \dfrac{1}{3}$, $P(\{e_2\}) = -\dfrac{1}{3}$, $P(\{e_3\}) = 1$

   (c) $P(\{e_1\}) = \dfrac{1}{5}$, $P(\{e_2\}) = \dfrac{3}{5}$, $P(\{e_3\}) = \dfrac{1}{5}$

   (d) $P(\{e_1\}) = \dfrac{1}{2}$, $P(\{e_2\}) = 0$, $P(\{e_3\}) = \dfrac{1}{2}$

6. A card is drawn at random from a standard deck of cards. Find the probability that the card is

   (a) a black card

   (b) a face card

   (c) a black card or a face card.

7. A letter of the English alphabet is chosen at random. Find the probability of the event that the letter selected

   (a) is a vowel

   (b) is a consonant

   (c) follows the letter p

   (d) follows the letter p and is a vowel

   (e) follows the letter p or is a vowel.

8. An urn contains 10 tubes of which 4 are known to be defective. If one tube is picked at random, find the probability that it is

   (a) defective   (b) nondefective.

9. Four hundred people attending a party are each given a number, 1 to 400. One number is called at random. Find the probability that the number called

   (a) is 123   (b) has the same three digits   (c) ends in 9.

10. A fair die is tossed once.

    (a) List all the outcomes in the event ''an even number or a number less than 3 shows up.''

    (b) What is the probability that an even number or a number less than 3 shows up?

11. Bertha has the following four books on a shelf: history, calculus, statistics, and English. If she picks two books at random, list all the outcomes in the following events:

    (a) The sample space.

    (b) She picks the statistics book as one of the two books.

12. A die is rolled twice. List the outcomes in each of the following events and then find the probability of each event:

    (a) The sample space.

    (b) The same number is on the two tosses.

    (c) The total score is 7.

    (d) The total score is even.

    (e) The total score is less than 6.

    (f) The outcome on the first toss is greater than 4.

    (g) Each toss results in the same even number.

13. A letter is chosen at random from the letters of the English alphabet. Find the probability of each of the following events:

    (a) The letter is j.

    (b) The letter is one of the letters in the word *board*.

    (c) The letter is not in the word *board*.

    (d) The letter is between d and m, both inclusive.

(e) The letter is in the word *card* or the word *board*.

(f) The letter is in the both the words *card* and *board*.

(g) The letter is in the word *board* but not in the word *card*.

14. Suppose probabilities are assigned to the simple events of $S = \{e_1, e_2, e_3, e_4\}$ as follows:
$$P(\{e_1\}) = 3P(\{e_2\}), \; P(\{e_3\}) = P(\{e_1\}), \; P(\{e_4\}) = 4P(\{e_2\})$$

(a) Find the probabilities assigned to the simple events.

(b) Find $P(A)$, where $A = \{e_1, e_3, e_4\}$.

15. A person decides to take a vacation and has three places in mind: Las Vegas, the Bahamas, and Hawaii. She is twice as likely to go to Hawaii as to Las Vegas, and three times as likely to go to Las Vegas as to the Bahamas. Find the probability that she vacations in Las Vegas.

16. The probabilities that a student will receive an A, B, C, D, F, or incomplete in a course are, respectively, 0.05, 0.15, 0.20, 0.25, 0.30, and x. Find the probability that the student will

   (a) receive an incomplete   (b) get at most a C   (c) get a B or a C.

17. Suppose a coin is tossed three times yielding the sample space
$S = \{HHH, HHT, HTH, THH, HTT, TTH, THT, TTT\}$, where HTH means heads on the first toss, tails on the second toss, and heads on the third toss. Assume the assignment of probabilities to the simple events as follows:

$$P(\{HHH\}) = 0.064, \; P(\{HHT\}) = 0.096, \; P(\{HTH\}) = 0.096,$$
$$P(\{THH\}) = 0.096, \; P(\{HTT\}) = 0.144, \; P(\{TTH\}) = 0.144,$$
$$P(\{THT\}) = 0.144, \; P(\{TTT\}) = 0.216.$$

Find the probability of

   (a) heads on the first toss   (b) heads on the second toss
   (c) exactly one head   (d) at least one head.

3-3 SOME RESULTS OF ELEMENTARY PROBABILITY

In this section we shall present some results that follow from our definition of probability.

THE ADDITION LAW:

If A and B are two events, then the probability that A or B (or both)* occur is equal to the sum of their probabilities minus the probability of their simultaneous occurrence. That is,

$$P(A \text{ or } B) = P(A) + P(B) - P(A \text{ and } B)$$

FIGURE 3-5

Venn diagram in which shaded region represents the event A or B. Darker shaded region represents the event A and B.

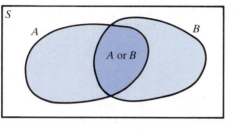

To justify this result consider the Venn diagram in Figure 3-5. On the left side of the equation above, $P(A$ or $B)$ stands for the probability assigned to the sample points in the shaded portion in the figure.

As a first step in justifying the right side of the equation, we might add $P(A)$, the probability assigned to the sample points in A, to $P(B)$, the probability assigned to the sample points in B, to get $P(A) + P(B)$. This is certainly not the same as $P(A$ or $B)$ because we realize immediately that the sample points in the overlap "A and B" are included twice; once with A and once with B. Therefore, to compensate for counting these points twice, we must subtract $P(A$ and $B)$ from $P(A) + P(B)$ to obtain $P(A$ or $B)$.

EXAMPLE 1   On a TV quiz show a contestant is asked to pick an integer at random from the first 100 consecutive positive integers, that is, the integers 1 through 100. If the number picked is divisible by 12 or 9, the contestant will win a free trip to the Bahamas. What is the probability that the contestant will win the trip?

SOLUTION   Since one integer is picked at random from the one hundred integers, the sample space consists of 100 equally likely outcomes and is given by $S = \{1, 2, 3, \ldots, 100\}$. Let two events be defined

   A: The number is divisible by 12

   B: The number is divisible by 9.

Then

   $A = \{12, 24, 36, 48, 60, 72, 84, 96\}$

   $B = \{9, 18, 27, 36, 45, 54, 63, 72, 81, 90, 99\}$.

Now the event "A and B" consists of the set of integers that are divisible by 12

---

*In set theory the event "A or B" is denoted as $A \cup B$ and is called the *union* of events A and B. Also, the event "A and B" is denoted as $A \cap B$ and is called the *intersection* of the events.

and 9. This is, of course, the set of integers divisible by 36 and, therefore, consists of the set $\{36, 72\}$. Hence

$$P(A) = \frac{8}{100}, \qquad P(B) = \frac{11}{100}, \qquad P(A \text{ and } B) = \frac{2}{100}.$$

Applying the addition law, we now get

$$P(A \text{ or } B) = \frac{8}{100} + \frac{11}{100} - \frac{2}{100} = \frac{17}{100}.$$

The probability of winning a free trip is 0.17. ▬▬

EXAMPLE 2   In a certain area, television channels 4 and 7 are affiliated to the same national network. The probability that channel 4 carries a particular sports program is 0.5, that channel 7 carries it is 0.7, and the probability that they both carry it is 0.3. What is the probability that José will be able to watch the program on either of the channels?

SOLUTION   Let the events be defined

   A: Channel 4 carries the program

   B: Channel 7 carries the program.

We wish to find the probability that channel 4 carries the program *or* channel 7 carries the program, that is, $P(A \text{ or } B)$. Since $P(A) = 0.5$, $P(B) = 0.7$, and $P(A \text{ and } B) = 0.3$, from the addition law we get

   $P(A \text{ or } B) = 0.5 + 0.7 - 0.3 = 0.9.$

Hence the probability that José will be able to watch the program is 0.9. ▬▬

EXAMPLE 3   The town of Metroville has two ambulance services: the city service and a citizens' service. In an emergency, the probability that the city service responds is 0.6, the probability that the citizens' service responds is 0.8, and the probability that either of the services responds is 0.9. Find the probability that both services will respond to an emergency.

SOLUTION   If A represents the event that the city service responds, and B the event that the citizens' service responds, then we are interested in finding $P(A \text{ and } B)$. We are given the following information:

   $P(A) = 0.6 \qquad P(B) = 0.8 \qquad P(A \text{ or } B) = 0.9$

By the addition law, we have

   $P(A \text{ or } B) = P(A) + P(B) - P(A \text{ and } B).$

Therefore, transposing appropriately, we have

$$P(A \text{ and } B) = P(A) + P(B) - P(A \text{ or } B)$$
$$= 0.6 + 0.8 - 0.9$$
$$= 0.5.$$

In the long run, about 50 percent of the times both services will respond to an emergency. ▪

We now consider a special case of the addition law. Suppose we draw a card from a standard deck. Consider the two events A and B, where A stands for getting a spade and B, for getting a red card. We know that if a card is a spade it cannot be red and, conversely, if it is a red card then it cannot be a spade. Two events of this type are said to be mutually exclusive and are defined as follows:

MUTUALLY EXCLUSIVE EVENTS

Two events A and B are **mutually exclusive events** if they do not have any outcome in common and, consequently, cannot occur simultaneously. It follows, therefore, that

$P(A \text{ and } B) = 0$ **for mutually exclusive events A and B.**

By the addition law of probability, we then have the following consequence:

$P(A \text{ or } B) = P(A) + P(B)$ **if A and B are mutually exclusive.**

This result can be extended to any finite number of mutually exclusive events, that is, events that cannot occur together in pairs. For instance, if A, B, and C are three events, then

$$P(A \text{ or } B \text{ or } C) = P(A) + P(B) + P(C)$$

if A, B, and C are mutually exclusive. Here note that A and B cannot occur simultaneously. Similarly, A and C cannot occur together, nor can B and C.

EXAMPLE 4 A box contains 3 red, 4 green, and 5 white balls. One ball is picked at random. What is the probability that it will be red or white?

SOLUTION Let us write the events

R: A red ball is picked.

W: A white ball is picked.

The events R and W are mutually exclusive, because if a ball is red it cannot be white and vice versa. Therefore, $P(R \text{ and } W) = 0$ and

$$P(R \text{ or } W) = P(R) + P(W)$$

$$= \frac{3}{12} + \frac{5}{12} = \frac{8}{12} = \frac{2}{3}. \qquad \blacksquare$$

EXAMPLE 5   How is Jane prepared for her test? Well, from past experience she thinks there is a probability of 0.1 that she will get an A, a probability of 0.4 that she will get a B, and a probability of 0.3 that she will get a C. What is the probability that she will get at least a C?

SOLUTION   Jane will get at least a C if she gets an A *or* she gets a B *or* she gets a C. Now the events ''getting an A,'' ''getting a B,'' and ''getting a C'' are mutually exclusive. One cannot get two grades on one test at the same time. Therefore,

$$P(\text{Jane gets at least a C}) = P(\text{Jane gets an A}) + P(\text{Jane gets a B})$$

$$+ P(\text{Jane gets a C})$$

$$= 0.1 + 0.4 + 0.3 = 0.8. \qquad \blacksquare$$

EXAMPLE 6   Consider a family with four children. A survey indicates that it is reasonable to believe that the probabilities of having 0, 1, 2, 3, or 4 sons in such a family are, respectively, 0.1, 0.2, 0.35, 0.2, and 0.15. Find the probability that in a family with four children there will be

(a) at least two sons
(b) at most two sons.

SOLUTION   (a) There will be at least two sons if there are two sons *or* three sons *or* four sons. (Here, when we say two sons, for example, we mean *exactly* two sons. In what follows we shall adopt this usage of the phrase in similar contexts.) The three events are mutually exclusive since, for example, a family that has three sons cannot, at the same time, have two sons or four sons. Therefore,

$$P(\text{at least two sons}) = P(2 \text{ sons}) + P(3 \text{ sons}) + P(4 \text{ sons})$$

$$= 0.35 + 0.2 + 0.15$$

$$= 0.7.$$

(b) Arguing as in Part (a), we have

$$P(\text{at most two sons}) = P(0 \text{ sons}) + P(1 \text{ son}) + P(2 \text{ sons})$$

$$= 0.1 + 0.2 + 0.35$$

$$= 0.65. \qquad \blacksquare$$

EXAMPLE 7 From her long experience, a car salesperson has found that the probability that a customer will buy a sports car is 0.1, that the customer will buy a station wagon is 0.2, and that the customer will buy a truck is 0.15. What is the probability that a prospective customer will buy one of the given types of vehicles?

SOLUTION Consider the three events "buying a sports car," "buying a station wagon," and "buying a truck." It is, of course, possible that a customer will buy both a sports car *and* a station wagon. In that case, certainly, the events are not mutually exclusive. We shall assume that our car salesperson knows that this is not the case, and that the three events are indeed mutually exclusive. Therefore,

$$P\begin{pmatrix} \text{customer buys a sports car} \\ \textit{or} \text{ a station wagon} \\ \textit{or} \text{ a truck} \end{pmatrix} = 0.1 + 0.2 + 0.15 = 0.45.$$

This example illustrates that events may not be assumed to be mutually exclusive. Always satisfy yourself that you have not made this assumption blindly.

LAW OF THE COMPLEMENT

The probability that an event A will not occur is equal to 1 minus the probability that it will occur. That is, for any event A

$$P(not\ A) = 1 - P(A)$$

where *not* A means the event "nonoccurrence of A." The event *not* A is called the *complement* of the event A.

Let us prove the result above. It is obvious that, in any situation, of the two events A and *not* A, one of them is bound to occur. Hence,

$$P(A \text{ or } not\ A) = 1$$

Also, if A occurs, then the event *not* A cannot occur and vice versa. In other words, the events A and *not* A are mutually exclusive. Consequently,

$$P(A) + P(not\ A) = 1$$

or $$P(not\ A) = 1 - P(A).$$

As a few examples of the applications of this law: (a) If the probability that the stock goes up is 0.6, then the probability that it will not go up is $1 - 0.6 = 0.4$; (b) if the probability of precipitation is 0.8, then the probability of no precipitation is $1 - 0.8 = 0.2$; (c) if the probability of at least three sons in a family of five children is 0.4, then the probability of at most two sons is $1 - 0.4 = 0.6$.

EXAMPLE 8 A person owns five stocks. The probabilities that exactly 0, 1, 2, 3, 4, or 5 of his stocks go up in price on a given day are, respectively, 0.1, 0.2, 0.3, 0.22, 0.1, and 0.08. Find the probability that

(a) at least one stock goes up
(b) at most four stocks go up.

SOLUTION Let us write A_0, A_1, A_2, A_3, A_4, and A_5, respectively, for the six events that exactly 0, 1, 2, 3, 4, or 5 of the stocks go up in price. Observe that the six events are mutually exclusive.

(a) P(at least one stock goes up) $= P(A_1 \text{ or } A_2 \text{ or } A_3 \text{ or } A_4 \text{ or } A_5)$

$$= P(A_1) + P(A_2) + P(A_3) + P(A_4) + P(A_5)$$
$$= 0.2 + 0.3 + 0.22 + 0.1 + 0.08$$
$$= 0.9.$$

Alternatively, notice that the event "at least one stock goes up" negates the event "no stock goes up." Hence, by the law of the complement,

$$P\text{(at least one stock goes up)} = 1 - P\text{(no stock goes up)}$$
$$= 1 - 0.1 = 0.9.$$

The latter approach is much more efficient especially in a case where, for instance, there were twenty stocks involved instead of five.

(b) P(at most four stocks go up) $= P(A_0 \text{ or } A_1 \text{ or } A_2 \text{ or } A_3 \text{ or } A_4)$

$$= P(A_0) + P(A_1) + P(A_2)$$
$$+ P(A_3) + P(A_4)$$
$$= 0.1 + 0.2 + 0.3 + 0.22 + 0.1$$
$$= 0.92.$$

An alternate method is to note that the event of interest negates the event that all five stocks go up. Hence,

$$P\text{(at most four stocks go up)} = 1 - P(A_5)$$
$$= 1 - 0.08 = 0.92. \quad \blacksquare$$

EXAMPLE 9 Two employees on a university campus, Tom from Plant Operations and Becky from Public Safety, are assigned to check that a certain building on the campus is locked over the weekends. Suppose that on any weekend the probability that Tom checks is 0.96, that Becky checks is 0.98, and that they both check is 0.95. What is the probabilty that neither Tom nor Becky check on a weekend?

SOLUTION Let us write the events

T: Tom checks

B: Becky checks.

The event "neither Tom nor Becky check" can be considered as the event that it is *not* the case that either Tom *or* Becky check. Thus, the event that we are interested in is "not (T or B)". Now

$$P(\text{not } (T \text{ or } B)) = 1 - P(T \text{ or } B)$$
$$= 1 - [P(T) + P(B) - P(T \text{ and } B)]$$
$$= 1 - [0.96 + 0.98 - 0.95]$$
$$= 0.01. \quad \blacksquare$$

SECTION 3-3   EXERCISES

1. If $P(A) = 0.4$, $P(B) = 0.5$, $P(A \text{ and } B) = 0.25$, find $P(A \text{ or } B)$.

2. If $P(A) = 0.7$, find $P(\text{not } A)$.

3. If $P(A) = 0.7$, $P(\text{not } B) = 0.4$, and $P(A \text{ and } B) = 0.5$, find $P(A \text{ or } B)$.

4. If $P(\text{not } A) = 0.5$, $P(\text{not } B) = 0.4$, and $P(A \text{ and } B) = 0.3$, find $P(A \text{ or } B)$.

5. If $P(A) = 0.15$, $P(B) = 0.65$, and $P(A \text{ or } B) = 0.72$, find $P(A \text{ and } B)$.

6. If $P(\text{not } A) = 0.3$, $P(B) = 0.5$, and $P(A \text{ or } B) = 0.9$, find $P(A \text{ and } B)$.

7. If $P(A \text{ and } B) = 0.25$, $P(A \text{ or } B) = 0.85$, and $P(B) = 0.45$, find $P(\text{not } A)$.

8. The probability that a radioactive substance emits at least one particle during a one-hour period is 0.008. What is the probability that it does not emit any particle during the period?

9. When is the following a true statement: $P(A \text{ or } B) = P(A) + P(B)$?

10. Suppose the probability that a person chosen at random is over 6 feet in height or weighs more than 160 pounds is 0.3. What is the probability that he is neither over 6 feet in height nor weighs more than 160 pounds.

11. Suppose α represents the probability of concluding that a drug is good when in fact it is not good.

    (a) Give the maximum and minimum values that α can have.

    (b) Describe an event for which $1 - \alpha$ is the probability.

12. If A and B are events for which $P(A) = 0.6$, $P(B) = 0.7$, and $P(A \text{ or } B) = 0.9$, find $P(\text{not}(A \text{ and } B))$.

13. Is it possible to have two events A and B such that $P(A) = 0.6$, $P(B) = 0.7$, and $P(A \text{ and } B) = 0.2$? Explain.

14. Explain why it is *not* possible to have two events A and B with the following assignments of probabilities:

    (a) $P(A) = 0.6$ and $P(A \text{ and } B) = 0.8$

    (b) $P(A) = 0.7$ and $P(A \text{ or } B) = 0.6$

15. Two fair dice are rolled. Find the probability that the total on the two dice is not equal to 5.

16. The probability that an adult male is a Democrat is 0.6; the probability that he belongs to a labor union is 0.5; and the probability that he is a Democrat or belongs to a labor union is 0.75. Find the probability that a randomly picked adult male is

    (a) a Democrat and belongs to a labor union

    (b) not a Democrat nor does he belong to a labor union.

17. From past experience, little Scotty knows that the probability that Mommy will serve ice cream for dessert is 0.5, the probability that she will serve pie is 0.7, and the probability that she will serve ice cream or pie (or both) is 0.9. Find the probability that Mommy will serve

    (a) both ice cream and pie

    (b) neither ice cream nor pie.

18. In a certain club, the probability that a member picked at random is a lawyer is 0.64, the probability that the member is male is 0.75, and the probability that the member is a male lawyer is 0.50. Find the probability that

    (a) the member is a lawyer or male

    (b) the member is neither a lawyer nor male.

19. The probability that a person goes to a concert on Saturday is 2/3, and the probability that the person goes to a baseball game on Sunday is 4/9. If the probability of going to either program is 7/9, find the following probabilities:

    (a) The person goes to both programs.

    (b) The person goes to neither program.

20. A student is taking two courses, history and English. If the probability that she will pass either of the courses is 0.7, that she will pass both courses is 0.2, and that she will fail in history is 0.6, find the probability that

    (a) she will pass history.

    (b) she will pass English.

21. Past records have led a bank to believe that the probability that a customer opens a savings account is 0.4, the probability that a customer opens a checking account is 0.7, and the probability that a customer opens a savings and a checking account is 0.25.

    (a) What is the probability that a customer will open either a savings account or a checking account?

    (b) What is the probability that a customer will open neither a savings account nor a checking account?

22. The probability that an American tourist traveling to Europe will visit Paris is 0.8, the probability that the tourist will visit Rome is 0.6, and the probability that the tourist will visit either of the cities is 0.9. Find the probability that an American tourist will visit

    (a) both cities

    (b) neither city.

23. The probabilities that a typist makes 0, 1, 2, 3, 4, 5, or 6 mistakes on a page are, respectively, 0.05, 0.1, 0.3, 0.25, 0.15, 0.1, and 0.05. Find the probability that the typist makes

(a) a least one mistake

(b) at least three mistakes

(c) at most five mistakes

(d) an even number of mistakes.

24. The probabilities that an office will receive, in a half-hour period, 0, 1, 2, 3, 4, or 5 calls are 0.05, 0.1, 0.25, 0.3, 0.1, and 0.2, respectively. Find the probability that in a half-hour period

(a) more than two calls will be received

(b) at most four calls will be received

(c) four or more calls will be received.

25. The portion of university students who own an automobile is 0.3 and the portion of university students who live in the dorm is 0.6. If the portion of university students who either own a car or live in a dorm is 0.8, find the portion of students who live in a dorm and own a car. (Assume that the portions are obtained from a large student body.)

3-4 COUNTING TECHNIQUES AND APPLICATIONS TO PROBABILITY (OPTIONAL)

COUNTING TECHNIQUES: PERMUTATIONS AND COMBINATIONS

As we have seen, the classical definition of the probability of an event E requires that we know $n(E)$, the number of outcomes in E, and $n(S)$, the number of outcomes in the sample space S. A straightforward method to find the number of outcomes would be to list them explicitly. However, if there are too many outcomes, this method certainly cannot be recommended. In such situations we often use certain rules based on counting techniques. The basic rule for counting is the multiplication rule, which we shall refer to as the basic counting principle. It can be stated as follows:

THE BASIC COUNTING PRINCIPLE

If a certain experiment can be performed in r ways, and *corresponding to each of these ways* another experiment can be performed in k ways, then the combined experiment can be performed in rk ways.

The essential reasoning behind this principle can be illustrated through the following example. Suppose a coin and a die are tossed. We could represent the outcomes as ordered pairs, where the first component refers to the experiment involving a coin and the second to the one with a die, as follows:

FIGURE 3-6

Tree diagrams
showing 12
branches when
a coin and a
die are tossed.

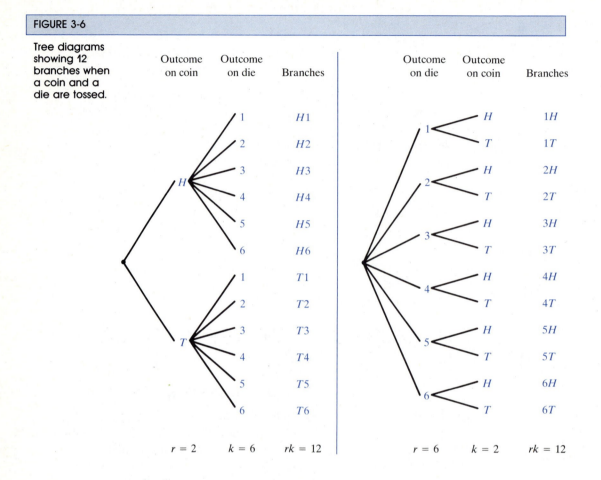

$(H, 1), (H, 2), (H, 3), (H, 4), (H, 5), (H, 6)$

$(T, 1), (T, 2), (T, 3), (T, 4), (T, 5), (T, 6)$

There are $6 \cdot 2$ outcomes corresponding to the two rows, each with six outcomes.

The principle can also be illustrated by means of a **tree diagram,** as in Figure 3-6. First we list the r outcomes of one experiment and then, corresponding to each of these, the outcomes of the other experiment. The total number of branches (namely, rk) gives the number of all the combined possibilities.

Thus, applying the basic counting principle, we have the following examples:

(a) The number of possible outcomes when a die is rolled twice is $6 \cdot 6$, or 36.

(b) The number of different ways for a man to get dressed if he has 8 different shirts and 6 different ties is $8 \cdot 6$, or 48.

The basic counting principle can be extended, in an obvious way, to cases where the combined experiment consists of three or more steps. Thus,

(c) There are $6 \cdot 6 \cdot 2 \cdot 2$, or 144, possibilities when two dice and two coins are tossed.

(d) The number of different ways for a man to get dressed if he has 8 different shirts, 6 different ties, and 5 different jackets is $8 \cdot 6 \cdot 5$, or 240.

(e) If a student has a choice of 3 math courses, 2 social sciences courses, 4 physics courses, and 3 biology courses, and she can take only one course in each discipline, then she can make her schedule in $3 \cdot 2 \cdot 4 \cdot 3$, or 72, different ways.

Next we apply the basic counting rule to find the number of possible ways when some fixed number of items are picked from a lot containing a certain number of items. We shall concern ourselves with the special case of sampling *without replacement,* where once an object is picked, it is not returned to the lot. We are thus led to the discussion of *permutations* and *combinations.*

Permutations

Suppose there are four contestants, Ann, Betty, Carl, and Dick, who are competing for three prizes, a first prize, a second prize, and a third prize. In how many ways can the prizes be awarded to the contestants so that any person gets at most one prize?

Notice that Ann getting first prize, Betty second prize, and Dick third prize is quite different from Dick getting first prize, Betty second prize, and Ann third prize, although the same three people are involved in both cases. *Order does make a difference.*

In Table 3-1 we give all the possible ways of awarding the prizes by listing the individuals by their initials A, B, C, D. For example, by the outcome *ACB* we mean that Ann gets the first prize, Carl the second prize, and Betty the third prize.

TABLE 3-1			
Permutations of letters A, B, C, D taken three at a time			
ABC	*ABD*	*ACD*	*BCD*
ACB	*ADB*	*ADC*	*BDC*
BCA	*BAD*	*CAD*	*CBD*
BAC	*BDA*	*CDA*	*CDB*
CAB	*DAB*	*DAC*	*DBC*
CBA	*DBA*	*DCA*	*DCB*

As mentioned earlier, the order in which the letters are written is important. Any particular arrangement is called a permutation. In Table 3-1 there are twenty-four permutations altogether. The reason is simple. We are choosing three letters out of four. There are 4 choices for the first letter. Corresponding to any of these ways, the second letter can be chosen in 3 ways (remember we are sampling without replacement). Last, for any way of drawing the first two letters, the third letter can be chosen in 2 ways. Therefore, after applying the basic counting principle, the total number of ways is $4 \cdot 3 \cdot 2$, or 24.

In general, suppose there are n distinct objects and we choose r of these without replacement. We cannot choose more objects than there are in the lot. Hence r cannot exceed n. Any particular ordered arrangement consisting of r objects is called a **permutation.** We shall denote the total number of permutations by the symbol $_nP_r$ and call it *the number of permutations of n objects taken r at a time.*

> The number of permutations (or ordered arrangements) of n distinct objects taken r at a time is given by
>
> $$_nP_r = n(n - 1)(n - 2) \cdots (n - r + 1).$$

On the right hand side of the formula for $_nP_r$, remember that there are r consecutively decreasing factors in the product with the first term as n. Thus, for example, the number of permutations when 4 objects are picked without replacement from 8 distinct objects is found by

$$_8P_4 = 8 \cdot 7 \cdot 6 \cdot 5 = 1680.$$

Notice that starting with 8 as the first term, we have four factors in the product.

In particular, $_nP_n$, the number of permutations of n objects *taken all together,* is $n(n - 1)(n - 2) \cdots 3 \cdot 2 \cdot 1$. We shall denote such a product of consecutive integers by $n!$. This is read n **factorial.** Thus, for example, $4! = 4 \cdot 3 \cdot 2 \cdot 1 = 24$. Also, for mathematical purposes, it is convenient to define $0! = 1$. Using the factorial symbol we can write $_nP_r$ in a compact way as

$$_nP_r = \frac{n!}{(n - r)!}$$

Thus, in summary,

> $$_nP_r = n(n - 1) \cdots (n - r + 1) = \frac{n!}{(n - r)!}$$
> $$_nP_n = n(n - 1) \cdots 3 \cdot 2 \cdot 1 = n!$$
> $$0! = 1$$

EXAMPLE 1   How many numbers with three distinct digits are possible using the digits 3, 4, 5, 6, 7, and 8? Write five of these numbers.

SOLUTION   There are six distinct digits and we are picking three. Observe that order is important, since, for example, the numbers 356 and 536 are different even though they have the same digits. Applying the formula, there are

$$_6P_3 = 6 \cdot 5 \cdot 4 = 120$$

distinct numbers. Five of these numbers are 345, 456, 465, 678, and 567.   ▬

EXAMPLE 2   Five people arrive at the checkout counter at the same time. In how many different ways can these people line up?

SOLUTION   The number of ways of arranging five people in a line is five factorial, that is,

$$5! = 5 \cdot 4 \cdot 3 \cdot 2 \cdot 1 = 120.$$ ▬

EXAMPLE 3   (a) Find the number of ways of arranging the letters in the word *object*.
                 (b) Find the number of ways if each arrangement in Part (a) starts with the letter *j*.

SOLUTION   (a) There are six distinct letters in the word. Hence, the number of arrangements is 6!, that is,

$$6! = 6 \cdot 5 \cdot 4 \cdot 3 \cdot 2 \cdot 1 = 720.$$

(b) Since the first letter is fixed as *j*, the total number of possibilities are found by arranging the remaining five letters in all possible ways. As we know, this can be done in five factorial ways, that is, 120 ways.   ▬

Combinations

As noted, the order in which objects are arranged matters in permutations. However, order does not matter under *combinations*. Let us consider the following example where we invite three volunteers from a group of four students, Ann, Betty, Carl, and Dick. If Ann, Betty, and Carl volunteer, this is in no way different from Carl, Ann, and Betty volunteering. We are interested in who volunteered and not in what order they volunteered. We get just four possibilities, namely, *ABC*, *ABD*, *ACD*, and *BCD*. Each of these possibilities is called a combination. Thus, there are four combinations.

Turning our attention to the general case, suppose we pick *r* objects from *n* distinct objects without replacement, and *order is not important*. Each such selection is a **combination.** Symbolically, we shall denote the number of these unordered arrangements by the symbol $\binom{n}{r}$ and call it *the number of combinations of n objects taken r at a time*.

> The number of combinations of n things taken r at a time is given by
> $$\binom{n}{r} = \frac{n(n-1)(n-2)\cdots(n-r+1)}{r!} = \frac{n!}{r!(n-r)!}.$$

Observe that the numerator in the first expression for $\binom{n}{r}$ is simply equal to $_nP_r$.

EXAMPLE 4   Evaluate the following:

(a) $\binom{8}{4}$ (b) $\binom{7}{7}$ (c) $\binom{10}{3}$ (d) $\binom{4}{0}$.

SOLUTION   (a) $\binom{8}{4} = \frac{8\cdot7\cdot6\cdot5}{4!} = \frac{8\cdot7\cdot6\cdot5}{4\cdot3\cdot2\cdot1} = 70$

(b) $\binom{7}{7} = \frac{7!}{7!(7-7)!} = \frac{7!}{7!0!} = 1$

Notice that $0! = 1$, and $7!$ cancels out in the numerator and the denominator of the second fraction.

(c) $\binom{10}{3} = \frac{10\cdot9\cdot8}{3!} = \frac{10\cdot9\cdot8}{3\cdot2\cdot1} = 120$

(d) $\binom{4}{0} = \frac{4!}{0!(4-0)!} = \frac{4!}{1\cdot4!} = 1$ ▬

The results of Example 4(b) and (d) illustrate that in general, for any positive integer n,

$$\binom{n}{n} = 1 \qquad \text{and} \qquad \binom{n}{0} = 1.$$

A commonly raised question is, When do we want permutations and when do we want combinations? To answer this we must first satisfy ourselves that objects are drawn from the lot without replacement. This done, the total number of ways will be $_nP_r$ if the order in which the objects are drawn is important, and $\binom{n}{r}$ if it is not important. Of course, whether order is relevant or not will depend on the nature of the experiment and the situation of interest. For instance, if we are dealing cards from a deck for a game of bridge, the order in which we deal the cards does not matter. What matters is the kinds of cards. Therefore, in such a situation we would be interested in the number of combinations. However, in seating people in a row, the order in which they are seated could be quite important. (People tend to be protocol conscious!) If we are interested in the different seating arrangements, we have a candidate for permutations.

EXAMPLE 5 Find the number of ways in which a hand of 5 cards can be dealt from a deck of 52 cards.

SOLUTION When cards are dealt from a deck, we are interested in what cards are dealt in a hand and not in the order in which they are dealt. Therefore, we want the number of combinations, which is

$$\binom{52}{5} = \frac{52 \cdot 51 \cdot 50 \cdot 49 \cdot 48}{5 \cdot 4 \cdot 3 \cdot 2 \cdot 1} = 2{,}598{,}960. \quad \blacksquare$$

EXAMPLE 6 How many ways are there of choosing a set of 3 books from a set of 8 books?

SOLUTION Since the order of choosing the books is not important, the answer is

$$\binom{8}{3} = \frac{8 \cdot 7 \cdot 6}{3 \cdot 2 \cdot 1} = 56. \quad \blacksquare$$

EXAMPLE 7 A company vice president has to visit 4 of the 12 subsidiaries that the company owns.

(a) How many sets of 4 companies are there from which the vice president can pick one set to visit?

(b) In how many different ways can she plan her itinerary in visiting 4 of the 12 subsidiaries?

SOLUTION (a) Here, we are simply picking 4 subsidiaries out of 12. What matters is which subsidiaries are picked. The order of picking them is not relevant. We have a case where combinations are appropriate. The answer is $\binom{12}{4}$, or 495.

(b) Let us identify the subsidiaries as $s_1, s_2, \ldots, s_{12}$. Suppose the vice president decides to visit the subsidiaries s_1, s_3, s_6, s_7. Then, for example, the route $s_1 \rightarrow s_3 \rightarrow s_6 \rightarrow s_7$ is different from the route $s_3 \rightarrow s_1 \rightarrow s_7 \rightarrow s_6$ though the same 4 subsidiaries are visited in both cases. Order is relevant and we use permutations. The answer is $_{12}P_4$, or 11,880. $\quad \blacksquare$

APPLICATIONS OF COUNTING TO FINDING PROBABILITIES

In the examples that follow, we shall show how counting techniques can be employed to solve problems in probability.

EXAMPLE 8 Three digits are picked at random from the digits 1 through 9. Find the probability that the digits are consecutive digits.

SOLUTION We are picking three digits out of nine. The number of ways this can be done is

$$\binom{9}{3} = \frac{9 \cdot 8 \cdot 7}{3 \cdot 2 \cdot 1} = 84.$$

The event of interest consists of the set of outcomes 123, 234, 345, 456, 567, 678, and 789. There are seven outcomes in the event. Therefore, the probability that the digits are consecutive is 7/84, or 1/12. ■

EXAMPLE 9   If the letters of the word *volume* are arranged in all possible ways, find the probability that

(a) the word ends in a vowel
(b) the word starts with a consonant and ends in a vowel.

SOLUTION   The problem clearly states that order is relevant. Since there are 6 distinct letters in the word *volume,* there are 6!, or 720 possible ways of arranging these letters.

(a) There are 3 vowels in the word *volume*. Since the word must end in a vowel, there are 3 choices for the last letter. This done, there are 5 choices for the first letter, 4 for the second, 3 for the third, 2 for the fourth, and 1 for the fifth. By the basic counting rule, we get $3 \cdot 5 \cdot 4 \cdot 3 \cdot 2 \cdot 1$, or 360 ways. Therefore, the probability is 360/720, or 1/2.
(b) Since we have 3 consonants and 3 vowels, there are 3 choices for the first letter and 3 choices for the last. Observe that we now have 4 choices for the second letter, 3 for the third, 2 for the fourth, and 1 for the fifth. Hence there are $3 \cdot 3 \cdot 4 \cdot 3 \cdot 2 \cdot 1$, or 216 possible ways. Therefore, the probability is 216/720, or 3/10. ■

An important type of probability problems can be couched in the following format:

Items in a lot represent a mixture of two or more kinds of items. Suppose we select from these items, without replacement, a sample of a certain number of items. We are interested in finding the number of ways of selecting these items from the lot so that there are so many of each kind.

For example, a person might have 9 stocks of which 4 are utility stocks and 5 are industrial stocks. If 3 stocks are selected, we might want to find the number of ways of doing this so that there are 2 utility stocks and 1 industrial stock in the selection.

We can represent the situation schematically as follows:

```
            9 stocks
          ↙         ↘
       utility    industrial
          4           5          ← composition of stocks
          ↓           ↓
          2           1          ← composition of selection
```

There are $\binom{4}{2}$ ways of selecting 2 utility stocks out of 4. *Corresponding to any one way of doing this,* there are $\binom{5}{1}$ ways of selecting 1 industrial stock out of 5. Applying the basic counting rule, we have $\binom{4}{2}\binom{5}{1}$ ways in which the selection can be made. (Notice that, in effect, we have simply put parentheses around the vertical arrows, omitting the arrows.)

As a further extension of the above example, suppose there are 15 stocks of which 3 are utility stocks, 7 industrial stocks, and 5 transportation stocks. If 6 stocks are selected, then the number of ways in which there are 2 utility, 1 industrial, and 3 transportation stocks is $\binom{3}{2}\binom{7}{1}\binom{5}{3}$ as can be seen from the following scheme.

$$
\begin{array}{ccc}
 & \text{15 stocks} & \\
\swarrow & \downarrow & \searrow \\
\text{utility} & \text{industrial} & \text{transportation} \\
3 & 7 & 5 \qquad \leftarrow \text{composition of stocks} \\
\downarrow & \downarrow & \downarrow \\
2 & 1 & 3 \qquad \leftarrow \text{composition of selection}
\end{array}
$$

EXAMPLE 10 Mrs. Dexter has 8 stocks in her portfolio of which 3 are utility and 5 are industrial. She selects 3 stocks at random to give to her nephew. Find the probability that the nephew receives:

(a) exactly 2 utility stocks
(b) only utility stocks
(c) no utility stock.

SOLUTION We are selecting 3 stocks out of 8, and since order is not important, there are

$$
\binom{8}{3} = \frac{8 \cdot 7 \cdot 6}{3 \cdot 2 \cdot 1} = 56
$$

outcomes in the sample space.

(a) "Exactly 2 utility stocks" means 2 utility stocks and 1 industrial stock. Schematically we have the following situation:

$$
\begin{array}{cc}
 & \text{8 stocks} \\
\swarrow & \searrow \\
\text{utility} & \text{industrial} \\
3 & 5 \qquad \leftarrow \text{composition of stocks} \\
\downarrow & \downarrow \\
2 & 1 \qquad \leftarrow \text{composition of selection}
\end{array}
$$

The number of ways of selecting 2 utility stocks and 1 industrial stock is

$$\binom{3}{2}\binom{5}{1} = \frac{3\cdot 2}{2\cdot 1}\cdot\frac{5}{1} = 15.$$

Consequently, the probability of exactly 2 utility stocks in the nephew's gift is 15/56.

(b) If the nephew receives only utility stocks, then he receives 3 utility stocks and 0 industrial. We have the followng scheme:

8 stocks

utility   industrial

3   5   ← composition of stocks

↓   ↓

3   0   ← composition of selection

The number of ways is $\binom{3}{3}\binom{5}{0} = 1$. Hence the probability of only utility stocks in the gift is 1/56.

(c) In this case we have the following scheme:

8 stocks

utility   industrial

3   5   ← composition of stocks

↓   ↓

0   3   ← composition of selection

The number of ways is $\binom{3}{0}\binom{5}{3} = 10$ and, consequently, the probability of no utility stock in the gift is 10/56. ■

EXAMPLE 11 A geology professor has 5 silicates, 7 pyrites, and 8 carbonates in a rock collection. He picks 6 rocks at random for a student to analyze. Find the probability that the professor picked

(a) 2 silicates, 1 pyrite, and 3 carbonates
(b) 2 silicates
(c) 2 silicates and 3 pyrites.

SOLUTION If there are 20 rocks in all, the number of ways of picking 6 of these is

$$\binom{20}{6} = \frac{20\cdot 19\cdot 18\cdot 17\cdot 16\cdot 15}{6\cdot 5\cdot 4\cdot 3\cdot 2\cdot 1} = 38{,}760.$$

(a) We have the following schematic representation:

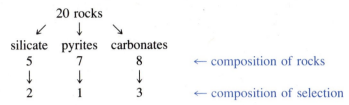

Therefore, the number of ways of picking 2 silicates, 1 pyrite, and 3 carbonates is

$$\binom{5}{2}\binom{7}{1}\binom{8}{3} = \frac{5\cdot 4}{2\cdot 1}\cdot\frac{7}{1}\cdot\frac{8\cdot 7\cdot 6}{3\cdot 2\cdot 1} = 3920.$$

Hence the probability of picking these rocks is 3920/38,760, or approximately 0.101.

(b) We are picking 6 rocks. If 2 rocks are silicates, then 4 rocks will have to be "nonsilicates." Hence for this situation, we have the following scheme:

20 rocks

silicates   nonsilicates
5   15   ← composition of rocks
↓   ↓
2   4   ← composition of selection

The number of ways is

$$\binom{5}{2}\binom{15}{4} = \frac{5\cdot 4}{2\cdot 1}\cdot\frac{15\cdot 14\cdot 13\cdot 12}{4\cdot 3\cdot 2\cdot 1} = 13{,}650.$$

Consequently, the probability of picking 2 silicate rocks is 13,650/38,760, or approximately 0.352.

(c) If among 6 rocks picked there are 2 silicates and 3 pyrites, then there must be 1 carbonate. From the following scheme

20 rocks

silicate   pyrites   carbonates
5   7   8   ← composition of rocks
↓   ↓   ↓
2   3   1   ← composition of selection

the number of ways is

$$\binom{5}{2}\binom{7}{3}\binom{8}{1} = \frac{5\cdot 4}{2\cdot 1}\cdot\frac{7\cdot 6\cdot 5}{3\cdot 2\cdot 1}\cdot\frac{8}{1} = 2800.$$

As a result, the probability of picking 2 silicates and 3 pyrites is 2800/38,760, or approximately 0.072. ▬

EXAMPLE 12   Eight vice presidents—Mr. Cox, Mr. Evans, Mrs. Fisher, Mr. Vegotsky, Mr. Pritchard, Ms. McKie, Mr. Alvarez, and Mr. Olson—are being considered for four vacancies on the board of directors in a company. If the four candidates are picked at random to fill the vacancies, what is the probability that Mr. Cox and Mrs. Fisher are among the four selected?

SOLUTION   There are $\binom{8}{4}$, or 70 ways of picking 4 vice presidents out of 8. Thus the sample space will have 70 possible outcomes. We are interested in the probability that Mr. Cox and Mrs. Fisher are included. The basic scheme can be represented as:

$$8 \text{ vice presidents}$$

$$\swarrow \qquad \downarrow \qquad \searrow$$

Mr. Cox	Mrs. Fisher	the rest
1	1	6	← composition of the group
↓	↓	↓
1	1	2	← composition of selection

The number of ways is $\binom{1}{1}\binom{1}{1}\binom{6}{2}$, or 15. Hence the probability that Mr. Cox and Mrs. Fisher are among the four selected to fill the vacancies is 15/70, or approximately 0.214.   ▪

SECTION 3-4   EXERCISES

1. Compute the following:

   (a) $_5P_2$ (b) $_8P_5$ (c) $_6P_6$ (d) $6!$

   (e) $4!0!$ (f) $\dfrac{20!}{18!}$ (g) $\binom{8}{2}$ (h) $\binom{6}{4}$

   (i) $\binom{14}{3}$ (j) $\binom{20}{20}$ (k) $\binom{16}{0}$ (l) $\binom{125}{1}$

2. Evaluate $\binom{8}{3}$ and $\binom{8}{5}$. Provide a rationale justifying why the answer is the same. Generalize by writing a rule that is illustrated by this result.

3. A menu lists two soups, three meat dishes, and five desserts. How many different meals are possible consisting of one soup, one meat dish, and a dessert?

4. A customer who wants to buy an automobile has a choice of 3 makes, 5 body styles, and 6 colors. Find how many choices he has in the selection.

5. A die is rolled four times and a coin is tossed twice. Find how many outcomes there are in the sample space. List three outcomes in the sample space.

6. An individual can be of genotype AA, Aa, or aa at one locus on the chromosome, of genotype BB, Bb, or bb at a second locus, and of genotype CC, Cc, or cc at a third

locus. How many possibilities are there for the genotypic composition of an individual with respect to the three loci?

7. A farmer has 3 concentrations of a nitrogen fertilizer, 4 concentrations of a phosphate fertilizer, and 5 concentrations of an organic fertilizer. How many ways are there for the farmer to combine the three types of fertilizers containing one and only one concentration of each type?

8. A hi-fi store has in its inventory six similar speakers, one of which is defective. Find how many ways there are

   (a) of picking two speakers

   (b) that would not include the defective speaker.

9. Ten people have gathered at a party. Find the number of handshakes when each person shakes hands with everyone else at the party.

10. An IRS agent has 200 income-tax returns on his desk. If he has decided to scrutinize 20 of the returns, find how many sets of 20 returns are possible. (Do not simplify your answer.)

11. A high school wants to buy 6 minicomputers for its computer laboratory from a local supplier. The supplier has 10 minicomputers in stock of which 4 are foreign-made.

    (a) Find how many ways there are to buy 6 minicomputers from the supplier.

    (b) Find how many ways there are if the high school wants 4 domestic and 2 foreign-made computers.

12. A superintendent of education has 16 schools in her district and wishes to visit 4 of the schools.

    (a) How many different ways are there for her to pick a group of 4 schools to visit?

    (b) Suppose the superintendent is also interested in the order in which she will visit the 4 schools. How many possible routes are there?

13. Sixteen college graduates have applied for six vacancies in a company. How many ways are there in which the company can make six offers?

14. Sixteen college graduates, of which five are women, have applied for six vacancies in a company. If the company has decided to hire two women, how many ways are there in which the company can make six offers of this type?

15. If there are one hundred entries in a contest, find the number of ways in which three different prizes—first, second, and third—can be awarded if no contestant can win more than one prize.

16. The letters of the word *problem* are arranged in all possible ways. If an arrangement is picked at random, find the probability that

    (a) the two vowels are next to each other

    (b) the arrangement will end with a consonant

    (c) the arrangement will start with *p* and end with *m*

17. A six-digit number is formed with the digits 1, 2, 3, 4, 5, 7 with no repetitions. Find the probability that

    (a) the number is even

    (b) the number is divisible by 5, that is, the units digit is 5.

18. Ted, Carol, Bob, and Alice are invited to a party. If they arrive at different times and at random, find the probability that

    (a) they will arrive in the order Ted, Carol, Bob, and Alice

    (b) Ted will arrive first

    (c) Ted will arrive first and Carol last.

19. If each license plate contains 3 different nonzero digits followed by 3 different letters, find the probability that, if a license plate is picked at random, the first digit will be odd and the first letter will be a vowel.

20. In a group there are 3 lawyers, 2 professors, and 3 doctors. If they are seated in a row, find the probability that those of the same profession sit together.

21. From a group of 10 lawyers, 15 accountants, and 12 doctors, a committee of six is selected at random. What is the probability that the committee consists of 3 lawyers, 2 accountants, and 1 doctor?

22. From six married couples, five people are selected at random. Find the probability of selecting two men and three women.

23. A box contains 4 white balls and 5 red balls. If three balls are drawn from the box, find the probability that

    (a) one ball is white

    (b) no ball is white

    (c) at least one ball is white

    (d) at most one ball is white.

24. If a committee of five is selected from a group of 8 seniors, 6 juniors, and 4 sophomores, find the probability that the committee has

    (a) 2 seniors, 2 juniors, and 1 sophomore

    (b) no seniors

    (c) at least one senior.

25. There are eight people at a picnic and their ages are as follows: 15, 5, 2, 20, 7, 30, 40, 23. If three people are picked at random, what is the probability that their combined ages will exceed 27 years?

26. A box contains eight balls marked 1, 2, 3, . . . , 8. If four balls are picked at random, find the probability that the balls marked 1 and 5 are among the four selected balls.

27. A box contains 8 good transistors, 4 transistors with minor defects, and 3 transistors with major defects. Four transistors are picked at random without replacement. Find the probability that

    (a) no transistor is good; that is, all transistors are defective

    (b) at least one transistor is good

    (c) exactly 2 transistors are good

    (d) one transistor is good, one has a minor defect, and 2 have a major defect

    (e) one transistor is good and 3 have major defects

3-5 CONDITIONAL PROBABILITY AND INDEPENDENT EVENTS

CONDITIONAL PROBABILITY

If we are interested in the probabilities of events, given some condition, we are concerned with *conditional probability*. When certain information is available about the outcome of the underlying experiment, we are led to make an appropriate adjustment of the probabilities of the associated events. As an example, consider the city of Burlington, Vermont. Suppose it snows on about 20 percent of the days each year; that is, if we pick a random calendar day, the probability of snow on that day is 0.2. Now suppose the random calendar day is a winter day. With this additional information, we would most likely say that the probability of snow is much higher than 20 percent. Similarly, if we know that the random day is a summer day, we would say that the probability of snow on that day is just about zero. In both situations, information about the day has led us to reassess the probability of the event that it snows on that day.

To make our ideas more concrete, let us now consider the following example. Suppose a certain sports club has 35 members of whom 20 are adults and 15 are minors. Among the adults, 8 are basketball players whereas among the minors, 5 are basketball players. The composition of the club membership is shown in the Venn diagram in Figure 3-7.

Suppose a person picked at random is a basketball player. Given this information, what is the probability that the person is an adult?

Let us write the events involved symbolically as follows:

A: The person is an adult

B: The person is a basketball player.

We now introduce the notation $P(A|B)$ to denote the desired probability. Here the vertical line is to be read *given*. Thus $P(A|B)$ is read "probability of A given B."

FIGURE 3-7

A Venn diagram depicting the club membership

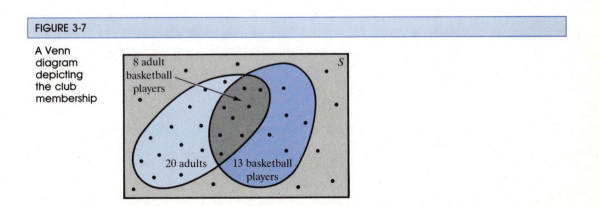

8 adult basketball players

20 adults   13 basketball players

Given the condition that the person picked is a basketball player, our concern is with the 13 basketball players. Of these 8 are adults. (See the Venn diagram in Figure 3-7.) Therefore, we can say

$$P(A|B) = \frac{8}{13} \quad \left(\text{that is, } \frac{n(A \text{ and } B)}{n(B)}\right).$$

We can rewrite $P(A|B)$ obtained above to give

$$P(A|B) = \frac{8/35}{13/35} \quad \left(\text{that is, } \frac{n(A \text{ and } B)/n(S)}{n(B)/n(S)}\right).$$

In other words, in the general context,

$$P(A|B) = \frac{P(A \text{ and } B)}{P(B)}.$$

Motivated by the discussion above, we give the following definition:

CONDITIONAL PROBABILITY

The **conditional probability** of an event A given the occurrence of an event B is defined by

$$P(A|B) = \frac{P(A \text{ and } B)}{P(B)}, \text{ provided } P(B) \neq 0.$$

$P(A|B)$ is not defined if $P(B) = 0$.

EXAMPLE 1

Suppose a card is picked at random from a deck of cards. Given that the card is a black card, find the probability that it is a face card.

SOLUTION

Given that the card is a black card, we concentrate our attention on the 26 black cards. Of these, there are 6 face cards. Hence the desired probability is 6/26, or 3/13.

Alternatively, we note that the probability that the card is a black face card (a black card *and* a face card) is 6/52, and the probability that it is a black card is 26/52. By applying the formula, the probability of the desired event is $\frac{6/52}{26/52}$, or 3/13. ▬

EXAMPLE 2

It is known that the probability that a bulb will last beyond 100 hours is 0.7 and the probability that it will last beyond 150 hours is 0.28. Given that a bulb lasts beyond 100 hours, find the probability that it will last beyond 150 hours.

SOLUTION Let us describe the events involved as follows:

L_{100}: the bulb lasts beyond 100 hours

L_{150}: the bulb lasts beyond 150 hours.

We want to find

$$P(L_{150}|L_{100}) = \frac{P(L_{150} \text{ and } L_{100})}{P(L_{100})}.$$

Now notice that the event "L_{150} and L_{100}" states that a bulb lasts beyond 150 hours and beyond 100 hours. This is simply the event that the bulb lasts beyond 150 hours. Hence

$$P(L_{150}|L_{100}) = \frac{P(L_{150})}{P(L_{100})} = \frac{0.28}{0.7} = 0.4. \quad \blacksquare$$

If we multiply both sides of the formula $P(A|B) = P(A \text{ and } B)/P(B)$ by $P(B)$, we get the *general multiplication law of probabilities*.

THE GENERAL MULTIPLICATION LAW

If A and B are two events, then $P(A \text{ and } B)$ is given by the formula

$$P(A \text{ and } B) = P(B) \cdot P(A|B).$$

This law is extremely useful in many instances to find the probability that two events will occur simultaneously.

EXAMPLE 3 The probability that the stock market goes up on Monday is 0.6. Given that it goes up on Monday, the probability that it goes up on Tuesday is 0.3. Find the probability that the market goes up on both days.

SOLUTION We are interested in the event that the market goes up on Monday *and* it goes up on Tuesday. Let the events be defined

M: market goes up on Monday

T: market goes up on Tuesday.

We are given that $P(M) = 0.6$ and $P(T|M) = 0.3$, and we want to find $P(T \text{ and } M)$. Applying the general multiplication law of probabilities, we get

$$P(T \text{ and } M) = P(M) \cdot P(T|M)$$
$$= (0.6)(0.3) = 0.18. \quad \blacksquare$$

EXAMPLE 4   Tom has an instructor who gives a midterm exam and a final exam. The probability that a student passes the midterm exam is 0.6. Given that a student fails the midterm exam, the probability that the student passes the final exam is 0.8. What is the probability that Tom fails the midterm and passes the final?

SOLUTION   We have

P(Tom fails the midterm *and* passes the final)

    = P(Tom fails the midterm) $\cdot$ P(Tom passes the

    final | Tom fails the midterm)

    = $(0.4)(0.8) = 0.32$

since the probability that Tom fails the midterm is $1 - 0.6$, or 0.4, and the probability of his passing the final, given that he fails the midterm, is 0.8.　◼

EXAMPLE 5   An advertising agency makes house calls to different households in a drive to sell a product. The probability that a resident is home is 0.7. Given that a resident is home, the probability that the resident buys the product is 0.3. Find the probability that a resident is home when called and buys the product as a result.

SOLUTION   The probability is equal to the product of the two probabilities P(a resident is home) and P(a resident buys a product | a resident is home) and, consequently, is equal to $(0.7)(0.3) = 0.21$.　◼

INDEPENDENT EVENTS

In some situations the fact that an event B has occurred may not influence the probability of the occurrence (or nonoccurrence) of the event A, so that we have $P(A|B) = P(A)$. (It turns out that if $P(A|B) = P(A)$, then $P(B|A) = P(B)$). As might be expected, such events are called *independent events*.

Suppose A and B are independent events. From the general multiplication law of probabilities, namely $P(A \text{ and } B) = P(B) \cdot P(A|B)$, it follows then that

$$P(A \text{ and } B) = P(A) \cdot P(B).$$

INDEPENDENT EVENTS

Two events A and B are said to be **independent events** if the probability of the simultaneous occurrence of A and B is equal to the product of the respective probabilities, that is,

$$P(A \text{ and } B) = P(A) \cdot P(B).$$

Two events that are not independent are said to be **dependent events**.

EXAMPLE 6   Past attendance records show that the probability that the chairman of the board attends a meeting is 0.7, that the president of the company attends a meeting is 0.8, and that they both attend a meeting is 0.4. Would you say that the chairman and the president act independently regarding their attendance of the meetings of the board of directors?

SOLUTION   Since $0.4 \neq (0.7)(0.8)$, we see that

P(chairman attends a meeting *and* president attends it)

$\neq P$(chairman attends a meeting) $\cdot P$(president attends a meeting).

Therefore, the two events 'the chairman attends a meeting' and 'the president attends a meeting' are not independent.   ▬

EXAMPLE 7   Victor flies from San Francisco to New York via Chicago. He takes Unified Airlines from San Francisco to Chicago, and Beta Airlines from Chicago to New York. The probability that a Unified plane lands safely is 0.95, and the probabilty that a Beta plane lands safely is 0.98. Find the probability that

(a) Victor lands safely in Chicago and New York
(b) Victor lands safely in Chicago, but has a mishap in New York.

SOLUTION   We can assume that independence is inherent in the situation since the safety on one airline should not influence that on the other.

(a) Since the events are independent,

   P(Victor lands safely in Chicago *and* New York)

   $= P$(Victor lands safely in Chicago)

   $\cdot P$(Victor lands safely in New York)

   $= (0.95)(0.98) = 0.931$.

(b) Here we want

   P(a safe landing in Chicago *and* mishap in New York)

   $= P$(a safe landing in Chicago) $\cdot P$(mishap in New York).

Now the probability of a mishap in New York is $1 - 0.98 = 0.02$. Hence the probability of the desired event is $(0.95)(0.02) = 0.019$.   ▬

The multiplication rule for two independent events given above can be extended to the case of three or more independent events. Thus, if A, B, C, and D are *independent events*, we can write

$$P(A \text{ and } B \text{ and } C \text{ and } D) = P(A) \cdot P(B) \cdot P(C) \cdot P(D).$$

EXAMPLE 8   A number is picked at random from the digits 1, 2, . . . , 9, and a coin and a die are tossed. Find the probability of picking an odd digit, getting a head on the coin, and getting a multiple of 3 on the die.

SOLUTION   Notice that

$$P(\text{odd digit}) = \frac{5}{9}$$

$$P(\text{head}) = \frac{1}{2}$$

$$P(\text{multiple of 3 on the die}) = \frac{1}{3}.$$

Therefore, since the events are independent,

$P(\text{odd digit } and \text{ a head } and \text{ a multiple of 3 on the die})$

$\quad = P(\text{odd digit}) \cdot P(\text{head}) \cdot P(\text{a multiple of 3 on the die})$

$\quad = \dfrac{5}{9} \cdot \dfrac{1}{2} \cdot \dfrac{1}{3} = \dfrac{5}{54}.$ ▬

EXAMPLE 9   A message will be transmitted from terminal A to terminal B if the relays R_1, R_2, R_3, and R_4 in Figure 3-8 are closed. If the probabilities that these relays are closed are, respectively, 0.5, 0.7, 0.8, and 0.8, find the probability that a message will get through. Assume that the relays function independently.

SOLUTION   The message will get through if relay R_1 is closed *and* relay R_2 is closed *and* relay R_3 is closed *and* relay R_4 is closed. Since the relays function independently, it follows that

$P(\text{message gets through}) = P(R_1 \text{ is closed}) \cdot P(R_2 \text{ is closed})$

$\qquad\qquad\qquad\qquad\qquad \cdot P(R_3 \text{ is closed}) \cdot P(R_4 \text{ is closed})$

$\qquad\qquad\qquad\qquad\quad = (0.5)(0.7)(0.8)(0.8) = 0.224.$ ▬

Remark: The two concepts, independent events and mutually exclusive events, are divergent concepts in the following sense: Suppose A and B are two events with positive probabilities. If the events are independent, they cannot be mutually exclusive, and vice versa.

FIGURE 3-8

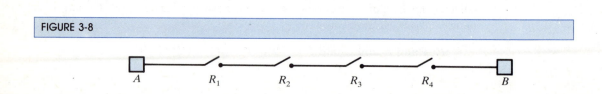

This follows simply from the fact that if A and B are mutually exclusive, they are not compatible and

$$P(A \text{ and } B) = 0$$

whereas, if they are independent

$$P(A \text{ and } B) = P(A) \cdot P(B) > 0 \qquad (\text{since } P(A) > 0, P(B) > 0)$$

and the two cannot be reconciled.

As an intuitive example, suppose Mr. Arnold owns a motel. Consider the two events

A: Mr. Arnold takes off on a weekend

B: Mr. Arnold's son, Bill, takes off on a weekend.

Suppose Mr. Arnold or Bill has to be on duty at the motel on every weekend. Since they cannot both take off on the same weekend the two events are obviously mutually exclusive. But since what one does depends on what the other does, that fact makes the two events dependent events.

On the other hand, if the two decide independently, then it is very likely that they will both take off sometimes on the same weekend and, then, the events are not mutually exclusive.

SECTION 3-5 EXERCISES

1. If A and B are two events, find $P(A|B)$ if
   (a) $P(B) = 0.6$, $P(A \text{ and } B) = 0.2$
   (b) $P(\text{not } B) = 0.7$, $P(A \text{ and } B) = 0.1$
   (c) $P(B) = 0.8$, $P(A \text{ and } B) = 0.5$.

2. If A and B are two events, find $P(A \text{ and } B)$ if
   (a) $P(A) = 0.5$, $P(B) = 0.7$, and $P(A|B) = 0.4$
   (b) $P(A) = 0.8$, $P(B) = 0.6$, and $P(B|A) = 0.55$
   (c) $P(A) = 0.6$, $P(B) = 0.8$, and $P(A|B) = 0.7$
   (d) $P(\text{not } A) = 0.3$, $P(\text{not } B) = 0.5$, and $P(B|A) = 0.6$.

   [*Note:* Some of the information given is superfluous.]

3. In each of the cases in Exercise 2 determine $P(A \text{ or } B)$.

4. In the table below, 400 individuals are classified according to whether or not they were vaccinated against influenza and whether or not they were attacked by influenza:

	Vaccinated	*Not vaccinated*
Attacked	60	85
Not attacked	190	65

If a person is chosen at random, find the probability that

(a) the person was vaccinated and attacked

(b) the person was attacked, given the person was vaccinated

(c) the person was not attacked, given the person was not vaccinated

(d) the person was vaccinated, given the person was attacked.

5. In a particular geographic region, 60 percent of the people are regular smokers, and evidence indicates that 10 percent of the smokers have lung cancer. Find the percent of people who are smokers and have lung cancer in this region.

6. A petroleum company exploring for oil has decided to drill two wells, one after the other. The probability of striking oil in the first well is 0.2. Given that the first attempt is successful, the probability of striking oil on the second attempt is 0.8. What is the probability of striking oil in both wells?

7. The probability that an adult male is over 6 feet tall is 0.3. Given that he is over 6 feet tall, the probability that he weighs less than 150 pounds is 0.2. Find the probability that a randomly chosen man will be taller than 6 feet and will weigh less than 150 pounds.

8. The probability that a person contracts a certain type of influenza is 0.2. Past experience has revealed that the probability that a person who gets the influenza does not recover is 0.05. Find the probability that a randomly chosen person

(a) contracts the influenza and does not recover

(b) contracts the influenza and recovers.

9. The probability that it will snow is 0.3. Given that it snows, the probability is 0.4 that the plane will take off on time, and given that it does not snow, the probability is 0.9 that the plane will take off on time. Find the probability that

(a) it will snow and the plane will take off on time

(b) it will not snow and the plane will take off on time

(c) the plane will take off on time.

10. The probability that a student studies is 0.7. Given that she studies, the probability is 0.8 that she will pass a course. Given that she does not study, the probability is 0.3 that she will pass the course. Find the probability that

(a) she will study and pass the course

(b) she will not study and will pass the course

(c) she will pass the course.

11. Suppose A and B are independent events. Find $P(A \text{ and } B)$ if

(a) $P(A) = 0.5, P(B) = 0.4$

(b) $P(A) = 0.3, P(\text{not } B) = 0.6$

(c) $P(A) = 0.8, P(B) = 0.7$

(d) $P(\text{not } A) = 0.3, P(\text{not } B) = 0.6$.

12. In each of the cases given in Exercise 11 determine $P(A \text{ or } B)$.

13. If A and B are independent events with $P(A) = 0.8$ and $P(A \text{ and } B) = 0.24$, find
(a) $P(B)$ and (b) $P(A \text{ or } B)$.

14. Suppose A and B are two events with $P(A) = 0.35$, and $P(B) = 0.6$. Find $P(A \text{ or } B)$ if the two events are

(a) mutually exclusive

(b) independent.

15. Suppose A and B are two events for which $P(A) = 0.4$, $P(B) = 0.6$, and $P(A \text{ or } B) = 0.76$.

    (a) Are A and B mutually exclusive?

    (b) Are A and B independent?

16. Orwell wants to buy a special ribbon for his typewriter. The probability that the university bookstore carries it is 0.3 and the probability that Village Stationers carries it is 0.3. Assuming the two stores stock the item independently, find the probability that neither store carries the item.

17. The probability that a plane arrives at the airport before 11:00 A.M. is 0.7. The probability that a passenger leaving from downtown by bus arrives at the airport by 11:00 A.M. is 0.6. Find the probability that

    (a) the plane and the passenger both arrive by 11:00 A.M.

    (b) the plane arrives by 11:00 A.M. and the passenger arrives after 11:00 A.M.

18. A genetic population consists of a large number of alleles that are of two types, type A and type a. It is known that 60 percent of the alleles in the population are of type A. If two alleles are picked at random, assuming independence, find the probability that

    (a) both alleles are of type A

    (b) one allele is of type A and the other is of type a

    (c) both alleles are of type a.

19. The probability that a person picked at random is an octogenarian is 0.05. If two people are picked at random, assuming independence, find the probability that

    (a) they are both octogenarians

    (b) neither is an octogenarian

    (c) at least one is an octogenarian.

20. It is known from past experience that the probability that a person has cancer is 0.02 and the probability that a person has heart disease is 0.1. If we assume that incidence of cancer and heart disease are independent, find the probability that

    (a) a person has both ailments

    (b) a person does not have both ailments

    (c) a person does not have either ailment.

21. From the records of the Public Health Division, the percentages of A, B, O, and AB blood types are, respectively, 40 percent, 20 percent, 30 percent, and 10 percent. If two individuals are picked at random from this population, assuming independence, find the probability that

    (a) they are both of blood type A

    (b) they are both of the same blood type

    (c) neither is of blood type A

    (d) one is of blood type A and the other is of blood type O

    (e) they are both of different blood types.

22. Suppose a penny, a dime, and a nickel with respective probabilities of heads equal to 0.2, 0.3, and 0.4 are tossed simultaneously. Find the probability that there will be

    (a) no heads

    (b) three heads

    (c) exactly one head.

23. A mouse caught in a maze has to maneuver through three escape hatches in order to escape. If the hatches operate independently and the probabilities of maneuvering through them are 0.6, 0.4, and 0.2, find the probability that

    (a) the mouse will make good his escape

    (b) the mouse will not be able to escape.

24. The genetic composition of a population is such that if a person is chosen at random, the probabilities that the person is of genotype *AA*, *Aa*, or *aa* at a locus on a chromosome are, respectively, 0.36, 0.48, and 0.16. The probabilities that the individual is of genotype *BB*, *Bb*, or *bb* at another locus are, respectively, 0.49, 0.42, and 0.09. If the genotypic distributions at the two loci are independent, find the probability that a randomly drawn person is

    (a) of genotype *AA* at one locus and of genotype *Bb* at the other

    (b) of genotype *Aa* at one locus and *Bb* at the other.

25. The probability that an adult male is shorter than 66 inches is 0.3, that he is between 66 and 70 inches is 0.5, and that he is taller than 70 inches is 0.2. If two males (for example, Tom and Dick) are picked at random, find the probability that

    (a) the heights of both are between 66 and 70 inches

    (b) Tom is shorter than 66 inches and Dick is taller than 70 inches

    (c) Tom is shorter than 70 inches and Dick is taller than 66 inches.

26. The probabilities that an office receives 0, 1, 2, or 3 calls during a half-hour period are, respectively, 0.1, 0.2, 0.4, and 0.3. It is safe to assume that the number of phone calls received during any two nonoverlapping time periods are independent. Find the probability that during the two time periods 11:00–11:30 and 3:00–3:30, together,

    (a) no calls will be received

    (b) three calls will be received

    (c) four calls will be received.

27. A nuclear power plant has a fail-safe mechanism consisting of six protective devices that function independently of each other. The probabilities of failure for the six devices are given as 0.3, 0.2, 0.2, 0.2, 0.1, and 0.1. If an accident will take place if all the devices fail, find the probability of an accident.

28. On a day in December, the probability that it will snow in Burlington, Vermont, is 0.6. Given that it snows in Burlington, the probability that it will snow in Boston is 0.7. Find the probability that on a random December day it will snow in both cities.

29. On a day in December, the probability that it will snow in Boston is 0.4 and the probability that it will snow in Moscow is 0.7. Assuming independence, find the probability that it will snow

    (a) in both cities

    (b) in neither city

(c) only in Moscow

(d) in exactly one city.

30. Suppose current flows through two switches A and B to a radio and back to the battery as shown in the following figure. If the probability that switch A is closed is 0.8, and given that switch A is closed the probability that switch B is closed is 0.7, find the probability that the radio is playing.

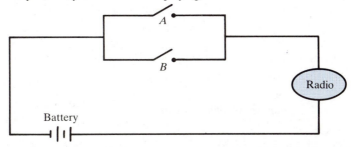

31. Suppose current flows through two switches A and B to a radio and back to the battery as shown in the figure for Exercise 30. If the probability that switch A is closed is 0.8 and the probability that switch B is closed is 0.9, assuming that the switches function independently, find the probability that the radio is playing.

32. Suppose current flows through two switches A and B to a radio and back to the battery as shown in the following figure. Suppose the probability that switch A is closed is 0.8, the probability that switch B is closed is 0.6, and given that switch A is closed the probability that switch B is closed is 0.7. Find the probability that the radio is playing.

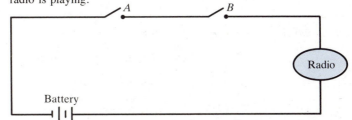

33. Suppose current flows through two switches A and B to a radio and back to the battery as shown in the figure for Exercise 32. If the probability that switch A is closed is 0.8 and the probability that switch B is closed is 0.9, and given that the switches function independently, find the probability that the radio is playing.

KEY TERMS AND EXPRESSIONS

empirical concept of probability	infinite sample space
set	event
subset	simple event
Venn diagram	impossible event
experiment	sure event
sample space	probability of a simple event
sample points	probability of an event
finite sample space	equally likely outcomes

mutually exclusive events   combination*
tree diagram*   conditional probability
permutation*   independent events

KEY FORMULAS

probability $P(A) =$ sum of the probabilities assigned to the simple events that comprise
of an event   event A, always

$$= \frac{\text{number of outcomes in } A}{\text{number of outcomes in } S}, \text{ if the outcomes are finite and equally likely.}$$

law of the $P(\text{not } A) = 1 - P(A).$
complement

the addition $P(A \text{ or } B) = P(A) + P(B) - P(A \text{ and } B), \text{ always}$
law (with
special cases) $\qquad\qquad = P(A) + P(B), \text{ if } A \text{ and } B \text{ are mutually exclusive}$

$\qquad\qquad = P(A) + P(B) - P(A) \cdot P(B), \text{ if } A \text{ and } B \text{ are independent.}$

conditional $P(A|B) = \dfrac{P(A \text{ and } B)}{P(B)}, \text{ provided } P(B) \neq 0$
probability

general $P(A \text{ and } B) = P(B) \cdot P(A|B), \text{ always}$
multiplication
law (with $\qquad\qquad = P(B) \cdot P(A), \text{ if } A \text{ and } B \text{ are independent}$
special cases) $\qquad\qquad = 0, \text{ if } A \text{ and } B \text{ are mutually exclusive.}$

r factorial $r! = r(r - 1) \cdots 3 \cdot 2 \cdot 1$

$0! = 1$

number of $_nP_r = n(n - 1) \cdots (n - r + 1) = \dfrac{n!}{(n - r)!}$
permutations

number of $\dbinom{n}{r} = \dfrac{n!}{r!(n - r)!}$
combinations

---

* From optional Section 3-4.

CHAPTER 3   TEST

[*Note:* Starred (*) exercises are solved using counting techniques of Section 3-4.]

1. Give an empirical interpretation of the following probability statement: The probability that an electric bulb will last beyond one hundred hours is 0.12.

2. What does it mean to say that an event has occurred?

*3. An agriculturist wants to test 3 fertilizers, each at 4 levels of concentration, on 5 different types of soil. How many different tests must he make?

4. Find $P(A \text{ and } B)$ if
   (a) $P(A) = 0.65$ and $P(B|A) = 0.4$
   (b) $P(B) = 0.3$ and $P(A|B) = 0.6$
   (c) $P(A) = 0.4$, $P(B) = 0.6$, and the events are independent.
   (d) $P(A) = 0.3$, $P(B) = 0.55$, and $P(A \text{ or } B) = 0.7$.

5. Find $P(A \text{ or } B)$ if
   (a) $P(A) = 0.4$, $P(B) = 0.7$, and $P(A \text{ and } B) = 0.3$
   (b) $P(A) = 0.65$, $P(B) = 0.35$, and $P(B|A) = 0.4$
   (c) $P(A) = 0.4$, $P(B) = 0.3$, and $P(A|B) = 0.6$
   (d) $P(A) = 0.4$, $P(B) = 0.6$, and the events are independent.

6. If $P(A) = 0.4$, $P(B) = 0.3$, and $P(A|B) = 0.6$ find $P(B|A)$.

7. Suppose A and B are two mutually exclusive events and the event B is not an impossible event. Find $P(A|B)$.

8. In the following cases, state whether the events A and B are independent or mutually exclusive or neither.
   (a) $P(A) = 0.4$, $P(B) = 0.3$, and $P(A \text{ or } B) = 0.56$
   (b) $P(A) = 0.5$, $P(B) = 0.3$, and $P(A \text{ or } B) = 0.8$
   (c) $P(A) = 0.4$, $P(B) = 0.5$, and $P(A \text{ or } B) = 0.7$

9. The probability that the propulsion system of a missile functions properly is 0.90, and the probability that its guidance system functions properly is 0.75. If the two systems operate independently, what is the probability that a launch is successful?

10. Henry is sick in the hospital. The probability that his brother Bill will come to visit him on a certain day is 0.3, and the probability that his friend Janice will come to visit him on that day is 0.8. If Bill and Janice decide independently, what is the probability that
    (a) at least one of them will pay a visit on that day?
    (b) neither will pay a visit on that day?

*11. A box contains three 15-watt bulbs, four 30-watt bulbs, two 60-watt bulbs, and one 100-watt bulb. If three bulbs are picked at random, find the probability that there are
    (a) three 15-watt bulbs
    (b) one 15-watt bulb, one 60-watt bulb, and one 100-watt bulb

(c) one 15-watt bulb and one 60-watt bulb

(d) no 30-watt bulb.

12. From past information it is known that machine *A* produces 10 percent defective tubes and machine *B* produces 5 percent defective tubes. Two tubes are picked at random, one produced by machine *A* and the other produced by machine *B*. Find the probability that

    (a) both tubes are defective

    (b) only the tube from machine *A* is defective

    (c) exactly one tube is defective

    (d) neither tube is defective.

13. The probability that a driver on a freeway will be speeding is 0.3. The probability that the driver will be speeding and will be detected is 0.1. Find the probability that a driver will be detected, given that the driver is speeding.

14. A radar complex consists of two units operating independently. The probability that one of the units detects an incoming missile is 0.95, and the probability that the other unit detects it is 0.90. Find the probability that

    (a) both will detect a missile

    (b) at least one will detect it

    (c) neither will detect it.

*15. Suppose the National Aeronautics and Space Administration (NASA) is interested in recruiting five schoolteachers in the astronaut training program and, by region, the following number of teachers have applied: East 10, West 12, North 6, South 8, and Midwest 5. If the five trainees are picked at random what is the probability that each region is represented?

16. A high-tech company manufacturing floppy discs has a two-stage inspection plan. Each disc is first subjected to visual inspection and is then checked electronically. The probability that a disc is rejected in the first stage is 0.15. If it is accepted at this stage, the probability of rejecting it under electronic inspection is 0.12. What is the probability that a randomly picked floppy disc will pass the inspection plan?

4

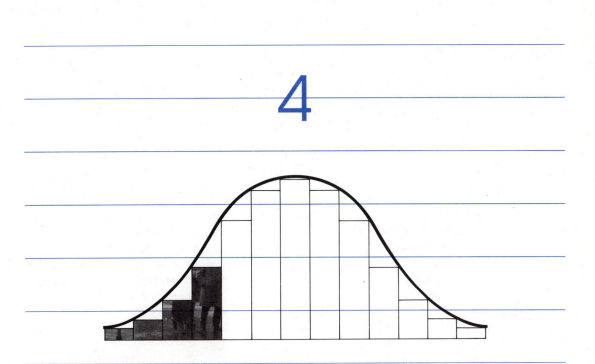

PROBABILITY DISTRIBUTIONS

4-1 Random Variables—The Definition

4-2 Discrete Random Variables

4-3 Continuous Random Variables

INTRODUCTION

It is not uncommon to hear someone claim to have "developed a system" for a game of chance at Las Vegas. That is, given idealized conditions in such a game, this person has built a probability model that incorporates the factors that have a bearing on the particular game, such as, for example, the game of blackjack. Based on this, the individual has developed a strategy that will optimize his or her winnings. In actually trying the system, the person will play the game many times, thus getting a mass of data. Using the techniques of Chapters 1 and 2, it would be possible to obtain the frequency distribution, the arithmetic mean, and the variance of the actual winnings. This is, of course, the empirical aspect. The person should also take into account the theoretical counterparts of these while building such a system. The theoretical counterpart of a frequency distribution is called the *probability distribution,* which is the main object of our study in this chapter. This important concept is at the very heart of the theoretical study of populations. ▄▄

4-1 RANDOM VARIABLES—THE DEFINITION

Often when an experiment is performed, our interest is not so much in the outcomes of the experiment as it is in the numerical values assigned to these outcomes. For instance, a gambler who will win $5 if heads shows up and lose $4 if tails shows up will think in terms of the numbers 5 and -4 rather than in terms of heads and tails. Technically this person is interested in the winnings and is assigning the value 5 to heads and -4 to tails. We shall refer to ''the winnings of the gambler on a toss'' as a *random variable*. Its value depends on which outcome shows up when a coin is tossed and, consequently, depends on chance. In general, we give the following definition of a random vairable.

A **random variable** assigns numerical values to the outcomes of a chance experiment. Therefore, mathematically a random variable is a function defined on the outcomes of the sample space.

It is customary to denote random variables by uppercase letters such as X, Y, and Z and their typical values by the corresponding lowercase letters x, y, and z. Thus suppose we let X represent the gambler's winnings on a toss. Then instead of making the long-winded statement, ''the winnings of the gambler on a toss are $5,'' we would simply write $X = 5$. In general $X = x$ will represent the statement that the random variable X assumes a value x.

Also, if a and b are any given numbers, then, for instance, $a < X \leq b$ will represent the statement that the random variable X assumes a value greater than a and less than or equal to b. Thus, if X represents the height of a person in inches, then $68 < X \leq 72$ will mean that the height of the person is greater than 68 inches and less than or equal to 72 inches. Similar interpretations for $X \geq b$, $X \leq a$, and so on, are obvious.

EXAMPLE 1 Suppose 3 satellites are launched into space. Let X denote the number of satellites that go into orbit. Describe the random variable X.

SOLUTION Writing h if a satellite attains an orbit successfully, and m if it does not, we have the following assignment, where the first row lists the outcomes of the sample space, and the second row lists corresponding numerical values assigned to X.

Sample point	hhh	hhm	hmh	mhh	mmh	mhm	hmm	mmm
Number of satellites attaining orbit	3	2	2	2	1	1	1	0

The random variable assumes four distinct values, namely, 0, 1, 2, and 3. ▬

EXAMPLE 2 Suppose a bowl contains 5 beads, of which 3 are white and 2 are black, and two beads are picked at random without replacement. If X represents the number of white beads in the sample, describe the random variable X.

SOLUTION Identifying the white beads as w_1, w_2, and w_3 and the black beads as b_1 and b_2, the outcomes of the sample space and the corresponding values that X assumes are given as follows:

Sample point	w_1w_2	w_1w_3	w_2w_3	w_1b_1	w_1b_2	w_2b_1	w_2b_2	w_3b_1	w_3b_2	b_1b_2
Number of white beads	2	2	2	1	1	1	1	1	1	0

The random variable assigns the values 0, 1, and 2 to the outcomes of the sample space. ▬

EXAMPLE 3 Suppose a person throws a dart at a circular dart board of radius 3 feet. (We shall assume that he is bound to hit somewhere on the board.) Let X denote the distance (in feet) by which he misses the center of the board. Describe the random variable X.

SOLUTION In this case the sample space consists of all the points on the board and X measures the distance (in feet) from the point of impact to the center of the board. Thus X assumes any value x where $0 \leq x < 3$. ▬

Additional examples of random variables are the maximum temperature during a day, the life of an electric bulb, the amount of annual rainfall, the salary that a new college graduate makes, the number of cavities a child has, the number of bacteria in a culture, the number of telephone calls received in an office between 9:00 A.M. and 10:00 A.M., and so on.

A random variable is classified as discrete or continuous, depending upon the range of its values. (The range of a random variable is the set of values it can assume.) As will be recalled, the concept of continuous and discrete variables was introduced in Chapter 1.

For our intuitive grasp, a **discrete random variable** is one which has a definite gap between any two possible values in its range. Formally the number of values that such a random variable assumes can be counted, and there can be as many values as there are positive integers. Thus, we often say that a discrete random variable can assume at most a *countably infinite* number of values.

The salary that a new college graduate earns, the number of bacteria in a culture, the number of cavities that a child has, the price of gold on the exchange on a particular day—these are all examples of discrete random variables.

A **continuous random variable** is one that can assume any value over an interval or intervals.

The maximum daily temperature, the life of an electric bulb, the distance by which a sharpshooter misses a target, the height of a person, and the length of time a person has to wait at a bus stop provide examples of continuous random variables.

SECTION 4-1   EXERCISES

1. A fair die is cast. Describe the random variable X in each of the following cases:

   (a) A person wins as many dollars as the number that shows up and X represents the person's winnings.

   (b) The same as in Part (a), but this time the person pays \$1 for the privilege of playing and X represents the net winnings.

   (c) A person wins \$4 if an even number shows up and loses \$3 if an odd number shows up, and X represents the person's winnings.

In Exercises 2–15, drawing on your past experience, give the range, that is, the set of values that the given random variable assumes. Also, classify the random variables as discrete or continuous.

2. Burning time of an experimental rocket

3. Speed of a car on a freeway

4. Tensile strength of an alloy

5. Typing speed of a typist

6. Number of home runs scored in a baseball game

7. Voltage of a battery

8. Balance of payments during a fiscal year

9. Reaction time of a frog to a given electric impulse

10. Time taken by a biological cell to divide into two

11. Number of accidents during a random day

12. Number of attempts a billiards player has to make to sink a ball

13. Number of stoplights that are green at intersections when a car driver encounters six stoplights on a drive

14. The amount of chlorophyll in a rubber plant leaf

15. The number of cavities a child has

4-2 DISCRETE RANDOM VARIABLES

We shall consider here only discrete random variables that assume a finite number of values.

The **probability distribution** of a discrete random variable can be described by listing all the values that the random variable can take, together with the corresponding probabilities. Such a listing is called the **probability function** of the random variable.

To understand a probability function, let us consider the random variable X described in Example 2 of Section 4-1. The random variable X represents the number of white beads in a sample of 2 picked at random from a bowl containing 5 beads of which 3 are white and 2 are black. As we have seen, the random variable assumes values 0, 1, and 2. Let us, for example, find the probability that both beads picked are white; that is, find $P(X = 2)$. We have

$$P(X = 2) = P(\{w_1w_2, w_1w_3, w_2w_3\})$$
$$= P(\{w_1w_2\}) + P(\{w_1w_3\}) + P(\{w_2w_3\}).$$

Now since there are ten outcomes in the sample space, each sample point is assigned a probability of $1/10$. Therefore,

$$P(X = 2) = \frac{3}{10} = 0.3.$$

(The reader, using counting techniques, should be able to justify this result by noting that

$$P(X = 2) = \frac{\binom{3}{2}\binom{2}{0}}{\binom{5}{2}} = 0.3.)$$

Similarly, it can be shown that $P(X = 1) = 6/10$ and $P(X = 0) = 1/10$. We can thus write the values that the random variable assumes and their respective

TABLE 4-1	
The probability function of the number of white beads	
Number of white beads	*Probabilities*
0	0.1
1	0.6
2	0.3
Sum	1.0

TABLE 4-2		
The empirical distribution when the experiment was performed 200 times		
Number of white beads	*Actual frequency*	*Relative frequency*
0	24	0.12
1	126	0.63
2	50	0.25

probabilities, obtaining the probability function given in Table 4-1. Notice that the total of all the probabilities in column 2 equals 1, which is a requirement that every probability function must satisfy.

A probability function can be given graphically by representing the values of the random variable along the horizontal axis and erecting ordinates to represent probabilities at these values, as is done in Figure 4-1.

The distribution given in Table 4-1 is a theoretical distribution. If we actually perform the experiment for perhaps 200 times—that is, if we repeat the experiment of picking 2 beads without replacement from the 5 beads (3 white and 2 black) 200 times—we might get the empirical distribution given in Table 4-2.

The agreement between the relative frequencies in Table 4-2 and the corresponding probabilities in Table 4-1 will improve if the experiment is repeated a greater number of times.

FIGURE 4-1

Probability function of the number of white beads.

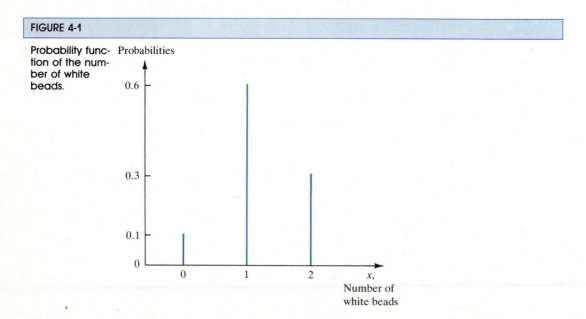

In general, suppose a random variable X assumes the values $x_1, x_2, \ldots, x_n$. Let us represent the probability that X assumes the value x_i by $p(x_i)$. That is,

$$P(X = x_i) = p(x_i).$$

The probability function can then be given in the form of a table as in Table 4-3.

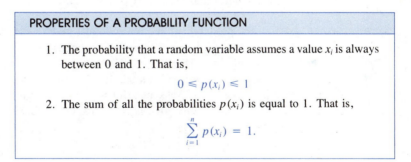

TABLE 4-3	
The probability function of a random variable.	
Value x	Probability $p(x)$
x_1	$p(x_1)$
x_2	$p(x_2)$
.	.
.	.
.	.
x_n	$p(x_n)$
Sum	$\displaystyle\sum_{i=1}^{n} p(x_i) = 1$

TABLE 4-4	
Probability function of the number of telephone calls.	
Number of calls x	Probability $p(x)$
0	0.05
1	0.20
2	0.25
3	0.20
4	0.10
5	0.15
6	0.05

In summary, we have the following properties:

PROPERTIES OF A PROBABILITY FUNCTION

1. The probability that a random variable assumes a value x_i is always between 0 and 1. That is,

$$0 \leq p(x_i) \leq 1$$

2. The sum of all the probabilities $p(x_i)$ is equal to 1. That is,

$$\sum_{i=1}^{n} p(x_i) = 1.$$

EXAMPLE 1   The number of telephone calls received in the math office between 12:00 noon and 1:00 P.M. has the probability function given in Table 4-4 above.

(a) Verify that it is in fact a probability function.
(b) Find the probability that there will be three or more calls.
(c) Find the probability that there will be an even number of calls.

SOLUTION   (a) Since all the probabilities in column 2 of Table 4-4 are between 0 and 1, and since their sum is 1, we have a genuine probability function.
(b) Writing X for the random variable "the number of telephone calls," we want $P(X \geq 3)$.

$$P(X \geq 3) = P(X = 3, \text{ or } X = 4, \text{ or } X = 5, \text{ or } X = 6)$$
$$= P(X = 3) + P(X = 4) + P(X = 5) + P(X = 6) \quad \text{(Why?)}$$
$$= p(3) + p(4) + p(5) + p(6)$$
$$= 0.20 + 0.10 + 0.15 + 0.05$$
$$= 0.5$$

(c) Here we want $P(X \text{ is even})$, which can be written as

$$P(X \text{ is even}) = P(X = 0, \text{ or } X = 2, \text{ or } X = 4, \text{ or } X = 6)$$
$$= P(X = 0) + P(X = 2) + P(X = 4) + P(X = 6)$$
$$= p(0) + p(2) + p(4) + p(6)$$
$$= 0.05 + 0.25 + 0.10 + 0.05$$
$$= 0.45. \quad \blacksquare$$

EXAMPLE 2   Mrs. Jones has owned 5 stocks for a number of years. She has concluded during the years she has owned the stocks that the probabilities that exactly 1, 2, 3, 4, or 5 of her stocks go up in price during a trading session are, respectively, 0.18, 0.30, 0.20, 0.15, and 0.08. Let X represent the number of stocks of Mrs. Jones that go up in price on a trading session.

(a) Describe the probability function of X.
(b) Find the probability that at most two stocks go up during a trading session.

SOLUTION   (a) To describe the probability function we must give the values that the random variable assumes and the corresponding probabilities. We are given the probabilities $p(1) = 0.18$, $p(2) = 0.30$, $p(3) = 0.20$, $p(4) = 0.15$, and $p(5) = 0.08$. But since it is possible that none of the stocks' prices will go up, there is a value that X assumes for which no probability is given, namely, 0. The probability $p(0)$ is found using the fact

$$p(0) + p(1) + p(2) + p(3) + p(4) + p(5) = 1.$$

That is,

$$p(0) + 0.18 + 0.30 + 0.20 + 0.15 + 0.08 = 1.$$

Hence

$$p(0) = 1 - 0.91 = 0.09.$$

The probability function is presented in Table 4-5.

(b) The probability that at most 2 stocks go up in price can be written as $P(X \leq 2)$ and can be found as follows:

$$P(X \leq 2) = P(X = 0, \text{ or } X = 1, \text{ or } X = 2)$$
$$= p(0) + p(1) + p(2)$$
$$= 0.09 + 0.18 + 0.30$$
$$= 0.57$$

TABLE 4-5	
The probability function of the number of Mrs. Jones' stocks going up in price	
Number of stocks going up in price *x*	*Probability* *p(x)*
0	? ← 0.09.
1	0.18
2	0.30
3	0.20
4	0.15
5	0.08
Sum	1.00

In Chapter 2 we calculated the arithmetic mean and the variance of a set of data. Just as a probability distribution is the theoretical counterpart of the relative frequency distribution, the mean and the variance of the probability distribution are, respectively, the theoretical counterparts of the empirical mean and the empirical variance.

THE MEAN OF A DISCRETE RANDOM VARIABLE

The **mean** of the probability distribution of a discrete random variable, or simply, the mean of a discrete random variable, is a number obtained by multiplying each possible value of the random variable by the corresponding probability and then adding these terms. Writing μ_X for the mean of a random variable X, or simply μ if the context is clear, in the case of a discrete random variable we have the following:

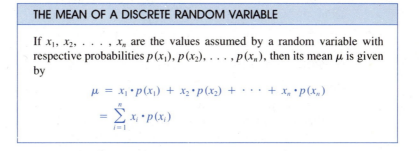

THE MEAN OF A DISCRETE RANDOM VARIABLE

If $x_1, x_2, \ldots, x_n$ are the values assumed by a random variable with respective probabilities $p(x_1), p(x_2), \ldots, p(x_n)$, then its mean μ is given by

$$\mu = x_1 \cdot p(x_1) + x_2 \cdot p(x_2) + \cdots + x_n \cdot p(x_n)$$

$$= \sum_{i=1}^{n} x_i \cdot p(x_i)$$

It is also common in the literature to refer to the mean as the *mathematical expectation* or the *expected value* of the random variable X, and to denote it as $E(X)$ (read *the expected value of X*).

EXAMPLE 3   Suppose the hourly earnings X of a self-employed landscape gardener are given by the following probability function:

Gardener's hourly earnings x	0	6	12	16
Probability $p(x)$	0.3	0.2	0.3	0.2

Find the gardener's mean hourly earnings.

SOLUTION   Applying the formula, we have

$$\mu = 0(0.3) + 6(0.2) + 12(0.3) + 16(0.2) = 8.0 \quad \blacksquare$$

Empirical Interpretation of $E(X)$

In Example 3, what does the number 8.0 mean in practical terms? Suppose we keep track of the gardner's earnings during N hours, and suppose he earns nothing during f_1 hours, \$6 during f_2 hours, \$12 during f_3 hours, and \$16 during f_4 hours, (so that $f_1 + f_2 + f_3 + f_4 = N$). Then, using the definition in Chapter 2, the arithmetic mean of his *actual hourly earnings* will be

$$\frac{0 \cdot f_1 + 6 \cdot f_2 + 12 \cdot f_3 + 16 \cdot f_4}{N} = 0\left(\frac{f_1}{N}\right) + 6\left(\frac{f_2}{N}\right) + 12\left(\frac{f_3}{N}\right) + 16\left(\frac{f_4}{N}\right)$$

Now the empirical notion of probability tells us that if N is very large, then f_1/N, f_2/N, f_3/N, and f_4/N should be close to the corresponding theoretical probabilities 0.3, 0.2, 0.3, and 0.2. That is,

$$0\left(\frac{f_1}{N}\right) + 6\left(\frac{f_2}{N}\right) + 12\left(\frac{f_3}{N}\right) + 16\left(\frac{f_4}{N}\right) \approx 0(0.3) + 6(0.2) + 12(0.3) + 16(0.2)$$

where $\approx$ means *is approximately equal to*. Hence if we keep track of the gardener's earnings over a very long period of time, then the arithmetic mean of his actual earnings would be close to 8.0, the theoretical mean. In general, the empirical concept of expectation entails repeating the experiment a large number of times and taking the grand mean.

Physical Interpretation of Mean μ

The mean is a numerical measure of the center of a probability distribution. As such, it can be interpreted in a physical sense as the *center of gravity* of a probability distribution. This can be seen best if we imagine weights located on a weightless seesaw at the various values that the random variable assumes. If the weights correspond to the probabilities with which these values are assumed, then a fulcrum located precisely at the mean will balance the seesaw. Figure 4-2 illustrates this interpretation with reference to Example 3.

FIGURE 4-2

Mean, viewed as the center of gravity

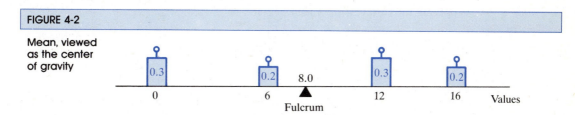

EXAMPLE 4   For the probability distribution of the number of telephone calls given in Example 1, find the mean number of telephone calls between 12:00 noon and 1:00 P.M.

SOLUTION   The distribution of the random variable and related computations are presented in Table 4-6. The sum in column 3 gives the mean number of telephone calls as 2.75.

TABLE 4-6

Computations to find the mean number of telephone calls (Example 4)

Number of calls x	Probability $p(x)$	$x \cdot p(x)$
0	0.05	0.00
1	0.20	0.20
2	0.25	0.50
3	0.20	0.60
4	0.10	0.40
5	0.15	0.75
6	0.05	0.30
Sum	1.00	2.75

TABLE 4-7

Probability function of net winnings in a raffle (Example 5)

Net winnings x	Probability $p(x)$
2998.5	1/10,000
998.5	5/10,000
98.5	20/10,000
−1.5	9974/10,000

EXAMPLE 5   In a raffle the prizes include 1 first prize of $3000, 5 second prizes of $1000 each, and 20 third prizes of $100 each. In all, ten thousand tickets are sold at $1.50 each. What are the expected net winnings of a person who buys one ticket?

SOLUTION   We are interested in the expected net winnings. So let X represent the net winnings of the person. The possible values of X are $(3000 - 1.5)$, $(1000 - 1.5)$, $(100 - 1.5)$, and $(0 - 1.5)$; that is, 2998.5, 998.5, 98.5, and −1.5. The respective probabilities are $\dfrac{1}{10,000}$, $\dfrac{5}{10,000}$, $\dfrac{20}{10,000}$, and $\dfrac{9974}{10,000}$. (Why?) The probability function of X is given in Table 4-7.

Applying the formula, the mean is given by

$$\mu = (2998.5)\frac{1}{10,000} + (998.5)\frac{5}{10,000} + (98.5)\frac{20}{10,000} + (-1.5)\frac{9974}{10,000}$$

$$= -0.5.$$

The interpretation of this result is that if a person were to participate in many raffles of this type, this person would lose approximately half a dollar per raffle. ▬

THE VARIANCE OF A DISCRETE RANDOM VARIABLE

The variance of a probability distribution provides a numerical measure of the spread and is defined as follows:

VARIANCE OF A DISCRETE RANDOM VARIABLE

The variance of the probability distribution of a discrete random variable X, or simply, the variance of a discrete random variable X, written $\text{Var}(X)$, is defined as

$$\text{Var}(X) = \sum_{i=1}^{n} (x_i - \mu)^2 \cdot p(x_i).$$

The *positive* square root of the variance is called the **standard deviation** of the distribution (or of the random variable).

The variance is commonly denoted as σ_X^2 or, if the context is clear, simply as σ^2 (read *sigma squared*). Here σ is positive. Hence the standard deviation equals σ.

The quantity $x_i - \mu$ represents the deviation of the value x_i from the mean μ. The variance is found by multiplying the squared deviation $(x_i - \mu)^2$ by the corresponding probability $p(x_i)$ and then adding the terms so obtained. A large value of σ^2 will indicate that extreme values on either side of μ are assigned high probabilities, that is, large deviations from the mean are highly probable. Thus the magnitude of σ^2 reflects the extent of spread about the mean μ. The larger the variance, the larger the spread.

EXAMPLE 6   The distribution of the number of raisins in a cookie is given as follows:

Number of raisins x	0	1	2	3	4	5
Probability p(x)	0.05	0.1	0.2	0.4	0.15	0.1

Find the mean and the variance of the number of raisins in a cookie.

SOLUTION   The necessary computations are presented in Table 4-8. The sum in column 3 gives the mean as 2.8, and the sum in column 6 gives the variance as 1.56. The standard deviation is equal to $\sqrt{1.56}$, or about 1.249. ▬

TABLE 4-8

Computations to find the mean μ and the variance σ^2

Number of raisins x	Probability $p(x)$	$x \cdot p(x)$	$(x - \mu)$	$(x - \mu)^2$	$(x - \mu)^2 \cdot p(x)$
0	0.05	0.0	-2.8	7.84	0.392
1	0.10	0.1	-1.8	3.24	0.324
2	0.20	0.4	-0.8	0.64	0.128
3	0.40	1.2	0.2	0.04	0.016
4	0.15	0.6	1.2	1.44	0.216
5	0.10	0.5	2.2	4.84	0.484
Sum	1.0	$\displaystyle\sum_{i=1}^{n} x_i \cdot p(x_i)$ $= 2.8$			$\displaystyle\sum_{i=1}^{n} (x_i - \mu)^2 \cdot p(x_i)$ $= 1.56$

It is often possible to simplify the computations involved in finding σ^2 by using the following version of the formula:

A SHORTCUT FORMULA FOR COMPUTING σ^2

$$\sigma^2 = \sum_{i=1}^{n} x_i^2 \cdot p(x_i) - \mu^2$$

For example, using this formula, let us compute the variance of the number of raisins in a cookie whose distribution is described in Example 6. The necessary computations are shown in Table 4-9.

TABLE 4-9

Computations to find σ^2 using the alternate formula

Number of raisins x	Probability $p(x)$	$x^2 \cdot p(x)$
0	0.05	0
1	0.10	0.1
2	0.20	0.8
3	0.40	3.6
4	0.15	2.4
5	0.10	2.5
Sum	1.0	$\displaystyle\sum_{i=1}^{6} x_i^2 \cdot p(x_i)$ $= 9.4$

Now in Example 6 we found the mean μ to be 2.8. Therefore,

$$\sigma^2 = \sum_{i=1}^{n} x_i^2 \cdot p(x_i) - \mu^2$$
$$= 9.4 - (2.8)^2$$
$$= 1.56.$$

This is the same answer that we found in Example 6.

EXAMPLE 7   Suppose X denotes the number that shows up when a fair die is rolled. Find the mean, the variance, and the standard deviation of X.

SOLUTION   The probability function of X and the necessary computations are shown in Table 4-10.

TABLE 4-10			
Computations to find the mean and the variance of the number when a die is rolled			
x	$p(x)$	$x \cdot p(x)$	$x^2 \cdot p(x)$
1	1/6	1/6	1/6
2	1/6	2/6	4/6
3	1/6	3/6	9/6
4	1/6	4/6	16/6
5	1/6	5/6	25/6
6	1/6	6/6	36/6
Sum	1	$21/6 = 3.5$	91/6

From column 3 of the table we find that the mean $\mu = 3.5$. Since

$$\sum_{i=1}^{n} x_i^2 \cdot p(x_i) = \frac{91}{6}$$

we get the variance as

$$\sigma^2 = \sum_{i=1}^{n} x_i^2 \cdot p(x_i) - \mu^2$$
$$= \frac{91}{6} - \left(\frac{21}{6}\right)^2$$
$$= \frac{35}{12}.$$

Taking the square root, we get the standard deviation

$$\sigma = \sqrt{35/12} \approx 1.71.$$

EXAMPLE 8   Suppose a population consists of N items with the values of a certain variable, for example, X, as $x_1, x_2, \ldots, x_N$. Find

(a) the population mean μ (b) the population variance σ^2.

SOLUTION   When a population of this type is mentioned, there is an implied assumption that the items are *equally likely*. In other words, since there are N values, $p(x_i) = \dfrac{1}{N}$ for any value x_i.

(a) Using the formula for the mean we get

$$\mu = \sum_{i=1}^{N} x_i \cdot p(x_i) = \sum_{i=1}^{N} x_i \cdot \frac{1}{N}$$

which can be rewritten to yield

$$\mu = \frac{\displaystyle\sum_{i=1}^{N} x_i}{N}.$$

Thus the population mean is simply the arithmetic mean of all the values in the population if there are a finite number of values, all equally likely.

(b) Using the formula for the variance, we have

$$\sigma^2 = \sum_{i=1}^{N} (x_i - \mu)^2 \cdot p(x_i) = \sum_{i=1}^{N} (x_i - \mu)^2 \cdot \frac{1}{N}$$

which may be rewritten

$$\sigma^2 = \frac{\displaystyle\sum_{i=1}^{N} (x_i - \mu)^2}{N}.$$

Notice that this result shows that the sum of the squares of the deviations from the mean is divided by N. ▪

There are several important discrete distributions. One of these, namely, the binomial distribution, will be considered in detail in the next chapter.

SECTION 4-2   EXERCISES

[*Note:* Starred (*) exercise is solved using counting techniques of Section 3-4.]

1. A box contains 3 defective bolts and 3 nondefective bolts. Suppose 3 bolts are picked at random and X represents the number of defective bolts in the sample. Describe a suitable sample space, the value that X assumes at each of the sample points, and the corresponding probabilities.

2. Jimmy tosses an unbiased coin three times in succession. He wins $1 for heads on the first toss, $2 for heads on the second toss, and $3 for heads on the third toss. For each tails on a toss he loses $1.50. (For example, if he gets the sequence *HTH* he will win $1 − $1.50 + $3, or $2.50). If X represents Jimmy's winnings, find the probability function of X.

*3. A group consists of 15 Democrats and 10 Republicans. A committee is formed by picking six people at random from the group. Let X represent the number of Republicans on the committee.

   (a) What does the event "$X = 2$" mean? Using the combinatorial formula, find the probability $P(X = 2)$.

   (b) What does the event "$2 \leq X < 5$" mean? Find $P(2 \leq X < 5)$.

   (c) Express in terms of X the event "Republicans have a majority". Find the probability of this event.

4. A random variable X has the following probability function:

x	-15	5	10	25	30	40
$p(x)$	0.25	0.15	0.10	0.25	0.15	0.10

Find the following probabilities:

   (a) $P(X \geq 15)$ (b) $P(5 \leq X < 28)$ (c) $P(X < 25)$

   (d) $P(X \leq 5)$ (e) $P(X \geq 30)$ (f) $P(X \geq 8)$

   (g) $P(5 < X < 30)$ (h) $P(5 \leq X \leq 30)$

5. The number of passengers in a car on a freeway has the following probability function:

Number of passengers x	1	2	3	4	5
$p(x)$	0.40	0.30	0.15	0.10	0.05

Find

   (a) the expected number of passengers in a car, and provide an interpretation to the value obtained

   (b) the variance of the number of passengers in a car.

6. Maria, a substitute teacher, lives 10 miles from Freshwater school, 15 miles from Sweetwater school, and 6 miles from Clearwater school. On any school day the probability that she goes to teach at Freshwater school is 0.4, at Sweetwater school is 0.3, and at Clearwater school is 0.2. Find the expected distance that Maria drives to school on a school day. (Assume there are only three schools in the area.)

7. Tom pays $2 for the privilege of tossing a coin once. He will win $5 if heads shows up and will lose $3 if tails shows up. If the probability of heads on a toss is 0.6, find Tom's expected net winnings. In practical terms how would you interpret the expected winnings?

8. Tom, John, Bob, and Mary weigh 200, 170, 150, and 120 pounds, respectively. A person in the group is picked at random. Find

(a) the expected weight of the person

(b) the variance of the weight.

9. For an experiment, prepare four chips marked 200, 170, 150, and 120. Pick a chip at random, record the number on it, and return it to the lot. Repeat the process forty times so that you have a set of forty values. Obtain

   (a) a frequency distribution of the data collected

   (b) the arithmetic mean, and compare it with the theoretical answer in Exercise 8(a)

   (c) the variance, and compare it with the theoretical answer in Exercise 8(b).

10. A fair die is rolled. If an even number shows up, Dolores will win as many dollars as the number that shows up on the die. If an odd number shows up, she will lose $2.20. Find

    (a) the probability function of Dolores's winnings, giving the values of the random variable and the corresponding probabilities

    (b) Dolores's expected winnings and the standard deviation of these winnings.

11. Conduct an experiment where you actually play the game in Exercise 10 sixty times. Prepare a table like the one below:

Game number	Number showing up	Winnings
1	4	4
2	5	−2.2
.	.	.
.	.	.
.	.	.
60	.	.

    Find the arithmetic mean and the standard deviation of the winnings from the empirical data. Compare these values with the theoretical answers obtained in Exercise 10.

12. A building contractor pays $250 to bid on a contract. If he gets the contract, the probability of which is 0.2, he will make a net profit of $10,000. On the other hand, if he does not get the contract, the amount of $250 is forfeited. Find the contractor's expected net profit.

13. Suppose, in an eight-round match between two boxers, the number of rounds that the match lasts has the following distribution:

Number of rounds	1	2	3	4	5	6	7	8
Probability	0.05	0.15	0.20	0.20	0.15	0.10	0.10	0.05

    Find the probability that a match will last

    (a) more than four rounds

    (b) less than four rounds

    (c) more than two rounds and less than or equal to five rounds.

14. For Exercise 13, find

(a) the expected number of rounds that a match lasts

(b) the variance of the number of rounds.

15. Suppose a random variable X assumes the values 1, 2, and 4. If $P(X = 1) = 0.3$ and the expected value of X is 2.7, find $P(X = 2)$ and $P(X = 4)$.

16. On each day a confectionery bakes four fancy cakes at the cost of $8 each, and prices them at $25 each. (Any cake not sold at the end of the day is discarded.) The demand for cakes on any day has the following probability distribution.

Number of cakes sold x	0	1	2	3	4
Probability $p(x)$	0.1	0.2	0.5	0.1	0.1

(a) If the confectionery sells two cakes on a day, find the profit on that day. Find the probability of making this profit.

(b) Continuing as in (a), find the profits if the number of cakes sold on a day is 0, 1, 3, and 4. Find the corresponding probabilities.

(c) Use the probability function derived above to find the expected profit of the confectionery when four cakes are baked on a day.

17. In Exercise 16, find the expected profit if instead of baking four cakes the confectionery bakes

(a) three cakes

(b) two cakes.

In view of the answers in Exercise 16(c) and in (a) and (b) of this exercise, how many cakes should the confectionery bake in order to make it most profitable in the long run?

18. The number of traffic accidents in a certain town during a week is a random variable with the following probability function:

Number of accidents x	0	1	2	3
Probability $p(x)$	0.50	0.30	0.15	0.05

Find the probability that during a week there will be

(a) no accident

(b) at least one accident

(c) at most one accident.

19. For Exercise 18, compute

(a) the expected number of accidents per week and interpret the figure

(b) the standard deviation of the number of accidents during a week.

20. Mr. Jones wants to invest money in a venture where the probability that he earns $10,000 is 0.38, that he earns $8000 is 0.26, and that he loses $15,000 is 0.36. Describe the probability function of Mr. Jones' earnings on his investment and find his expected earnings. Interpret the answer.

4-3 CONTINUOUS RANDOM VARIABLES

A clue to understanding probability distributions of continuous random variables is found in the histogram. Let us consider the following example, where we are interested in a random variable X, which represents the delay in the departure time of a plane scheduled to depart at a certain time, say 8 A.M. We shall assume that time is measured in units of hours. Suppose records are kept over 200 days and the information is summarized in a relative frequency distribution given in columns 1 and 2 of Table 4-11.

TABLE 4-11		
Relative frequency distribution of the delay times.		
Delay-in-departure time*	*Relative frequency*	*Relative frequency density*
$0.0 \leqslant t < 0.5$	0.10	0.20
$0.5 \leqslant t < 1.0$	0.35	0.70
$1.0 \leqslant t < 1.5$	0.20	0.40
$1.5 \leqslant t < 2.0$	0.18	0.36
$2.0 \leqslant t < 2.5$	0.10	0.20
$2.5 \leqslant t < 3.0$	0.05	0.10
$3.0 \leqslant t < 3.5$	0.02	0.04

*In column 1, $1.5 \leqslant t < 2.0$, for example, means that the delay-in-departure time is at least 1.5 hours but less than 2 hours.

Let us draw a histogram of the distribution as a modified version of the type presented in Chapter 1. Instead of using frequency densities, we use what are called the *relative frequency densities*. The **relative frequency density** of a class is obtained by dividing the relative frequency of the class by the class size. Thus

$$\left(\begin{array}{c}\textbf{relative frequency density}\\ \textbf{of a class}\end{array}\right) = \frac{\textbf{relative frequency of the class}}{\textbf{size of the class}}.$$

In our example, the width of any class interval is 0.5 and hence the relative frequency density column in Table 4-11 is obtained by dividing the entries in column 2 by 0.5. We now erect rectangles in the usual way with their bases centered at the class marks, except that the heights of the rectangles are taken to correspond to the relative frequency densities. The histogram so obtained is called a *relative frequency density histogram*. For the distribution in Table 4-11, the relative frequency density

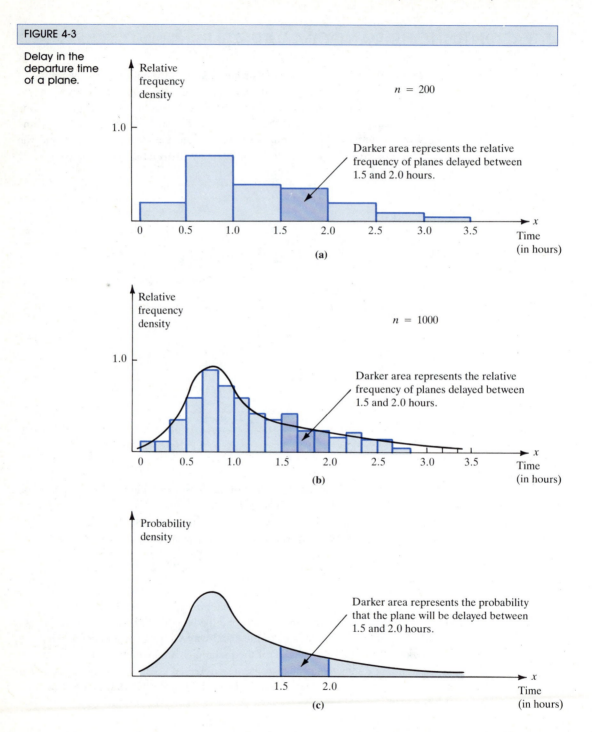

FIGURE 4-3

Delay in the departure time of a plane.

(a)

$n = 200$

Darker area represents the relative frequency of planes delayed between 1.5 and 2.0 hours.

(b)

$n = 1000$

Darker area represents the relative frequency of planes delayed between 1.5 and 2.0 hours.

(c)

Darker area represents the probability that the plane will be delayed between 1.5 and 2.0 hours.

histogram is shown in Figure 4-3(a). Notice that

the area of a rectangle = base × height

$$= \left(\begin{array}{c} \text{size of the} \\ \text{class interval} \end{array}\right) \times \left(\begin{array}{c} \text{relative frequency} \\ \text{density} \end{array}\right)$$

$$= \left(\begin{array}{c} \text{size of the} \\ \text{class interval} \end{array}\right) \times \dfrac{\left(\begin{array}{c} \text{relative frequency} \\ \text{of the class} \end{array}\right)}{\left(\begin{array}{c} \text{size of the} \\ \text{class interval} \end{array}\right)}$$

$$= \text{relative frequency of the class.}$$

Thus, in geometric terms, the area of any rectangle erected over an interval represents the relative frequency of that class. Now, for instance, the relative frequency of the flights of the plane delayed between 1.5 hours and 2.0 hours is precisely equal to the area of the shaded rectangle in Figure 4-3(a). Also, since the sum of the relative frequencies is 1, it follows that *the area under a relative frequency density histogram is always equal to 1.*

With a larger sample we are in a position to take shorter and shorter class intervals, obtaining a refinement of the relative frequency density histogram as in Figure 4-3(b), bringing into clearer focus the nature of the delay time. In Figure 4-3(b), where the histogram is obtained with 1000 observations, the relative frequency of the flights delayed between 1.5 hours and 2.0 hours is equal to the sum of the areas of the three darker rectangles. In the ultimate, as we take larger and larger samples an idealization of the situation is the rendering of the histogram into a smooth curve, as in Figure 4-3(c). This curve is superimposed on the histogram of Figure 4-3(b). We call this curve the *probability density curve* and it represents the graph of a function called the **probability density function.** The shaded area in Figure 4-3(c) represents the *probability* that the plane will be delayed between 1.5 hours and 2.0 hours.

A few comments are in order at this point.

1. Since the area under each relative frequency density histogram is equal to 1, it can be seen, intuitively at least, that in the limiting situation, the area under the probability density curve is also equal to 1.
2. The graph of the probability density function is always above the horizontal axis.
3. In the graph of the probability density function, the height at any point has *no* probability interpretation. It is the *areas* under the graph that give the probabilities. Specifically if *a* and *b* are any two numbers, then *the probability that the random variable assumes a value between a and b (written P(a < X < b)) is given as the area under the probability density curve between a and b, as shown in Figure 4-4 (page 150).*
4. For a continuous random variable, the probability that the random variable assumes any particular value is zero, that is, $P(X = a) = 0$ for any number a. This is because we can look at $P(X = a)$ as equal to

FIGURE 4-4

The shaded area
under the curve
is $P(a < X < b)$.

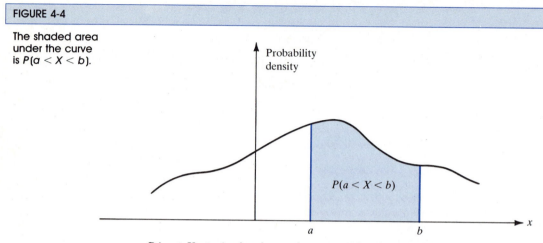

$P(a \leq X \leq a)$, that is, as the area under the curve between a and a.
But this area is zero. Hence $P(X = a) = 0$. Thus, in our example, the
probability that the plane will be delayed exactly 2.5 hours is zero.

5. Since $P(X = a) = 0$ for a continuous random variable, it does not
matter whether we include or exclude equalities at the endpoints of the
intervals, and we have

$$P(c < X < d) = P(c \leq X < d) = P(c < X \leq d)$$
$$= P(c \leq X \leq d)$$

Caution: In the discrete case this need not be true.

EXAMPLE 1   It is known that an executive arrives in his office between 8 A.M. and 10 A.M.
Records are kept of his arrival times (measured in hours) over 1000 days and the
relative frequency density histogram given in Figure 4-5 is obtained.

(a) Give an idealization that might best provide a probability model describing
the distribution of the arrival times.
(b) What is the probability that the executive will arrive between 8:45 A.M. and
9:25 A.M.?
(c) What is the probability that he will arrive after 9:36 A.M.?

SOLUTION   (a) Since the vertical bars are approximately the same height, it seems reasonable that the graph given in Figure 4-6 provides an appropriate probability
model. Notice that the length of the interval 8 A.M. to 10 A.M. is 2 units,
and since we want the total area of the rectangle equal to 1, the probability
density function at every point in the interval is 0.5. The area of a rectangle
with these dimensions is 2(0.5), or 1.
(b) We want to find the probability that the executive will arrive between 8:45
A.M. and 9:25 A.M. Converting into hours, we want $P(8.75 < X < 9.42)$.
This is equal to the area of the darker rectangle in Figure 4-7 on page 152.

FIGURE 4-5

The relative fre-
quency density
histogram for
the arrival times
of an executive

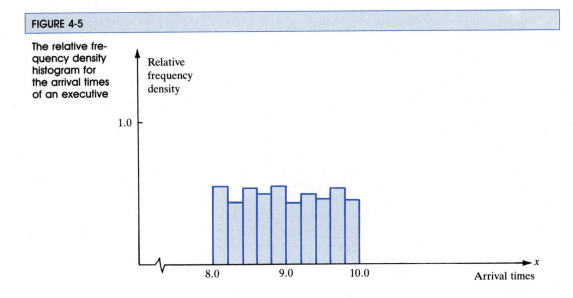

FIGURE 4-6

The probability
density function
of the arrival
times of an
executive

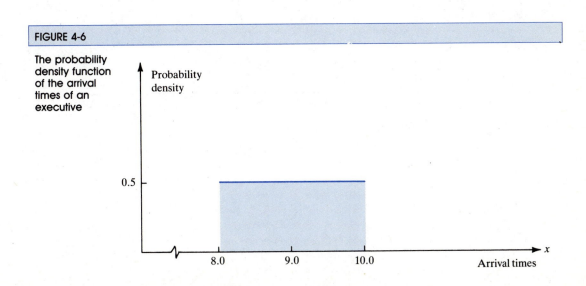

FIGURE 4-7

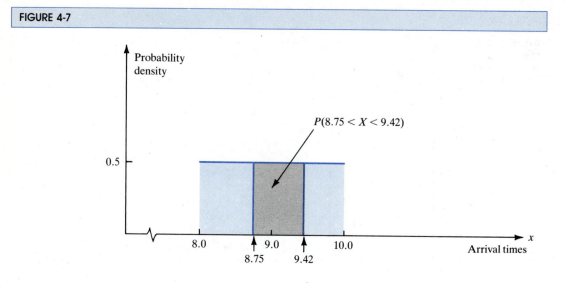

FIGURE 4-8

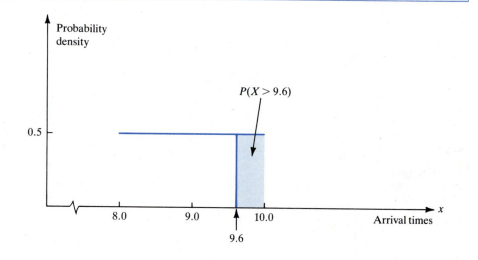

FIGURE 4-9

In Figure (a) the fulcrum is at the balance point; in Figure (b) it is not.

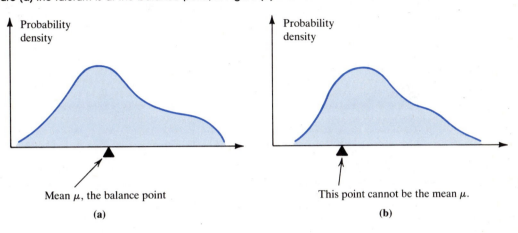

(a)   (b)

Since the length of the rectangle is $9.42 - 8.75 = 0.67$, the darker area is equal to $(0.67)(0.5) = 0.335$.

(c) Converting into hours, the probability of arriving after 9:36 A.M. is equal to $P(X > 9.6)$ and is obtained as the shaded area in Figure 4-8. As can be seen, this area is equal to $(10 - 9.6)(0.5)$, or 0.2. Hence the probability that the executive will arrive after 9:36 A.M. is equal to 0.2. That is, over a long period of time, approximately 20 percent of the time the executive will arrive between 9:36 A.M. and 10:00 A.M.

In Example 1 it was easy to compute the areas since the graph of the probability density function was a familiar geometric figure, the rectangle. In many situations, computation of the areas is an impossible task without using techniques of integral calculus or numerical analysis. Extensive tables of areas are available for important continuous distributions and we shall have occasion to use them later in our study. One such distribution is the so-called normal distribution, which we shall study in the next chapter. Also, in the continuous case, the definitions of the mean and the variance of a probability distribution require a background in calculus, and so we shall steer clear of this subject.

However, it is worth mentioning that, if we regard the probability distribution (the shaded portion in Figure 4-9) as a thin plate of some material of uniform density, then the position of the mean along the horizontal axis will be where the plate balances.

FIGURE 4-10

The curve in Figure (a) has smaller variance than the one in Figure (b).

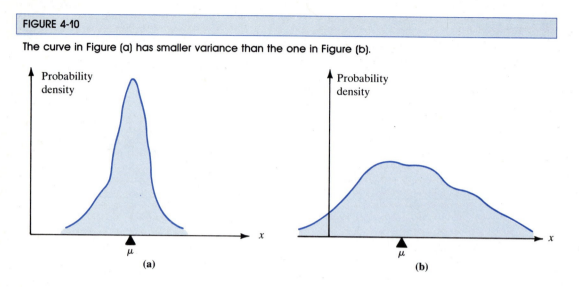

(a)   (b)

Also the magnitude of the variance reflects the spread of the distribution about the mean. Distributions with small variance σ^2 have high, narrow curves while those with large variance σ^2 have low, flatter curves. Thus, for instance, the curve in Figure 4-10(a) has smaller variance than the one in Figure 4-10(b).

SECTION 4-3 EXERCISES

1. The figures given below show three different relative frequency density histograms. Figure (a) shows the histogram of the distribution of the life of 10,000 electric bulbs. Along the horizontal axis we plot time (in hours), and the vertical bar at any point represents the relative frequency density of bulbs that stopped functioning during that

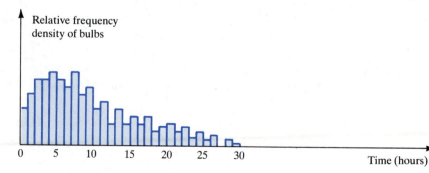

(a)

time period. Figure (b) shows the histogram of the distribution of time needed to perform a task. Figure (c) shows the histogram of the distribution of the heights of 1000 adult individuals. For each case, approximate the histogram by a theoretical probability density curve that you think is best representative of the data.

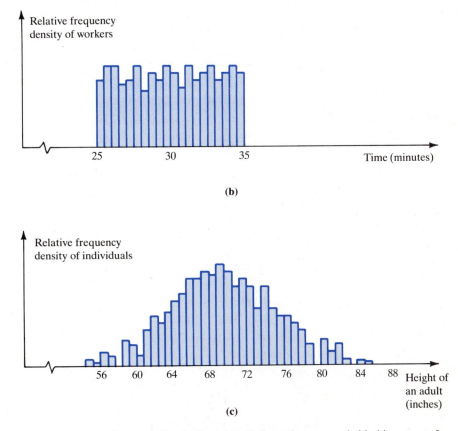

(b)

(c)

For the continuous random variables in Exercises 2–8, conjecture a suitable histogram of each distribution if you were to obtain a large number of measurements on the variable. From this infer the nature of the theoretical probability density curve.

2. The amount of rainfall during a year

3. The weight of an adult male

4. The burning time of an experimental rocket

5. The reaction time of a frog to a given electric impulse

6. The time taken by a biological cell to divide into two

7. The length of time that a person has to wait in line in order to be served

8. After applying brakes, the distance that a car going at 40 miles per hour covers before it comes to a stop

9. The curves in the figure that follows represent theoretical probability density curves.

   (a) In Figure (a) on page 156, shade the appropriate region to represent the probability $P(175 < X < 250)$.

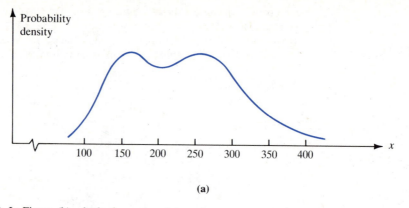

(a)

(b) In Figure (b), shade the appropriate region to represent the probability $P(X \geq -2.5)$.

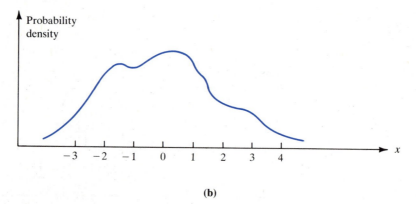

(b)

(c) In Figure (c), shade the appropriate region to represent the probability that X does *not* lie between 10 and 20.

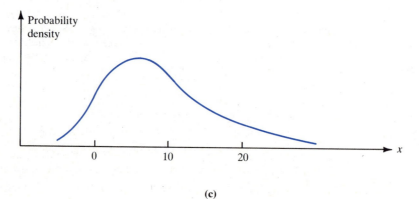

(c)

10. In the following figure, probabilities over some intervals are given as appropriate areas under the curve. Use this information to find the following probabilities:

(a) $P(X \leq -6)$

(b) $P(-3.5 < X \leq 3)$

(c) $P(X \geq 1)$

(d) Probability that X lies between -6 and -3.5, or between 1 and 3

(e) Probability that X does not lie between -6 and 1

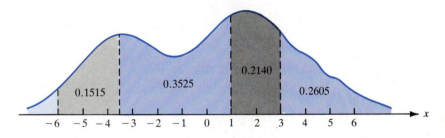

11. The probability distribution of the height (in feet) attained by a balloon is given by the probability density function shown in the figure below. Find the probability that a balloon will rise

(a) beyond 1800 feet

(b) not more than 1400 feet

(c) between 1200 and 1600 feet

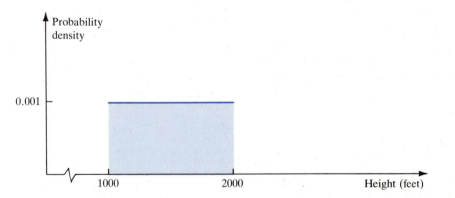

12. If the distribution of the height attained by a balloon is given by the probability density curve in the figure on page 158, find the probability that a balloon will rise

(a) beyond 1800 feet

(b) not more than 1400 feet.

(c) between 1200 and 1600 feet.

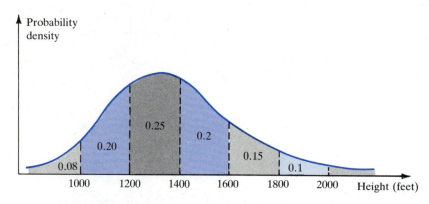

13. Of the two probability density curves in Exercises 11 and 12 describing the height attained by a balloon, which, would you say, depicts a more realistic situation? Explain.

14. Referring to the probability density curve in the figure for Exercise 12, find the value c for which

(a) $P(1200 < X < c) = 0.60$

(b) $P(X > c) = 0.27$

(c) $P(X < c) = 0.53$.

KEY TERMS AND EXPRESSIONS

random variable
discrete random variable
continuous random variable
probability distribution
probability function
mean of a random variable
  (or its distribution)

variance of a random variable
  (or its distribution)
standard deviation of a random variable
  (or its distribution)
relative frequency density
probability density function

KEY FORMULAS

mean of a discrete random variable

$$\mu = \sum_{i=1}^{n} x_i \cdot p(x_i)$$

variance of a discrete random variable

$$\sigma^2 = \sum_{i=1}^{n} (x_i - \mu)^2 \cdot p(x_i) = \sum_{i=1}^{n} x_i^2 \cdot p(x_i) - \mu^2$$

CHAPTER 4   TEST

[*Note:* Starred exercise is solved using counting techniques of Section 3-4.]

1. Give examples of two random variables that are discrete and two that are continuous.

*2. Four defective transistors are mixed with five good transistors. Suppose three transistors are picked at random without replacement. If X represents the number of defective transistors, find the values X assumes together with the corresponding probabilities.

3. The number of cars that a salesman sells on any day has the following probability function:

Number of cars	0	1	2	3	4	5
Probability	0.3	0.2	0.3	0.1	0.08	0.02

If the salesman earns $80 for every car he sells, find

(a) the probability that on any day he earns

(i)   more than $260

(ii) $240

(iii) between $120 and $280

(iv) $200

(v)   less than $280

(vi) no money

(b) the expected earnings of the salesman on any day

(c) the variance of the earnings of the salesman on any day.

4. A lottery is set up with a first prize of $500 and a second prize of $100. If a total of 2000 tickets are sold what should the price of a ticket be so that the lottery can be considered fair? (A game is considered fair if the expectation is zero.)

5. A lottery is set up with 1 first prize of $5000, 3 second prizes of $2000, and 10 third prizes of $1000 each. If 10,000 tickets are sold in all, what should the price of a ticket be so that the lottery can be considered fair?

6. Mrs. Wilson sells vacuum cleaners. A table including some information about the number of sales she makes during a day with the corresponding probabilities follows:

Number of vacuum cleaners	0	1	2	3
Probability	0.05	0.15	0.30	0.40

Mrs. Wilson's commission is $100, $75, $55, $30, $0, respectively, if she sells 4 or more, 3, 2, 1, 0 vacuum cleaners. Find the expected amount of Mrs. Wilson's commission during a day.

7. The graph at the top of page 160 gives the probability density curve of a continuous random variable X.

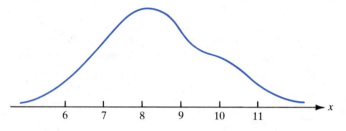

Copy the graph and shade appropriate regions to describe the following probabilities:

(a) $P(7 < X \leq 9)$ (b) $P(X \leq 6)$

(c) $P(X > 10)$.

8. The curve shown below is the probability distribution of annual rainfall in a certain region.

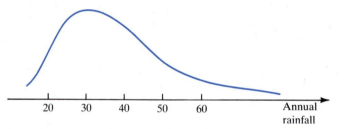

Copy the graph and shade appropriate regions to describe the following probabilities that

(a) the annual rainfall will exceed 40 inches

(b) the annual rainfall will be between 20 and 40 inches

(c) the annual rainfall will not be between 20 and 50 inches.

9. When two fruit flies are crossed, the probabilities that there are 100, 110, 120, 140, or 150 flies in the progeny are, respectively, 0.15, 0.15, 0.20, 0.30, and 0.20. Find the expected number of flies in a cross and the variance of the number of flies.

10. The following figure represents the probability density curve describing the distribution of the error (in ounces) in filling a one-pound bag of flour. (Error is positive if the bag is overfilled, negative if underfilled.) Find the probability that a bag is

(a) overfilled by more than 0.2 ounces

(b) underfilled by more than 0.3 ounces

(c) underfilled by more than 0.2 ounces or overfilled by more than 0.3 ounces.

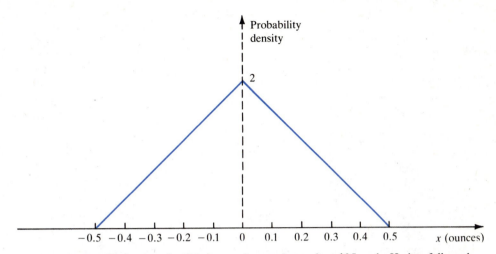

11. Mr. Roberts has bought 100 shares of a certain stock at $25 each. He has followed the stock closely over a long period of time for its price fluctuations and has determined that the price of the stock on any day is $20, $22, $25, $32, $36 with respective probabilities 0.1, 0.35, 0.3, 0.15, 0.1. If Mr. Roberts intends to sell the stock at the next trading session, what is his expected profit?

5

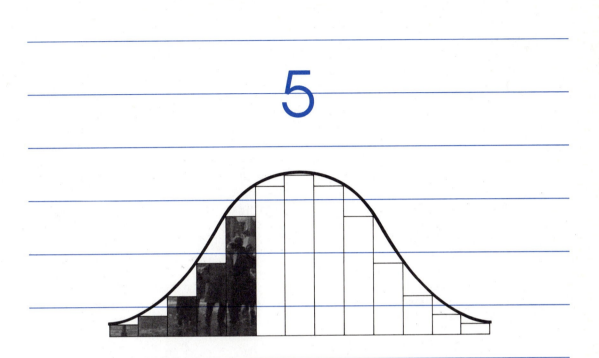

THE BINOMIAL AND NORMAL PROBABILITY DISTRIBUTIONS

5-1 The Binomial Distribution

5-2 The Normal Distribution

5-3 The Normal Approximation to the Binomial Distribution

INTRODUCTION

Our goal in the preceding chapter was to become acquainted with some basic ideas about probability distributions. We were introduced to two types of distributions, discrete and continuous. In the discrete case the distribution can be described by giving the values that the random variable assumes along with the corresponding probabilities. In the continuous case the distribution is described by means of a smooth curve, and the probabilities are given in terms of appropriate areas under this curve. We now single out two distributions that are of special importance in statistics, since probability distributions of many physical phenomena can be explained through them. One of these is a discrete distribution called the *binomial distribution** and the other, a continuous distribution called the *normal distribution.* The normal distribution is also called the *Gaussian* distribution in honor of one of the greatest mathematicians, Karl Gauss (1777–1855). In the last section of this chapter, we will see a connection between the two distributions and we will show how in some situations the normal distribution can be used to compute, approximately, the probabilities of a binomial distribution. ■

5-1 THE BINOMIAL DISTRIBUTION

THE PROBABILITY DISTRIBUTION

The starting point in discussing a **binomial distribution** is to consider a basic experiment in which the possible outcomes can be classified into one of two categories, a **success** or a **failure.** For instance, a tossed coin lands heads or tails, a basketball player either makes a free throw or does not make it, a new vaccine is effective or is not effective, a missile is fired successfully or is not fired successfully, and so on. The use of the word *success* does *not* carry the usual connotation of a desirable outcome. The basic experiment, often called a trial, is repeated a given number of times, perhaps *n* times, thus giving *n* trials which constitute the **binomial experiment.** In performing the trials, a tacit assumption is that the outcome of one trial does not influence the outcome of any other trial. Thus we say that the trials are *independent.* This assumption, together with others characterizing a binomial distribution, can be stated as follows:

1. The experiment called the binomial experiment consists of a fixed number *n* of independent and identical trials.
2. The performance of a trial results in an outcome that can be classified either as a *success* or a *failure.*

---

*The Poisson distribution, another well-known and important discrete distribution, is considered in Appendix I.

3. The probability of success is known and remains the same from trial to trial. Furthermore, if $P(\text{success}) = p$ and $P(\text{failure}) = q$, then

$p + q = 1$.

It is common in statistical literature to refer to each trial as a *Bernoulli* trial, in honor of the Swiss mathematician Jacob Bernoulli (1654–1705). The **binomial random variable** X is defined as follows:

$X = $ the number of successes in n trials.

It is easily seen that the possible values of X consist of integers 0, 1, 2, . . . , n. Thus if X denotes the number of successful missile launches in 10 attempts, then X could assume any value 0, 1, . . . , 10. Our objective will be to obtain the probabilities with which these values are assumed. The values that X assumes together with the corresponding probabilities will give us the distribution of X. Before treating the general case, we shall indicate the approach through the following example.

Suppose a man fires 4 shots at a target. If he scores a hit on any shot, we will denote the outcome as h and call it a success; if he misses, we will denote the outcome as m and call it a failure. Let X represent the number of hits. Since the man can score 0 hits, 1 hit, 2 hits, 3 hits, or 4 hits, the possible values of X are 0, 1, 2, 3, and 4. Assuming that the probability of scoring a hit on any shot is p, we want to find the probabilities with which these values are assumed. The sample space, which represents the results of firing 4 shots, consists of 16 (or 2^4) outcomes, and these are listed in Table 5-1 according to the number of hits scored.

TABLE 5-1				
Outcomes of the sample space according to the number of hits scored in four shots.				
0 hits	*1 hit*	*2 hits*	*3 hits*	*4 hits*
mmmm	*hmmm*	*hhmm*	*hhhm*	*hhhh*
	mhmm	*hmhm*	*hhmh*	
	mmhm	*hmmh*	*hmhh*	
	mmmh	*mhhm*	*mhhh*	
		mhmh		
		mmhh		

Let us first find the probability of scoring one hit. From Table 5-1, the possible outcomes of this event are *hmmm*, *mhmm*, *mmhm*, and *mmmh* given in column 2. Hence,

$P(1 \text{ hit}) = P(\{hmmm\}) + P(\{mhmm\}) + P(\{mmhm\}) + P(\{mmmh\})$

Now, *hmmm* stands for "hit on the first shot" *and* "miss on the second shot" *and* "miss on the third shot" *and* "miss on the fourth shot." Since the trials are independent, we can therefore write

$$P(\{hmmm\}) = P\left(\begin{array}{c} \text{hit on} \\ \text{the first} \\ \text{shot} \end{array}\right) \cdot P\left(\begin{array}{c} \text{miss on} \\ \text{the second} \\ \text{shot} \end{array}\right) \cdot P\left(\begin{array}{c} \text{miss on} \\ \text{the third} \\ \text{shot} \end{array}\right) \cdot P\left(\begin{array}{c} \text{miss on} \\ \text{the fourth} \\ \text{shot} \end{array}\right)$$

$$= pqqq$$
$$= pq^3.$$

As a matter of fact, a similar argument shows that the probabilities $P(\{mhmm\})$, $P(\{mmhm\})$, and $P(\{mmmh\})$ are all equal to pq^3. Hence

$$P(1 \text{ hit}) = 4pq^3.$$

probability of each outcome
number of outcomes

Next, to find the probability of 2 hits, we note from Table 5-1 that there are six outcomes describing the event, and each of these outcomes has the same probability p^2q^2. For example,

$$P(\{hhmm\}) = ppqq = p^2q^2.$$

Hence

$$P(2 \text{ hits}) = 6p^2q^2.$$

probability of each outcome
number of outcomes

Proceeding in this way, it is now simple to construct Table 5-2, which gives the probability distribution of X.

TABLE 5-2	
The probability distribution of the number of hits in four shots.	
Number of hits x	*Probability of* x *hits*
0	q^4
1	$4pq^3$
2	$6p^2q^2$
3	$4p^3q$
4	p^4

*Turning to the general case of n trials, the sample space consists of 2^n outcomes. Writing S for success and F for failure, these outcomes are indicated in Table 5-3 according to the number of successes.

---

*At this point, the reader who is not familiar with counting techniques can go directly to the paragraph following Example 2 where tables to find binomial probabilities are explained.

TABLE 5-3

Outcomes of the sample space when *n* trials are performed, arranged according to the number of successes.

0 successes	1 success ...	x successes	...	(n − 1) successes	n successes
$FF \ldots F$	$SFF \ldots F$	$SS \ldots SF \ldots F$		$FSS \ldots S$	$SS \ldots S$
	$FSF \ldots F$	.		$SFS \ldots S$	
	.	.		.	
	.	$SFS \ldots F \ldots SF$		.	
	.	.		.	
	.	.		.	
	.	.		.	
	$FF \ldots FS$	.		$SS \ldots SF$	
↑	↑	↑		↑	↑
1 outcome	*n* outcomes	$\binom{n}{x}$ outcomes in this listing. Each outcome has *x* successes and *n* − *x* failures.		*n* outcomes	1 outcome

We shall find the probability of *x* successes in *n* trials, where, of course, *x* could be any integer 0, 1, 2, . . . , *n*. One of the ways in which we can have *x* successes is to have successes on the first *x* trials and failures on the remaining *n* − *x* trials, as

$$\underbrace{SS \ldots S}_{x \text{ trials}}\underbrace{FF \ldots F}_{(n-x) \text{ trials}}$$

Arguing as in our earlier example, the probability of this outcome is

$$P(\{SS \ldots SFF \ldots F\}) = \underbrace{pp \ldots p}_{x \text{ times}}\underbrace{qq \ldots q}_{(n-x) \text{ times}}$$
$$= p^x q^{n-x}$$

Now there are $\binom{n}{x}$ outcomes, each having *x* successes and *n* − *x* failures, since this is simply the number of ways of picking *x* trials out of *n* for successes to occur on them. Hence we have the formula

$$P(x \text{ successes}) = \frac{n!}{x!(n-x)!} p^x q^{n-x}.$$

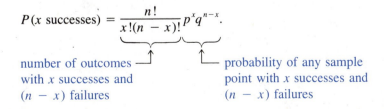

number of outcomes with *x* successes and (*n* − *x*) failures

probability of any sample point with *x* successes and (*n* − *x*) failures

In summary we have the following result:

BINOMIAL PROBABILITY DISTRIBUTION

In a binomial experiment with a constant probability p of success at each trial, the probability of x successes in n trials is given by

$$P(x \text{ successes}) = \frac{n!}{x!(n - x)!} p^x q^{n-x}$$

where x is any integer $0, 1, 2, \ldots, n$.

The probability distribution can be presented as in Table 5-4.

TABLE 5-4

The probability distribution of the number of successes in n trials.

Number of successes x	Probability
0	q^n
1	npq^{n-1}
.	.
.	.
.	.
x	$\dfrac{n!}{x!(n - x)!} p^x q^{n-x}$
.	.
.	.
.	.
n	p^n

EXAMPLE 1   The probability that a person who undergoes a kidney operation will recover is 0.6. Find the probability that of the 6 patients who undergo similar operations,

(a) none will recover   (b) all will recover
(c) half will recover   (d) at least half will recover.

SOLUTION   Since there are 6 patients, $n = 6$. If a patient recovers, we shall call the outcome a success. Hence $p = P(\text{success}) = 0.6$ and $q = 1 - 0.6 = 0.4$. We therefore have the following probabilities:

(a) $P(\text{none recover}) = P(0 \text{ successes})$

$$= q^n$$

$$= (0.4)^6 \qquad \text{Use Table 5-4 with } n = 6, x = 0.$$

$$= 0.0041$$

(b) $P(\text{all recover}) = P(6 \text{ successes})$

$$= p^n$$

$$= (0.6)^6 \qquad \text{Use Table 5-4 with } n = 6, \, x = 6.$$

$$= 0.0467$$

(c) $P(\text{half recover}) = P(3 \text{ successes})$

$$= \frac{6!}{3!3!} (0.6)^3 (0.4)^3 \qquad \text{Use Table 5-4 with } n = 6, \, x = 3.$$

$$= \frac{(6)(5)(4)(3)(2)(1)}{(3)(2)(1)(3)(2)(1)} (0.6)^3 (0.4)^3$$

$$= 20(0.6)^3 (0.4)^3$$

$$= 0.2765$$

(d) $P(\text{at least half recover}) = P(\text{at least 3 successes})$

$$= P(3 \text{ successes}) + P(4 \text{ successes})$$

$$+ P(5 \text{ successes}) + P(6 \text{ successes})$$

$$= 0.2765 + \frac{6!}{4!2!} (0.6)^4 (0.4)^2$$

$$+ \frac{6!}{5!1!} (0.6)^5 (0.4)^1 + 0.0467 \qquad \begin{array}{l} \text{Use the formula} \\ \text{and the results} \\ \text{of (c) and (b).} \end{array}$$

$$= 0.2765 + 15(0.0207) + 6(0.0311) + 0.0467$$

$$= 0.8203 \quad \blacksquare$$

EXAMPLE 2   A test consists of 5 questions, and to pass the test, a student has to answer at least 4 questions correctly. Each question has three possible answers, of which only one is correct. If a student guesses on each question what is the probability that the student will pass the test?

SOLUTION   Let $p = P(\text{correct answer})$. Since the student guesses on each question, we have $p = \frac{1}{3}$ and, consequently, $q = 1 - p = \frac{2}{3}$. Hence, with $n = 5$, we get

$$P(\text{student passes the test}) = P(\text{at least 4 correct answers})$$

$$= P(4 \text{ correct answers}) + P(5 \text{ correct answers})$$

$$= \frac{5!}{4!1!} \left(\frac{1}{3}\right)^4 \left(\frac{2}{3}\right)^1 + \frac{5!}{5!0!} \left(\frac{1}{3}\right)^5 \left(\frac{2}{3}\right)^0$$

$$= 5\left(\frac{2}{3^5}\right) + \frac{1}{3^5}$$

$$= \frac{10 + 1}{3^5}$$

$$= 0.0453. \quad \blacksquare$$

The reader will have realized by now that the computations involving binomial probabilities can become very tedious. Extensive tables giving these probabilities for various values of n and p are available. Each pair of n and p gives a different binomial distribution. For ease of reference, we shall denote the probability of x successes in n trials with $P(\text{success}) = p$ by $b(x; n, p)$. In Appendix II, Table A-2 we give $b(x; n, p)$ for $n = 2, 3, \ldots, 15$ and various values of p. For example, to find $b(5; 12, 0.4)$, we go to the part of Table A-2 where $n = 12$ and look under the column headed by 0.4 corresponding to $x = 5$, obtaining the value 0.227. Hence the probability of 5 successes in 12 trials when the probability of success on a trial is 0.4 is equal to 0.227.

EXAMPLE 3   Suppose that 80 percent of all families own a television set. If 10 families are interviewed at random, use the table of binomial probabilities (Table A-2) to find the probability that

(a) 7 families own a television set
(b) at least 7 families own a television set
(c) at most 3 families own a television set.

SOLUTION   Let $p = P(\text{a family owns a television})$. Then $p = 0.8$.

(a) Since $n = 10$, referring to the binomial table with $n = 10$, $p = 0.8$, and $x = 7$, we get

$$P(7 \text{ families own television}) = b(7; 10, 0.8)$$
$$= 0.201.$$

(b) "At least 7 families own a television set" means that either 7 families *or* 8 families *or* 9 families *or* 10 families own a television set. Hence

$$P\binom{\text{at least 7 families}}{\text{own television}} = b(7; 10, 0.8) + b(8; 10, 0.8)$$
$$+ b(9; 10, 0.8) + b(10; 10, 0.8)$$
$$= 0.201 + 0.302 + 0.268 + 0.107$$
$$= 0.878.$$

(c) "At most 3 families own a television set" means that either 0 families *or* 1 family *or* 2 families *or* 3 families own a television set. Hence

$$P\binom{\text{at most 3 families}}{\text{own television}} = b(0; 10, 0.8) + b(1; 10, 0.8)$$
$$+ b(2; 10, 0.8) + b(3; 10, 0.8)$$
$$= 0 + 0 + 0 + 0.001$$
$$= 0.001. \quad \blacksquare$$

EXAMPLE 4   A machine works on 14 identical components that function independently. It will stop working if 3 or more components fail. If the probability that a component fails is equal to 0.1, find the probability that the machine will be working.

SOLUTION   The machine will be working if at most 2 components fail. Since $n = 14$ and $p = 0.1$, we get

$$P(\text{machine works}) = P(0\text{ failures}) + P(1\text{ failure}) + P(2\text{ failures})$$
$$= b(0; 14, 0.1) + b(1; 14, 0.1) + b(2; 14, 0.1)$$
$$= 0.229 + 0.356 + 0.257$$
$$= 0.842. \quad \blacksquare$$

EXAMPLE 5   **Risks accompanying decision making.** A drug manufacturing company is debating whether a vaccine is safe enough to be marketed. The company claims that the vaccine is 90 percent effective; that is, when tried on a person, the chance for that person to develop immunity is 0.9. The federal drug agency, however, believes that the claim is exaggerated and that the drug is 40 percent effective. To test the company claim, the following procedure is devised: The vaccine will be tried on 10 people. If 8 or more people develop immunity, the company claim will be granted. Find the probability that

(a) the company claim will be granted incorrectly, that is, the federal drug agency is correct in its assertion;
(b) the company claim will be denied incorrectly, that is, indeed, the vaccine is 90 percent effective.

SOLUTION   Let X denote the number of people among ten who develop immunity. If the company claim is valid, then the distribution of X is binomial with $n = 10$, $p = 0.9$. On the other hand, if the federal agency claim is valid, then X has a binomial distribution with $n = 10$, $p = 0.4$.

(a) The company claim will be granted precisely when 8 or more people develop immunity. The probability that the claim will be granted *incorrectly* is equal to the probability that 8 or more people develop immunity when p = 0.4. Hence the probability is obtained from Table A-2 as

$$b(8; 10, 0.4) + b(9; 10, 0.4) + b(10; 10, 0.4)$$
$$= 0.011 + 0.002 + 0$$
$$= 0.013.$$

(b) The company claim will be denied precisely when less than eight people develop immunity. The probability that the claim is denied *incorrectly* is equal to the probability that less than 8 people develop immunity when $p = 0.9$. Hence the probability in question is

$$\sum_{k=0}^{7} b(k; 10, 0.9) = 0.001 + 0.011 + 0.057$$
$$= 0.069. \quad \blacksquare$$

THE MEAN AND THE VARIANCE OF A BINOMIAL DISTRIBUTION

Since the binomial distribution is a discrete distribution, the mean and the variance of the distribution can be found by following the methods presented in Chapter 4. Since the algebraic manipulations become cumbersome, we shall illustrate the method for $n = 3$ and, from the results obtained, inductively state the general formula for arbitrary n. The probability distribution and related computations are presented in Table 5-5.

TABLE 5-5

Binomial probability distribution, $n = 3$.

Number of successes x	Probability $p(x)$	$x \cdot p(x)$	$x^2 \cdot p(x)$
0	q^3	$0 \cdot q^3$	$0^2 \cdot q^3$
1	$3pq^2$	$1 \cdot 3pq^2$	$1^2 \cdot 3pq^2$
2	$3p^2q$	$2 \cdot 3p^2q$	$2^2 \cdot 3p^2q$
3	p^3	$3 \cdot p^3$	$3^2 \cdot p^3$

Following the general rule described in Chapter 4, the mean μ of the distribution can be found by multiplying each of the values 0, 1, 2, 3 by the corresponding probability and then adding the terms. Hence

$$\begin{aligned} \mu &= 0(q^3) + 1(3pq^2) + 2(3p^2q) + 3(p^3) \\ &= 3p(q^2 + 2pq + p^2) \\ &= 3p(p + q)^2 \\ &= 3p \qquad \text{since } p + q = 1. \end{aligned}$$

In the general case, the formula for μ reduces to np. This should not come to us as a surprise. Wouldn't we expect a basketball player to make 12 out of 20 free throws, if we knew from past experience that he makes 60 percent of the free throws? Notice that $20(0.6) = 12$.

To find the variance σ^2, we employ the formula

$$\sigma^2 = \sum_{x=0}^{n} x^2 \cdot p(x) - \mu^2.$$

Now, $\sum_{x=0}^{n} x^2 \cdot p(x)$ is the sum of the entries in column 4 of Table 5-5. Hence

$$\begin{aligned} \sum_{x=0}^{n} x^2 \cdot p(x) &= 0^2(q^3) + 1^2(3pq^2) + 2^2(3p^2q) + 3^2(p^3) \\ &= 3pq^2 + 12p^2q + 9p^3. \end{aligned}$$

Therefore

$$\begin{aligned}
\sigma^2 &= 3pq^2 + 12\,p^2q + 9p^3 - (3p)^2 \qquad \text{since } \mu = 3p \\
&= 3pq^2 + 12p^2q - 9p^2(1 - p) \\
&= 3pq^2 + 3p^2q, \text{ since } 1 - p = q \\
&= 3pq(p + q) \\
&= 3pq.
\end{aligned}$$

It turns out that in the case of arbitrary n, the formula for the variance is given as npq.

In summary we have the following results:

MEAN AND VARIANCE OF A BINOMIAL DISTRIBUTION

The mean μ and the variance σ^2 of a binomial distribution consisting of n trials and probability of success p are given as

$$\mu = np \qquad \text{and} \qquad \sigma^2 = np(1 - p).$$

Also, the standard deviation σ is given as

$$\sigma = \sqrt{np(1 - p)}.$$

EXAMPLE 6   The probability that a freshman entering a university will graduate after a four-year program is 0.6. If 1000 freshmen enroll at a university, find the expected number of those who graduate. Also find the standard deviation of the number of freshmen who graduate.

SOLUTION   Since $n = 1000$ and $p = 0.6$, we have

$$\mu = 1000(0.6) = 600$$

and

$$\sigma = \sqrt{1000(0.6)(0.4)} = 15.49.$$

Hence the expected number of students who graduate is 600 with a standard deviation of 15.49. ∎

Remark: Suppose a finite population consists of a mixture of two kinds of items. When a random sample of items is picked with replacement from such a population, all the conditions of the binomial experiment are satisfied. However, if the sampling is carried out without replacement, then the condition of independence of trials is not met. In such cases, if the population is very large compared to the sample size, sampling without replacement does not make an appreciable difference and we can work the problem as if it were a binomial experiment.

SECTION 5-1   EXERCISES

1. Suppose the probability that a person picked at random feels that the President is doing a satisfactory job is 0.4. If Tom, John, Mary, Brice, Julie, Nancy, and Bob are interviewed, and respond independently, what is the probability that they will answer, respectively, agree, disagree, agree, disagree, disagree, agree, disagree?

2. The probability that a player makes a free throw is 0.6. In 4 free throws, find the probability that the player makes

   (a) exactly 3 free throws   (b) at least 1 free throw

   (c) at most 3 free throws   (d) an even number of free throws.

3. A person who claims to have powers of extrasensory perception (ESP) is shown 6 cards, each of which could be black or red with equal probability. If the person really does not have ESP and guesses on each card, what is the probability that all the cards will be identified correctly?

4. The Associated Press wire service reported that a poll of former U.S. prisoners of war in Vietnam found 67 percent of them were unhappy with the design of a national memorial honoring Americans who fought in the war. If this figure is correct and reflects the probability that a former P.O.W. is unhappy with the design, what is the probability that if 4 former P.O.W.s are interviewed and they respond independently, exactly one will say that he is unhappy with the design?

5. Use the table of binomial probabilities (Table A-2) to find

   (a) $b(3; 5, 0.4)$ (b) $b(8; 10, 0.7)$

   (c) $b(5; 12, 0.3)$ (d) $b(4; 8, 0.5)$.

6. Use Table A-2 to find the value of p, the probability of success, if

   (a) $b(5; 10, p) = 0.201$

   (b) $b(2; 15, p) = 0.267$

   (c) $b(8; 12, p) = 0.133$.

7. A machine manufactures a large number of bolts of a certain type. From past experience, it is known that if 10 bolts are tested, the probability of finding 4 defective bolts is 0.111. Find the probability of finding 6 defective bolts among 10 bolts tested assuming that defects on bolts are independent.

   [*Hint:* Use Table A-2 with the appropriate n to find p and then proceed.]

8. The probability that a child is a son is 0.4. If there are 8 children in a family, find the most probable number of sons and the corresponding probability.

   [*Hint:* Use the appropriate binomial table.]

9. The Center for Disease Control has determined that when a person is given a vaccine, the probability that the person will develop immunity to a virus is 0.8. If 8 people are given this vaccine, find the probability that

   (a) none will develop immunity

   (b) exactly 4 will develop immunity

   (c) all will develop immunity.

10. Suppose 10 percent of the tubes produced by a machine are defective. If 6 tubes are inspected at random, determine the probability that

    (a) 3 tubes are defective

    (b) at least 2 tubes are defective

    (c) at most 5 tubes are defective

    (d) at least 3 tubes are defective.

11. It is known that 60 percent of adults favor capital punishment. Find the probability that on a jury of 12 members,

    (a) all will favor capital punishment

    (b) there will be a tie; that is, 6 members will favor capital punishment and 6 will oppose it.

12. **Quality control inspection.** In a factory that manufactures a large number of washers, the quality control department has adopted the following inspection plan for the production of the day: 10 washers are picked at random and inspected; if this sample contains less than 3 defective washers, the entire production is accepted as tolerable and shipped. What is the probability that the production of a day will be shipped when in fact it contains 50 percent defective washers?

13. A machine produces defective items at the rate of 1 in 10. If 100 items produced by this machine are inspected, what is the expected number of defective items? Also find the variance of the number of defective items.

14. The probability that a door-to-door salesman makes a sale when he visits a house is 0.2. If the salesman visits 60 houses, find

    (a) the expected number of houses where he makes sales

    (b) the variance and the standard deviation of the number of houses where he makes a sale.

15. A large genetic pool contains 60 percent alleles of type A and 40 percent of type a. If 14 alleles are picked at random, find

    (a) the probability that there are exactly 5 alleles of type a

    (b) the expected number of alleles of type A

    (c) the variance of the number of alleles of type a.

16. A company received 8 fluorescent tubes packed in a box. The probability that a tube is defective is 0.1. Assuming independence, find

    (a) the probability that there are 3 defective tubes

    (b) the expected number of defective tubes

    (c) the standard deviation of the number of defective tubes.

17. Suppose a company received a consignment of 200 boxes of fluorescent tubes, eight tubes to a box. If the probability that a tube is defective is 0.1, how many of the boxes would you expect to have three defective tubes?

    [*Hint:* See Part (a) in Exercise 16.]

18. If the probability that a child is a son is 0.4, find the probability that in a family of four children there are

    (a) two sons

    (b) at least two sons

    (c) all girls

    (d) three girls.

19. Suppose the probability that a child is a son is 0.4. If 100 families, each with four children, are interviewed, find how many families you would expect to have

(a) two sons

(b) at least two sons

(c) all girls

(d) three girls.

[*Hint:* Refer to Exercise 18.]

20. Suppose a box contains 100,000 beads, of which 30,000 are black and 70,000 are red. If 10 beads are picked at random from this box, find the probability that there are

(a) 4 black beads

(b) more than 4 black beads

(c) at most 6 red beads

(d) no black beads.

21. On the question of whether marijuana should be decriminalized, it is known that the probability that a person favors it is 0.4, that a person opposes it is 0.5, and that a person has no opinion is 0.1. If 6 persons are interviewed and they respond independently, find the probability that

(a) 5 favor decriminalization

(b) 5 oppose decriminalization

(c) 5 have no opinion

(d) more than 4 favor decriminalization

(e) less than 3 oppose decriminalization.

22. Suppose the probability is 0.8 that a person picked at random will support a constitutional amendment requiring an annual balanced budget. If 9 individuals are interviewed and if they respond independently, what is the probability that at least two-thirds of them will support the amendment?

23. A radar complex consists of 8 units that operate independently. The probability that a unit detects an incoming missile is 0.90. Find the probability that an incoming missile will

(a) not be detected by any unit

(b) be detected by at most 4 units.

5-2 THE NORMAL DISTRIBUTION

For a large number of phenomena, a smooth, bell-shaped curve serves as a mathematical model to describe their probability distribution. Such a bell-shaped curve is called a **normal curve,** and the probability distribution is called a **normal distribution.*** It is by far the most celebrated of the continuous distributions and plays a

---

*For the mathematically oriented, the shape of the curve is defined by

$$f(x) = \frac{1}{\sigma\sqrt{2\pi}}\, e^{-(x-\mu)^2/(2\sigma^2)}, \quad -\infty < x < \infty$$

where μ is the population mean and σ is the population standard deviation.

FIGURE 5-1

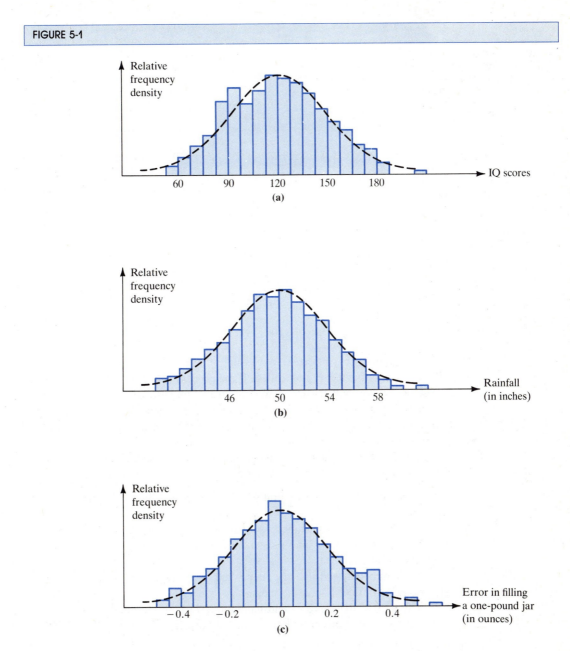

(a)

(b)

(c)

central role in statistical inference. The IQ of university students, the amount of annual rainfall, and the amount of error in filling a one-pound jar of coffee provide some examples. In Figure 5-1, we show how a cursory consideration of the histogram in each case gives an idealization by means of a smooth, symmetrical curve that tapers off at both ends and is peaked at the center. For example, consider the histogram and the corresponding smooth curve describing the distribution of error

FIGURE 5-2

Curves of normal distribution with mean 50 and $\sigma = \frac{1}{2}$, 1, 2. The greater the standard deviation, the flatter the curve.

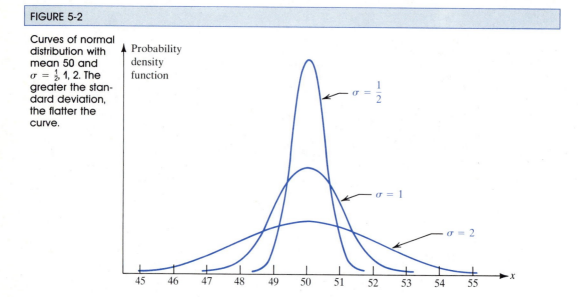

in filling one-pound jars of coffee. We assume that the error is positive if a jar is overfilled and negative if it is underfilled. If the process is under control, we expect that most of the jars will be neither overfilled nor underfilled by large amounts. Fewer and fewer jars will be overfilled by larger and larger amounts; hence the graph will taper off on the right. In the same way, fewer and fewer jars will be underfilled by larger and larger amounts; hence the graph will taper off on the left.

The mathematical equation of the normal curve was developed by DeMoivre in 1733. The following features characterize the normal distribution:

1. It is a continuous distribution which is described by a bell-shaped curve.
2. It is completely determined by its mean, which can be any number, positive or negative, and its standard deviation, which can be any positive number.
3. It is unimodal, and its curve is peaked at the center and symmetric about a vertical line at the mean.
4. The tails of the curve extend indefinitely in both directions from the center, getting closer and closer to the horizontal axis but never quite touching it.
5. The mean μ determines where the center of the curve is located and the standard deviation σ determines its flatness.

All of the curves in Figure 5-2 have the same mean of 50 but different variances, with $\sigma = \frac{1}{2}$, $\sigma = 1$, and $\sigma = 2$. Observe that the greater the standard deviation, the flatter the curve. For example, the curve with $\sigma = \frac{1}{2}$ is more peaked than the one with $\sigma = 2$.

FIGURE 5-3

Normal curves
with $\sigma = 2$ and
$\mu = -6, 0, 8.$

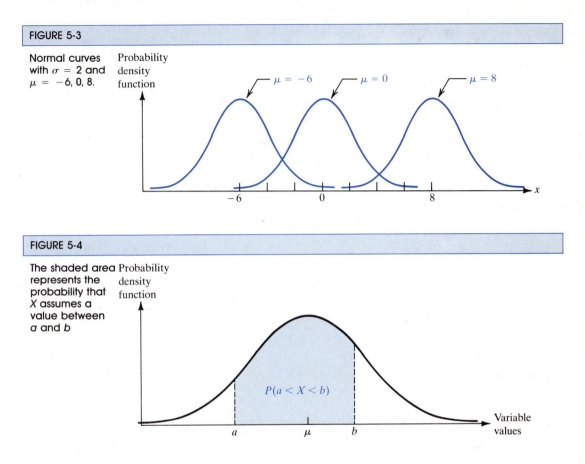

FIGURE 5-4

The shaded area
represents the
probability that
X assumes a
value between
a and b

In Figure 5-3 are given curves all having the same standard deviation, $\sigma = 2$, but different means, $\mu = -6$, $\mu = 0$, and $\mu = 8$. Since σ is the same, the curves have the same shape and size, but the one with $\mu = 8$ is to the right of the one with $\mu = -6$.

THE STANDARD NORMAL DISTRIBUTION

Since a normal distribution is a continuous distribution, the probabilities are given in terms of appropriate areas. Thus the probability that a random variable X having a normal distribution will assume a value between two numbers a and b is equal to the area under the curve between $x = a$ and $x = b$, indicated by the shaded region in Figure 5-4. Because of the shape of the curve, it is impossible to compute these areas geometrically. Tables of areas for normal distributions with all possible means and standard deviations are needed, which would of course be impractical, since there is an endless number of such curves. Fortunately, we need a table of areas for

FIGURE 5-5

Area between
$z = 0$ and
$z = 1.73$ is 0.4582.

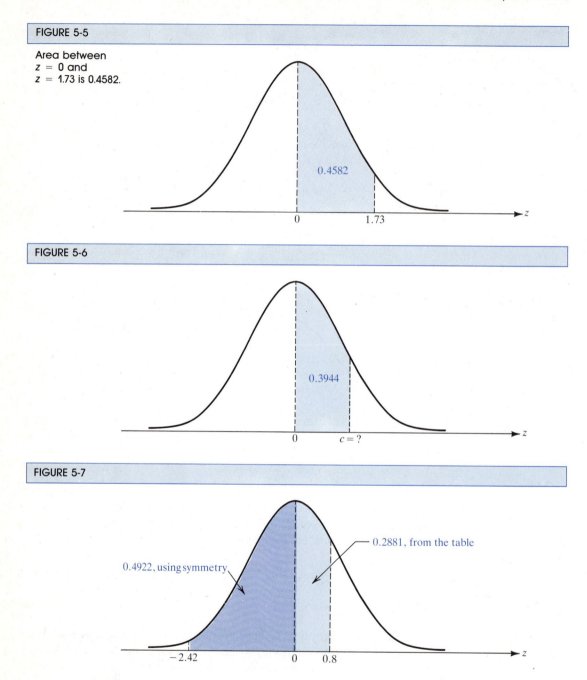

0.4582

0   1.73

z

FIGURE 5-6

0.3944

0          $c = ?$

z

FIGURE 5-7

0.2881, from the table

0.4922, using symmetry

-2.42 0   0.8

z

only one normal curve. The curve that is used is called the **standard normal curve** which describes the distribution of a normal random variable with mean 0 and standard deviation 1. The random variable itself is called the **standard normal**

variable. In the future we shall always denote this random variable by Z. Areas under any normal curve can be obtained by comparing the curve with the standard normal curve through an appropriate transformation, as we shall see later.

*Appendix II, Table A-3 gives areas under the standard normal curve from $z = 0$ to $z = a$, where a is any number 0.00, 0.01, 0.02, . . . , 3.09. To find the area between $z = 0$ and $z = 1.73$, for instance, we go to 1.7 in the column and 0.03 in the row, both headed by z, and read the corresponding entry in the body of the table as 0.4582, as indicated below:

z	0.00	0.01	0.02	0.03	0.04	0.05	0.06	0.07	0.08	0.09
0.0	0.0000	0.0040	0.0080	0.0120	0.0160	0.0199	0.0239	0.0279	0.0319	0.0359
0.1	0.0398	0.0438	0.0478	0.0517	0.0557	0.0596	0.0636	0.0675	0.0714	0.0753
0.2	0.0793	0.0832	0.0871	0.0910	0.0948	0.0987	0.1026	0.1064	0.1103	0.1141
1.6	0.4452	0.4463	0.4474	0.4484	0.4495	0.4505	0.4515	0.4525	0.4535	0.4545
1.7	0.4554	0.4564	0.4573	0.4582	0.4591	0.4599	0.4608	0.4616	0.4625	0.4633
1.8	0.4641	0.4649	0.4656	0.4664	0.4671	0.4678	0.4686	0.4692	0.4699	0.4706

Hence, the area between 0 and 1.73 is 0.4582 and

$$P(0 < Z < 1.73) = 0.4582$$

This is shown by means of the shaded region in Figure 5-5.

We shall now illustrate the use of Table A-3 to find areas under the standard normal curve in several examples. The reader should remember that the curve is symmetric with respect to the vertical axis through 0. It is helpful to draw the curves and identify the areas under the curve and the values along the horizontal axis. We recommend this very strongly.

EXAMPLE 1   If $P(0 < Z < c) = 0.3944$, find c.

SOLUTION   Here the entry in the body of the table is given as 0.3944 and we want to find the corresponding z value. Having found the row heading is 1.2 and the column heading is 0.05, we read $c = 1.25$. This reading is shown in Figure 5-6.   ▬

EXAMPLE 2   Find $P(-2.42 < Z < 0.8)$.

SOLUTION   This probability is shown in Figure 5-7 as the sum of the shaded regions be-

---

*In later chapters dealing with statistical inference the reader will have occasion to look up areas in the tails of the standard normal curve. For convenient reference we also provide Appendix II, Table A-4. It is read exactly as Table A-3 is read, except that the entries in the body of the table are areas in the right tail of the standard normal curve.

FIGURE 5-8

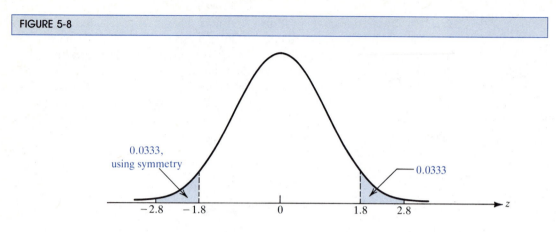

tween 0 and 0.8 and between -2.42 and 0. Notice that due to symmetry, the area between -2.42 and 0 is equal to the area between 0 and 2.42.

Hence

$$P(-2.42 < Z < 0.8) = P(0 < Z < 0.8) + P(0 < Z < 2.42)$$
$$= 0.2881 + 0.4922$$
$$= 0.7803. \quad \blacksquare$$

EXAMPLE 3   Find (a) $P(1.8 < Z < 2.8)$ (b) $P(-2.8 < Z < -1.8)$.

SOLUTION   (a) $P(1.8 < Z < 2.8)$ is the area between $z = 1.8$ and $z = 2.8$ and is equal to the area from 0 to 2.8 *minus* the area from 0 to 1.8, as can be seen from the shaded portion in the right tail of the curve in Figure 5-8.
Hence

$$P(1.8 < Z < 2.8) = 0.4974 - 0.4641$$
$$= 0.0333.$$

(b) Due to the symmetry of the normal curve, as can be seen from the left tail of the curve in Figure 5-8,

$$P(-2.8 < Z < -1.8) = P(1.8 < Z < 2.8)$$
$$= 0.0333. \quad \blacksquare$$

EXAMPLE 4   Find (a) $P(Z > -2.13)$ (b) $P(Z < -1.81)$.

SOLUTION   (a) The area representing $P(Z > -2.13)$ is shaded light blue in Figure 5-9. This area is equal to the area between -2.13 and 0 *plus* the area to the right of

FIGURE 5-9

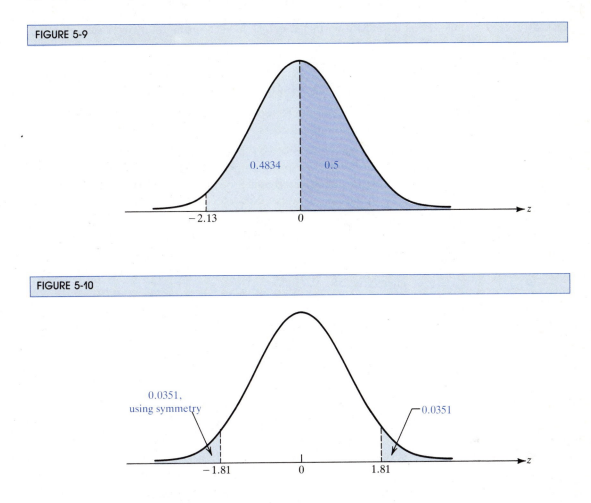

FIGURE 5-10

0. Now the area between -2.13 and 0 is the same as the area between 0 and 2.13 and is equal to 0.4834. Also, the area to the right of 0 is 0.5. Hence

$$P(Z > -2.13) = 0.4834 + 0.5$$
$$= 0.9834.$$

(b) $P(Z < -1.81)$ is equal to the area to the left of -1.81, which, due to symmetry, is equal to the area to the right of 1.81, as can be seen from Figure 5-10. This, in turn, is equal to the area to the right of 0 *minus* the area between 0 and 1.81. Hence

$$P(Z < -1.81) = 0.5 - 0.4649$$
$$= 0.0351.$$

EXAMPLE 5   Suppose Z is a standard normal variable. In each of the following cases, find c for which

(a) $P(Z \leq c) = 0.1151$ (b) $P(Z \leq c) = 0.8238$

(c) $P(1 \leq Z < c) = 0.1525$ (d) $P(-c < Z < c) = 0.8164$.

SOLUTION   In this example we are given the areas and want to read the values along the horizontal axis.

(a) Since the area to the left of c is 0.1151, and since it is less than 0.5, c must be located to the left of 0 as in Figure 5-11. Hence c is negative. Now, due to symmetry, the area between 0 and $-c$ is $0.5 - 0.1151 = 0.3849$. But from Table A-3, the area between 0 and 1.2 is 0.3849. Hence $-c = 1.2$, so that $c = -1.2$.

(b) Since $P(Z \leq c) = 0.8238$, c is to the right of 0; that is, c is positive. Also, as can be seen from Figure 5-12, the area between 0 and c is equal to $0.8238 - 0.5 = 0.3238$. From Table A-3, the area between 0 and 0.93 is 0.3238. Therefore, $c = 0.93$.

FIGURE 5-11

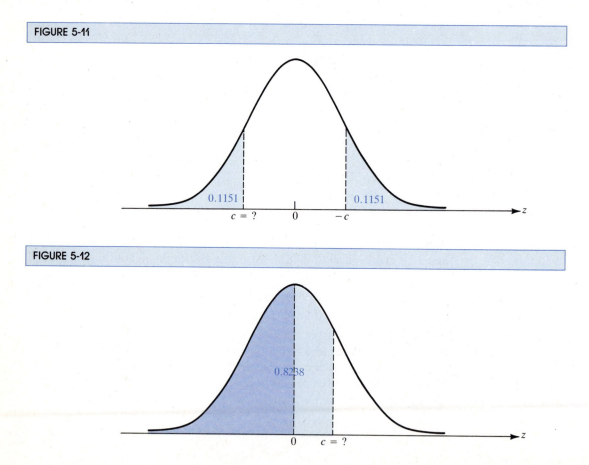

FIGURE 5-12

FIGURE 5-13

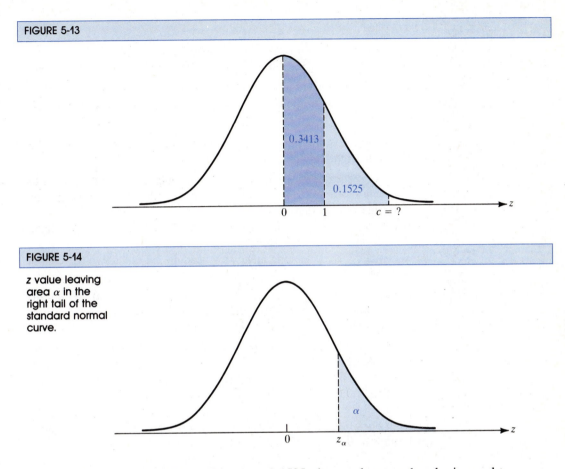

FIGURE 5-14

z value leaving
area α in the
right tail of the
standard normal
curve.

(c) Since $P(1 \leq Z < c) = 0.1525$, the area between 1 and c is equal to 0.1525. Now the area between 0 and 1 is 0.3413. Hence

$$P(0 < Z < c) = 0.3413 + 0.1525$$
$$= 0.4938.$$

We have the situation shown in Figure 5-13. Referring back to Table A-3, it follows that $c = 2.5$.

(d) $P(-c < Z < c) = 0.8164$. Therefore, on account of symmetry,

$$P(0 < Z < c) = \frac{1}{2}(0.8164)$$

$$= 0.4082.$$

Hence, from Table A-3, $c = 1.33$. ■

EXAMPLE 6   Suppose the Greek letter α (alpha) represents a number between 0 and 0.5 and z_α represents the z value such that to its right the area under the standard normal curve is α as shown in Figure 5-14.

Find

(a) α if (i) $z_\alpha = 1.24$ (ii) $z_\alpha = 1.99$ (iii) $z_\alpha = 2.73$

(b) (i) $z_{0.0043}$ (ii) $z_{0.0202}$ (iii) $-z_{0.0202}$ (iv) $z_{0.025}$.

SOLUTION   We could find all the answers in this example using Appendix II, Table A-3. However, since α is area in the right tail it will be more convenient to use Appendix II, Table A-4.

(a) In this part of the example we are given z values and are expected to find α, the area in the right tail.

  (i) To find α when $z_\alpha = 1.24$ we look for the entry in Table A-4 along the row with heading 1.2 and down the column with heading 0.04, getting the value of α as 0.1075.

  (ii) When $z_\alpha = 1.99$, the entry α with row heading 1.9 and column heading 0.09 is 0.0233.

  (iii) When $z_\alpha = 2.73$, looking along row heading 2.7 and down the column heading 0.03 we find $\alpha = 0.0032$.

(b) In this part we do exactly the reverse of what we did in Part (a). We locate the given number α in the body of the table and read the corresponding row and column headings to find the z value.

  (i) Since $\alpha = 0.0043$, the row heading is 2.6 and the column heading is 0.03. Hence $z_{0.0043} = 2.63$.

  (ii) When $\alpha = 0.0202$ we see that the row heading is 2.0 and the column heading is 0.05. Therefore $z_{0.0202} = 2.05$.

  (iii) Obviously $-z_{0.0202} = -2.05$.

  (iv) Here $\alpha = 0.025$. The corresponding row heading is 1.9 and column heading is 0.06. Hence $z_{0.025} = 1.96$. ▆▆

Remark: In dealing with inference problems, we shall have occasion to use the values $z_{0.05}$ and $z_{0.06}$. From Appendix II, Table A-4, $z_{0.05}$ is midway between 1.64 and 1.65. So we take $z_{0.05} = 1.645$. In the same way we take $z_{0.06} = 1.555$ since this value lies between 1.55 and 1.56.

NORMAL DISTRIBUTION WITH ANY MEAN μ AND VARIANCE σ^2

Having considered areas under the standard normal curve, we now turn our attention to the general case of a normal distribution with any mean μ and any standard deviation σ where, of course, $\sigma > 0$. It can be shown that if X is a normal variable with mean μ and standard deviation σ, then one can convert X into a standard normal variable Z by setting

$$Z = \frac{X - \mu}{\sigma}.$$

FIGURE 5-15

The solid curve in (a) is first shifted to the dashed curve. Then a change of scale is applied to produce the curve in (b)

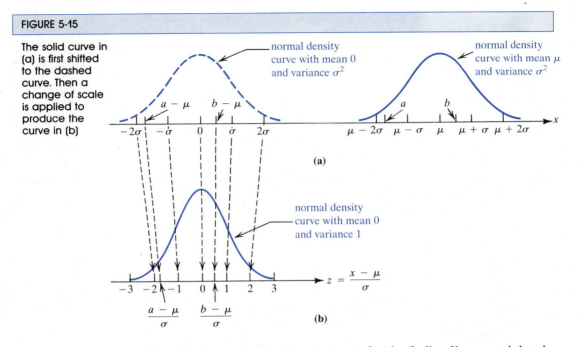

(a)

(b)

The result is accomplished in two steps, first by finding $X - \mu$ and then by dividing by σ. The effect of subtracting μ from X is that of shifting the curve with its center of symmetry at μ so that its new center is at 0. The new position is denoted by the dashed curve in Figure 5-15(a). This step gives us the deviation from the mean μ.

The effect of dividing by σ is to change the scale of the dashed curve in the figure so that it coincides with the standard normal curve as shown in Figure 5-15(b). This step gives us the number of standard deviations from the mean.

How do we find $P(a < X < b)$? Notice that

when $x = a$, we have $z = \dfrac{a - \mu}{\sigma}$

and when $x = b$, we have $z = \dfrac{b - \mu}{\sigma}$.

This means that when X is between a and b, Z is between $\dfrac{a - \mu}{\sigma}$ and $\dfrac{b - \mu}{\sigma}$. Hence we have the following result:

PROBABILITIES UNDER A GENERAL NORMAL CURVE

If X is a normal random variable with mean μ and variance σ^2, then

$$P(a < X < b) = P\left(\frac{a - \mu}{\sigma} < Z < \frac{b - \mu}{\sigma}\right)$$

In short, the area between a and b under a normal curve with mean μ and variance σ^2 is equal to the area under the standard normal curve between $\dfrac{a - \mu}{\sigma}$ and $\dfrac{b - \mu}{\sigma}$.

The resulting situation can be shown conveniently as in Figure 5-16. The important point is that *we subtract the mean of the random variable from the end points of the interval, divide by the standard deviation of the random variable, and refer to Table A-3 of areas under the standard normal curve.*

The conversion of x values into z values as described above is called **standardization.** If x is a given value, then its **standard score** or **z-score** is found by

$$z = \frac{x - \mu}{\sigma}.$$

FIGURE 5-16

The shaded regions under the two curves have the same area. The center of the upper curve is at μ; its variance σ^2. The center of the lower curve is at 0; its variance 1.

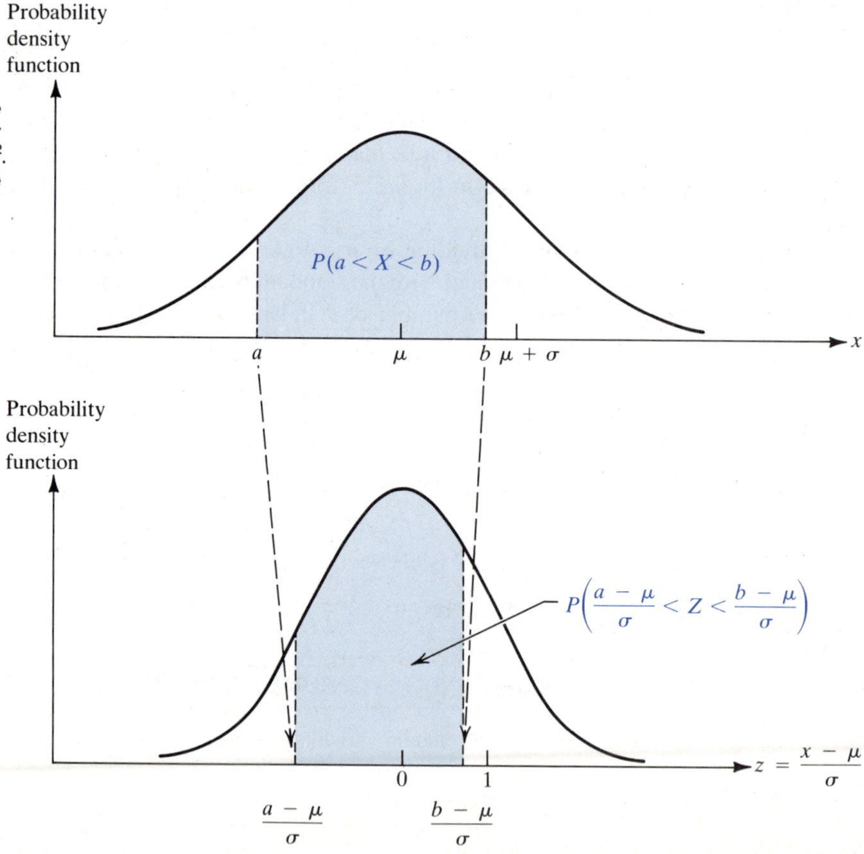

EXAMPLE 7   Suppose X has a normal distribution with $\mu = 30$ and $\sigma = 4$. Find

(a) $P(30 < X < 35)$ (b) $P(X > 40)$
(c) $P(X < 22)$ (d) $P(X > 20)$
(e) $P(X < 37)$ (f) $P(20 < X < 37)$.

SOLUTION   (a) $P(30 < X < 35) = P\left(\dfrac{30 - 30}{4} < Z < \dfrac{35 - 30}{4}\right)$

$$= P(0 < Z < 1.25)$$
$$= 0.3944$$

(b) $P(X > 40) = P\left(Z > \dfrac{40 - 30}{4}\right)$

$$= P(Z > 2.5)$$
$$= 0.5 - 0.4938$$
$$= 0.0062$$

(c) $P(X < 22) = P\left(Z < \dfrac{22 - 30}{4}\right)$

$$= P(Z < -2)$$
$$= P(Z > 2)$$
$$= 0.5 - 0.4772$$
$$= 0.0228$$

(d) $P(X > 20) = P\left(Z > \dfrac{20 - 30}{4}\right)$

$$= P(Z > -2.5)$$
$$= P(-2.5 < Z < 0) + P(Z \geqslant 0)$$
$$= 0.4938 + 0.5$$
$$= 0.9938$$

(e) $P(X < 37) = P\left(Z < \dfrac{37 - 30}{4}\right)$

$$= P(Z \leqslant 0) + P(0 < Z < 1.75)$$
$$= 0.5 + 0.4599$$
$$= 0.9599$$

(f) $P(20 < X < 37) = P\left(\dfrac{20 - 30}{4} < Z < \dfrac{37 - 30}{4}\right)$

$$= P(-2.5 < Z < 1.75)$$
$$= P(-2.5 < Z < 0) + P(0 < Z < 1.75)$$
$$= P(0 < Z < 2.5) + P(0 < Z < 1.75)$$
$$= 0.4938 + 0.4599$$
$$= 0.9537 \quad \blacksquare$$

FIGURE 5-17

Area within k standard deviations from the mean, $k = 1, 2, 3$.

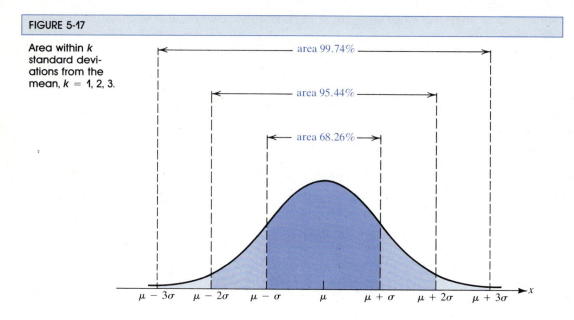

EXAMPLE 8   Suppose X has a normal distribution with mean μ and standard deviation σ. Find

(a) $P(\mu - \sigma < X < \mu + \sigma)$
(b) $P(\mu - 2\sigma < X < \mu + 2\sigma)$
(c) $P(\mu - 3\sigma < X < \mu + 3\sigma)$.

SOLUTION   Notice that for any number k

$$P(\mu - k\sigma < X < \mu + k\sigma) = P\left(\frac{(\mu - k\sigma) - \mu}{\sigma} < Z < \frac{(\mu + k\sigma) - \mu}{\sigma}\right)$$
$$= P(-k < Z < k).$$

(Notice: We subtract the mean μ and divide by the standard deviation σ.) Hence for $k = 1$, $k = 2$, $k = 3$, we find, respectively,

(a) $P(\mu - \sigma < X < \mu + \sigma) = P(-1 < Z < 1)$ since $k = 1$
$$= 2(0.3413)$$
$$= 0.6826$$

(b) $P(\mu - 2\sigma < X < \mu + 2\sigma) = P(-2 < Z < 2)$ since $k = 2$
$$= 2(0.4772)$$
$$= 0.9544$$

(c) $P(\mu - 3\sigma < X < \mu + 3\sigma) = P(-3 < Z < 3)$ since $k = 3$
$$= 2(0.4987)$$
$$= 0.9974.$$

These probabilities are given in Table 5-6 and displayed in Figure 5-17.

TABLE 5-6

Probability that X is within k standard deviations from the mean, $k = 1, 2, 3$.

k	Interval	Probability that X is in the interval	Area under the normal curve
1	$\mu - \sigma$ to $\mu + \sigma$	0.6826	68.26 percent
2	$\mu - 2\sigma$ to $\mu + 2\sigma$	0.9544	95.44 percent
3	$\mu - 3\sigma$ to $\mu + 3\sigma$	0.9974	99.74 percent

Remark: Theoretically a normal curve extends indefinitely in both directions in that the range of the random variable is any number from minus infinity to plus infinity. However, in practice we would approximate the distribution of such random variables as the heights of people by a normal distribution, although we are fully aware that the height of a person can never assume a negative value. In view of the fact that practically all the area under a normal curve is within three standard deviations from the mean (99.74 percent, as seen from Table 5-6), we need only be concerned with whether the normal curve provides a good model within three standard deviations from the mean. The behavior in the curve's tails beyond three standard deviations is inconsequential.

EXAMPLE 9   The amount of annual rainfall in a certain region is known from past experience to be a normally distributed random variable with a mean annual rainfall of 50 inches and a standard deviation of 4 inches. If the rainfall exceeds 57 inches during a year, it leads to floods. Find the probability that during a randomly picked year there will be floods.

SOLUTION   Let X denote the amount of annual rainfall (in inches). The distribution of X is shown in Figure 5-18 on page 192. The probability of a flood is equal to the probability that the rainfall exceeds 57 inches, that is, $P(X > 57)$. Now

$$P(X > 57) = P\left(Z > \frac{57 - 50}{4}\right)$$
$$= P(Z > 1.75)$$
$$= 0.5 - 0.4599$$
$$= 0.0401.$$

FIGURE 5-18

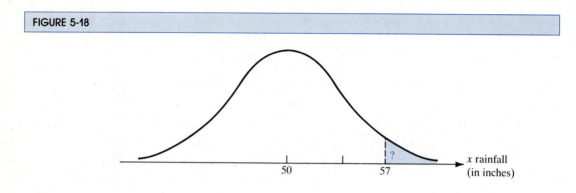

Over a long period of years, about once every 25 years there will be floods. ■

EXAMPLE 10   The weight of food packed in certain containers is a normally distributed random variable with a mean weight of 500 pounds and a standard deviation of 5 pounds. Suppose a container is picked at random. Find the probability that it contains

(a) more than 510 pounds
(b) less than 498 pounds
(c) between 491 and 498 pounds.

SOLUTION   Suppose X denotes the weight of food packed in the containers.

(a) We want to find $P(X > 510)$. We have

$$P(X > 510) = P\left(Z > \frac{510 - 500}{5}\right)$$
$$= P(Z > 2)$$
$$= 0.5 - 0.4772$$
$$= 0.0228.$$

(b) Here we want $P(X < 498)$. We get

$$P(X < 498) = P\left(Z < \frac{498 - 500}{5}\right)$$
$$= P(Z < -0.4)$$
$$= P(Z > 0.4)$$
$$= 0.5 - 0.1554$$
$$= 0.3446.$$

(c) We want $P(491 < X < 498)$. We get

$$P(491 < X < 498) = P\left(\frac{491 - 500}{5} < Z < \frac{498 - 500}{5}\right)$$

FIGURE 5-19

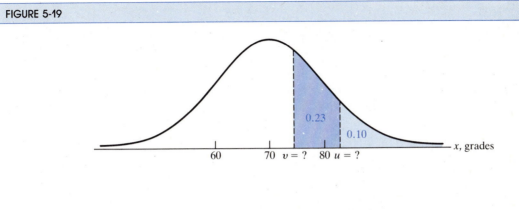

$$= P(-1.8 < Z < -0.4)$$
$$= P(0.4 < Z < 1.8) \quad \text{(Why?)}$$
$$= P(0 < Z < 1.8) - P(0 < Z < 0.4)$$
$$= 0.4641 - 0.1554 = 0.3087. \quad \blacksquare$$

EXAMPLE 11
Grading on the curve. Suppose the scores of students on a test are approximately normally distributed with a mean score of 70 points and a standard deviation of 10 points. It is decided to give A's to 10 percent of the students and B's to 23 percent of the students. Find what scores should be assigned A's and B's.

SOLUTION
"The scores on a test" is in reality a discrete random variable, but experience shows that the normal distribution provides a good approximation. Let us denote the random variable by X. Suppose scores above u are assigned the grade A and those between v and u are assigned the grade B. Then the problem is reduced to finding two numbers u and v for which

$$P(X \geq u) = 0.1$$

and

$$P(v \leq X < u) = 0.23$$

as shown in Figure 5-19. From the figure we see that the area between 70 and u is $0.5 - 0.1$, or 0.4, and that between 70 and v is $0.5 - (0.23 + 0.1)$, or 0.17. Hence

$$P(70 < X < u) = 0.4 \quad \text{and} \quad P(70 < X < v) = 0.17.$$

Since $\mu = 70$ and $\sigma = 10$, converting to the z scale we have

$$P\left(0 < Z < \frac{u - 70}{10}\right) = 0.4$$

and

$$P\left(0 < Z < \frac{v - 70}{10}\right) = 0.17.$$

From the standard normal table, it therefore follows that

$$\frac{u - 70}{10} = 1.28, \text{ that is, } u = 82.8$$

and

$$\frac{v - 70}{10} = 0.44, \text{ that is, } v = 74.4.$$

Thus students with a score of 83 points or over should be given an A, and those with a score between 74 and 82 points should receive a B. ▬

SECTION 5-2 EXERCISES

1. If Z is a standard normal variable, use Table A-3 and find the following probabilities:

(a) $P(0 \leqslant Z < 1.2)$ (b) $P(Z \geqslant 1.2)$

(c) $P(Z > -1.2)$ (d) $P(-1.2 < Z \leqslant 0)$

(e) $P(1 < Z < 2.5)$ (f) $P(-2.5 < Z < -1)$

(g) $P(-1 < Z < 2.3)$ (h) $P(0.23 < Z \leqslant 0.68)$

(i) $P(-0.18 < Z < 0.45)$ (j) $P(Z \geqslant -0.75)$

(k) $P(Z \leqslant 0.68)$

2. Suppose Z is a standard normal variable. In each of the following cases, find the value of c if

(a) $P(0 < Z \leqslant c) = 0.1844$ (b) $P(Z \leqslant c) = 0.8729$

(c) $P(Z \leqslant c) = 0.1038$ (d) $P(Z \geqslant c) = 0.7357$

(e) $P(Z \geqslant c) = 0.0154$ (f) $P(-c \leqslant Z \leqslant c) = 0.8926$

(g) $P(Z \geqslant 2c) = 0.0778$ (h) $P(0.5 \leqslant Z < c) = 0.2184$

(i) $P(c \leqslant Z \leqslant 0) = 0.4881$ (j) $P(-2 \leqslant Z \leqslant c) = 0.055$

(k) $P(c \leqslant Z \leqslant 2) = 0.64.$

3. Find the following values of z where z_α is defined as in Example 6:

(a) $z_{0.0918}$ (b) $z_{0.1446}$ (c) $-z_{0.0244}$

(d) $-z_{0.0104}$ (e) $z_{0.3745}$ (f) $-z_{0.3897}$

4. Find α if

(a) $z_\alpha = 2.25$ (b) $z_\alpha = 1.39$ (c) $z_\alpha = 1.93$

(d) $z_\alpha = 0.56$ (e) $z_{\alpha/2} = 1.74$ (f) $z_{\alpha/2} = 1.46.$

5. Suppose X is normally distributed with mean 50 and standard deviation 4. In each case indicate the given probabiity as an area under the normal curve, and determine its value:

   (a) $P(51 \leqslant X \leqslant 56)$ (b) $P(X \geqslant 47)$

   (c) $P(44 \leqslant X \leqslant 53)$ (d) $P(X \leqslant 48)$

   (e) $P(X \leqslant 52)$ (f) $P(X \geqslant 59)$

6. Suppose X has the normal distribution with mean 10 and standard deviation 2. Find the following probabilities:

   (a) $P(10 \leqslant X \leqslant 11.6)$ (b) $P(X \leqslant 12.8)$

   (c) $P(7 < X \leqslant 13)$ (d) $P(X \geqslant 8.6)$

   (e) $P(11 \leqslant X \leqslant 13)$ (f) $P(7.6 \leqslant X \leqslant 9.2)$

7. If X is normally distributed with mean -0.5 and standard deviation 0.02, find the following probabilities:

   (a) $P(-0.5 \leqslant X \leqslant -0.46)$ (b) $P(X \leqslant -0.47)$

   (c) $P(-0.53 \leqslant X \leqslant -0.45)$ (d) $P(X \geqslant -0.53)$

   (e) $P(-0.48 \leqslant X \leqslant -0.45)$ (f) $P(-0.55 \leqslant X \leqslant -0.52)$

8. The length of life of an instrument produced by a machine has a normal distribution with mean life of 12 months and standard deviation of 2 months. Find the probability that an instrument produced by the machine will last

   (a) less than 7 months

   (b) between 7 and 12 months

9. The number of orders that a mail-order service receives in a day has, approximately, a normal distribution with a mean number of orders equal to 500 and a standard deviation of 30 orders. Find the probability that on any day the number of orders received will be

   (a) less than 440 orders

   (b) between 530 and 575 orders

   (c) between 440 and 470 orders.

10. The height of an adult male is known to be normally distributed with a mean of 69 inches and a standard deviation of 2.5 inches. How high should a doorway be so that 96 percent of the adult males can pass through it without having to bend?

11. The diameter of a lead shot produced by a machine has normal distribution with a mean diameter equal to 2 inches and a standard deviation equal to 0.05 inch. Find approximately what the diameter should be of a circular hole through which only 3 percent of the lead shots will pass.

12. The nicotine content in a certain brand of king-size cigarettes has a normal distribution with a mean content of 1.8 mg and a standard deviation of 0.2 mg. Find the probability that the nicotine content of a randomly picked cigarette of this brand will be

    (a) less than 1.45 mg

    (b) between 1.45 and 1.65 mg

    (c) between 1.95 and 2.15 mg

    (d) more than 2.15 mg.

13. In Exercise 12, what value is such that 80 percent of the cigarettes will exceed it in their nicotine content?

14. The diameter of a bolt produced by a machine is normally distributed with a mean diameter of 0.82 cm and a standard deviation of 0.004 cm. What percent of the bolts will meet the specification that they be between 0.816 cm and 0.826 cm?

15. The amount of sap collected from a tree has a normal distribution with a mean amount of 12 gallons and a standard deviation of 2.5 gallons.

    (a) Find the probability that a tree will yield more than 11 gallons.

    (b) If 1000 trees are tapped, approximately how many trees will yield more than 11 gallons?

16. The amount of coffee (in ounces) filled in a jar by a machine has a normal distribution with a mean amount of 16 ounces and a standard deviation of 0.2 ounces.

    (a) Find the probability that a randomly picked jar will contain (i) less than 15.7 ounces, (ii) more than 16.3 ounces.

    (b) In a random sample of 200 jars, find approximately how many jars will contain (i) less than 15.7 ounces, (ii) more than 16.3 ounces.

17. The demand for meat at a grocery store during any week is approximately normally distributed with a mean demand of 5000 pounds and a standard deviation of 300 pounds.

    (a) If the store has 5300 pounds of meat, what is the probability that it is over-stocked?

    (b) How much meat should the store have in stock per week so as not to run short more than 10 percent of the time?

18. The time that a skier takes on a downhill course has a normal distribution with a mean of 12.3 minutes and a standard deviation of 0.4 minutes. Find the probability that on a random run the skier will take between 12.1 and 12.5 minutes.

Solved using the binomial formula

19. In Exercise 18, if the skier makes 6 independent runs, find the probability that

    (a) 2 of the runs will be between 12.1 and 12.5 minutes

    (b) all of the runs will be between 12.1 and 12.5 minutes.

20. Suppose X has a normal distribution with $\sigma = 4$. If $P(X \geqslant -8) = 0.0668$, find the mean μ.

21. Suppose the systolic resting blood pressure of an adult male is normally distributed with mean 138 mm of mercury and standard deviation 10 mm of mercury. If an adult male is picked at random, find the probability that his systolic blood pressure will be

    (a) greater than 160 mm

    (b) between 120 and 135 mm.

22. The mean annual income of a worker in a certain professional category is $18,200, with a standard deviation of $1,400. Assuming that the distribution of income is normal, what is the probabilty that two workers picked at random will each earn less than $17,500?

23. According to the local transit authority, the mean length of time that a passenger spends waiting for a bus is 5 minutes with a standard deviation of 1.2 minutes. It may be assumed that the waiting time is normally distributed. If a passenger waits for a bus at two different times, what is the probability that this person will have to wait for more than 7 minutes on each occasion?

5-3 THE NORMAL APPROXIMATION TO THE BINOMIAL DISTRIBUTION

The arithmetic computations for finding the binomial probabilities can be both involved and time consuming, even for moderately small values of n.

If n is large, there is a convenient way to obtain an approximate value for binomial probabilities. This is accomplished by approximating the binomial distribution with n trials and probability of success p by means of a normal distribution that has a mean np and a standard deviation $\sqrt{npq}$, where $q = 1 - p$. Let us consider the case when $n = 15$ and $p = 0.4$.

The *exact* probabilities $b(x; 15, 0.4)$ for various values of x, except for rounding, are reproduced in Table 5-7 from Table A-2 in Appendix II. Graphically these probabilities can be represented as in Figure 5-20 on page 198 by drawing line segments at each of the values 0, 1, 2, . . . , 15, or by means of a histogram, as in Figure 5-21 on page 198, by erecting vertical bars.

TABLE 5-7	
Binomial probabilities when $n = 15$ and $p = 0.4$.	
x	$b(x; 15, 0.4)$
0	– – –
1	0.005
2	0.022
3	0.063
4	0.127
5	0.186
6	0.207
7	0.177 $\longleftarrow$ $= \binom{15}{7}(0.4)^7(0.6)^8$
8	0.118
9	0.061
10	0.024
11	0.007
12	0.002
13	– – –
14	– – –
15	– – –

From Table 5-7 we see that the exact probability of getting 7 successes is equal to 0.177. In the discussion that follows, we shall show how to obtain the approximate value for this probability by normal approximation.

Notice first of all that in Figure 5-21 the vertical bar centered at $x = 7$ has a base length of 1 unit extending from 6.5 to 7.5, and has a height equal to 0.177. Thus the area of this vertical bar is equal to 0.177×1, or 0.177. In other words, $b(7; 15, 0.4)$ is *precisely* equal to the area of the darker rectangle in Figure 5-21. The important fact to realize is that the rectangle has the base extending from 6.5 to 7.5.

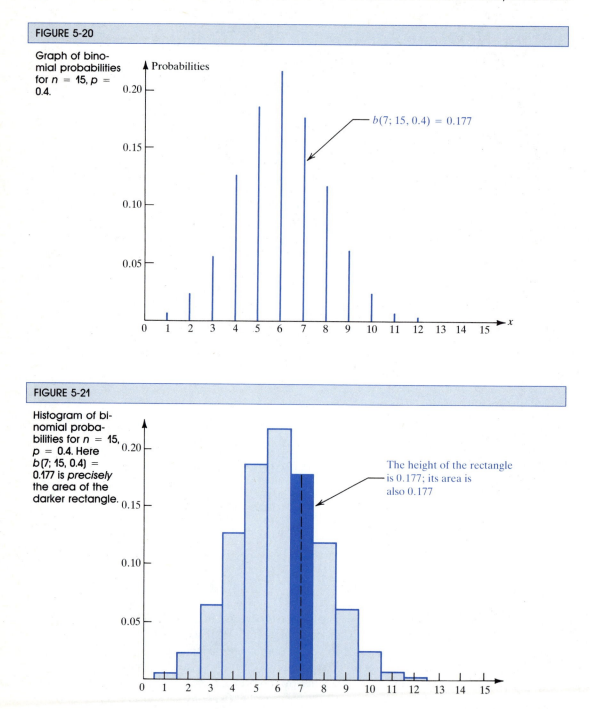

FIGURE 5-20

Graph of binomial probabilities for $n = 15$, $p = 0.4$.

Probabilities

$b(7; 15, 0.4) = 0.177$

FIGURE 5-21

Histogram of binomial probabilities for $n = 15$, $p = 0.4$. Here $b(7; 15, 0.4) = 0.177$ is *precisely* the area of the darker rectangle.

The height of the rectangle is 0.177; its area is also 0.177

FIGURE 5-22

The darker area under the normal curve is *approximately* the probability *b*(7; 15, 0.4).

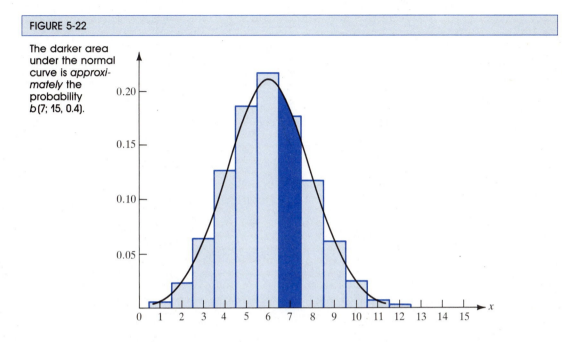

Under the normal approximation, the area of the rectangle will be approximated by the darker region in Figure 5-22, the area between 6.5 and 7.5 under the normal curve with $\mu = 15(0.4) = 6$ and $\sigma = \sqrt{15(0.4)(0.6)} = \sqrt{3.6} = 1.897$. Therefore, converting to the z scale, the probability of 7 successes is approximately equal to the area between $z = \dfrac{6.5 - 6}{1.897}$ and $z = \dfrac{7.5 - 6}{1.897}$ under the standard normal curve. Letting the symbol $\approx$ stand for "approximately equal to," we have

$$b(7; 15, 0.4) \approx P\left(\frac{6.5 - 6}{1.897} < Z < \frac{7.5 - 6}{1.897}\right)$$
$$= P(0.26 < Z < 0.79)$$
$$= P(0 < Z < 0.79) - P(0 < Z < 0.26)$$
$$= 0.2852 - 0.1026$$
$$= 0.1826.$$

This approximate value, as can be seen, is remarkably close to the exact value of 0.177.

It now follows, for instance, that

(i) $P(5 \leqslant X < 9)$ = sum of the areas of the rectangles *centered* at 5, 6, 7, and 8

$\approx$ area under the normal curve between 4.5 and 8.5

$$= P\left(\frac{4.5 - 6}{1.897} < Z < \frac{8.5 - 6}{1.897}\right)$$

$$= P(-0.79 < Z < 1.32)$$

$$= 0.4066 + 0.2852$$

$$= 0.6918.$$

(The exact value from Table 5-7 is $0.186 + 0.207 + 0.177 + 0.118 = 0.688$.)

(ii) $P(5 < X \leqslant 9) =$ sum of the areas of the rectangles centered at 6, 7, 8, and 9

$\approx$ area under the normal curve between 5.5 and 9.5

$$= P\left(\frac{5.5 - 6}{1.897} < Z < \frac{9.5 - 6}{1.897}\right)$$

$$= P(-0.26 < Z < 1.85)$$

$$= 0.4678 + 0.1026 = 0.5704$$

(iii) $P(X \geqslant 8) =$ sum of the areas of the rectangles centered at 8, 9, . . . , 15

$\approx$ area under the normal curve to the right of 7.5

$$= P\left(Z > \frac{7.5 - 6}{1.897}\right)$$

$$= P(Z > 0.79)$$

$$= 0.5 - 0.2852$$

$$= 0.2148$$

(iv) $P(X < 7) =$ sum of the areas of the rectangles centered at 0, 1, . . . , 6

$\approx$ areas under the normal curve to the left of 6.5

$$= P\left(Z < \frac{6.5 - 6}{1.897}\right)$$

$$= P(Z < 0.26)$$

$$= 0.5 + 0.1026$$

$$= 0.6026$$

and so on.

In general, the approximation is carried out as follows: If n is large and p is not close to 0 or 1, the binomial probability $b(x; n, p)$ can be obtained approximately as the area from $x - \frac{1}{2}$ to $x + \frac{1}{2}$ under a normal curve where the mean of the distribution is np and the standard deviation is $\sqrt{npq}$.

Converting to the z scale, we get the following result:

> **APPROXIMATION OF BINOMIAL PROBABILITIES**
>
> $$b(x; n, p) \approx P\left(\frac{(x - \frac{1}{2}) - np}{\sqrt{npq}} < Z < \frac{(x + \frac{1}{2}) - np}{\sqrt{npq}}\right)$$
>
> if n is large and p is not close to 0 or 1.

The approximation is rather close, even for n as low as 15 if p is in the neighborhood of 0.5. As a working rule, *the procedure will yield a satisfactory approximation to the binomial probabilities if both np and nq are greater than 5.* In the context of a normal approximation to the binomial distribution, this is what we shall mean when we say n is large. The addition and subtraction of $\frac{1}{2}$ is called the **continuity correction.**

EXAMPLE 1   Suppose the probability that a worker meets with an accident during a one-year period is 0.4. If there are 200 workers in a factory, find the probability that more than 75 workers will meet with an accident in one year. (Assume that whether a worker has an accident or not is independent of any other worker having one.)

SOLUTION   Let X denote the number of workers who meet with an accident during a year. Then X has a binomial distribution with

$$\text{mean} = 200(0.4) = 80$$

and

$$\text{standard deviation} = \sqrt{200(0.4)(0.6)} = 6.928.$$

Now the probability that more than 75 workers meet with an accident is

$$P(X > 75) = P(X = 76) + P(X = 77) + \cdots + P(X = 200)$$

$$= \frac{200!}{76!124!} (0.4)^{76}(0.6)^{124} + \cdots + \frac{200!}{200!0!} (0.4)^{200}(0.6)^{0}$$

using the formula for binomial probabilities. The computation of the value of the expression above without a high-speed computer is out of the question. Under normal approximation, we shall approximate $P(X > 75)$ as the area to the right of 75.5 under the normal curve with $\mu = 80$ and $\sigma = 6.928$; that is, as the area to the right of $\dfrac{75.5 - 80}{6.928}$ under the standard normal curve. Thus

$$P(X > 75) \approx P\left(Z > \frac{75.5 - 80}{6.928}\right)$$

$$= P(Z > -0.65)$$
$$= 0.5 + P(0 < Z < 0.65)$$
$$= 0.5 + 0.2422$$
$$= 0.7422 \quad \blacksquare$$

EXAMPLE 2   A new vaccine was tested on 100 persons to determine its effectiveness. If the claim of the drug company is that a random person who is given the vaccine will develop immunity with probability 0.8, find the probability that

(a) less than 74 people will develop immunity
(b) between 74 and 85 people, inclusive, will develop immunity.

SOLUTION   Let X represent the number of people who develop immunity. Since $n = 100$ and $p = 0.8$,

$$\mu = 100(0.8) = 80 \quad \text{and} \quad \sigma = \sqrt{100(0.8)(0.2)} = 4.$$

(a) We want $P(X < 74)$. The approximate value of this probability is equal to the area to the left of 73.5 under the normal curve with $\mu = 80$ and $\sigma = 4$. Converting to the z scale, this is equal to $P\left(Z < \dfrac{73.5 - 80}{4}\right)$. Thus

$$P(X < 74) \approx P\left(Z < \frac{73.5 - 80}{4}\right)$$
$$= P(Z < -1.62)$$
$$= P(Z > 1.62)$$
$$= 0.5 - P(0 < Z < 1.62)$$
$$= 0.5 - 0.4474$$
$$= 0.0526.$$

(b) $P(74 \le X \le 85)$ can be approximated as the area between 73.5 and 85.5 under the normal curve with $\mu = 80$ and $\sigma = 4$; that is, as the area between $\dfrac{73.5 - 80}{4}$ and $\dfrac{85.5 - 80}{4}$ under the standard normal curve. Hence

$$P(74 \le X \le 85) \approx P\left(\frac{73.5 - 80}{4} < Z < \frac{85.5 - 80}{4}\right)$$
$$= P(-1.63 < Z < 1.37)$$
$$= P(0 < Z < 1.37) + P(0 < Z < 1.63)$$
$$= 0.4147 + 0.4484$$
$$= 0.8631. \quad \blacksquare$$

Remark: The discrepancy resulting from not using the continuity correction in a normal approximation to a binomial distribution will not be appreciable if n is very large. With an eye to keeping computations simple, we shall not be concerned with using it in later chapters.

SECTION 5-3   EXERCISES

1. Suppose the probability that a college student favors strict environmental controls is 0.7. If 200 randomly picked students are interviewed and they respond independently, find

   (a) the expected number of students in the sample who favor strict controls

   (b) the variance of the number of students in the sample who favor strict controls

   (c) the approximate probability that 140 students favor strict controls

   (d) the approximate probability that between 130 and 146 students, inclusive, favor strict controls.

2. From past experience, a restaurant owner knows that about 40 percent of her customers order cold drinks. If 400 customers visit the restaurant, find the approximate probability that more than 180 customers will order cold drinks. (Assume that the customers order independently.)

3. The probability that a household owns a self-defrosting refrigerator is 0.3. If 200 households are interviewed, find the approximate probability that a self-defrosting refrigerator is owned by

   (a) 60 households

   (b) more than 60 households

   (c) less than 70 households

4. It is claimed that 80 percent of the people in the United States watched the first moon-landing on TV. If this claim is true, what is the approximate probability that in a sample of 100 polled at random, the number of people who saw the landing will be

   (a) less than 70?

   (b) greater than 86?

   (c) between 76 and 86, inclusive?

5. The city of Metroville has a population of 50,000. It is felt that in the mayoral election 60 percent of the voters are in favor of Mr. Nesbitt. What is the approximate probability that of the 200 people interviewed at random by a television network, less than 110 will favor Mr. Nesbitt. (Assume that the people respond independently.)

6. When a certain seed is planted, the probability that it will germinate is 0.1. If 1000 seeds are planted, find the approximate probability that

   (a) more than 130 seeds will germinate

   (b) between 90 and 95 seeds, inclusive, will germinate.

7. **Risks accompanying decision making**. If a sample of 80 washers produced by a machine yields less than 10 defective washers, the entire production of the day will be accepted as good. Otherwise it is rejected. Find the probability that the production will be

   (a) accepted when in fact the machine produces 20 percent defective washers

   (b) rejected when the machine produces 7 percent defective washers

8. The probability that an electronic component functions beyond one hour is 0.8. If a machine uses 120 such components functioning independently, find the approximate probability that at the end of one hour

(a) at least 90 components will be functioning

(b) between 70 and 90 components, inclusive, will be functioning.

9. **Risks accompanying decision making.** A drug company is debating whether a vaccine is effective enough to be marketed. The company claims that the vaccine is 90 percent effective. To test the company claim, the vaccine will be tried on 100 people. If 82 or more people develop immunity, the company claim will be granted. Find the approximate probability that the company claim will *not* be granted when in fact the drug is 90 percent effective, as the company claims.

10. The probability that a cathode tube manufactured by a machine is defective is 0.4. If a sample of 60 tubes is inspected, find the approximate probability that

(a) exactly half of the tubes will be defective

(b) over half of the tubes will be defective.

11. The probability that a mail-order service receives between 400 and 500 orders on a day is 0.3. Find the approximate probability that during 90 days the mail-order service will receive between 400 and 500 orders on

(a) 32 days

(b) more than 32 days.

12. Suppose the probability that a person favors decriminalization of marijuana is 0.45. If 200 people are interviewed, find the approximate probability that a majority will favor decriminalization.

13. A door-to-door salesman visits 100 households. If the probability that he makes a sale at a household is 0.4, find the approximate probability that he will make sales in more than 45 households.

14. The probability that a drunk driver meets with an accident is 0.2. If there are 80 drunk drivers on a freeway, assuming independence, find the approximate probability that

(a) fewer than 14 will meet with an accident

(b) between 14 and 17, inclusive, will meet with an accident.

15. Two hundred individuals who were picked at random and respond independently were asked whether food stamps should be restricted to families who earn below the level of $12,000 per year for a family of four. If it is believed that 70 percent of the population favors such a proposition, what is the approximate probability that in the sample of 200 less than 122 will support the proposition?

16. An advertising agency claims that 75 percent of all doctors recommend aspirin over any other pain reliever. If this claim is indeed valid, what is the approximate probability that in a sample of 96 doctors who were interviewed at random and who respond independently less than 60 will recommend aspirin? If in such a survey you actually found out that 58 out of 96 preferred aspirin, what would you think about the genuine nature of the claim?

17. If a genotype *Aa* is self-fertilized, then the probability that an offspring is of *AA* genotype is 1/4. When a genotype *Aa* was self-fertilized, there were 184 offsprings. Assuming independence, what is the approximate probability that there will be more than 60 offsprings of *AA* genotype?

KEY TERMS AND EXPRESSIONS

binomial experiment
binomial distribution
success
failure
binomial random variable
normal curve
normal distribution

standard normal distribution
standardization
standard score
z-score
normal approximation to the binomial
continuity correction

KEY FORMULAS

binomial probabilities

$$P(x \text{ successes}) = \frac{n!}{x!(n-x)!} p^x (1-p)^{n-x} \qquad x = 0, 1, \ldots, n.$$

binomial mean

$$\mu = np$$

binomial variance

$$\sigma^2 = np(1-p)$$

z-score for x

$$z = \frac{x - \mu}{\sigma}$$

approximation of binomial probabilties

$$b(x; n, p) \approx P\left(\frac{(x - \frac{1}{2}) - np}{\sqrt{np(1-p)}} < Z < \frac{(x + \frac{1}{2}) - np}{\sqrt{np(1-p)}}\right)$$

CHAPTER 5   TEST

1. Explain clearly what assumptions are made in applying the binomial distribution formula.

2. Describe the probability density curve of a normal distribution. Draw, on the same graph, probability density curves for two normal distributions that have the same variance but have means -3 and 5.

3. Ernesto is interested in the results of four baseball games (against the L.A. Dodgers, St. Louis Cardinals, Atlanta Braves, and New York Mets) that his home team, the San Francisco Giants, is scheduled to play. Suppose the following are the probabilities of his team winning: 0.6 against the L.A. Dodgers, 0.3 against the St. Louis Cardinals, 0.7 against the Atlanta Braves, and 0.5 against the New York Mets. What is the probability that Ernesto's team will win exactly two of the games? (You may assume that the outcome of one game does not influence that of the other.)

4. As in Exercise 3 above Ernesto is interested in the results of the four games. However, this time suppose the probability of his team winning against any team is the same and equal to 0.6. What is the probability that Ernesto's team will win exactly two of the games? Once again, assume independence.

5. In a poll the following response was obtained to the following question: Some people believe that President Reagan's proposal for strategic nuclear arms reduction talks

with the Soviet Union is a genuine effort to cut the size of U.S. and Soviet nuclear arsenals. Others believe the proposal is designed to stop criticism in the United States and Western Europe of his defense build-up. Which comes closer to your view?

Genuine effort	50 percent
Stop criticism	40 percent
Don't know	10 percent

Suppose these figures do indeed reflect the national sentiment. If seven randomly picked people are interviewed by a local TV station find the probability that

(a) exactly 4 will believe that the effort is genuine

(b) exactly 2 will believe that the effort is to stop criticism

(c) none will respond "I don't know."

6. From past experience, an automobile dealer has determined that 20 percent of his customers buy compact cars. Find the probability that of 10 customers who buy an automobile (independently)

(a) 3 will buy a compact

(b) at most 3 will buy a compact

(c) at least 3 will buy a compact.

7. A machine manufactures a very large number of resistors. It is known that 10 percent of the resistors produced by the machine are defective. What is the approximate probability that a sample of 400 will contain 43 defective resistors?

8. The probability that an electric sander will last between 500 and 600 hours before it needs repairs is 0.7. In a school district, 200 sanders are in use in different woodworking classes. What is the approximate probability that more than 150 sanders will last between 500 and 600 hours?

9. The distance (in miles) that a brand of radial tire lasts is normally distributed with a mean distance of 40,000 miles and a standard deviation of 4000 miles. If the company guarantees free replacement of any tire that lasts less than 33,000 miles, what fraction of the tires will the company have to replace?

10. The breaking strength of cables manufactured by a certain company has the normal distribution with a mean strength of 1200 pounds and a standard deviation of 64 pounds.

(a) Find the probability that a randomly picked cable will have breaking strength

    (i)   less than 1059 pounds

    (ii) greater than 1136 pounds

(b) If the breaking strengths of 500 randomly picked cables are measured, find approximately how many will have breaking strength

    (i)   less than 1059 pounds

    (ii) greater than 1136 pounds.

11. Dresswell company is a large apparel manufacturing company. A survey conducted by the company has revealed that collar sizes of men's shirts in a certan region are normally distributed with a mean collar size of 17 and a standard deviation of 1.6. Suppose that an individual whose collar size is between 17.5 and 18.5 will purchase a shirt with collar size 18. If Dresswell manufactures shirts of all collar sizes to meet the needs of its customers, what proportion of shirts should be of size 18?

12. A company administers a battery of tests in its recruiting procedure. The first test is a verbal test. Suppose it is known from past experience that scores on the verbal test have a normal distribution with a mean score μ of 150 points and standard deviation σ of 30 points. A candidate will be considered for further screening only if his or her score on this test is 160 points or more. What fraction of the candidates will be eliminated at this stage?

6

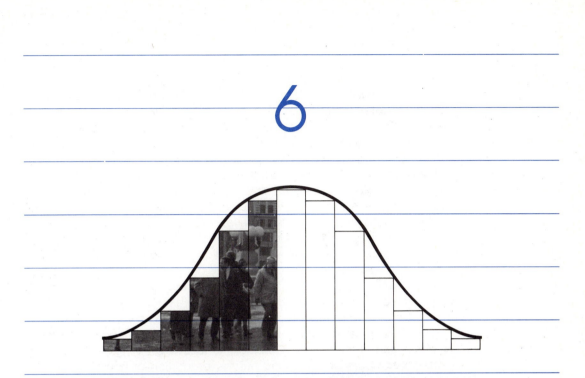

SAMPLING DISTRIBUTIONS

6-1 Sampling Distribution of the Mean

6-2 Random Samples and Random Number Tables

INTRODUCTION

This is the last chapter in our step-by-step development of probability theory, and in the chapters that follow, we will study the inferential, or inductive, aspects of statistics. The field of inductive statistics is concerned with studying facts about populations. Specifically, our interest is in learning about population parameters. We accomplish this by picking a sample and computing the values of appropriate statistics. Now, the value of a statistic will depend on the observed sample values, which, more than likely, will differ from sample to sample. Thus a statistic is a random variable. In keeping with the convention adopted in Chapter 4, we shall denote a statistic by an uppercase letter. For example, $\bar{X}$ will stand for the statistic *sample mean*. Its value for a given set of sample values will be denoted by $\bar{x}$, the lowercase letter x with a bar. Similarly, the sample variance S^2 is a statistic whose value for a given sample will be denoted as s^2. Although we shall try not to overemphasize this notational encumbrance, it is important to be aware of the distinction between a statistic, which is a random variable, and its particular value, which depends on a given set of sample values.

By the **sampling distribution** of a statistic we mean the theoretical probability distribution of the statistic. Consider the following example. Suppose an advertising company wants to find out the proportion of households in the United States that use Toothpaste A. The exact value of this proportion will not be known unless every household is interviewed. Suppose the company decides to question 100 households and use the proportion in this sample as an indicator of the proportion in the population. It is possible that when 100 households are actually interviewed, 62 will say they use Toothpaste A, thus giving a value of the sample proportion as 0.62. If another sample of 100 is taken, we might get a response that 55 use Toothpaste A, that is, a sample proportion of 0.55. Every time a new group of 100 is interviewed, a new value for the sample proportion is obtained. At this point we might envision the totality of all the values of the sample proportion when groups of 100 households are interviewed. This totality of values will describe the distribution of the sample proportion.

We are interested in sampling distributions for a reason that is easy to understand. From the first sample that was drawn, the sample proportion obtained was 0.62. Is 0.62 the true proportion in the population? Maybe not. But before drawing the sample, we intuitively feel that the chance is high that we will find a value near the true population value. Exactly how great is this chance? To be more specific, what is the probability that the sample proportion will differ from the true proportion by, perhaps, more than 0.05? We can answer this question and questions of this type only if we know the probability distribution of the sample proportions; for this reason we must learn about sampling distribution of sample proportion. In the same way, a need arises to discuss sampling distributions of other statistics, such as $\bar{X}$, S^2, and so on.

The problem of obtaining the sampling distribution of a statistic is essentially a mathematical problem. Informally the following concept is involved: We conceive of *all the possible samples* of a given size n that can be drawn from the population under study. For each of these samples we compute the value of the statistic of interest. The sampling distribution of the statistic is then the distribution of the totality of these values. Consequently a nonmathematical understanding of the theoretical distribution can be acquired by considering an experiment where a *large* number of samples of size *n* are drawn, every time computing the value of the statistic. From these values we might obtain the relative frequencies and the histogram of the empirical distribution as discussed in Chapter 1. If a fairly large number of samples are drawn, the histogram so obtained will provide a fairly close approximation to the theoretical sampling distribution of the statistic. ▬▬

6-1 SAMPLING DISTRIBUTION OF THE MEAN

In this section we shall concentrate entirely on the **sampling distribution of the mean,** or, to put it more descriptively, the distribution of the sample mean $\overline{X}$. For the purpose of illustration, let us consider the following rather artificial example where a sample of two (sample of size 2) is picked from a population consisting of four members whose weights (in pounds) are 148, 156, 176, and 184. The probability distribution of the weights in this population can be described as follows:

Weight x	148	156	176	184
Probability	1/4	1/4	1/4	1/4

Using the formulas in Chapter 4, we obtain the mean μ and the variance σ^2 of this population as follows:

$$\mu = (148)\frac{1}{4} + (156)\frac{1}{4} + (176)\frac{1}{4} + (184)\frac{1}{4}$$

$$= 166$$

$$\sigma^2 = (148 - 166)^2\,\frac{1}{4} + (156 - 166)^2\,\frac{1}{4} + (176 - 166)^2\,\frac{1}{4}$$

$$+ (184 - 166)^2\,\frac{1}{4}$$

$$= 212$$

The population described above is the parent population with which we start. Now suppose we pick a random sample of two. Sampling carried out in such a way that once an item is picked, it is not returned to the lot, is called **sampling without replacement.** On the other hand, if an item is returned to the lot before picking the next one, the sampling is called **sampling with replacement.** In an actual experiment the procedure will usually consist of picking items without replacement. However, to inject some important ideas into our discussion, we shall not only consider sampling without replacement but also sampling with replacement. We shall begin with the latter case.

SAMPLING WITH REPLACEMENT

In sampling with replacement from a population of four items, we get 4^2, or 16, possible samples. The 16 samples are listed below in Table 6-1 along with the corresponding sample means and probabilities. When we, as experimenters, actually pick our sample, we will end up with *one* of these. By a *random sample* we mean that the probability of picking any one of these samples is the same and equal to 1/16. (We could have written each sample on a slip of paper, mixed the 16 slips of paper thoroughly, and then picked one blindfolded. At the end of this chapter we will explain how random number tables can be used for this purpose.)

TABLE 6-1			
Samples of size 2 drawn with replacement, their means and probabilities.			
Sample number	Sample	Mean $\bar{x}$	Probability
1	148, 148	148	1/16
2	148, 156	152	1/16
3	148, 176	162	1/16
4	148, 184	166	1/16
5	156, 148	152	1/16
6	156, 156	156	1/16
7	156, 176	166	1/16
8	156, 184	170	1/16
9	176, 148	162	1/16
10	176, 156	166	1/16
11	176, 176	176	1/16
12	176, 184	180	1/16
13	184, 148	166	1/16
14	184, 156	170	1/16
15	184, 176	180	1/16
16	184, 184	184	1/16

TABLE 6-2	
Sampling distribution of $\bar{X}$ (sampling with replacement).	
Sample mean $\bar{x}$	Probability
148	1/16
152	2/16
156	1/16
162	2/16
166	4/16
170	2/16
176	1/16
180	2/16
184	1/16

Listing the distinct values of $\bar{x}$ and combining the corresponding probabilities, we get the probability distribution given in Table 6-2. It is the so-called sampling distribution of $\bar{X}$.

Now that we have obtained the sampling distribution of $\bar{X}$, let us find the mean and the variance of this distribution. We shall denote these quantities by using the subscript $\bar{x}$ as in $\mu_{\bar{x}}$ and $\sigma^2_{\bar{x}}$ respectively. This will enable us to distinguish them from the mean μ and variance σ^2 of the parent population, which are written without subscripts.

Applying the rule for finding the mean of a discrete random variable given in Chapter 4, we get

$$\mu_{\bar{x}} = (148)\frac{1}{16} + (152)\frac{2}{16} + (156)\frac{1}{16} + (162)\frac{2}{16} + (166)\frac{4}{16}$$

$$+ (170)\frac{2}{16} + (176)\frac{1}{16} + (180)\frac{2}{16} + (184)\frac{1}{16}$$

$$= 166.$$

Similarly, using the formula for the variance appropriately, we get

$$\sigma_{\bar{x}}^2 = (148 - 166)^2\frac{1}{16} + (152 - 166)^2\frac{2}{16} + (156 - 166)^2\frac{1}{16}$$

$$+ (162 - 166)^2\frac{2}{16} + (166 - 166)^2\frac{4}{16} + (170 - 166)^2\frac{2}{16}$$

$$+ (176 - 166)^2\frac{1}{16} + (180 - 166)^2\frac{2}{16} + (184 - 166)^2\frac{1}{16}$$

$$= 106.$$

Notice from the above computations that $\mu_{\bar{x}}$ and the mean μ of the original population are equal to each other, that is, both are 166. However, since $\sigma_{\bar{x}}^2 = 106$ and $\sigma^2 = 212$, we get

$$\sigma_{\bar{x}}^2 = \frac{\sigma^2}{2} = \frac{\text{variance of the parent population}}{\text{sample size}}.$$

The variance $\sigma_{\bar{x}}^2$ is smaller than the variance of the parent population. This is not difficult to understand. For instance, in the original population the value 184 at the upper extreme had a probability of 1/4 and the value 148 at the lower extreme had a probability of 1/4. But as can be seen from Table 6-2 of the distribution of $\bar{X}$, the values at the extremes of 184 and 148 carry a probability of only 1/16. This is reflected in the smaller value for the variance $\sigma_{\bar{x}}^2$.

The connection between μ and $\mu_{\bar{x}}$ and between σ^2 and $\sigma_{\bar{x}}^2$ that we have shown is always true irrespective of the size of the population. We can make the following statement in general:

MEAN AND VARIANCE—SAMPLING WITH REPLACEMENT

If samples of size n are drawn *with replacement* from a population with mean μ and variance σ^2, then

$$\mu_{\bar{x}} = \mu, \text{ that is, the population mean}$$

and

$$\sigma_{\bar{x}}^2 = \frac{\sigma^2}{n}, \text{ that is, } \frac{\text{the population variance}}{\text{the sample size}}.$$

SAMPLING WITHOUT REPLACEMENT

When sampling without replacement, we have 4×3, or 12, possible samples given in Table 6-3. By a *random sample,* we now mean that any sample has a probability of 1/12 of being drawn.

TABLE 6-3

Samples of size 2 drawn without replacement, their means and probabilities.

Sample number	Sample	Mean $\bar{x}$	Probability
1	148, 156	152	1/12
2	148, 176	162	1/12
3	148, 184	166	1/12
4	156, 148	152	1/12
5	156, 176	166	1/12
6	156, 184	170	1/12
7	176, 148	162	1/12
8	176, 156	166	1/12
9	176, 184	180	1/12
10	184, 148	166	1/12
11	184, 156	170	1/12
12	184, 176	180	1/12

Combining the values of $\bar{x}$, we obtain the sampling distribution of $\overline{X}$ given in Table 6-4.

TABLE 6-4

Sampling distribution of $\overline{X}$ (sampling without replacement).

Sample mean $\bar{x}$	Probability
152	1/6
162	1/6
166	1/3
170	1/6
180	1/6

The mean of the sampling distribution of $\overline{X}$ in this case is given as

$$\mu_{\bar{x}} = (152)\frac{1}{6} + (162)\frac{1}{6} + (166)\frac{1}{3} + (170)\frac{1}{6} + (180)\frac{1}{6}$$

$$= 166.$$

Thus we see that $\mu_{\bar{X}}$ is equal to the mean of the parent population even when the sampling is carried out without replacement. This is always the case. *Irrespective of whether the sampling is with or without replacement, the mean of the sampling distribution of $\overline{X}$ is always equal to the mean μ of the parent population.* Turning now to the variance $\sigma_{\bar{X}}^2$, since $\mu_{\bar{X}} = 166$, we have $\sigma_{\bar{X}}^2$ given as

$$\sigma_{\bar{X}}^2 = (152 - 166)^2 \frac{1}{6} + (162 - 166)^2 \frac{1}{6} + (166 - 166)^2 \frac{1}{3}$$

$$+ (170 - 166)^2 \frac{1}{6} + (180 - 166)^2 \frac{1}{6}$$

$$= \frac{424}{6}.$$

This value of $\sigma_{\bar{X}}^2$ can be rewritten as

$$\sigma_{\bar{X}}^2 = \frac{212}{2} \cdot \frac{4 - 2}{4 - 1}$$

which contains a clue as to what we might anticipate for the formula in the general case where we have a population of size N from which a sample of size n is drawn without replacement. It turns out that, in sampling without replacement,

$$\sigma_{\bar{X}}^2 = \frac{\sigma^2}{n} \cdot \frac{N - n}{N - 1}.$$

Notice that on the right-hand side of the equation there is not only the term σ^2/n, which we had in the case of sampling with replacement, but also a factor $(N - n)/(N - 1)$. This factor is the **finite population correction factor.** If the population is very large compared to the sample size, then this factor will be very close to 1. For example, if $N = 10,000$ and $n = 40$, then

$$\frac{N - n}{N - 1} = \frac{10,000 - 40}{10,000 - 1} = 0.9961$$

which for all practical purposes is equal to 1. Hence if the population size is large relative to the sample size, then $\sigma_{\bar{X}}^2 \approx \sigma^2/n$, and this is the same as the variance of $\overline{X}$ when sampling with replacement. Intuitively, this is just what one might expect. After all, if the population is extremely large, its composition is not going to change appreciably if we do not return the objects to the lot as they are picked. Even if the sampling is without replacement, for all practical purposes we can regard it as a case of sampling with replacement.

In summary, we have the following results:

MEAN AND VARIANCE—SAMPLING WITHOUT REPLACEMENT

When random samples of size n are drawn *without replacement* from a finite population of size N that has a mean μ and a variance σ^2, the mean and the variance of the sampling distribution of $\overline{X}$ are given by

$$\mu_{\overline{X}} = \mu$$

and

$$\sigma_{\overline{X}}^2 = \frac{\sigma^2}{n} \cdot \frac{N - n}{N - 1}.$$

If the population size is large compared to the sample size,

$$\sigma_{\overline{X}}^2 \approx \frac{\sigma^2}{n}.$$

The standard deviation of the sampling distribution of $\overline{X}$ is commonly known as the **standard error of the mean.** It is $\dfrac{\sigma}{\sqrt{n}}$ when sampling with replacement. For a sample drawn without replacement from a finite population of size N, the standard error of the mean is $\dfrac{\sigma}{\sqrt{n}} \sqrt{\dfrac{N - n}{N - 1}}$. In the latter case it is approximately $\dfrac{\sigma}{\sqrt{n}}$ if the population is very large compared to the sample size.

In our discussion, for the most part, we shall assume that the population is large enough that σ^2/n can be taken as the value of $\sigma_{\overline{X}}^2$ even when sampling without replacement. The standard error of the mean then depends on two quantities, σ^2 and n. It will be large if σ^2 is large, that is, if the scatter in the parent population is large, just as one would expect. On the other hand, since n appears in the denominator, the standard error will be small if the sample size is large. This would seem reasonable since with a larger sample we stand to acquire more information about the population mean μ and consequently less scatter of the sample mean about μ. The variance of the parent population is usually not under the experimenter's control. Therefore, one sure way of reducing the standard error of the mean is by picking a large sample—the larger the better.

EXAMPLE 1   A sample of 36 is picked at random from a population of adult males. If the standard deviation of the distribution of their heights is known to be 3 inches, find the standard error of the mean if

(a) the population consists of 1000 males
(b) the population is extremely large.

SOLUTION   (a) We are given that $n = 36$, $N = 1000$, and $\sigma = 3$. Therefore the standard error of the mean is

$$\sqrt{\frac{\sigma^2}{n} \cdot \frac{N-n}{N-1}} = \sqrt{\frac{9}{36} \cdot \frac{1000-36}{1000-1}}$$

$$= \sqrt{0.2412}$$

$$= 0.491.$$

(b) If the population size is extremely large (practically infinite), the standard error is given as $\sigma/\sqrt{n} = 3/\sqrt{36} = 0.5$. ■

So far we have concerned ourselves with the two parameters of the sampling distribution of $\overline{X}$, namely, its mean and variance. We now turn our attention to the distribution itself. The probability distribution of $\overline{X}$ will depend greatly on the distribution of the sampled population. In the particular example where we picked a sample of two from a population of four, it was fairly straightforward to obtain the distribution of $\overline{X}$. In general, the problem is not so simple as that. It is remarkable and most fortunate that if n, the sample size, is large, the distribution of $\overline{X}$ is close to a normal distribution, of course, with mean μ and variance σ^2/n. The statement of this result is contained in one of the most celebrated theorems of probability theory, called the **central limit theorem:**

> ### CENTRAL LIMIT THEOREM
>
> The distribution of the sample mean $\overline{X}$ of a random sample drawn from practically any population with mean μ and variance σ^2 can be approximated by means of a normal distribution with mean μ and variance σ^2/n, provided the sample size is large.

To realize the scope of the central limit theorem consider the parent population whose distribution is described in Figure 6-1 on page 218, together with the distributions of $\overline{X}$ for different values of n. (These distributions of $\overline{X}$ can be derived from mathematical theory.) The parent population in Figure 6-1(a) has the so-called exponential distribution which, as you will agree, can hardly be regarded as normal. The distribution is not even symmetric! But notice how strongly the tendency towards normality is pronounced as n increases. Even for n as low as 15, the approximation by a normal distribution would seem satisfactory.

The central limit theorem tells us that the shape of the distribution of the sample mean is approximately the shape of a normal distribution. From our previous discussion, we already know that if the population has mean μ and variance σ^2, then $\mu_{\overline{X}} = \mu$ and $\sigma_{\overline{X}}^2 = \sigma^2/n$. Converting to the z-scale, we can give an alternate version of the central limit theorem:

FIGURE 6-1

Tendency of the distribution of $\overline{X}$ towards normality as n increases when the parent population has the exponential distribution.

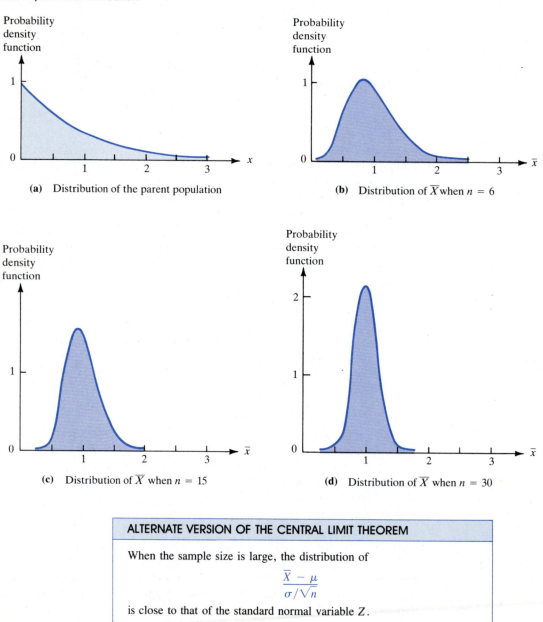

(a) Distribution of the parent population

(b) Distribution of $\overline{X}$ when $n = 6$

(c) Distribution of $\overline{X}$ when $n = 15$

(d) Distribution of $\overline{X}$ when $n = 30$

ALTERNATE VERSION OF THE CENTRAL LIMIT THEOREM

When the sample size is large, the distribution of

$$\frac{\overline{X} - \mu}{\sigma/\sqrt{n}}$$

is close to that of the standard normal variable Z.

(Recall that to convert to the z-scale the rule is: subtract the mean and divide by the standard deviation of the random variable in question.) Thus, for example,

$$P(a < \overline{X} < b) \approx P\left(\frac{a - \mu}{\sigma/\sqrt{n}} < Z < \frac{b - \mu}{\sigma/\sqrt{n}}\right)$$

if n is large.

Since the central limit theorem applies if the sample size is large, a natural question is, How large is large enough? This will depend on the nature of the sampled population. First of all, there is the following result of statistical theory:

> If the parent population is normally distributed, then the distribution of $\overline{X}$ is exactly normal for any sample size.

If the parent population has a symmetric distribution, the approximation to the normal distribution will be reached for a moderately small sample size, in many cases as low as 10. In most instances the tendency towards normality is so strong that the approximation is fairly satisfactory with a sample of about 30. With larger samples, of course, the approximation is even more satisfactory.

EXAMPLE 2 The records of an agency show that the mean expenditure incurred by a student during 1985 was $8000 and the standard deviation of the expenditure was $800. Find the approximate probability that the mean expenditure of 64 students picked at random was

(a) more than $7820
(b) between $7800 and $8120.

SOLUTION Here $\mu = 8000$, $\sigma = 800$, and $n = 64$. Therefore,

$$\mu_{\overline{X}} = 8000$$

and

$$\sigma_{\overline{X}} = \frac{800}{\sqrt{64}} = 100.$$

(a) We want to find $P(\overline{X} > 7820)$. Converting to the z-scale, we get

$$P(\overline{X} > 7820) \approx P\left(Z > \frac{7820 - 8000}{100}\right)$$

$$= P(Z > -1.8)$$

$$= 0.9641.$$

(b) We want $P(7800 < \overline{X} < 8120)$, which, after we have converted to the z-scale, is given by

$$P(7800 < \overline{X} < 8120) \approx P\left(\frac{7800 - 8000}{100} < Z < \frac{8120 - 8000}{100}\right)$$

$$= P(-2 < Z < 1.2)$$

$$= 0.3849 + 0.4772$$

$$= 0.8621.$$

Interpretation: If we considered a large number of samples, each having 64 students, in approximately 86.2 percent of the samples the mean expenditure would be between $7800 and $8120. ▬

EXAMPLE 3   The length of life (in hours) of a certain type of electric bulb is a random variable with a mean life of 500 hours and a standard deviation of 35 hours. What is the approximate probability that a random sample of 49 bulbs will have a mean life between 488 and 505 hours?

SOLUTION   Here $\mu = 500$, $\sigma = 35$, and $n = 49$. Hence

$$\mu_{\bar{X}} = 500$$

and

$$\sigma_{\bar{X}} = 35/\sqrt{49} = 5.$$

We are interested in finding $P(488 < \bar{X} < 505)$, which, after we have converted to the z-scale, is given by

$$P(488 < \bar{X} < 505) \approx P\left(\frac{488 - 500}{5} < Z < \frac{505 - 500}{5}\right)$$

$$= P(-2.4 < Z < 1)$$

$$= 0.3413 + 0.4918$$

$$= 0.8331. \quad ▬$$

EXAMPLE 4   The weight of food packed in certain containers is a random variable with a mean weight of 16 ounces and a standard deviation of 0.6 ounces. If the containers are shipped in boxes of 36, find, approximately, the probability that a randomly picked box will weigh over 585 ounces.

SOLUTION   Here $\mu = 16$, $\sigma = 0.6$, and $n = 36$. We want to find the probability that the total weight of 36 containers exceeds 585 ounces, that is, the mean weight $\bar{X}$ exceeds 585/36, or 16.25 ounces. Now

$$\mu_{\bar{X}} = 16$$

and

$$\sigma_{\bar{X}} = 0.6/\sqrt{36} = 0.1.$$

Conversion to the z-scale gives

$$P(\bar{X} > 16.25) \approx P\left(Z > \frac{16.25 - 16}{0.1}\right)$$

$$= P(Z > 2.5)$$

$$= 0.0062. \quad ▬$$

EXAMPLE 5   The amount of time (in minutes) devoted to commercials on a TV channel during any half-hour program is a random variable whose mean is 6.3 minutes and whose standard deviation is 0.866 minutes. What is the approximate probability that a person who watches 36 half-hour programs will be exposed to over 220 minutes of commercials?

SOLUTION   Here $\mu = 6.3$, $\sigma = 0.866$, and n represents the number of half-hour programs and is equal to 36. We shall assume that $n = 36$ is large enough that we can use the normal approximation for the distribution of $\overline{X}$, the mean amount of time devoted to commercials in 36 half-hour programs.

Now "the total exposure in 36 programs is over 220 minutes" implies that the mean exposure per program is over 220/36, or 6.11 minutes. Thus we want to find $P(\overline{X} > 6.11)$ where $\mu_{\overline{x}} = 6.3$ and $\sigma_{\overline{x}} = 0.866/\sqrt{36} = 0.1443$. We get

$$P(\overline{X} > 6.11) \approx P\left(Z > \frac{6.11 - 6.3}{0.1443}\right)$$

$$= P(Z > -1.32)$$

$$= 0.9066. \quad \blacksquare$$

SAMPLING DISTRIBUTION OF THE PROPORTION

A need to know the *sampling distribution of the proportion* might arise if, for instance, we were conducting a survey and wanted to find the probability that in a sample of 40 randomly picked people at least 70 percent will be in favor of a certain presidential candidate. Stated differently, we are interested in the probability that in a sample of 40 *the proportion of voters in favor of the candidate* will be greater than 0.7. To be able to answer this, we would need to know the distribution of the proportion of voters in favor of the candidate when samples of 40 are taken.

If p denotes the probability that a randomly picked person is in favor of the candidate, we already know that X, the number of voters among 40 who favor the candidate, has a binomial distribution. The proportion of voters in favor is simply the ratio of the number of voters in favor to the number of voters interviewed, and hence it is simply the random variable $X/40$. In general, when we pick a sample,

$$\binom{\text{sample proportion}}{\text{of an attribute}} = \frac{\binom{\text{number of items in the sample}}{\text{having the attribute}}}{\text{sample size}}.$$

So, if X is the number of items having a certain attribute in a sample of size n, then "the sample proportion having the attribute" is the random variable X/n. The probability distribution of this statistic is called the **sampling distribution of the proportion.** We can discuss the distribution of X/n, its mean, and its variance by using the known facts about the distribution of X that we learned in Chapter 5. However, we prefer to look at it here in the framework of the central limit theorem.

The method of counting votes essentially consists of recording a 1 if the voter is in favor of the candidate and a 0 if he or she is not in favor. In a typical survey, the sample proportion might be obtained as a value $(1 + 0 + 0 + 1 + \cdots + 0 + 1)/40$, or $29/40$, if there were twenty-nine 1's, that is, twenty-nine in favor. This shows that we can regard the sample proportion X/n as a mean of a sample of size n drawn from an infinite population consisting of 1's and 0's where the fraction of 1's is p and the fraction of 0's is $1 - p$. Thus the sampled population can be described by the following probability distribution:

x	Probability
0	$1 - p$
1	p

The mean of this population is

$$\mu = 0 \cdot (1 - p) + 1 \cdot p = p$$

and, substituting p for μ in the formula for the variance, we have

$$
\begin{aligned}
\sigma^2 &= (0 - p)^2 \cdot (1 - p) + (1 - p)^2 \cdot p \\
&= p^2(1 - p) + (1 - p)^2 p \\
&= p(1 - p)[p + (1 - p)] \\
&= p(1 - p).
\end{aligned}
$$

Now recall that if the population has mean μ and variance σ^2, then the distribution of the sample mean (based on a sample of size n) has mean μ and variance σ^2/n. In our present case X/n represents the mean of a sample of size n from a population with mean p and variance $p(1 - p)$. It follows, therefore, that the mean of the sampling distribution of X/n is p, the same as the parent population mean, and that the variance of its sampling distribution is $p(1 - p)/n$, the parent population variance divided by the sample size. Moreover, using the central limit theorem, we can assert that if the sample size is large, the distribution of X/n will be very near the normal distribution (provided p is not very near 0 or 1). We now summarize our findings:

DISTRIBUTION OF THE SAMPLE PROPORTION

If n items are picked independently from a population where the probability of success is p (not very near 0 or 1), and if n is large, then the distribution of the sample proportion X/n is approximately normal with mean p and variance $p(1 - p)/n$.

Converting to the z-scale, we find that

$$\frac{\dfrac{X}{n} - p}{\sqrt{\dfrac{p(1 - p)}{n}}}$$

has a distribution that is very near the standard normal distribution provided n is large.

EXAMPLE 6   According to the Mendelian law of segregation in genetics, when certain types of peas are crossed, the probability that the plant yields a yellow pea is 3/4 and that it yields a green pea is 1/4. For a plant yielding 400 peas, find

(a) the standard error of the proportion of yellow peas
(b) the approximate probability that the proportion of yellow peas will be between 0.71 and 0.78.

SOLUTION   Here $p = 3/4$, $n = 400$ and X = number of yellow peas.

(a) The standard error of the proportion of yellow peas is

$$\sqrt{\frac{p(1-p)}{n}} = \sqrt{\left(\frac{3}{4} \cdot \frac{1}{4}\right) \Big/ 400}$$

$$= 0.0217$$

(b) We want to find $P(0.71 < X/n < 0.78)$. Converting to the z-scale, we get

$$P(0.71 < X/n < 0.78) \approx P\left(\frac{0.71 - 0.75}{0.0217} < Z < \frac{0.78 - 0.75}{0.0217}\right)$$

$$= P(-1.85 < Z < 1.38)$$

$$= 0.4678 + 0.4162$$

$$= 0.8840. \quad \blacksquare$$

SECTION 6-1   EXERCISES

1. A sample of 8 items is picked at random from a population consisting of 100 items and a certain quantitative variable X is measured. If the population mean is 112.8, find the mean of the distribtuion of $\overline{X}$ if the sampling is carried out

   (a) with replacement   (b) without replacement.

2. Suppose 12 items are picked at random from a population having 80 items. If the population variance of a variable X is 67.24, find the variance of the distribution of $\overline{X}$ if the sampling is

   (a) with replacement   (b) without replacement.

3. A population consisting of 100 items has a mean of 60 and a standard deviation of 5. If a sample of 9 items is drawn without replacement from this population, what is the mean and the standard deviation of the distribution of the sample mean?

4. Answer the question in Exercise 3 if the sample is drawn from the population with replacement.

5. Suppose it is known that the annual income of a family has a distribution with a mean of $18,200 and a standard deviation of $3,500. If a sample of 64 families is

picked at random, find the standard error of the mean family income if the population

(a) consists of 400 families

(b) can be considered extremely large.

6. Assuming that the population is infinite, discuss the effect on the standard error of the mean if the sample size is changed

   (a) from 25 to 225   (b) from 320 to 20.

7. Suppose a population consists of six values (all equally likely): 10, 8, 5, 6, 9, 13. Find

   (a) the population mean   (b) the population variance.

8. A sample of 4 items is picked at random from the population described in Exercise 7. Use appropriate formulas to find the mean and the variance of $\overline{X}$ if the four items are picked

   (a) with replacement   (b) without replacement.

9. Suppose a population has the distribution given by the following probability function:

x	-3	0	5	12
$p(x)$	0.3	0.2	0.4	0.1

   (a) Find the mean μ and the variance σ^2 of the population.

   (b) Find the mean and the variance of $\overline{X}$, the sample mean based on a sample of 100 independent observations from the above population.

   (c) Approximately what is the distribution of $\overline{X}$?

   (d) Find the approximate probability that

      (i) $\overline{X}$ will lie between 1.366 and 2.767

      (ii) $\overline{X}$ will exceed 3.234

      (iii) $\overline{X}$ will be less than 1.366.

10. The annual rainfall in a region has a distribution with a mean rainfall of 100 inches and a standard deviation of 12 inches. What is the approximate probability that the mean rainfall during 36 randomly picked years will exceed 103.8 inches?

11. A cigarette manufacturing company claims that the mean nicotine content in their king-size cigarettes is 2 mg with the standard deviation of the nicotine content equal to 0.3 mg. If this claim is valid, what is the approximate probability that a sample of 900 cigarettes will yield a mean nicotine content exceeding 2.02 mg?

12. **Quality control inspection.** A hardware store receives a consignment of bolts whose diameters have a normal distribution with a mean diameter of 12 inches and a standard deviation of 0.2 inch. The consignment will be considered substandard and returned if the mean diameter of 100 bolts is less than 11.97 inches or greater than 12.04 inches. Find the probability that the consignment will not be returned.

13. A farmer has divided his entire farm into a large number of small plots of the same size. The amount of yield on a plot has a mean of 100 bushels with a standard deviation of 8 bushels. If 64 of the plots are picked at random, find the approximate probability that the mean yield will

   (a) exceed 102.5 bushels

(b) be less than 98.5 bushels

(c) be between 98 and 102 bushels.

14. The maximum load that a ferry can carry is 16,375 pounds. Suppose on any trip the ferry carries 100 passengers and the distribution of the weights of the passengers has a mean of 160 pounds and a standard deviation of 15 pounds. Find the approximate probability that the ferry will exceed the limit on a trip.

15. The distance that Jane, a substitute teacher, travels to teach on a schoolday has the following probability function:

Distance traveled	0	6	10	15
Probability	0.1	0.2	0.4	0.3

During an academic year consisting of 225 days, find the approximate probability that Jane will travel between 2082 and 2258 miles.

16. A fair die is rolled 100 times. On any roll, if an even number shows up, Tom will win as many dollars as the number that shows up on the die, and if an odd number shows up, he will lose $3.60. Find the approximate probability that Tom's total winnings will be

    (a) more than $21   (b) between $19.50 and $21.

17. A truck is loaded with 900 boxes containing books. If the total weight of the boxes exceeds 36,450 pounds, the driver will be fined at the weighing station. If the distribution of the weight of a box has a mean of 40 pounds and a standard deviation of 6 pounds, find the approximate probability that the driver will be fined.

18. The amount of ice cream in an ice-cream cone has a distribution with a mean amount of 3.2 ounces per cone and a standard deviation of 0.4 ounce. If there are 100 children at a birthday party, what is the approximate probability that more than 330 ounces of ice cream will be served?

19. A computer is programmed in such a way that the error involved in carrying out one computation has a distribution with mean 10^{-7} and standard deviation 10^{-8}. If 10,000 independent computations are made, find the approximate probability that the total error will be less than 0.001001.

20. The distribution of the test scores of students has an unknown mean μ and a standard deviation equal to 15 points. Suppose 100 students are picked at random. What is the approximate probability that the mean score of these students will differ from their true mean score μ by at most 3 points?

21. Suppose 10 percent of the tubes produced by a machine are defective. If a sample of 100 tubes is inspected at random, find

    (a) the expected proportion of defectives in the sample

    (b) the variance of the proportion of defectives in the sample

    (c) the approximate distribution of the sample proportion

    (d) the probability that the proportion of defectives will exceed 0.16.

22. It is believed that 40 percent of the people favor capital punishment. If 400 persons are interviewed at random, find the approximate probability that the proportion of individuals in the sample who favor capital punishment

    (a) will exceed 0.43   (b) will be between 0.35 and 0.42.

23. If 60 percent of the population feels that the President is doing a satisfactory job, find the approximate probability that in a sample of 900 people interviewed at random, the proportion who share this view will

   (a) exceed 0.65

   (b) be less than 0.56.

24. It is hypothesized that the proportion of individuals in the population with blood type O is 0.3. If this hypothesis is correct, what is the approximate probability that in a random sample of 400, the proportion of blood type O individuals will be less than 0.25 or greater than 0.35?

25. The probability that a signal that originates as a dash will be received as a dot is 0.20. If there are 400 dashes in a message, find the approximate probability that the proportion of dashes received as dots will exceed 0.21.

26. It is felt that 80 percent of the population in the United States favor some form of federal income-tax reform. What is the approximate probability that in a random sample of 64 individuals interviewed, less than 70 percent will favor some income-tax reform?

27. It is believed that 70 percent of the adult population feel that Congress should provide additional tax incentives for family-operated farms. Find the approximate probability that in a sample of 100 adults interviewed, the proportion of individuals who feel this way is

   (a) less than 0.6

   (b) greater than 0.65 and less than 0.73.

6-2 RANDOM SAMPLES AND RANDOM NUMBER TABLES

We conclude this chapter with a brief discussion of what is meant by a *random sample* and how *random number tables* can be used for this purpose. (In Chapter 13, Nonparametric Methods, we shall show how to test for randomness.)

When a sample is drawn from a population, it is our hope that it will be as representative of the population as possible. The concept of randomness implies that the investigation should be "honest" with no extraneous bias in the sampling procedure. Thus a **random sample** may be defined as a sample drawn in such a way that every member of the population has an equal chance of being included. This means that if the population is finite and if we list all the possible samples of a given size, the probability of picking any one sample is the same. The mechanical procedure might entail writing slips, one for each sample, mixing them well, and then drawing one slip. It is, of course, important to realize that the term *random* refers to the mode of drawing the sample and in no way guarantees that the sample will be representative. However, it does minimize any bias on the part of the experimenter.

As can be seen, the procedure above could become extremely tedious and time-consuming even for moderately small populations and samples. An alternative to this is provided through the use of **tables of random numbers.** Such tables consist of digits 0, 1, . . . , 9 generated by the computer using a random process and recorded randomly in rows and columns. Random number tables running literally into hundreds of pages are available. In Table A-8 in Appendix II, we reproduce a small portion of such a table.

How is the table used? Suppose we want to pick a sample of 4 from a population consisting of 150 households in a residential block. As a first step we write down the 150 households in a serial listing as 001, 002, . . . , 150. Then we pick any arbitrary page of the table of random numbers, for example, the third page, an arbitrary row, such as the fourth, and three contiguous columns, such as columns 18, 19, and 20. Starting with row four, we then read down the page in the three columns. We obtain successively the following 3-digit numbers:

691, **078,** 736, 272, 441, **093,** 227, 683, 317, **038,** 543, **002**

Discarding those that are over 150, we stop as soon as we have accomplished our objective of picking four numbers not exceeding 150. The first four such numbers that we obtain are 078, 093, 038, and 002. We would accordingly select the sample of households bearing these serial numbers.

If there were 5000 households in the population, the procedure would still be the same, except that we would make a serial listing as 0001, 0002, . . . , 5000, and we would, of course, use four columns reading down the page.

SECTION 6-2 EXERCISES

1. Suppose 250 students in a grade school are listed serially 1 through 250. Select a random sample of 8 students using columns 7, 8, and 9 of the table of random numbers, and starting in row 14 of the third page of the table.

2. Describe how, using the random number tables, you would pick a random sample of 6 members from a club consisting of 60 members.

3. There are 1200 members attending a convention and it is desired to form a committee by picking 10 members at random. How would you go about picking such a committee using a table of random numbers?

KEY TERMS AND EXPRESSIONS

sampling distribution

sampling distribution of the mean

sampling without replacement

sampling with replacement

finite population correction factor

standard error of the mean

central limit theorem

sampling distribution of the proportion

table of random numbers

random sample

KEY FORMULA

$$P(a < \overline{X} < b) \approx P\left(\frac{a - \mu}{\sigma/\sqrt{n}} < Z < \frac{b - \mu}{\sigma/\sqrt{n}}\right)$$

CHAPTER 6   TEST

1. The mean of the sampling distribution of $\overline{X}$ will depend on whether the sampling is with replacement or without replacement. Is this statement true or false?

2. It is said that the normal distribution forms the cornerstone of statistics. Explain this statement.

3. Suppose the population mean is μ. State whether the following statements are true or false:

    (a) The distribution of $\overline{X}$ is normal if the population is normally distributed.

    (b) The distribution of $\overline{X}$ is approximately normal if the population is large.

    (c) The distribution of $\overline{X}$ is approximately normal if the sample size n is large.

    (d) The mean of the sampling distribution of $\overline{X}$ is μ only if the sample size is large.

    (e) The mean of the sampling distribution of $\overline{X}$ is always μ.

    (f) The mean of the sampling distribution of $\overline{X}$ is approximately μ if the sample size n is large.

4. A sample of 10 is drawn from a population of 30. If the population variance is 16, find the variance of the distribution of $\overline{X}$ if the sampling is

    (a) with replacement   (b) without replacement

5. Which of the following statements is correct? The central limit theorem is important because it tells us that

    (a) the mean and the variance of the distribution of $\overline{X}$ are μ and σ^2/n, respectively

    (b) the distribution of $\overline{X}$ is close to a normal distribution for large n.

6. The amount of gasoline (in gallons) an automobile driver uses during a week has a distribution with a mean amount of 10 gallons per week and a standard deviation of 3.5 gallons. What is the approximate probability that during a 49-week period the driver will use more than 540 gallons?

7. A firm is manufacturing ball bearings with a diameter (in inches) that is normally distributed with a mean length of 0.5 inch and a standard deviation of 0.02 inch. If a sample of 25 is picked at random, what is the probability that the mean diameter will be less than 0.496 inch or greater than 0.512 inch?

8. According to available actuarial tables, 88 percent of the people of age 50 will reach age 60 and 12 percent will die before reaching that age. Suppose an insurance company has approved 80 applications for ten-year term insurance policies from people aged 50. Find the probability that the proportion of people reaching age 60 is

    (a) less than 0.78   (b) greater than 0.82.

9. Noskid Tire Company has introduced a new radial tire. The number of trouble-free miles it gives has a normal distribution with mean of 35,000 miles and standard deviation of 4,000 miles. If 16 tires are tested in a simulated experiment,

   (a) what is the distribution of the mean trouble-free miles $\overline{X}$ based on 16 tires? Is the distribution exact or approximate?

   (b) find the probability that the mean trouble-free mileage based on 16 tires will exceed 37,500 miles.

10. The time that Leonard takes to repair a machine has a certain probability distribution with mean time of 2.5 hours and standard deviation of 1.2 hours. If he has to repair 36 machines of this type, find the approximate probability that he will take

    (a) more than 100 hours

    (b) between 82 and 100 hours.

7

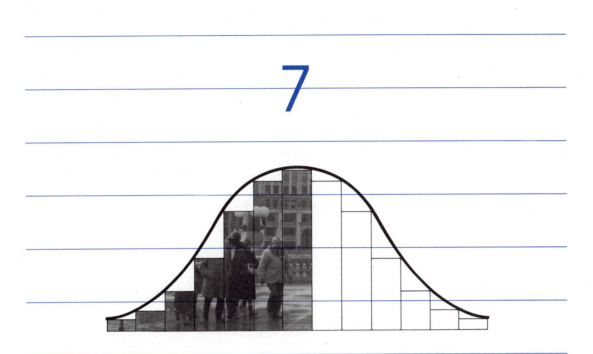

ESTIMATION—SINGLE POPULATION

INTRODUCTION

We are entering the most important phase of a statistical investigation, called *statistical inference,* which is a procedure whereby, on the basis of observed experimental data in a particular sample, we generalize and draw conclusions about the population from which the sample is drawn. Among other things, this procedure is concerned with estimation of unknown parameter values and tests of hypotheses regarding these values. We shall discuss the principles of estimation in this and the next chapter, deferring tests of hypotheses to Chapter 9.

The overall procedure associated with the problem of statistical inference is shown in the following diagram:

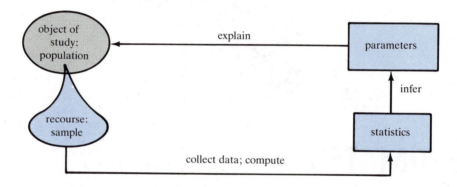

The object of the study of an investigator is the population. As we are aware, the investigation of the entire population of interest may not be feasible for several reasons. So recourse is taken to picking a sample of a certain size from the population. From the sample data that is collected, we compute the values of appropriate statistics. On the basis of the values of the statistics, we draw inferences about the corresponding population parameters which serve to explain the population.

The *Thorndike-Barnhart Advanced Dictionary* defines the word *estimate* as "judgment or opinion of the approximate size, amount, etc." The practice of estimating is not foreign to us. For instance, if we are taking a trip from San Diego to San Francisco, we might estimate the distance as 650 miles, the mileage per gallon as 22 miles, and the price per gallon as $1.20, and eventually we might put all the information together and estimate how much the entire trip might cost us. We might also estimate the distance as being between 600 and 700 miles, the mileage per gallon as being between 20 and 25 miles, and so on. In the first case we are estimating the distance, mileage, and price per gallon as specific values, or specific points. In the latter case we are estimating these quantities in terms of certain intervals. The first method of giving an estimate is referred to as *point estimation,* the second method as *interval estimation.*

The main objective of a statistical investigation is to acquire an understanding of the population, and this is often accomplished by studying the constants associated with the population. These constants are what we have called the population parameters. We have come across three parameters so far. One of these is p, the proportion of items of a certain type in the population, for example, the proportion of Democrats in a state, the proportion of defectives produced by a machine, the proportion of people of type A blood, and so on. While discussing quantitative data, we came across the population mean μ and another parameter, the population variance σ^2. For example, the mean IQ of students and variance of the IQ, the mean nicotine content in a certain brand of cigarettes and the variance of the nicotine content.

As in our example of the trip to San Francisco, there are two ways of giving an estimate of a parameter. One might give an estimate as a single value, in which case it is called a point estimate, or one might give an estimate stating that the parameter is enclosed in an interval, in which case the estimate is called an interval estimate. ▬▬

7-1 POINT ESTIMATION—DEFINITION AND CRITERIA

A *point estimate* provides a single numerical value as an assessed value of the parameter under investigation.

The procedure for obtaining a point estimate of a parameter involves the following steps:

1. Pick a random sample of a certain size.
2. Devise a random variable—a statistic.
3. Compute the value of the statistic from the observed sample data.

The statistic used for the purpose of estimation is called an **estimator.** It is a random variable which, in some sense, is relevant for estimating the parameter and is simply the form of a mathematical expression, that is, a formula. A numerical value of the estimator computed from a given set of sample values is called a **point estimate** of the parameter. Thus a point estimate is a single number.

For example, we might use the sample mean $\overline{X}$ based on four items as an estimator of the population mean μ. Its values are likely to differ from sample to sample. Suppose we pick a sample of size four yielding values 2, 3, 5, 8. Then the value of $\overline{X}$ computed from this sample is 4.5. We say that 4.5 is a point estimate of μ. In general, if a sample yields values $x_1, x_2, \ldots, x_n$ then the mean computed from this data, which we denote by $\overline{x}$, is an estimate of the true population mean, μ.

CRITERIA FOR "GOOD" ESTIMATORS

Many of the basic properties and criteria for characterizing point estimates were formulated and developed by Sir Ronald A. Fisher (1890–1962). They include (1) *unbiasedness*, (2) *efficiency*, (3) *consistency*, and (4) *sufficiency*.

FIGURE 7-1

Sampling distributions of estimators and the location of θ along the axis.

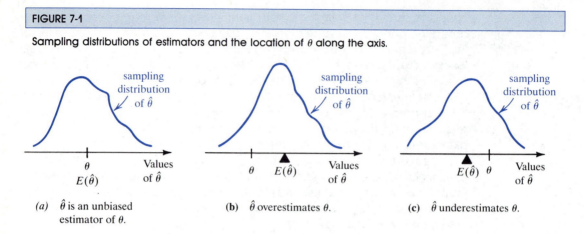

(a) $\hat{\theta}$ is an unbiased estimator of θ.

(b) $\hat{\theta}$ overestimates θ.

(c) $\hat{\theta}$ underestimates θ.

In the following discussion instead of referring to the specific parameters p, μ, σ^2 individually we shall sometimes use the Greek letter θ (theta) to refer to them collectively. The estimator of θ will be written as $\hat{\theta}$, where the hat "^" is read *estimator of*.

Unbiasedness

We are aware and have no misgivings that, when we take a specific sample and, from the observed data, find an estimate of the unknown parameter, it may well differ from the true value of the parameter. However, a practical and desirable estimator would be such that the mean value of several estimates of the parameter from a number of samples (of the same size) was near the true parameter value. Basically this is the notion of **unbiasedness.** Historically, this concept is attributed to Gauss and is defined as follows:

An estimator $\hat{\theta}$ is an *unbiased estimator* of a parameter θ if its sampling distribution has mean θ, the parameter being estimated. That is,

$$E(\hat{\theta}) = \theta.$$

Recall that $E(\hat{\theta})$ means the *expected value*, or *mean*, of $\hat{\theta}$.

An estimator that is not unbiased is said to be *biased*. A biased estimator will either underestimate or overestimate θ.

In Figure 7-1(a) $\hat{\theta}$ is an unbiased estimator of θ since $E(\hat{\theta})$ and θ coincide, whereas in Figures 7-1(b) and (c) $\hat{\theta}$ is a biased estimator.   .

We saw in Chapter 6 that if the population has mean μ, then $E(\overline{X}) = \mu$. Also, if p is the proportion of items having a certain attribute in a large population and X is the number of items with this attribute in a sample of n items, then $E\left(\dfrac{X}{n}\right) = p$.

In other words, the sample mean $\overline{X}$ is an unbiased estimator of μ and the sample proportion X/n is an unbiased estimator of p.

FIGURE 7-2

Sampling distri-
butions of two
estimators, $\hat{\theta}_1$
and $\hat{\theta}_2$, both unbiased estimators
of θ.

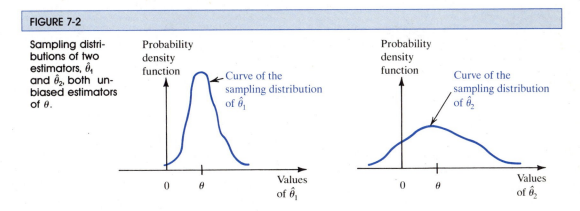

When it comes to estimating σ^2, the logical choice would be the statistic $\frac{1}{n} \sum_{i=1}^{n} (X_i - \bar{X})^2$. However, its sampling distribution has expected value $\sigma^2 - \frac{\sigma^2}{n}$ so that it underestimates the population variance by the amount $\frac{\sigma^2}{n}$. It turns out that $S^2 = \frac{1}{n-1} \sum_{i=1}^{n} (X_i - \bar{X})^2$ is an unbiased estimator of σ^2. It is for this reason, from the point of view of unbiasedness, that we prefer S^2 for the purpose of estimating σ^2.

Efficiency

When faced with the problem of estimating a certain parameter, it may happen that the criterion of unbiasedness is shared by many of its estimators. For example, suppose we have a normal population and we wish to estimate its unknown mean μ. It can be shown that the sample mean $\bar{X}$ and the sample median $\tilde{X}$ (read X *tilde*) are both unbiased estimators of μ. Which one should be preferred for estimating μ?

A reasonable basis for choosing one of several unbiased estimators that a parameter might have is provided through the concept of **efficiency.**

If $\hat{\theta}_1$ and $\hat{\theta}_2$ are two unbiased estimators of θ, then $\hat{\theta}_1$ is said to be *more efficient* than $\hat{\theta}_2$ if the variance of the sampling distribution of $\hat{\theta}_1$ is less than the variance of the sampling distribution of $\hat{\theta}_2$.

To get some idea about the concept involved, compare the sampling distributions of two unbiased estimators $\hat{\theta}_1$ and $\hat{\theta}_2$ of the parameter θ in Figure 7-2.

The estimator $\hat{\theta}_1$ has smaller variance than $\hat{\theta}_2$ and consequently has less spread around θ. We say that $\hat{\theta}_1$ is more efficient than $\hat{\theta}_2$ in that we have more faith that a value of $\hat{\theta}_1$ has a greater chance of being near θ, the parameter that is being estimated, than does a value of $\hat{\theta}_2$.

If at all possible, among all the unbiased estimators of a parameter, we should choose the one that has the smallest variance. Such an estimator, if it exists, is said to be the *most efficient estimator.*

We already know that the variance of the sample mean $\overline{X}$ is σ^2/n. Also, when sampling from a normal distribution, the variance of the sample median $\tilde{X}$ can be shown to be approximately $1.57\sigma^2/n$. Since the variance of $\overline{X}$ is less than that of $\tilde{X}$, $\overline{X}$ is more efficient than $\tilde{X}$ for estimating μ and is considered better.

Consistency

The third property that we would like an estimator to possess is **consistency.** As the sample size becomes extremely large, we would like the estimator to be close to the parameter with very high probability, almost to the point of being sure. Estimators with this property are said to be *consistent*.

The sample mean as an estimator of the population mean is consistent. The intuitive rationale is rather simple. We know that the sampling distribution of $\overline{X}$ has variance equal to σ^2/n, where σ^2 is the variance of the original parent population from which the sample is picked and n is the sample size. If the sample size is very large, then σ^2/n will be very small since n appears in the denominator. In other words the sampling distribution of $\overline{X}$ will have a negligible spread and will be narrower and more and more concentrated around μ. Hence if n is very large, the probability that $\overline{X}$ is close to μ will be high, near 1.

Sufficiency

Sufficiency is a rather involved concept mathematically. Essentially what it means is that an estimator should be such that it utilizes all the information contained in the sample for the purpose of estimating a given parameter. No other estimator should provide any more information.

Whether a particular estimator satisfies some or all of the above criteria may not always be an easy matter to establish. In selecting an estimator one additional point to consider is that whichever estimator we use in any given situation, in order to be able to make statements about its reliability, we should have some idea about its probability distribution.

As mentioned earlier, the sample mean $\overline{X}$ is an unbiased and a consistent estimator of μ. Also, by the central limit theorem we know that $\overline{X}$ is normally distributed for moderately large sample size. It is these features that make $\overline{X}$ an extremely appealing estimator of μ. Also, for our purpose, we will find it satisfactory to use the sample proportion to estimate p, the sample variance S^2 to estimate the population variance σ^2, and the sample standard deviation S to estimate σ. [*Note:* S is *not* an unbiased estimator of σ even though S^2 is an unbiased estimator of σ^2.]

EXAMPLE 1   The following readings give the weights of four bags of flour (in pounds) picked at random from a shipment: 102, 101, 97, and 96. Find estimates of

(a) the true (population) mean weight of all the bags

(b) the true variance of weights of all the bags

(c) the true standard deviation of weights.

SOLUTION

The sample mean is

$$\bar{x} = \frac{396}{4} = 99$$

and

$$\sum_{i=1}^{n} (x_i - \bar{x})^2 = 3^2 + 2^2 + (-2)^2 + (-3)^2 = 26.$$

(a) the estimate of the true mean is $\bar{x}$, that is, 99 pounds

(b) the estimate (in pounds^2) of the true variance is

$$\sum_{i=1}^{n} (x_i - \bar{x})^2/(4 - 1) = 26/3 = 8.67$$

(c) the estimate (in pounds) of the true standard deviation is

$$\sqrt{8.67} = 2.9. \quad \blacksquare$$

EXAMPLE 2

Antirrhinum majus (snapdragon) is a plant with saclike, two-lipped flowers. When a yellow-flowered plant which was known to be heterozygous was self-mated it was found that there were two types of flowers with 105 yellow flowers and 31 ivory flowers. Find a point estimate of the probability that a flower is yellow when the heterozygous variety is self-mated.

SOLUTION

We have qualitative information. Altogether there are $105 + 31$, or 136, flowers of which 105 are yellow. Hence $x = 105$ and $n = 136$. Consequently a point estimate of p is

$$\frac{x}{n} = \frac{105}{136} = 0.772. \quad \blacksquare$$

SECTION 7-1 EXERCISES

1. State which property of an estimator we are alluding to in each of the following:

    (a) If the sample size is large, we can be almost certain that the estimator will be very near the parameter it is supposed to estimate.

    (b) Estimator 1 and estimator 2 are two unbiased estimators of a parameter, and estimator 1 has a smaller variance. (Which estimator is to be preferred?)

    (c) Suppose $x_1, x_2, \ldots, x_n$ represent n values in a sample. Use $(x_1 + x_2 + \cdots + x_n)/n$ as an estimate of the population mean, rather than $(x_1 + x_2 + \cdots + x_{n-1})/(n - 1)$.

(d) The estimator is such that its theoretical distribution has a mean equal to the parameter it is supposed to estimate.

(e) The estimator has the following property: if we were to take repeated samples of a given size a large number of times, each time computing the value of the estimator and taking the arithmetic mean of these values, then the value so obtained would be approximately equal to the parameter.

(f) $\sum_{i=1}^{n} (x_i - \bar{x})^2/(n - 1)$, and not $\sum_{i=1}^{n} (x_i - \bar{x})^2/n$, is a better estimate of the population variance.

2. An aptitude test was administered to 100 of the 400 students in a freshman math class. The mean score for the 100 students was found to be 71.8 points. Which of the following statements are inferential?

   (a) The mean score of the 100 students is 71.8.

   (b) The mean score for the freshman math class is 71.8.

   (c) This year's freshmen are better than last year's.

3. The student records of a school extend from the day it opened to the present. State whether the mean height of all the students when they enrolled is a parameter or a statistic. (You are interested only in this school.)

4. In each of the following, identify the quantities as μ, p, $\bar{x}$, or x/n, if with reference to Exercise 3 it was found that

   (a) 42 percent of the students enrolled in the 1950s graduated

   (b) the mean weight of the students enrolled in 1960 was 115 pounds

   (c) 25 percent of all the students enrolled graduated as biology majors

   (d) 28 percent of all the students enrolled were blonde

   (e) the mean IQ of all the students was 110.

5. In each of the following, obtain point estimates of the population mean, variance, and standard deviation if

   (a) $n = 12$, $\sum_{i=1}^{n} x_i = 1080$, $\sum_{i=1}^{n} (x_i - \bar{x})^2 = 6875$

   (b) $n = 10$, $\sum_{i=1}^{n} x_i = 185$, $s^2 = 1.764$.

6. The following readings were obtained for the chlorophyll content (in mg/gm) of eight wheat leaves:

   2.3,   2.1,   2.6,   2.2,   1.9,   2.7,   2.4,   2.2.

   Obtain point estimates of the true mean, variance, and standard deviation of the chlorophyll content of wheat leaves.

7. Sylvia's grandmother deposits in a box whatever change she collects each day. In this way she has collected 50,000 coins of different denominations—pennies, nickels, dimes, and quarters. A random sample of 200 coins from this collection was found to contain 85 pennies, 52 nickels, 38 dimes, and 25 quarters. Estimate how much the collection of coins is worth.

8. A small company wants to know how many neckties it should make for the next year. When 100 people were interviewed, it was found that they had bought 165 ties produced by the company during the current year. If there are 100,000 people in the region, estimate how many ties the company should produce.

9. A manuscript has 5000 pages. You have only one hour in which to come up with an approximate figure for the number of words in it. Devise a suitable approach.

10. Describe a suitable scheme to estimate the number of grains in a container that holds 100 pounds of rice.

11. Estimate the proportion of people who wear glasses if a random sample of 370 yielded 76 individuals who wore them.

12. When 15 randomly picked people were asked about whether they smoked cigarettes, the following responses were obtained; yes, no, yes, no, no, yes, no, yes, no, yes, no, yes, no, no, no. Obtain a point estimate of the proportion of smokers in the population.

13. Of 100 fluorescent tubes tested in a laboratory, 68 lasted beyond 250 hours. Find a point estimate of the true proportion of fluorescent tubes that will last beyond 250 hours.

14. A fisherman cast his net five times. The figures below refer to the number of fish caught each time and the proportion of salmon in each catch:

Proportion of salmon	0.2	0.32	0.18	0.12	0.28
Number of fish caught	150	350	200	140	190

(a) How many salmon did he catch altogether?

(b) What was the proportion of salmon in the entire catch?

(c) Obtain a *pooled estimate* of the true proportion of salmon, that is, an estimate by pooling together the results of the five catches.

15. Suppose a box contains 100,000 beads, of which some are red and others are blue. If you have limited time at your disposal and are certainly short on patience, how would you go about estimating the number of red beads?

7-2 CONFIDENCE INTERVAL FOR μ WHEN σ^2 IS KNOWN

GENERAL APPROACH TO INTERVAL ESTIMATION

In the preceding section we devoted our attention to estimating a parameter with a single value. However, when we find a point estimate, we certainly do not expect that it will be *exactly* equal to the parameter value. Also, if we take two samples from the same population, we do not expect the two estimates computed from these samples to be exactly equal either. This is due to the sampling error involved. The method of point estimation has some serious drawbacks because, for one thing, unless some additional information is given, there is no way to decide how good the estimate is. Certainly we would like to have some idea about the chances of departure of the estimator from the parameter it is supposed to estimate.

For instance, suppose we want to estimate the mean weight of miniature poodles in the population. When a sample is picked, suppose the sample mean is found to be 20.5 pounds. We shall accept 20.5 as a point estimate of μ. How reliable is this estimate? Suppose we are told that the population standard deviation is 8 pounds. In that case we would not have very much confidence in our estimate because of the large spread indicated by the fact that σ is 8. Our confidence in the estimate would be improved considerably if, on the other hand, we knew that the population standard deviation was, for example, 1 pound. Again, suppose we are told that the mean of 20.5 pounds was obtained on the basis of the weights of two poodles. We do not put much faith in an estimate that is computed with such a small sample. But compare this with an estimate obtained by weighing 100 poodles. Naturally we would have much more faith in this latter estimate.

An alternate approach of estimating a parameter which takes into account considerations of the nature given above is the method of **interval estimation**, also called the method of **confidence intervals**. It provides an interval—a range of numerical values—which is believed to contain the unknown parameter. The general procedure for constructing a confidence interval for a parameter involves the following steps:

1. Pick a random sample of a certain size.
2. Devise two random variables, which we shall call L and U, thereby getting a **random interval** (L, U).
3. Compute the values of L and U from the sample data.

The random interval (L, U) is arrived at in such a way that the probability that it covers the parameter θ is $1 - \alpha$, where α (alpha) is a number between 0 and 1. That is,

$$P(L < \theta < U) = 1 - \alpha.$$

If the value of L from the specific sample data is l and that of U is u, then we give (l, u), also written $l < \theta < u$, as our interval estimate of θ based on the sample. Further we say that it is a $(1 - \alpha)100$ percent confidence interval for θ.

If, for example, $\alpha = 0.05$, then we get a $(1 - 0.05)100$ percent, that is, 95 percent confidence interval; with $\alpha = 0.02$ we get a $(1 - 0.02)100$ percent, that is, 98 percent confidence interval, and so on.

The fraction $1 - \alpha$ associated with the confidence interval is called the **confidence coefficient**. When the confidence coefficient is expressed as a percent, we get the **level of confidence**. The left endpoint of the interval (l, u) is called the **lower confidence limit** and the right endpoint, the **upper confidence limit**. For the interpretation of a confidence interval, read the discussion given later in this section.

We shall limit our discussion to a single population in this chapter and shall consider two populations in Chapter 8.

Confidence Interval

We shall now develop the general notions by specifically showing in detail how to construct a confidence interval for μ when the population variance σ^2 is known. The assumption that σ^2 is known may seem rather artificial since in practice it is more

FIGURE 7-3

Probability density curve of normal distribution with mean μ and variance σ^2/n.

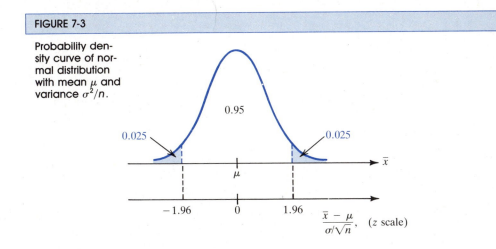

than likely that it will not be known. We shall handle this situation in the next section. In the meantime, the ideas developed here will serve to guide us.

To set a confidence interval for μ, *we make a basic assumption that the population has a normal distribution*. As we saw in Chapter 6, the distribution of $\overline{X}$ is then normal with mean μ and variance σ^2/n.

Conversion from the $\bar{x}$-scale to the z-scale, that is, into standard units, is accomplished by subtracting the mean of $\overline{X}$, namely, μ, and then dividing by the standard deviation of $\overline{X}$, namely, $\sigma/\sqrt{n}$. Thus the resulting random variable is

$$\frac{\overline{X} - \mu}{\sigma/\sqrt{n}}$$

and it has the standard normal distribution.

Now, for instance, we know from the table for the standard normal distribution (Appendix II, Table A-4) that the area to the right of 1.96 is 0.025 and, due to its symmetry about 0, that the area to the left of -1.96 is also 0.025. So the area between -1.96 and 1.96 is 0.95 as shown in Figure 7-3.

As a result we can write

$$P\left(-1.96 < \frac{\overline{X} - \mu}{\sigma/\sqrt{n}} < 1.96\right) = 0.95.$$

More generally, let $z_{\alpha/2}$ represent the value on the z-scale such that there is an area of $\alpha/2$ to its right. For example, if $\alpha = 0.05$, then $\alpha/2 = 0.025$ and $z_{\alpha/2} = z_{0.025} = 1.96$; if $\alpha = 0.02$, then $\alpha/2 = 0.01$ and $z_{\alpha/2} = z_{0.01} = 2.33$. Due to symmetry about 0 of the standard normal distribution, the z-value to the left of

FIGURE 7-4

Probability density curve of normal distribution with mean μ, variance σ^2/n showing area $\alpha/2$ in each of the two tails.

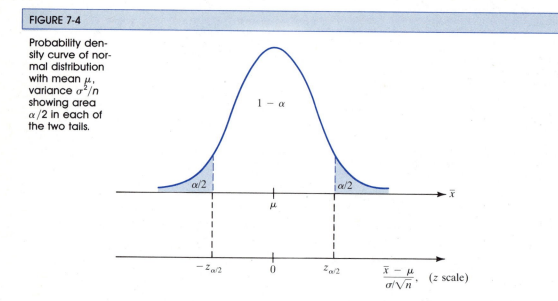

which there is an area of $\alpha/2$ is $-z_{\alpha/2}$. The balance of the total area under the curve, namely, $1 - \alpha$, is in the center between $-z_{\alpha/2}$ and $z_{\alpha/2}$ as shown in Figure 7-4.

Hence we can make the following probability statement.

$$P\left(-z_{\alpha/2} < \frac{\bar{X} - \mu}{\sigma/\sqrt{n}} < z_{\alpha/2}\right) = 1 - \alpha$$

With this as the starting point, through a series of algebraic steps, it can be established that

$$P\left(\bar{X} - z_{\alpha/2} \frac{\sigma}{\sqrt{n}} < \mu < \bar{X} + z_{\alpha/2} \frac{\sigma}{\sqrt{n}}\right) = 1 - \alpha.$$

The general procedure indicated that this is what we had to do, namely, isolate the parameter in the middle, between the two inequalities. Here, proper identification shows that

$$L = \bar{X} - z_{\alpha/2} \frac{\sigma}{\sqrt{n}}$$

and

$$U = \bar{X} + z_{\alpha/2} \frac{\sigma}{\sqrt{n}}.$$

Our premise is that σ is known. In any problem we will be given the level of confidence from which we should be able to find $z_{\alpha/2}$. When a sample of a given size is picked we will obtain $\bar{x}$, the mean for this data. Thus n and $\bar{x}$ will also be known to us. All this information will give us the resulting interval

$$\bar{x} - z_{\alpha/2} \frac{\sigma}{\sqrt{n}} < \mu < \bar{x} + z_{\alpha/2} \frac{\sigma}{\sqrt{n}}.$$

Often it is convenient to give this interval simply by specifying the limits as $\bar{x} \pm z_{\alpha/2} \frac{\sigma}{\sqrt{n}}$. We will say that it is a $(1 - \alpha)100$ percent confidence interval for μ.

We have established the following result given in summary:

CONFIDENCE INTERVAL FOR μ; σ KNOWN

If the population has a normal distribution and σ is known, then a $(1 - \alpha)100$ percent confidence interval for μ is given by

$$\bar{x} - z_{\alpha/2} \frac{\sigma}{\sqrt{n}} < \mu < \bar{x} + z_{\alpha/2} \frac{\sigma}{\sqrt{n}}$$

where $\bar{x}$ is the sample mean based on n observations.

We have constructed the confidence interval above for μ with the assumption that the population is normally distributed. This assumption was necessary since it allowed us to proceed by stating that $\bar{X}$ $\left(\text{consequently, } \frac{\bar{X} - \mu}{\sigma/\sqrt{n}}\right)$ has a normal distribution. If n is large (usually greater than 30), the assumption of a normal distribution is not crucial, because the central limit theorem permits us to proceed by stating that $\frac{\bar{X} - \mu}{\sigma/\sqrt{n}}$ has *approximately* a normal distribution. We still get the same confidence limits given above, but now they are approximate limits. This should be understood by the reader while working the exercises, even if the word *approximate* is omitted.

EXAMPLE 1 A gas station sold a total of 8019 gallons of gas on 9 randomly picked days. Suppose the amount sold on a day is normally distributed with a standard deviation σ of 90 gallons. Construct confidence intervals for the true mean amount sold on a day with the following confidence levels:

(a) 98 percent (b) 80 percent

SOLUTION We are given that $\sum_{i=1}^{9} x_i = 8019$, $n = 9$, and $\sigma = 90$. Therefore,

$$\bar{x} = \frac{8019}{9} = 891.$$

(a) Since $\alpha = 0.02$, $z_{\alpha/2} = z_{0.01} = 2.33$ from the standard normal table. Therefore, a 98 percent confidence interval is given by

$$\bar{x} - 2.33 \frac{\sigma}{\sqrt{n}} < \mu < \bar{x} + 2.33 \frac{\sigma}{\sqrt{n}}.$$

That is,

$$891 - 2.33\left(\frac{90}{\sqrt{9}}\right) < \mu < 891 + 2.33\left(\frac{90}{\sqrt{9}}\right)$$

which can be simplified to give

$$821.1 < \mu < 960.9.$$

(b) Here $\alpha = 0.20$. Therefore, $z_{\alpha/2} = z_{0.1} = 1.28$. Consequently, with 80 percent confidence,

$$891 - 1.28\left(\frac{90}{\sqrt{9}}\right) < \mu < 891 + 1.28\left(\frac{90}{\sqrt{9}}\right)$$

that is, $852.6 < \mu < 929.4.$ ▬

EXAMPLE 2   When 10 vehicles were observed at random for their speeds (in mph) on a freeway, the following data were obtained:

61,   53,   57,   55,   62,   58,   60,   54,   62,   60.

Suppose that from past experience it may be assumed that vehicle speed is normally distributed with $\sigma = 3.2$ mph. Construct a 90 percent confidence interval for μ, the true mean speed.

SOLUTION   From the available data we can compute $\bar{x} = 58.2$. Further we know that $n = 10$ and $\sigma = 3.2$. For a 90 percent confidence interval, $\alpha = 0.10$. Hence $z_{\alpha/2} = z_{0.05} = 1.645$ and the confidence interval for μ is given by

$$58.2 - 1.645\left(\frac{3.2}{\sqrt{10}}\right) < \mu < 58.2 + 1.645\left(\frac{3.2}{\sqrt{10}}\right)$$

which can be simplified to $56.54 < \mu < 59.86.$ ▬

EXAMPLE 3   Basing her findings on 50 successful pregnancies involving natural birth, Dr. Vivien McNutt found that the mean pregnancy term was 274 days. Suppose that from previous studies the standard deviation σ of a pregnancy term may safely be assumed to be 14 days. Construct a 98 percent confidence interval for the true mean pregnancy term μ.

SOLUTION   The problem does not state that pregnancy term is normally distributed. However, $n = 50$, which is large enough that we can apply the central limit theorem. Of course, we realize that the confidence interval will be approximate.

From the given information $\bar{x} = 274$, $\sigma = 14$, and $n = 50$. For a 98 percent confidence interval, $\alpha = 0.02$. Hence $z_{\alpha/2} = z_{0.01} = 2.33$ so that, with 98 percent confidence, the approximate interval is

$$274 - 2.33\left(\frac{14}{\sqrt{50}}\right) < \mu < 274 + 2.33\left(\frac{14}{\sqrt{50}}\right).$$

That is, $269.4 < \mu < 278.6$.   ▬▬

Empirical Interpretation

As an example of an empirical interpretation of a confidence interval statement, suppose we wish to construct a 95 percent confidence interval for the mean weight of a certain strain of monkeys. Based on past experience, we will assume that σ equals 4 pounds. Also, let us agree to construct an interval by weighing 64 monkeys; that is, $n = 64$. Substituting for σ and n, the random interval becomes $\left(\bar{X} - 1.96\,\frac{4}{\sqrt{64}}, \bar{X} + 1.96\,\frac{4}{\sqrt{64}}\right)$, that is, $(\bar{X} - 0.98, \bar{X} + 0.98)$. Thus

$$P(\bar{X} - 0.98 < \mu < \bar{X} + 0.98) = 0.95.$$

Suppose we weigh a sample of 64 monkeys and find their mean weight to be 20.8 pounds. The random interval then reduces to $(20.8 - 0.98, 20.8 + 0.98)$, that is, $(19.82, 21.78)$. Does this interval include μ? We do not know.

Next, suppose we take a second sample, once again consisting of 64 monkeys, and obtain a mean of 19.7 pounds. This time the random interval reduces to $(18.72, 20.68)$. Here, too, we do not know whether the interval encloses μ.

We can continue this process of taking samples of 64 and computing intervals. In Table 7-1 we give results for five samples of 64. These intervals corresponding to each sample are plotted in Figure 7-5 on page 246.

TABLE 7-1		
Sample number	*Sample mean* $\bar{x}$	*Intervals* $(\bar{x} - 0.98, \bar{x} + 0.98)$
1	20.8	(19.82, 21.78)
2	19.7	(18.72, 20.68)
3	21.3	(20.32, 22.28)
4	20.2	(19.22, 21.18)
5	21.0	(20.02, 21.98)
.	.	.
.	.	.
.	.	.

FIGURE 7-5

Confidence intervals for 5 samples each of size 64. μ is unknown but is shown in the figure for illustration. Only sample 2 excludes this value for μ.

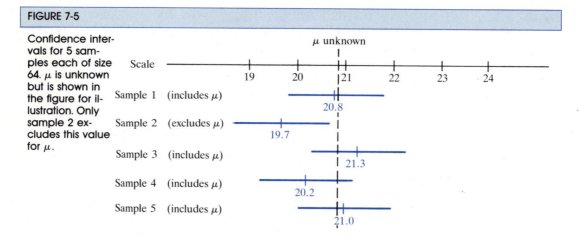

Which of these intervals enclose μ and which ones do not we cannot actually say. Recall at this point the empirical notion of probability: *The probability of an event is interpreted as the relative frequency with which the event occurs in a large series of identical trials of an experiment.* Therefore, the probability statement

$$P(\overline{X} - 0.98 < \mu < \overline{X} + 0.98) = 0.95$$

states that approximately 95 percent of the intervals constructed in the manner above will encompass the population mean.

Of course, we pick samples of size n repeatedly only in theory. When we construct a confidence interval, no repetitive process is involved. We pick just one sample and compute the endpoints. That's it! Now the population mean is either included in this interval or it is not. We will never really know.

Since the first sample that we picked gave us an interval (19.82, 21.78), we claim that the population mean is contained in the interval 19.82 to 21.78 and write

$$19.82 < \mu < 21.78.$$

We don't contemplate taking any more samples. We attach a 95 percent confidence to the claim $19.82 < \mu < 21.78$ with the following rationale: The approach that we adopted for computing the confidence interval is such that of all the intervals obtained by computing $(\overline{X} - 0.98, \overline{X} + 0.98)$ by repeatedly drawing samples of 64 monkeys, 95 percent will include μ and 5 percent will not. We have a strong feeling and hope that the one interval that we have obtained, namely (19.82, 21.78), is from the former category. We quantify this hope by saying that we have a 95 percent confidence in the claim.

Caution: It is definitely wrong to say

$$P(19.82 < \mu < 21.78) = 0.95.$$

However, it would not be inappropriate to say

$$\text{conf}(19.82 < \mu < 21.78) = 0.95$$

which means that the confidence is 95 percent.

FIGURE 7-6

A graphical presentation showing how the length of the confidence interval increases with the level of confidence.

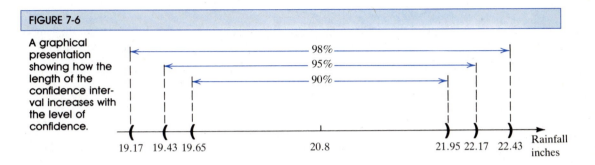

THE LENGTH OF A CONFIDENCE INTERVAL

Let us consider the following example as it pertains to the **length of a confidence interval**.

EXAMPLE 4   From past records, the seasonal rainfall in a county when observed over 16 randomly picked years yielded a mean rainfall of 20.8 inches. If it can be assumed from past experience that rainfall during a season is normally distributed with $\sigma = 2.8$ inches, construct confidence intervals for the true mean rainfall μ with the following confidence levels.

(a) 90 percent   (b) 95 percent   (c) 98 percent.

SOLUTION   We are given that $\bar{x} = 20.8$, $\sigma = 2.8$, and $n = 16$. Also, we are granted the assumption that rainfall is normally distributed.

(a) For a 90 percent confidence interval, $\alpha = 0.10$. Hence $z_{\alpha/2} = z_{0.05} = 1.645$ and the confidence interval is given by

$$20.8 - 1.645\left(\frac{2.8}{\sqrt{16}}\right) < \mu < 20.8 + 1.645\left(\frac{2.8}{\sqrt{16}}\right)$$

which can be simplified to $19.648 < \mu < 21.951$.

(b) Since $\alpha = 0.05$, $z_{\alpha/2} = 1.96$ and a 95 percent confidence interval is given by

$$20.8 - 1.96\left(\frac{2.8}{\sqrt{16}}\right) < \mu < 20.8 + 1.96\left(\frac{2.8}{\sqrt{16}}\right).$$

This, upon simplification, gives $19.428 < \mu < 22.172$.

(c) Since $\alpha = 0.02$, $z_{\alpha/2} = 2.33$ and we get, with 98 percent confidence,

$$20.8 - 2.33\left(\frac{2.8}{\sqrt{16}}\right) < \mu < 20.8 + 2.33\left(\frac{2.8}{\sqrt{16}}\right).$$

that is, $19.169 < \mu < 22.431$.   ■

If we represent the confidence intervals in Example 4 graphically as in Figure 7-6 we observe that they become wider as the level of confidence increases.

What are the other factors that affect the length of a confidence interval? Notice that the length of any interval from a to b is $b - a$. For instance, the length of the interval from 4 to 10 is $10 - 4$, or 6 units. Consequently, the length of a confidence interval $\left(\bar{x} - z_{\alpha/2} \dfrac{\sigma}{\sqrt{n}}, \bar{x} + z_{\alpha/2} \dfrac{\sigma}{\sqrt{n}} \right)$ is

$$\left(\bar{x} + z_{\alpha/2} \frac{\sigma}{\sqrt{n}} \right) - \left(\bar{x} - z_{\alpha/2} \frac{\sigma}{\sqrt{n}} \right) = 2 z_{\alpha/2} \frac{\sigma}{\sqrt{n}}.$$

We see that the length of the confidence interval for μ does not depend on $\bar{x}$ but it does depend on $z_{\alpha/2}$, σ, and n, in the manner indicated below:

1. The length of the interval will be large or small according as $z_{\alpha/2}$ is large or small, which, in turn, will be large or small depending on whether the level of confidence, $1 - \alpha$, is large or small. In short, for a higher confidence level, the interval will be wider than it is for a lower confidence level. (For example, you can be most confident in saying that μ is between $-\infty$ and ∞.)
2. The length of the interval depends on σ and will be large if σ is large and small if σ is small. Notice, however, that σ is a population characteristic and an experimenter has no control over it.
3. The third factor affecting the length of the interval is the sample size n. Since n appears in the denominator of $2 z_{\alpha/2} \dfrac{\sigma}{\sqrt{n}}$, the length will be small if n is large. Thus, if a very high accuracy is desired, that is, a small confidence interval with a specified degree of confidence, then a way to accomplish this goal is by picking an appropriately large sample. We shall discuss this aspect next.

How Large a Sample for Estimating μ?

Suppose we want to decide how large a sample should be picked so that we can be $(1 - \alpha)100$ percent confident that the estimate $\bar{x}$ does not differ from μ by more than some given amount. For example, if an environmental engineer wants to estimate the mean dissolved oxygen concentration in a stream, he might want to know how many measurements he should make so that he can be $(1 - 0.05)100$, or 95, percent confident that his estimate $\bar{x}$ will not differ from the true mean concentration μ by, for example, more than 1 mg/liter. We shall refer to the amount by which the estimate $\bar{x}$ differs from μ as the *error in estimation* and shall denote the maximum error allowed by e.

For a $(1 - \alpha)100$ percent confidence interval, we have observed previously that the left extremity of the interval is obtained by subtracting $z_{\alpha/2} \dfrac{\sigma}{\sqrt{n}}$ from $\bar{x}$, and the right extremity by adding $z_{\alpha/2} \dfrac{\sigma}{\sqrt{n}}$ to $\bar{x}$. This is shown in Figure 7-7.

The midpoint of the interval is $\bar{x}$. If the interval so constructed includes μ, then we have one of the two situations shown in Figure 7-8 (a) and (b). In either

FIGURE 7-7

The confidence limits are located $z_{\alpha/2}\dfrac{\sigma}{\sqrt{n}}$ units from $\bar{x}$.

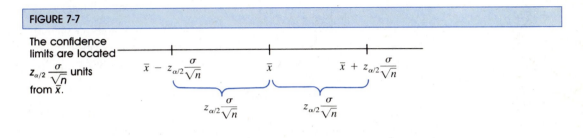

FIGURE 7-8

Error in estimating μ by $\bar{x}$.

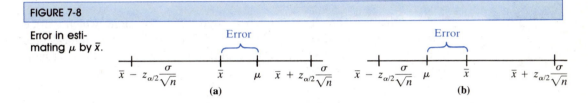

case, the error is less than $z_{\alpha/2}\dfrac{\sigma}{\sqrt{n}}$. Thus the maximum error is $z_{\alpha/2}\dfrac{\sigma}{\sqrt{n}}$ and we have the following result:

When $\bar{x}$ is used to estimate μ with $(1 - \alpha)100$ percent confidence level, the error in estimation will not exceed $z_{\alpha/2}\dfrac{\sigma}{\sqrt{n}}$ so that the maximum error e is given by

$$e = z_{\alpha/2}\frac{\sigma}{\sqrt{n}}.$$

A slight simplification then gives

$$\sqrt{n} = \frac{z_{\alpha/2}\sigma}{e}.$$

Squaring, we get

$$n = \left(\frac{z_{\alpha/2}\sigma}{e}\right)^2.$$

With a sample size as large as this, we can be $(1 - \alpha)100$ percent confident that the estimate will not be in error by more than an amount e. If the experimenter can take a sample larger than $\left(\dfrac{z_{\alpha/2}\sigma}{e}\right)^2$, so much the better. We should not forget that with a larger sample, we always stand to gain more information about the population. Of course, one might have reached a point of diminishing returns, in which case the additional information acquired may not be commensurate with the added expenditure in time and money needed to investigate more individuals. In summary, we have established the following result:

SAMPLE SIZE FOR ESTIMATING μ

The sample size needed so as to be $(1 - \alpha)100$ percent confident that the estimate $\bar{x}$ does not differ from μ by more than a preassigned quantity e is

$$n = \left(\frac{z_{\alpha/2}\sigma}{e}\right)^2$$

where the population is normally distributed with known variance σ^2.

Notice that n will be large if (1) σ is large, (2) $z_{\alpha/2}$ is large, which will be the case if we want a high level of confidence, and (3) e is small (that is, if the degree of accuracy desired is high). This is in keeping with what we might expect intuitively.

EXAMPLE 5 An electrical firm which manufactures a certain type of bulb wants to estimate its mean life. Assuming that the standard deviation σ is known to be 40 hours, find how many bulbs should be tested so as to be

(a) 95 percent confident that the estimate $\bar{x}$ will not differ from the true mean life μ by more than 10 hours
(b) 98 percent confident to accomplish accuracy in Part (a).

[Assume a normal distribution of the life of a bulb.]

SOLUTION We are given that $\sigma = 40$ and $e = 10$.

(a) Since $\alpha = 0.05$, $z_{\alpha/2} = z_{0.025} = 1.96$. Therefore

$$n = \left(\frac{1.96 \times 40}{10}\right)^2 = 61.5 \approx 62.$$

Approximately 62 bulbs should be tested.

(b) Since $\alpha = 0.02$, $z_{\alpha/2} = z_{0.01} = 2.33$. Therefore

$$n = \left(\frac{2.33 \times 40}{10}\right)^2 = 86.9 \approx 87.$$

Approximately 87 bulbs should be tested.

Notice that we need to test more bulbs if we increase the level of confidence. Is this not what you would expect intuitively? ▬

SECTION 7-2 EXERCISES

1. For the given level of confidence in each case determine α.

   (a) 96 percent   (b) 92 percent
   (c) 84 percent   (d) 97 percent

2. Determine the indicated value for the given level of confidence.

   (a) $z_{\alpha/2}$; 90 percent   (b) $z_{\alpha/2}$; 93.84 percent

   (c) $-z_{\alpha/2}$; 91.64 percent   (d) $-z_{\alpha/2}$; 96.16 percent

In each of the Exercises 3–5, assuming that the population is normally distributed, determine a confidence interval for the population mean at the specified level of confidence. State whether the interval is exact or approximate.

3. $n = 9$, $\bar{x} = 15.2$, $\sigma = 2.8$; 95 percent

4. $n = 16$, $\bar{x} = 0.8$, $\sigma = 0.12$; 90 percent

5. $n = 13$, $\bar{x} = 152.3$, $\sigma = 9.8$; 91.64 percent.

6. A random sample of size $n = 16$ is drawn from a population having a normal distribution with variance σ^2 of 36. If the sample mean is $\bar{x} = 46$, find a 90 percent confidence interval for the population mean μ.

7. The time (in minutes) taken by a biological cell to divide into two cells has a normal distribution. From past experience, the standard deviation σ can be assumed to be 3.5 minutes. When 16 cells were observed, the mean time taken by them to divide was 31.2 minutes. Estimate the true mean time for a cell division using a 98 percent confidence interval.

8. The mean yield of grain on ten randomly picked experimental plots of farm was found to be 150 bushels. If the yield per plot can be assumed to be normally distributed with a standard deviation of yield σ equal to 20 bushels, construct an 85 percent confidence interval on the true mean yield per plot.

9. A random sample of nine full-grown crabs had a mean weight of 24 ounces. If the population standard deviation σ of the weight of a crab, can be assumed to be 3 ounces, construct confidence intervals for the true mean weight of a crab with the following confidence levels:

   (a) 90 percent   (b) 95 percent   (c) 98 percent

   Comment on the lengths of the confidence intervals obtained. You may assume that the weight of a full-grown crab has a normal distribution.

10. A population has a normal distribution with variance 225. Find how large a sample must be drawn in order to be 95 percent confident that the sample mean will not differ from the population mean by more than 2 units.

11. In each case find the length of the given confidence interval.

    (a) $10.87 < \mu < 12.63$ (b) $-4.88 < \mu < 8.44$

12. The following confidence intervals are obtained for μ using the formula

    $\bar{x} \pm z_{\alpha/2}\dfrac{\sigma}{\sqrt{n}}$. In each case determine the sample mean $\bar{x}$.

    (a) $10.87 < \mu < 12.63$ (b) $-4.88 < \mu < 8.44$

    [*Hint:* The sample mean is the midpoint of the interval.]

13. In computing a confidence interval for a parameter for which of the following situations should we strive?

    (a) To have as wide an interval as possible with a high level of confidence

    (b) To have as narrow an interval as possible with a high level of confidence

14. Suppose the breaking strength of cables (in pounds) is known to have a normal distribution with a standard deviation σ equal to 6 pounds. Find how large a sample must

be taken so as to be 90 percent confident that the sample mean breaking strength will not differ from the true mean breaking strength by more than 0.75 pound.

15. To determine the mean weekly idle time of machines, the management has decided to study a sample of machines from the total number of machines in the factory. (It may be assumed that the total number of machines is large.) The mean idle time of a machine during a week is to be estimated within 1.12 hours of the true mean idle time with a 95 percent level of confidence. From past experience, it is safe to assume that the idle time of a machine during a week is normally distributed with a standard deviation of 4 hours. Find what size sample should be drawn.

16. Based on a random sample of 100 cows of a certain breed, a confidence interval for estimating the true mean yield of milk is given by $41.6 < \mu < 44.0$. If the yield of milk of a cow may be assumed to be normally distributed with σ equal to 6, what was the level of confidence used?

7-3 CONFIDENCE INTERVAL FOR THE POPULATION MEAN WHEN THE POPULATION VARIANCE IS UNKNOWN

The case where σ^2 is unknown is a more realistic situation since in practice σ^2 will usually not be known to the experimenter. *The method we are about to develop is particularly important when the sample size is small.*

We shall assume that the parent population has a normal distribution. Now recall that for setting a confidence interval for μ when σ^2 is known, we start with the fact that the sampling distribution of $\overline{X}$ is normal with mean μ and variance σ^2/n or, equivalently, converting to the z-scale, that

$$Z = \frac{\overline{X} - \mu}{\sigma/\sqrt{n}}$$

has the standard normal distribution. In the present context, since σ^2 is not known, we do the next best thing and use its point estimate. In other words, we start with the random variable

$$T = \frac{\overline{X} - \mu}{S/\sqrt{n}}.$$

Now it turns out that the random variable T does not have the standard normal distribution but has a distribution called Student's t distribution, which we shall study next.

STUDENT'S *t* DISTRIBUTION

We notice immediately that for Z only the numerator changes from sample to sample (assuming n is fixed). But for T both the numerator and the denominator change from sample to sample, since $\overline{X}$ and S both depend on sample values. Hence it is not

FIGURE 7-9

t distributions with 2, 4, and 10 df. Also, the figure shows how the *t* distribution approaches the standard normal distribution as the degrees of freedom become large.

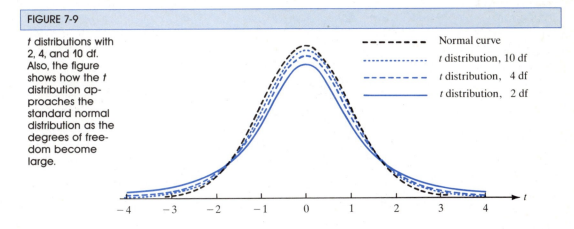

surprising that the exact distribution of T is not the same as that of Z. If n is small, the departure of the distribution of T from the standard normal distribution is rather marked. William Gosset, an employee of an Irish brewery, realized this fact in his research and derived the exact distribution of T. Gosset published his work under the pseudonym of Student. Hence the distribution is commonly known as **Student's *t* distribution,** or, simply, as the ***t* distribution.** There are many other statistics that also have the *t* distribution. A typical value of a random variable that has the *t* distribution is denoted by t.

The following features characterize the *t* distribution.

1. It is a continuous distribution with mean 0.
2. It is symmetric about the vertical axis at 0.
3. Its shape depends on a parameter ν (which is related in some way to the sample size) called the **number of degrees of freedom,** usually abbreviated "df." There is one *t* distribution curve for each degree of freedom. (*t* distributions with 2, 4, and 10 df are shown in Figure 7-9.)
4. The curves for the *t* distribution are lower at the center and higher at the tails than the standard normal curve (see Figure 7-9).
5. As the number of degrees of freedom becomes larger, the *t* distribution looks more and more like the standard normal distribution; the curves for the *t* distributions rise at the center, getting closer to the standard normal distribution (see Figure 7-9).

The particular statistic T that we have described above, namely,

$$T = \frac{\bar{X} - \mu}{S/\sqrt{n}}$$

has a *t* distribution with $n - 1$ degrees of freedom where n is the size of the sample.

Because the *t* distribution is continuous, probabilities are given as areas under the curve. Appendix II, Table A-5 gives such areas for different numbers of degrees of freedom. Since there is one curve for each number of degrees of freedom, for

FIGURE 7-10

For ν df, a t value such that area in the right tail is α.

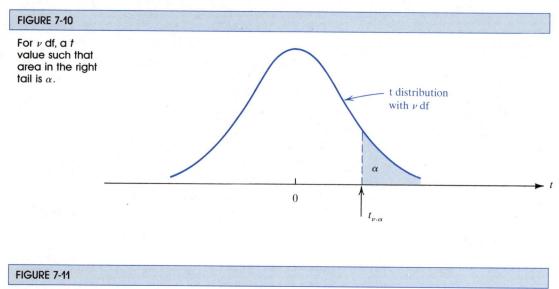

FIGURE 7-11

Probability density curve for the t distribution with 10 df. Area to the right of 1.372 is 0.1, and so on.

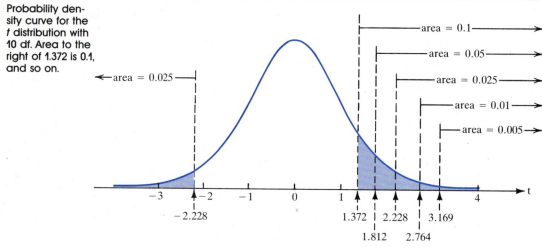

economy of space, the table is not as exhaustive as the one giving areas for the standard normal distribution. (We would need 30 tables for the first 30 df!) For each number of degrees of freedom ν, given in the first column, the *t-values are given along the row for a select set of probabilities contained in the right tail of the distribution.* These probabilities, α, are given as the corresponding column headings. For ν degrees of freedom, a t-value such that an area α is in the right tail is usually denoted as $t_{\nu,\alpha}$ as shown in Figure 7-10.

We shall show how to read Table A-5 specifically for $\nu = 10$. The relevant entries for $\nu = 10$ are included here in the partial reproduction of that table.

Degrees of freedom	α, area in the right tail				
	0.1	0.05	0.025	0.01	0.005
ν	$t_{\nu,0.1}$	$t_{\nu,0.05}$	$t_{\nu,0.025}$	$t_{\nu,0.01}$	$t_{\nu,0.005}$
1	3.078	6.314	12.706	31.821	63.657
2	1.886	2.920	4.303	6.965	9.925
3	1.638	2.353	3.182	4.541	5.841
.					
.					
.					
9	1.383	1.833	2.262	2.821	3.250
10	1.372	1.812	2.228	2.764	3.169
11	1.363	1.796	2.201	2.718	3.106
.					
.					

The values 1.372, 1.812, 2.228, 2.764, 3.169 represent the *t* values that we would ordinarily mark along the horizontal axis. The column headings, α, are the areas (that is, the probabilities) under the curve to the right of the corresponding *t* value as shown in Figure 7-11. Thus to the right of 1.812 there is an area of 0.05, to the right of 2.764 it is 0.01, and so on. Since the *t* distribution is symmetric about the vertical axis at $t = 0$, it is possible to obtain areas for certain other *t* values as well. For example, the area to the left of -1.812 is also 0.05, the area between -1.812 and 1.812 is 0.90, and so on.

EXAMPLE 1 Find

(a) $t_{4,0.05}$ (b) $t_{21,0.025}$ (c) $-t_{17,0.1}$ (d) $t_{5,\alpha/2}$ if $\alpha = 0.01$
(e) $-t_{16,\alpha/2}$ if $\alpha = 0.05$.

SOLUTION A rough sketch of the graph of the appropriate *t* distribution would help.

(a) To find $t_{4,0.05}$, we consider *t* distribution with 4 degrees of freedom. Notice that the area is 0.05 to the right of the value. From Table A-5, in the row corresponding to 4 degrees of freedom under column heading 0.05, we find $t_{4,0.05} = 2.132$ (see Figure 7-12(a) below).

FIGURE 7-12(a)

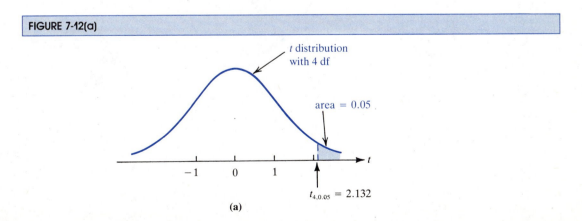

(a)

(b) As in (a), we look in Table A-5 in the row corresponding to $\nu = 21$, and under column heading 0.025. The answer is $t_{21,0.025} = 2.080$ (see Figure 7-12(b) below).

(c) To find $-t_{17,0.1}$, we first find $t_{17,0.1}$. From Table A-5 we see that $t_{17,0.1} = 1.333$. Hence $-t_{17,0.1} = -1.333$ (see Figure 7-12(c) below). Notice, incidentally, that $-t_{17,0.1}$ has area 0.9 to its right and area 0.1 to its left.

(d) Since $\alpha = 0.01$, $\alpha/2 = 0.005$. From Table A-5, we immediately get $t_{5,\alpha/2} = t_{5,0.005} = 4.032$.

(e) $\alpha = 0.05$; hence $\alpha/2 = 0.025$. As a result, $t_{16,\alpha/2} = t_{16,0.025} = 2.12$, and $-t_{16,\alpha/2} = -2.12$. ■

FIGURE 7-12(b)

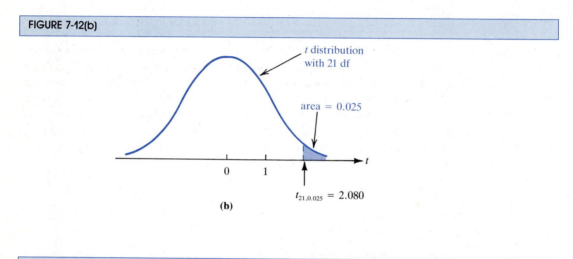

(b)

FIGURE 7-12(c)

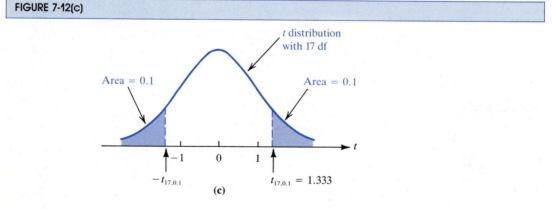

(c)

FIGURE 7-13

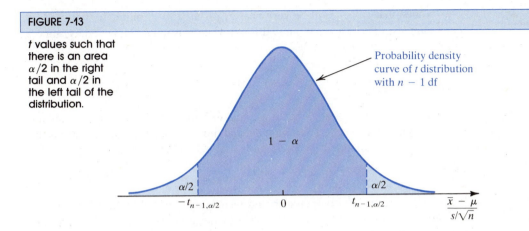

t values such that there is an area $\alpha/2$ in the right tail and $\alpha/2$ in the left tail of the distribution.

Probability density curve of *t* distribution with $n - 1$ df

$1 - \alpha$

$\alpha/2$ $\alpha/2$

$-t_{n-1,\alpha/2}$ 0 $t_{n-1,\alpha/2}$ $\dfrac{\bar{x} - \mu}{s/\sqrt{n}}$

Confidence Interval

Having discussed the *t* distribution, let us now return to our primary goal—that of setting a confidence interval for μ when σ is not known.

Since the statistic *T* has Student's *t* distribution with $n - 1$ degrees of freedom, we can write

$$P\left(-t_{n-1,\alpha/2} < \frac{\bar{X} - \mu}{S/\sqrt{n}} < t_{n-1,\alpha/2}\right) = 1 - \alpha$$

where $t_{n-1,\alpha/2}$ is the value from the table for a *t* distribution with $n - 1$ degrees of freedom such that the area in its right tail is $\alpha/2$. The value $-t_{n-1,\alpha/2}$ is the value located symmetrically on the other side of zero and is such that the area in its left tail is $\alpha/2$, as shown in Figure 7-13. Through some algebraic steps, this leads to

$$P\left(\bar{X} - t_{n-1,\alpha/2} \frac{S}{\sqrt{n}} < \mu < \bar{X} + t_{n-1,\alpha/2} \frac{S}{\sqrt{n}}\right) = 1 - \alpha.$$

Hence we get the following result:

CONFIDENCE INTERVAL FOR μ; σ UNKNOWN

A $(1 - \alpha)100$ percent confidence interval for μ when the population is normally distributed and σ is *not* known is given by

$$\bar{x} - t_{n-1,\alpha/2} \frac{s}{\sqrt{n}} < \mu < \bar{x} + t_{n-1,\alpha/2} \frac{s}{\sqrt{n}}$$

where $\bar{x}$ is the sample mean and *s* the sample standard deviation based on *n* observations.

Notice that $t_{n-1,\alpha/2}$ will be very near $z_{\alpha/2}$ if n is 30 or more. In that case, *for large samples,* the confidence interval for μ when σ is unknown becomes, approximately,

$$\bar{x} - z_{\alpha/2} \frac{s}{\sqrt{n}} < \mu < \bar{x} + z_{\alpha/2} \frac{s}{\sqrt{n}}.$$

Actually, if n is large, we can use this formula even if the parent population is not normally distributed. This comment and those made earlier make a strong case for planning an experiment properly and, if at all possible, taking a large enough sample.

EXAMPLE 2   When 16 cigarettes of a particular brand were tested in a laboratory for the amount of tar content, it was found that their mean content was 18.3 milligrams with $s = 1.8$ milligrams. Set a 90 percent confidence interval for the mean tar content μ in the population of cigarettes of this brand. (Assume that the amount of tar in a cigarette is normally distributed.)

SOLUTION   Here $n = 16$, $\bar{x} = 18.3$, $s = 1.8$, and $\alpha = 0.10$. Hence the t distribution has $16 - 1 = 15$ degrees of freedom and $t_{15,\alpha/2} = t_{15,0.05} = 1.753$. As a result, a 90 percent confidence interval is given by

$$18.3 - 1.753\left(\frac{1.8}{\sqrt{16}}\right) < \mu < 18.3 + 1.753\left(\frac{1.8}{\sqrt{16}}\right)$$

which can be simplified to

$$17.51 < \mu < 19.09. \quad \blacksquare$$

EXAMPLE 3   In order to estimate the amount of time (in minutes) that a teller spends on a customer, a bank manager decided to observe 64 customers picked at random. The amount of time the teller spent on each customer was recorded. It was found that the sample mean was 3.2 minutes with $s^2 = 1.44$. Find a 98 percent confidence interval for the mean amount of time μ.

SOLUTION   From the given information $\bar{x} = 3.2$, $s^2 = 1.44$, $n = 64$, and $\alpha = 0.02$. The population variance σ^2 is not known; so we use its estimate s^2. Since n is large, the confidence interval is constructed, approximately, using the formula

$$\bar{x} - z_{\alpha/2} \frac{s}{\sqrt{n}} < \mu < \bar{x} + z_{\alpha/2} \frac{s}{\sqrt{n}}.$$

Now $z_{\alpha/2} = z_{0.01} = 2.33$. Therefore, a 98 percent confidence interval for μ is approximately

$$3.2 - 2.33 \frac{\sqrt{1.44}}{\sqrt{64}} < \mu < 3.2 + 2.33 \frac{\sqrt{1.44}}{\sqrt{64}}$$

which, simplifying, gives

$$2.85 < \mu < 3.55. \quad \blacksquare$$

EXAMPLE 4 The following data represent the amount of sugar (in pounds) consumed in a household during five randomly picked weeks:

$$3.8, 4.5, 5.2, 4.0, 5.5.$$

Construct a 90 percent confidence interval for the true mean consumption μ. (Assume a normal distribution for the amount of sugar consumed.)

SOLUTION Routine calculations give $\bar{x} = 4.6$ and

$$\sum_{i=1}^{5} (x_i - \bar{x})^2 = (-0.8)^2 + (-0.1)^2 + (0.6)^2 + (-0.6)^2 + (0.9)^2$$

$$= 2.18.$$

Therefore, $s^2 = \dfrac{2.18}{5-1} = 0.545$, and $s = 0.738$.

The t distribution has $5 - 1 = 4$ degrees of freedom, and since $\alpha = 0.10$, $t_{4,0.05} = 2.132$. Consequently, a 90 percent confidence interval is given by

$$4.6 - 2.132\left(\frac{0.738}{\sqrt{5}}\right) < \mu < 4.6 + 2.132\left(\frac{0.738}{\sqrt{5}}\right).$$

Simplifying, we get

$$3.896 < \mu < 5.304. \quad \blacksquare$$

SECTION 7-3 EXERCISES

1. Find

   (a) $t_{18,0.05}$ (b) $t_{13,0.025}$

   (c) $-t_{6,0.01}$ (d) $-t_{9,0.025.}$

2. In each case find the indicated quantity for the given level of confidence.

   (a) $t_{6,\alpha/2}$; 95 percent   (b) $t_{10,\alpha/2}$; 98 percent

   (c) $-t_{14,\alpha/2}$; 98 percent   (d) $-t_{26,\alpha/2}$; 80 percent

   (e) $t_{16,\alpha/2}$; 99 percent   (f) $-t_{12,\alpha/2}$; 99 percent

   In Exercises 3–6, assuming that the population is normally distributed, determine a confidence interval for the corresponding population mean at the specified level of confidence. State whether the interval is exact or approximate.

3. $n = 9$, $\bar{x} = 15.2$, $s = 3.3$; 95 percent

4. $n = 16, \bar{x} = 0.8, s = 0.1$; 98 percent

5. $n = 13, \bar{x} = 152.3, s = 10.82$; 99 percent

6. $n = 21, \bar{x} = -1.23, s = 2.34$; 90 percent

7. A random sample of size $n = 16$ is drawn from a population having a normal distribution. If the sample mean and the sample variance from the data are given, respectively, as $\bar{x} = 23.8$ and $s^2 = 10.24$, find a 98 percent confidence interval for μ, the population mean.

8. When a skier took ten runs on the downhill course, the mean time was found to be 12.3 minutes with $s = 1.2$ minutes. Assuming a normal distribution, estimate the true mean time for a run of the skier using a 90 percent confidence interval.

9. In ten half-hour programs on a TV channel, Mary found that the number of minutes devoted to commercials were: 6, 5, 5, 7, 5, 4, 6, 7, 5, and 5. Set a 95 percent confidence interval on the true mean time devoted to commercials during a half-hour program. Assume that the amount of time devoted to commercials is normally distributed.

10. A sample of eight workers in a clothing manufacturing company gave the following figures for the amount of time (in minutes) needed to join a collar to a shirt: 10, 12, 13, 9, 8, 14, 10, and 11. Assuming that the length of time is normally distributed, set a 90 percent confidence interval on the true mean time μ.

In Exercises 11–13, construct a confidence interval for the corresponding population mean if the given information is available. State whether the interval is exact or approximate. (Notice n is large in each case; so you may use z-values.)

11. $n = 40, \bar{x} = 18.6, s = 9.486$; 95 percent

12. $n = 64, \bar{x} = 0.82, s = 0.12$; 90 percent

13. $n = 100, \bar{x} = 148.9, s^2 = 13.83$; 98 percent

14. In a laboratory experiment, 64 crosses of a certain type of beetle were made and the number of pupae produced were counted. It was found that the mean number of pupae were 85 with $s = 15$ pupae. Find a 95 percent confidence interval for the true mean number of pupae in a cross.

15. A laboratory tested 36 chicken eggs and found that the mean amount of cholesterol was 230 milligrams with $s = 20$ milligrams. Find a 90 percent confidence interval for the true mean cholesterol content of an egg.

7-4 CONFIDENCE INTERVAL FOR THE POPULATION VARIANCE

While discussing point estimation, we agreed to accept S^2 as providing a good estimate of σ^2. *For setting a confidence interval on σ^2, we shall assume that the population has a normal distribution* and consider the statistic

$$U = \frac{(n-1) S^2}{\sigma^2}.$$

FIGURE 7-14

Probability density curves for chi-square distributions with 5, 10, 15, 30 df.

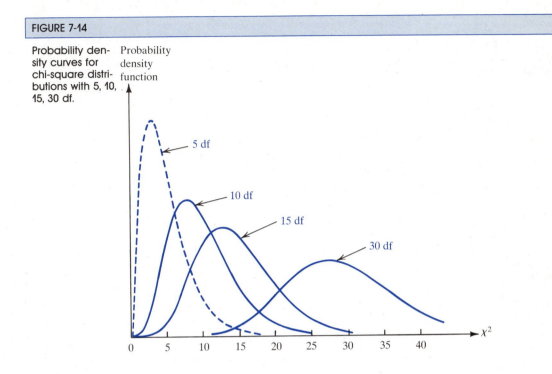

The value of U for any set of sample data $x_1, x_2, \ldots, x_n$ is given by

$$\frac{(n-1)s^2}{\sigma^2} = \frac{\sum_{i=1}^{n} (x_i - \bar{x})^2}{\sigma^2}.$$

Notice that only the numerator of the statistic U changes from sample to sample. The denominator is a fixed number σ^2.

A quick analysis of the statistic reveals that we are comparing the product of $(n-1)$ and the sample variance with the population variance. It might seem more relevant that we compare the sample variance itself with the population variance, that is, study the statistic S^2/σ^2. This would be fine. It just happens that the statistic U not only allows us to accomplish our primary goal of inference, but also it has a well-known distribution called the **chi-square distribution,** with $n-1$ degrees of freedom.

There are other statistics which also have the chi-square distribution. We are thus led to the study of yet another important continuous distribution.

THE CHI-SQUARE DISTRIBUTION

A typical value of a random variable that has a chi-square distribution is denoted as χ^2, the square of the Greek letter chi.

The following features characterize the chi-square distribution.

FIGURE 7-15

For ν df, χ^2 value such that area in the right tail is α.

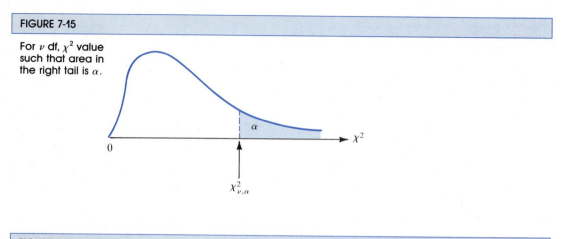

FIGURE 7-16

Probability density curve of the chi-square distribution with 10 df. Area to the right of 2.156 is 0.995, and so on.

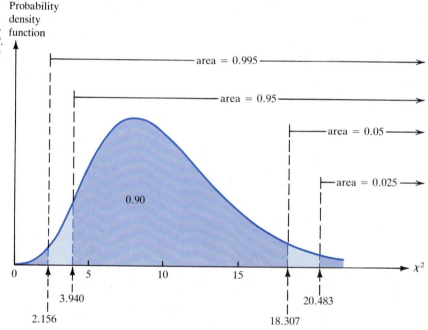

1. It is a continuous distribution.
2. χ^2 values cannot be negative. As a result, curves describing the distributions are always to the right of the vertical axis at 0.
3. Its shape depends on a parameter ν (which is related to the sample size) called the **number of degrees of freedom** usually abbreviated df. There is one chi-square distribution curve for each number of degrees of freedom. (Chi-square distributions with 5, 10, 15, and 30 df are shown in Figure 7-14 on page 261.)

4. For a small number of degrees of freedom, the distribution is markedly skewed to the right (see Figure 7-14).
5. Skewness disappears rapidly as degrees of freedom increase. For $\nu >$ 30 the distribution is approximately normal (see Figure 7-14).

Since a chi-square distribution is continuous, the probabilities are given as areas under the curve. Appendix II, Table A-6 gives χ^2 values for specified areas in the right tail of the distribution, indicated by the corresponding column headings α. The degrees of freedom are given in the first column. It is customary to represent the χ^2 value for which there is an area α in the right tail of the distribution with ν df as $\chi^2_{\nu,\alpha}$ as shown in Figure 7-15.

As with t distribution, since there is one curve for each degree of freedom, for economy of space, χ^2 *values in our table are given only for a certain set of probabilities contained in the right tail.* This table is read exactly as the one for the t distribution. We reproduce, in the following table, a part of Table A-6 which includes $\nu = 10$ degrees of freedom:

Degrees of freedom ν	α, area in the right tail							
	0.995	0.99	0.975	0.95	0.05	0.025	0.01	0.005
	$\chi^2_{\nu,0.995}$	$\chi^2_{\nu,0.99}$	$\chi^2_{\nu,0.975}$	$\chi^2_{\nu,0.95}$	$\chi^2_{\nu,0.05}$	$\chi^2_{\nu,0.025}$	$\chi^2_{\nu,0.01}$	$\chi^2_{\nu,0.005}$
1	0.0^4393	0.0^3157	0.0^3982	0.0^2393	3.841	5.024	6.635	7.879
2	0.0100	0.0201	0.0506	0.103	5.991	7.378	9.210	10.597
3	0.0717	0.115	0.216	0.352	7.815	9.348	11.345	12.838
.								
.								
.								
9	1.735	2.088	2.700	3.325	16.919	19.023	21.666	23.589
10	2.156	2.558	3.247	3.940	18.307	20.483	23.209	25.188
11	2.603	3.053	3.816	4.575	19.675	21.920	24.725	26.757
.								
.								
.								

For instance, when $\nu = 10$, the area to the right of 2.156 is 0.995, to the right of 3.940 is 0.95, to the right of 20.483 is 0.025, and so on. The locations of these χ^2 values along the horizontal axis together with the corresponding probabilities are shown in Figure 7.16. It will be noticed that, unlike the table for the t distribution, the column headings α in the table include not only the probabilities 0.05, 0.025, 0.01, and 0.005, but also the probabilities 0.950, 0.975, 0.990, and 0.995. Because the t distribution is symmetric, there was no need to provide the t values for this latter set of probabilities.

The chi-square distribution finds applications in many statistical tests, as the reader will soon discover. It will form the mainstay of the tests carried out in Chapter 10, where we discuss goodness of fit and contingency tables.

EXAMPLE 1   Find

(a) $\chi^2_{4,0.05}$ (b) $\chi^2_{21,0.025}$ (c) $\chi^2_{17,0.95}$ (d) $\chi^2_{5,\alpha/2}$ if $\alpha = 0.1$
(e) $\chi^2_{16,1-\alpha/2}$ if $\alpha = 0.05$

SOLUTION   It will help us to draw a rough sketch of the graph of the appropriate chi-square distribution.

(a) To find $\chi^2_{4,0.05}$, we consider the χ^2 distribution with 4 degrees of freedom. Notice that the area to the right of the desired value is 0.05. From Table A-6 in the row for 4 degrees of freedom under column heading 0.05, we find, in the body of the table, $\chi^2_{4,0.05} = 9.488$. (see Figure 7-17(a) below).

(b) As in (a), we look in Table A-6 for the row corresponding to $\nu = 21$, under column heading 0.025. The value we find is $\chi^2_{21,0.025} = 35.479$ (see Figure 7-17(b) below).

(c) To find $\chi^2_{17,0.95}$ we are interested in a chi-square distribution with 17 degrees of freedom where the area to the right of the value is 0.95. From Table A-6

FIGURE 7-17(a)

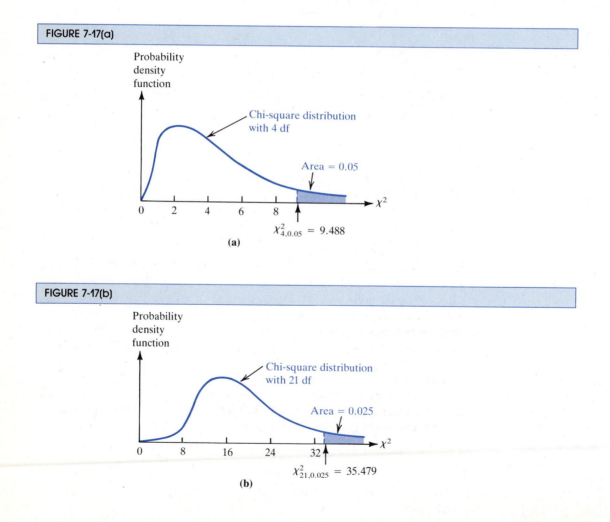

(a)

FIGURE 7-17(b)

(b)

FIGURE 7-17(c)

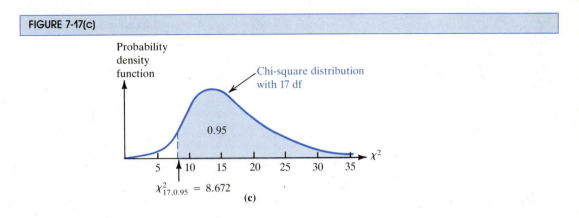

(c)

in the row corresponding to $\nu = 17$ under column heading 0.95, such a value is 8.672. Thus $\chi^2_{17,0.95} = 8.672$. (Notice that the value is in the left tail of the distribution in Figure 7-17(c).)

(d) Since $\alpha = 0.1$, $\alpha/2 = 0.05$. From Table A-6, we immediately find
$$\chi^2_{5,\alpha/2} = \chi^2_{5,0.05} = 11.07$$

(e) Here $\alpha = 0.05$. Hence
$$1 - \alpha/2 = 1 - 0.025 = 0.975$$

so that $\chi^2_{16,1-\alpha/2} = \chi^2_{16,0.975} = 6.908$ (in the row corresponding to $\nu = 16$, under column heading 0.975). ▄▄▄

Confidence Interval

Having considered the chi-square distribution at length, we return to our main problem of setting a confidence interval for σ^2. Since the statistic U has a chi-square distribution with $n - 1$ degrees of freedom, we can write

$$P\left(\chi^2_{n-1,1-\alpha/2} < \frac{(n-1)S^2}{\sigma^2} < \chi^2_{n-1,\alpha/2}\right) = 1 - \alpha$$

where $\chi^2_{n-1,1-\alpha/2}$ and $\chi^2_{n-1,\alpha/2}$ represent values from the chi-square distribution with $n - 1$ degrees of freedom such that they leave, respectively, areas of $1 - \alpha/2$ and $\alpha/2$ to their right, as shown in Figure 7-18 on page 266.

While setting a confidence interval on μ, we isolated it in the center between the inequalities. In the present case, we isolate σ^2. Remember that this is what the general procedure stipulates.

It can be shown through a series of algebraic steps that the probability statement given above leads to

$$P\left(\frac{(n-1)S^2}{\chi^2_{n-1,\alpha/2}} < \sigma^2 < \frac{(n-1)S^2}{\chi^2_{n-1,1-\alpha/2}}\right) = 1 - \alpha.$$

FIGURE 7-18

Chi-square values such that areas $1 - \alpha/2$ and $\alpha/2$ are to their right.

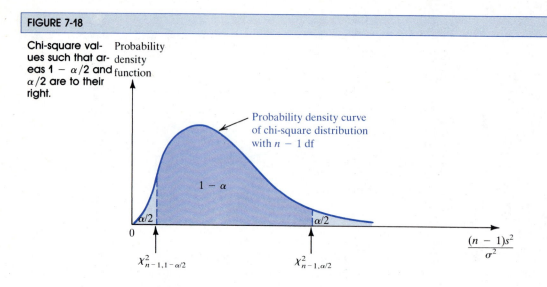

Hence we arrive at the following result:

CONFIDENCE INTERVAL FOR σ^2

A $(1 - \alpha)100$ percent confidence interval for σ^2 is given by

$$\frac{(n - 1)s^2}{\chi^2_{n-1,\alpha/2}} < \sigma^2 < \frac{(n - 1)s^2}{\chi^2_{n-1,1-\alpha/2}}$$

where s^2 is the sample variance based on n observations from a normal population.

Caution: Keep in mind that $\chi^2_{n-1,1-\alpha/2}$ is *not* equal to $-\chi^2_{n-1,\alpha/2}$. This is because the chi-square distribution is not symmetric.

EXAMPLE 2   The following data represent the amount of enamel coating (in ounces) needed to paint six plates:

.   8.1,   8.7,   7.6,   7.8,   8.5,   7.9.

Find

(a) a point estimate of σ^2

(b) a confidence interval for σ^2 with the following confidence coefficients:

(i) 95 percent   (ii) 98 percent.

(Assume that the amount of enamel coating per plate is normally distributed.)

SOLUTION   (a) It can be easily checked that $\sum_{i=1}^{6} (x_i - \bar{x})^2 = 0.9$. Hence a point estimate of σ^2 is

$$s^2 = \frac{\sum_{i=1}^{6} (x_i - \bar{x})^2}{6 - 1} = \frac{0.9}{5} = 0.18.$$

(b) Since $n = 6$, the chi-square distribution has 5 degrees of freedom.

(i) To find a 95 percent confidence interval we observe that $\alpha = 0.05$ so that $\alpha/2 = 0.025$ and $1 - \alpha/2 = 0.975$. Now $\chi^2_{5,0.025} = 12.832$ and $\chi^2_{5,0.975} = 0.831$. Since $(n - 1)s^2 = 0.9$, a 95 percent confidence interval is given by

$$\frac{0.9}{12.832} < \sigma^2 < \frac{0.9}{0.831}$$

that is,

$$0.07 < \sigma^2 < 1.08.$$

(ii) In this case $\alpha = 0.02$. We note that $\chi^2_{5,0.01} = 15.086$ and $\chi^2_{5,0.99} = 0.554$, and we get a 98 percent confidence interval given by

$$\frac{0.9}{15.086} < \sigma^2 < \frac{0.9}{0.554}$$

that is,

$$0.06 < \sigma^2 < 1.62. \quad ■$$

SECTION 7-4   EXERCISES

1. Find
   (a) $\chi^2_{7,0.05}$ (b) $\chi^2_{12,0.01}$
   (c) $\chi^2_{14,0.95}$ (d) $\chi^2_{7,0.025}$
   (e) $\chi^2_{22,0.99}$ (f) $\chi^2_{19,0.975}$.

2. In each case find the indicated quantity for the given level of confidence.
   (a) $\chi^2_{5,\alpha/2}$; 90 percent   (b) $\chi^2_{15,\alpha/2}$; 99 percent
   (c) $\chi^2_{7,\alpha/2}$; 98 percent   (d) $\chi^2_{10,\alpha/2}$; 90 percent.

3. Suppose you are constructing a 90 percent confidence interval. Determine $\alpha/2$, $1 - \alpha/2$ and find the indicated chi-square value.
   (a) $\chi^2_{16,1-\alpha/2}$ (b) $\chi^2_{16,\alpha/2}$ (c) $\chi^2_{8,1-\alpha/2}$ (d) $\chi^2_{7,1-\alpha/2}$

4. In each case, find the chi-square value for the indicated level of confidence.
   (a) $\chi^2_{14,\alpha/2}$; 95 percent   (b) $\chi^2_{14,1-\alpha/2}$; 95 percent
   (c) $\chi^2_{8,1-\alpha/2}$; 98 percent   (d) $\chi^2_{10,1-\alpha/2}$; 99 percent

In Exercises 5–10, assuming that the population is normally distributed, construct a confidence interval for the corresponding population variance σ^2 at the indicated confidence level if the given data summary is available:

5. $n = 10$, $s = 1.22$; 98 percent

6. $n = 14$, $s = 0.4$; 95 percent

7. $n = 23$, $s^2 = 12.8$; 90 percent

8. $n = 12$, $\sum_{i=1}^{12} (x_i - \bar{x})^2 = 16.38$; 95 percent

9. $n = 18$, $\sum_{i=1}^{18} (x_i - \bar{x})^2 = 4.597$; 90 percent

10. $n = 7$, $\sum_{i=1}^{7} (x_i - \bar{x})^2 = 797.645$; 99 percent

11. A grade-school teacher has obtained the following figures for the time (in minutes) that ten randomly picked students in her class took to complete a task:

    15,   10,   12,   8,   11,   10,   10,   13,   12,   9.

Find a 95 percent confidence interval for

(a) the true mean time taken to complete the task

(b) the true variance of the time.

Assume a normal distribution for the time needed to complete the task.

12. A machine filled six randomly picked jars with the following amounts of coffee (in ounces):

    15.7,   15.9,   16.3,   16.2,   15.7,   15.9.

Assuming that the amount of coffee in a jar is normally distributed,

(a) set a 95 percent confidence interval on the true mean weight of the amount filled in a jar.

(b) set a 95 percent confidence interval on σ^2, the population variance of the amount filled in a jar.

7-5 CONFIDENCE INTERVAL FOR THE POPULATION PROPORTION

If p represents the proportion of an attribute in the population, we have already seen that the sample proportion X/n provides a good estimate of p. We shall now give an interval estimate for p under the following assumptions. *We shall assume that the population proportion is not too close to zero or one, and that the sample size is large (at least 30).* Under these assumptions, we know that the sampling distribution of X/n can be approximated by a normal distribution that has mean p and standard deviation $\sqrt{p(1 - p)/n}$.

To construct a confidence interval for p, we can now adopt the same argument that was used in finding a confidence interval for μ and write

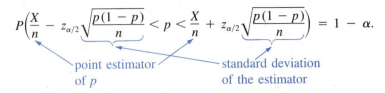

$$P\left(\frac{X}{n} - z_{\alpha/2}\sqrt{\frac{p(1-p)}{n}} < p < \frac{X}{n} + z_{\alpha/2}\sqrt{\frac{p(1-p)}{n}}\right) = 1 - \alpha.$$

point estimator of p standard deviation of the estimator

Hence one might give a $(1 - \alpha)100$ percent confidence interval as

$$\frac{x}{n} - z_{\alpha/2}\sqrt{\frac{p(1-p)}{n}} < p < \frac{x}{n} + z_{\alpha/2}\sqrt{\frac{p(1-p)}{n}}.$$

But you may notice that there is something incongruous about this formula. We are giving a confidence interval for estimating p, but that very unknown quantity is used in computing the endpoints of the interval. Luckily it turns out that we can substitute x/n for p in $\sqrt{p(1-p)}/n$. If the sample size is large, this will not bring in much discrepancy. Thus we get the following result:

LARGE SAMPLE CONFIDENCE INTERVAL FOR p

An approximate $(1 - \alpha)100$ percent confidence interval for the population proportion p is given by

$$\frac{x}{n} - z_{\alpha/2}\sqrt{\frac{\frac{x}{n}\left(1 - \frac{x}{n}\right)}{n}} < p < \frac{x}{n} + z_{\alpha/2}\sqrt{\frac{\frac{x}{n}\left(1 - \frac{x}{n}\right)}{n}}$$

if the sample size is large (usually $n > 30$). Here x/n is the proportion of an attribute in a sample of n independent observations.

EXAMPLE 1   In a sample of 400 people who were questioned regarding their participation in sports, 160 said that they did participate. Set an approximate 98 percent confidence interval for p, the proportion of people in the population who participate in sports.

SOLUTION   We are given that $x = 160$, $n = 400$, and $\alpha = 0.02$. Hence $z_{\alpha/2} = z_{0.01} = 2.33$ and since $x/n = 0.4$,

$$\sqrt{\frac{\frac{x}{n}\left(1 - \frac{x}{n}\right)}{n}} = \sqrt{\frac{(0.4)(0.6)}{400}} = 0.0245.$$

As a result, an approximate 98 percent confidence interval for p is given by

$$0.4 - 2.33(0.0245) < p < 0.4 + 2.33(0.0245)$$

that is,

$$0.343 < p < 0.457. \quad \blacksquare$$

How Large a Sample for Estimating p?

The question of how large a sample to pick in order to attain a certain degree of precision was considered earlier in Section 7-2 in the context of constructing a confidence interval for μ when σ is known. A question of a similar nature also comes up when estimating a proportion p. For example, a pollster might wish to know how many people he should interview so that he can be 95 percent confident that his estimate will not be in error of the true proportion p by more than 0.01.

Recall from Chapter 6 that we can regard a sample proportion X/n as a mean of a sample of size n drawn from an infinite population consisting of 1's and 0's with the proportion of 1's equal to p and the proportion of 0's equal to $1 - p$. As we saw, the standard deviation σ of such a population is $\sqrt{p(1 - p)}$. Hence, substituting in the formula for n on page 250, we get

$$n = \left(\frac{z_{\alpha/2}\sigma}{e}\right)^2 = \left(\frac{z_{\alpha/2}\sqrt{p(1 - p)}}{e}\right)^2 = z_{\alpha/2}^2\,\frac{p(1 - p)}{e^2}.$$

How can we use this formula for finding n if we do not know p in the first place? There are two ways to get around this impasse. We can either use a preliminary estimate of p if it is available, or, just to be on the safe side, we can give the maximum n that will "do the job." The latter estimate is a conservative estimate and is obtained when $p(1 - p)$ is maximum. Now $p(1 - p)$ is maximum, equal to $1/4$, when $p = 1/2$. (We shall accept this fact.) Therefore, a conservative estimate of n is

$$n = z_{\alpha/2}^2\,\frac{1/4}{e^2} = \frac{z_{\alpha/2}^2}{4e^2}.$$

In conclusion we obtain the following result:

SAMPLE SIZE FOR ESTIMATING p

A conservative sample size needed so as to be $(1 - \alpha)100$ percent confident that the estimate x/n is not in error of p by more than a quantity e is given by

$$n = \frac{z_{\alpha/2}^2}{4e^2}.$$

Thus, in the case of the pollster who wants to be 95 percent confident that his estimate will not be in error of p by more than 0.01, a conservative value of n will be

$$n = \frac{(1.96)^2}{4(0.01)^2} = 9604.$$

A much smaller sample might give an estimate that is in error by less than 0.01 with 95 percent confidence. But this will depend on the actual value of p. In being conservative, we have given a value of n that will do the job no matter what p is.

SECTION 7-5   EXERCISES

In Exercises 1–4 construct a confidence interval for the corresponding population proportion p at the indicated level of confidence if the following information is given. (State whether the interval is exact or approximate.)

1. $n = 60$, $x = 18$; 95 percent

2. $n = 78$, $x = 30$; 90 percent

3. $n = 100$, $x = 62$; 88 percent

4. $n = 80$, $x = 52$; 95 percent

5. Of 100 people who were given a vaccine, 80 developed immunity to a disease. Obtain a 98 percent confidence interval on the true proportion of people developing immunity.

6. In a sample of 100 cathode tubes inspected, 12 were found to be defective. Set a 95 percent confidence interval on the true proportion of defectives in the entire production.

7. Of the 200 individuals interviewed, 80 said that they were concerned about fluorocarbon emissions in the atmosphere. Obtain a 99 percent confidence interval for the true proportion of individuals who are concerned.

8. A new vaccine is to be tested on the market. Find how large a sample should be drawn if we want to be 95 percent confident that the estimate will not be in error of the true proportion of success by more than (a) 0.1, (b) 0.02, (c) 0.001. Comment on your findings.

9. A survey is to be conducted to estimate the proportion of citizens who favor imposing trade restraints on Japan. Determine how large the sample should be so that, with 98 percent confidence, the sample proportion will not differ from the true proportion by more than 0.02.

10. Should a certain nuclear power plant be reopened? A random sample of 500 adults gave the following results: 240 in favor, 180 opposed, 80 undecided. Using a 90 percent confidence interval, estimate the true proportion of individuals in the population

    (a) who favor reopening the plant

    (b) who are undecided.

11. A student obtained the following 95 percent confidence interval for the population proportion p: $1.38 < p < 2.33$. Comment on its acceptability.

KEY TERMS AND EXPRESSIONS

estimator
point estimate
unbiasedness
efficiency
consistency
sufficiency
interval estimation
confidence interval

random interval
confidence coefficient
level of confidence
lower (upper) confidence limits
length of a confidence interval
Student's t distribution
chi-square distribution

KEY FORMULAS

Summary of Confidence Intervals (Single Population)

Nature of the population	Parameter on which confidence interval is set	Procedure	Limits of confidence interval with confidence coefficient $1 - \alpha$
Quantitative data; variance σ^2 is *known;* population normal.	μ, the population mean	Draw a sample of size n and compute the value of $\bar{x}$, the estimate of μ.	$\bar{x} \pm z_{\alpha/2} \dfrac{\sigma}{\sqrt{n}}$
Quantitative data; variance σ^2 is *not* known; population normal	μ, the population mean	Draw a sample of size n and compute $\bar{x}$ *and* $$s = \sqrt{\frac{1}{n-1} \sum_{i=1}^{n} (x_i - \bar{x})^2}$$	$\bar{x} \pm t_{n-1,\alpha/2} \dfrac{s}{\sqrt{n}}$ $t_{n-1,\alpha/2}$ is the value obtained from the t distribution with $n - 1$ degrees of freedom.
Quantitative data; variance σ^2 is *not* known; population *not* necessarily normal; *sample size is large*	μ, the population mean	Draw a sample of size n and compute $\bar{x}$ and s.	$\bar{x} \pm z_{\alpha/2} \dfrac{s}{\sqrt{n}}$ Confidence interval is approximate.
Qualitative data; binomial case	p, the probability of success (the population proportion)	Draw a sample of size n and note x, the number of successes; obtain x/n, the estimate of p. Here n is assumed large.	$\dfrac{x}{n} \pm z_{\alpha/2} \sqrt{\dfrac{\dfrac{x}{n}\left(1 - \dfrac{x}{n}\right)}{n}}$ Confidence interval is based on the central limit theorem, hence is approximate.
Quantitative data; population normal	σ^2, the population variance	Draw a sample of size n and compute s^2	$\left(\dfrac{(n-1)s^2}{\chi^2_{n-1,\alpha/2}}, \dfrac{(n-1)s^2}{\chi^2_{n-1,1-\alpha/2}} \right)$

CHAPTER 7 TEST

In Exercises 1–4, obtain point estimates of the (a) population mean, (b) variance, and (c) standard deviation.

1. $n = 15$, $\sum_{i=1}^{n} x_i = 24.3$, $\sum_{i=1}^{n} (x_i - \bar{x})^2 = 0.824$.

2. $n = 6$, $\sum_{i=1}^{n} x_i = -16$, $s^2 = 12.67$

3. $x_1 = 3.2$, $x_2 = 3.9$, $x_3 = 2.8$, $x_4 = 4.3$, $x_5 = 3.4$

4. $n = 7$, $\sum_{i=1}^{n} x_i = 26.3$, $\sum_{i=1}^{n} x_i^2 = 128.03$

5. In a survey conducted to estimate the percentage of population in the 20–40 years bracket engaged in some kind of physical exercise, it was found that, of the 150 randomly picked individuals, 86 did exercise. Estimate the percent in the population.

6. Eight randomly picked shoppers at a shopping mall were asked the question, "Do you fear that communists are getting a stronghold in Central America?" The following response was obtained: yes, no, no, no, yes, no, yes, no. Based on this response, estimate the proportion in the population who share the fear.

In Exercises 7–11, find a confidence interval for the indicated parameter at the specified level of confidence. Mention the assumptions that would be considered necessary in each case.

7. $\bar{x} = 13.8$, $n = 17$, $s = 6.2$; parameter μ; 98 percent

8. $n = 14$, $s = 1.3$; parameter σ^2; 95 percent

9. $\bar{x} = -3.4$, $n = 65$, $s = 7.6$; parameter μ; 90 percent

10. $n = 80$, $x = 27$ successes; parameter p; 88 percent

11. $\bar{x} = 7.6$, $n = 16$, $\sigma = 2.8$; parameter μ; 95 percent

12. In each case find the length of the given confidence interval.

(a) $6.85 < \mu < 17.92$ (b) $1.8 < \sigma^2 < 3.84$

(c) $-3.13 < \mu < 6.84$ (d) $0.36 < p < 0.84$

13. A random sample of 16 servings of canned pineapple has a mean carbohydrate content of 49 grams. If it can be assumed that the population is normally distributed and that σ^2, the variance of the carbohydrate content of all the servings from which the sample was taken, is 4 (grams)^2, find a 98 percent confidence interval for the true mean carbohydrate content of a serving.

14. In a survey in which 100 randomly picked middle-class families were interviewed, it was found that their mean medical expenses during a year were $760 with a standard deviation s of $120. Find an 88 percent confidence interval for the true mean of the medical expenses of a middle-class family.

15. An archaeologist found that the mean cranial width of 17 skulls was 5.3 inches with s equal to 0.5 inch. Using a 90 percent confidence level, set a confidence interval on

(a) the true mean cranial width

(b) the true variance of cranial width.

You may assume that cranial width has a normal distribution.

16. Do you think the president is doing a satisfactory job? The following response was obtained among 40 randomly picked citizens (Y = yes, N = no, U = undecided):

    N U U N Y Y N Y U Y N Y Y N

    U Y Y Y U Y N Y Y U N Y Y Y

    U Y Y N Y Y U N N N Y U

    Estimate the true proportion of citizens who think the president is doing a satisfactory job.

17. It is important for timberland owners and managers to estimate the mean population site index with a high degree of precision since site index represents the only basis for the prediction of the future yields from the trees. The following figures were reported for site index from 10 one-fifth acre permanent plots in a relatively homogeneous tract of young-growth redwood trees.*

    166,   160,   150,   156,   170,

    176,   166,   143,   165,   175

    Estimate the mean population site index using a 98 percent confidence interval. Assume a normal distribution for site index.

18. It is suspected that a substance called actin is linked to various movement phenomena of nonmuscle cells. In a laboratory experiment when eight fertilized eggs were incubated for 14 days the following amounts (mg) of total brain actin were obtained:

    1.2,   1.4,   1.5,   1.2,   1.4,   1.7,   1.5,   1.7

    Assuming that brain-actin amount after 14 days of incubation is normally distributed find a 95 percent confidence interval for

    (a) the true mean brain-actin amount

    (b) the true variance of the brain-actin amount.

---

*James Lundquist and Marshal Palley, Bulletin 796, California Agricultural Station.

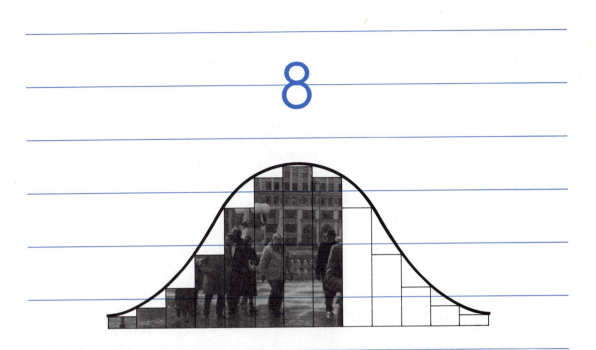

8

ESTIMATION—TWO POPULATIONS

INTRODUCTION

In a statistical investigation, it is often necessary to study two populations and compare their parameters. This might be done by comparing means of two populations, or by comparing proportions of objects having a given attribute in two populations, or by comparing variances of two populations. For example, we might wish to compare the mean diameter of bolts produced by one machine with the mean of those produced by another machine, or we might want to compare the proportion of smokers among men with that among women, or we might want to compare the variance of altimeter readings on one instrument with the variance of readings on another instrument.

There are two common ways to determine how near two quantities are to each other. One way is to take the difference: if the difference is nearly zero, then the quantities are very near each other; if the difference is very far removed from zero, then the two quantities are far apart. The other way is to take the ratio of the two quantities: if the ratio is nearly equal to 1, then the two quantities are near each other; departure of the ratio from 1 will provide a measure of how far apart the two quantities are. In this chapter we shall compare means of two populations and also proportions in two populations. For considerations in theoretical statistics, we use the difference between quantities as a measure. It is for this reason that we will construct confidence intervals for the **difference of means** and **difference of proportions** in two populations.

Whether we are comparing population means or population proportions, our basic approach will be to take two random samples, one from each population, and then compare corresponding sample analogues. We will be making a very important assumption throughout in this context. We shall assume that the two samples are independent; that is, the outcomes in one sample have no bearing on the outcomes in the other in that they do not influence each other.

8-1 CONFIDENCE INTERVAL FOR THE DIFFERENCE OF POPULATION MEANS WHEN POPULATION VARIANCES ARE KNOWN

Let us consider two populations, which we might designate as Population 1 and Population 2, both of which will be assumed to be normally distributed. Suppose the means of these two populations are, respectively, μ_1 and μ_2, and their variances are σ_1^2 and σ_2^2. Our objective is to set a confidence interval for the difference $\mu_1 - \mu_2$. For this purpose, we pick two random, independent samples, one from each population. The sizes of these samples need not be the same. Let us denote the size of the sample from Population 1 by m, and let us denote its mean by $\bar{x}$. Let n be the

FIGURE 8-1

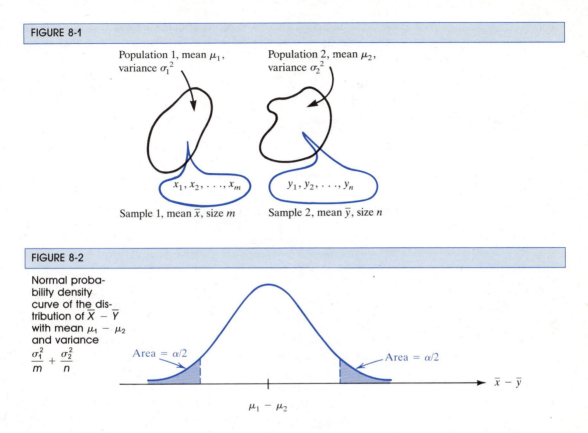

Population 1, mean μ_1, variance σ_1^2

Population 2, mean μ_2, variance σ_2^2

$x_1, x_2, \ldots, x_m$ $y_1, y_2, \ldots, y_n$

Sample 1, mean $\bar{x}$, size m Sample 2, mean $\bar{y}$, size n

FIGURE 8-2

Normal probability density curve of the distribution of $\overline{X} - \overline{Y}$ with mean $\mu_1 - \mu_2$ and variance
$$\frac{\sigma_1^2}{m} + \frac{\sigma_2^2}{n}$$

Area = $\alpha/2$ Area = $\alpha/2$

$\bar{x} - \bar{y}$

$\mu_1 - \mu_2$

sample size and $\bar{y}$ the mean of the sample from Population 2. (Since alphabetically the letter m precedes n, and x precedes y, we associate m and x with Population 1, and n and y with Population 2.) Schematically we can show the situation as in Figure 8-1.

We know that the point estimate of μ_1 is $\bar{x}$ and that the point estimate of μ_2 is $\bar{y}$. Intuitively, then, we would feel that a point estimate of $\mu_1 - \mu_2$ would be $\bar{x} - \bar{y}$. This turns out to be true. (If we were interested in the point estimate of $\mu_2 - \mu_1$, then we would take $\bar{y} - \bar{x}$.)

To construct a confidence interval for $\mu_1 - \mu_2$, we have to know the distribution of $\overline{X} - \overline{Y}$, the statistic that is used to estimate $\mu_1 - \mu_2$.

If two populations are normally distributed with respective means μ_1 and μ_2 and variances σ_1^2 and σ_2^2, then the sampling distribution of $\overline{X} - \overline{Y}$ is normal with mean $\mu_1 - \mu_2$ and variance $\dfrac{\sigma_1^2}{m} + \dfrac{\sigma_2^2}{n}$ when two random, independent samples are drawn, one from each population.

Let us accept this fact without any elaboration. The nature of the distribution is indicated in Figure 8-2.

Now recall how, in Section 7-2, we set a confidence interval for μ using the statistic $\overline{X}$ once we established that it had a certain distribution. Using that as the guide, we obtain the following result:

$$P\left(\underbrace{\overline{X} - \overline{Y}} - z_{\alpha/2}\underbrace{\sqrt{\frac{\sigma_1^2}{m} + \frac{\sigma_2^2}{n}}} < \mu_1 - \mu_2 < \overline{X} - \overline{Y} + z_{\alpha/2}\sqrt{\frac{\sigma_1^2}{m} + \frac{\sigma_2^2}{n}}\right) = 1 - \alpha.$$

point estimator of $\mu_1 - \mu_2$ standard deviation of the point estimator

In summary we have the following result:

CONFIDENCE INTERVAL FOR $\mu_1 - \mu_2$; VARIANCES KNOWN

If the populations are normally distributed and the variances are known, then a $(1 - \alpha)100$ percent confidence interval for $\mu_1 - \mu_2$ is given by

$$(\overline{x} - \overline{y}) - z_{\alpha/2}\sqrt{\frac{\sigma_1^2}{m} + \frac{\sigma_2^2}{n}} < \mu_1 - \mu_2$$

$$< (\overline{x} - \overline{y}) + z_{\alpha/2}\sqrt{\frac{\sigma_1^2}{m} + \frac{\sigma_2^2}{n}}$$

where $\overline{x}$ and $\overline{y}$ are the sample means based, respectively, on samples of sizes m and n, and where σ_1^2 and σ_2^2 are the population variances.

In particular, if the two populations have the same variance σ^2, that is, $\sigma_1^2 = \sigma_2^2 = \sigma^2$, then a $(1 - \alpha)$ 100 percent confidence interval for $\mu_1 - \mu_2$ is given by

$$(\overline{x} - \overline{y}) - z_{\alpha/2}\sigma\sqrt{\frac{1}{m} + \frac{1}{n}} < \mu_1 - \mu_2 < (\overline{x} - \overline{y}) + z_{\alpha/2}\sigma\sqrt{\frac{1}{m} + \frac{1}{n}}.$$

EXAMPLE 1

Tests were carried out to compare strengths of two types of yarn. When 15 random tests were carried out with Yarn A, it was found that the mean strength was 140 pounds/square inch. Also, 10 random tests carried out with Yarn B yielded a mean of 145 pounds/square inch. Assume that the strength of yarn has a normal distribution for both yarns and that the variance σ_1^2 for Yarn A is 25 (pounds/square inch)^2 while the variance σ_2^2 for Yarn B is 30 (pounds/square inch)^2. Construct an 80 percent confidence interval for the difference $\mu_1 - \mu_2$, where μ_1 and μ_2 represent, respectively, the true means for Yarn A and Yarn B.

SOLUTION

We are given that $\overline{x} = 140$, $\sigma_1^2 = 25$, $m = 15$, $\overline{y} = 145$, $\sigma_2^2 = 30$, $n = 10$, and $\alpha = 0.2$. Therefore, $z_{\alpha/2} = z_{0.1} = 1.28$. Consequently, an 80 percent confidence interval for $\mu_1 - \mu_2$ is given by

$$(140 - 145) - 1.28\sqrt{\frac{25}{15} + \frac{30}{10}} < \mu_1 - \mu_2 < (140 - 145) + 1.28\sqrt{\frac{25}{15} + \frac{30}{10}}$$

$$-5 - 1.28(2.16) < \mu_1 - \mu_2 < -5 + 1.28(2.16)$$

$$-7.765 < \mu_1 - \mu_2 < -2.235. \quad \blacksquare$$

We have assumed that the populations are normally distributed. This assumption is not vital if the sample sizes are large (as a rule of thumb, if m and n are both greater than 30). Actually, in this case, even the population variances need not be known. We can use s_1^2 and s_2^2, the sample variances, to provide the respective population variances σ_1^2 and σ_2^2. We have the following result:

LARGE SAMPLES CONFIDENCE INTERVAL FOR $\mu_1 - \mu_2$

If the sample sizes m and n are both large and s_1^2 and s_2^2 represent the sample variances, then an approximate $(1 - \alpha)100$ percent confidence interval for $\mu_1 - \mu_2$ is

$$(\bar{x} - \bar{y}) - z_{\alpha/2}\sqrt{\frac{s_1^2}{m} + \frac{s_2^2}{n}} < \mu_1 - \mu_2 < (\bar{x} - \bar{y}) + z_{\alpha/2}\sqrt{\frac{s_1^2}{m} + \frac{s_2^2}{n}}$$

where $\bar{x}$ and $\bar{y}$ are the sample means.

EXAMPLE 2   At the end of a crash dieting program administered to 50 men and 40 women, the following information was obtained about the loss of weight (in pounds):

	Men	Women
Mean loss of weight	$\bar{x} = 16.5$	$\bar{y} = 13.2$
s^2, *sample variance*	$s_1^2 = 25$	$s_2^2 = 16$

Suppose μ_1 and μ_2 represent, respectively, the mean losses for men and women. Obtain an approximate 85 percent confidence interval for the difference in the mean losses, $\mu_1 - \mu_2$.

SOLUTION   Since the sample sizes are large, we can treat s_1^2 and s_2^2 as though they are σ_1^2 and σ_2^2.

Since $\alpha = 0.15$, $z_{\alpha/2} = z_{0.075} = 1.44$. Therefore, an approximate 85 percent confidence interval for $\mu_1 - \mu_2$ is given by

$$(16.5 - 13.2) - 1.44\sqrt{\frac{25}{50} + \frac{16}{40}} < \mu_1 - \mu_2 < (16.5 - 13.2) + 1.44\sqrt{\frac{25}{50} + \frac{16}{40}}$$

$$3.3 - 1.44(0.949) < \mu_1 - \mu_2 < 3.3 + 1.44(0.949)$$

$$1.933 < \mu_1 - \mu_2 < 4.667. \quad \blacksquare$$

SECTION 8-1   EXERCISES

1. Find a point estimate of $\mu_1 - \mu_2$ if
   (a) $\bar{x} = 13.75$ and $\bar{y} = 10.7$
   (b) $\sum_{i=1}^{10} x_i = 105$, $\sum_{i=1}^{15} y_i = 150$. (Of course, $m = 10$ and $n = 15$.)

In Exercises 2–4, find a point estimate of $\mu_2 - \mu_1$.

2. $x_1 = 2.2$, $x_2 = 3.2$, $x_3 = 4.1$, $x_4 = 1.8$, $x_5 = 2.4$; and
   $y_1 = 3.5$, $y_2 = 4.1$, $y_3 = 3.0$, $y_4 = 3.7$

3. $\sum_{i=1}^{6} x_i = 36.18$ and $\sum_{i=1}^{8} y_i = 42.96$

4. $\bar{x} = 1.85$ and $\bar{y} = -0.63$.

Suppose there are two normally distributed populations. Assuming that two independent random samples are picked from the two populations, determine a confidence interval for estimating $\mu_1 - \mu_2$ in Exercises 5–7 at the indicated level of confidence.

5. $\bar{x} = 8.0$, $m = 20$, $\sigma_1^2 = 10$; and $\bar{y} = 7.2$, $n = 24$, $\sigma_2^2 = 12$; 95 percent

6. $\bar{x} = 1.85$, $m = 14$, $\sigma_1^2 = 20$; and $\bar{y} = -0.63$, $n = 9$, $\sigma_2^2 = 14$; 90 percent

7. $x_1 = 2.2$, $x_2 = 3.2$, $x_3 = 4.1$, $x_4 = 1.8$, $x_5 = 2.5$ with $\sigma_1^2 = 0.78$; and
   $y_1 = 3.5$, $y_2 = 4.1$, $y_3 = 3.0$, $y_4 = 3.7$ with $\sigma_2^2 = 0.28$; 98 percent.

8. Suppose there are two normally distributed populations. One population has variance $\sigma_1^2 = 10$, and when a sample of size 20 was picked from it, the mean was $\bar{x} = 8$. The second population has variance $\sigma_2^2 = 12$, and a sample of 24 yielded a mean $\bar{y} = 7.2$. Assuming that the samples are independent, determine a 95 percent confidence interval for estimating $\mu_1 - \mu_2$.

9. The mean height of ten adult Japanese women was found to be $\bar{x} = 58.8$ inches, and that of 15 adult Chinese women was found to be 57.2 inches. The heights of adult women for the two nationalities may be assumed to be normally distributed with a standard deviation σ_1 of 2.8 inches for Japanese women and σ_2 of 3.1 inches for Chinese women. Let μ_J represent the (population) mean height for Japanese women and μ_C for Chinese women.

   (a) Find a 92 percent confidence interval for $\mu_J - \mu_C$.

   (b) Find a 98 percent confidence interval for $\mu_C - \mu_J$.

10. An engineering student measured the daily dissolved oxygen concentration (in milligrams/liter) of two streams. The mean daily concentration of Stream 1 when observed during 20 days was found to be $\bar{x} = 4.58$ mg/l and that of Stream 2 when observed during 14 days was found to be $\bar{y} = 3.68$ mg/l. Suppose it is reasonable to assume that daily concentration is normally distributed for each stream and that the concentration of one stream does not influence that of the other. Also, it may be assumed that the standard deviation of the concentration for Stream 1 is $\sigma_1 = 0.8$ mg/l and that for Stream 2 is $\sigma_2 = 1.1$ mg/l. Determine a 98 percent confidence interval for $\mu_1 - \mu_2$ where μ_1 is the (population) mean concentration for Stream 1 and μ_2 that for Stream 2.

Suppose two independent random samples are picked from two populations. In Exercises 11–13, determine a confidence interval for estimating $\mu_1 - \mu_2$ at the indicated level of confidence.

[*Hint:* Notice m and n are large in each case.]

11. $\bar{x} = 6.2$, $m = 120$, $s_1^2 = 1.26$; and $\bar{y} = 3.8$, $n = 80$, $s_2^2 = 2.32$; 95 percent

12. $\bar{x} = 0.19$, $m = 60$, $s_1^2 = 1.69$; and $\bar{y} = -0.32$, $n = 68$, $s_2^2 = 2.56$; 90 percent

13. $\bar{x} = 123.6$, $m = 48$, $s_1^2 = 8.82$; and $\bar{y} = 98.7$, $n = 64$, $s_2^2 = 5.62$; 98 percent

14. When 100 king-size cigarettes of Brand A and 152 king-size cigarettes of Brand B were tested for their nicotine content (in milligrams), the following information was obtained:

	Sample mean	Sample standard deviation
Brand A	1.2	0.2
Brand B	1.4	0.25

Estimate the difference in the true means using a 90 percent confidence interval.

15. In a test to measure the performance of two comparable models of cars, Cheetah and Pioneer, 128 cars of each model were driven on the same terrain with ten gallons of gasoline in each car. The mean number of miles for Cheetah was 340 miles with s_1 of 10 miles; the mean number of miles for Pioneer was 355 miles with s_2 of 16 miles. Set a 98 percent confidence interval for the difference between the true mean numbers of miles for the two models of cars.

16. A rancher tried Feed A on 256 cattle and Feed B on 144 cattle. The mean weight of cattle given Feed A was found to be 1350 pounds with s_1 of 180 pounds. On the other hand, the mean weight of the cattle given Feed B was found to be 1430 pounds with s_2 of 210 pounds. Set a 95 percent confidence interval on the difference in the true mean weights of cattle given the two feeds.

17. A company has two branches, one on the East Coast and the other on the West Coast. The mean daily sales of the company on the East Coast, when observed on 100 days, were found to be $32,000 with s_1 of $2400. For the company on the West Coast, when observed on 200 days, the mean daily sales were $36,000 with s_2 of $2700. Obtain a 95 percent confidence interval for the difference in true mean daily sales of the two branches.

8-2 CONFIDENCE INTERVAL FOR THE DIFFERENCE OF POPULATION MEANS WHEN THE VARIANCES ARE UNKNOWN BUT ARE ASSUMED EQUAL

The method we develop will be found particularly useful when the *sample sizes are small*. We assume that both populations are normally distributed. An additional assumption will be that their variances σ_1^2 and σ_2^2, though unknown, are the same and each equals σ^2. Our immediate task is to find an estimate of this common variance σ^2.

To find s_1^2 we divide $\sum\limits_{i=1}^{m} (x_i - \bar{x})^2$ by $m - 1$, and to find s_2^2 we divide $\sum\limits_{i=1}^{n} (y_i - \bar{y})^2$ by $n - 1$. Since we are assuming that the variances of the two

populations are equal, it seems reasonable that we pool together whatever information the two samples can contribute toward estimating σ^2. Such a **pooled estimate
of variance** is obtained by pooling together the squares of deviations for the two
samples, that is, by finding $\sum\limits_{i=1}^{m} (x_i - \bar{x})^2 + \sum\limits_{i=1}^{n} (y_i - \bar{y})^2$ and then dividing the sum
by $(m - 1) + (n - 1)$, or $m + n - 2$.

Denoting this pooled estimate by s_p^2, we get

$$s_p^2 = \frac{\sum\limits_{i=1}^{m} (x_i - \bar{x})^2 + \sum\limits_{i=1}^{n} (y_i - \bar{y})^2}{m + n - 2}.$$

Since

$$\sum_{i=1}^{m} (x_i - \bar{x})^2 = (m - 1)s_1^2 \text{ and } \sum_{i=1}^{n} (y_i - \bar{y})^2 = (n - 1)s_2^2$$

the pooled estimate may also be written as

$$s_p^2 = \frac{(m - 1)s_1^2 + (n - 1)s_2^2}{m + n - 2}.$$

Now recall that if $\sigma_1^2 = \sigma_2^2 = \sigma^2$, the confidence interval for the difference
in population means becomes

$$(\bar{x} - \bar{y}) - z_{\alpha/2}\sigma\sqrt{\frac{1}{m} + \frac{1}{n}} < \mu_1 - \mu_2 < (\bar{x} - \bar{y}) + z_{\alpha/2}\sigma\sqrt{\frac{1}{m} + \frac{1}{n}}.$$

In the context of our present discussion, σ is to be replaced by its pooled
estimate s_p. But as soon as we do this, we cannot use z values from the normal table.
Rather we should use the values from the t distribution with $m + n - 2$ degrees of
freedom. The rationale for using $m + n - 2$ degrees of freedom is that in obtaining
s_p^2, we divided the pooled squares of deviations by $m + n - 2$. In conclusion, we
have the following result:

CONFIDENCE INTERVAL FOR $\mu_1 - \mu_2$; EQUAL UNKNOWN VARIANCES

The upper and lower limits of a $(1 - \alpha)100$ percent confidence interval
for $\mu_1 - \mu_2$ when two populations are normally distributed and have the
same (but unknown) variance are given by

$$(\bar{x} - \bar{y}) \pm t_{m+n-2,\alpha/2}s_p\sqrt{\frac{1}{m} + \frac{1}{n}}$$

where $\bar{x}$ and $\bar{y}$ are the sample means based, respectively, on m and n
observations and s_p is the pooled estimate of σ.

[*Note:* We shall omit the case involving *small* samples picked from two normal
populations having *unknown, unequal variances,* which involves a complicated
formula for computing the degrees of freedom.]

EXAMPLE 1 Two diets, Diet X and Diet Y, were fed to two groups of randomly picked dogs of a certain breed. Diet X was fed to 10 dogs. Their mean weight $\bar{x}$ was 50 pounds, and the sample standard deviation s_1 was 6 pounds. Diet Y was fed to 14 dogs. Their mean weight $\bar{y}$ was 56 pounds, and the sample standard deviation s_2 was 8 pounds. Find a confidence interval for the difference of the means with a 90 percent confidence level. (Assume that the weights of the dogs given Diet X and the weights of dogs given Diet Y are normally distributed with the same variance.)

SOLUTION Since $s_1 = 6$, $s_2 = 8$, $m = 10$, and $n = 14$,

$$s_p^2 = \frac{(m-1)s_1^2 + (n-1)s_2^2}{m+n-2} = \frac{9(6)^2 + 13(8)^2}{10 + 14 - 2} = \frac{1156}{22} = 52.55$$

and therefore $s_p = 7.249$. Also, there are $m + n - 2 = 22$ degrees of freedom. Since $\alpha = 0.1$, $t_{m+n-2,\alpha/2} = t_{22,0.05} = 1.717$. Hence a 90 percent confidence interval is

$$(\bar{x} - \bar{y}) - 1.717 s_p \sqrt{\frac{1}{m} + \frac{1}{n}} < \mu_1 - \mu_2 < (\bar{x} - \bar{y}) + 1.717 s_p \sqrt{\frac{1}{m} + \frac{1}{n}}.$$

Since $\bar{x} = 50$ and $\bar{y} = 56$, the upper and lower limits of the confidence interval are given by

$$(50 - 56) \pm 1.717(7.249)\sqrt{\frac{1}{10} + \frac{1}{14}}$$

which, upon simplification, gives

$$-11.15 < \mu_1 - \mu_2 < -0.85. \quad \blacksquare$$

EXAMPLE 2 A phosphate fertilizer was applied to 5 plots and a nitrogen fertilizer to 6 plots. The yield of grain on each plot was recorded (in bushels) as presented in Table 8-1.

TABLE 8-1	
Yield of grain on each plot.	
Phosphate	*Nitrogen*
40	50
49	41
38	53
48	39
40	40
	47

Construct a 98 percent confidence interval for $\mu_1 - \mu_2$. (μ_1 represents the mean for all the conceivable plots to which the phosphate fertilizer might be applied, with a similar interpretation for μ_2.) Suppose it may be assumed that the populations are normally distributed and the variances are equal.

SOLUTION   In Table 8-2, we find $\sum_{i=1}^{5} (x_i - \bar{x})^2$ and $\sum_{i=1}^{6} (y_i - \bar{y})^2$, which will be used to obtain a pooled estimate of the common variance.

TABLE 8-2

Computations of $\sum_{i=1}^{5} (x_i - \bar{x})^2$ and $\sum_{i=1}^{6} (y_i - \bar{y})^2$.

x	$(x - \bar{x})^2$	y	$(y - \bar{y})^2$
40	9	50	25
49	36	41	16
38	25	53	64
48	25	39	36
40	9	40	25
		47	4
$\bar{x} = \dfrac{215}{5}$	$\sum_{i=1}^{5} (x_i - \bar{x})^2$	$\bar{y} = \dfrac{270}{6}$	$\sum_{i=1}^{6} (y_i - \bar{y})^2$
$= 43$	$= 104$	$= 45$	$= 170$

Since $m = 5$ and $n = 6$, the pooled estimate of σ^2 is

$$s_p^2 = \frac{104 + 170}{5 + 6 - 2} = \frac{274}{9} = 30.44$$

Therefore, $s_p = 5.517$.

Now $\alpha = 0.02$ and the t distribution has $5 + 6 - 2 = 9$ degrees of freedom. Therefore, $t_{m+n-2,\alpha/2} = t_{9,0.01} = 2.821$.

Hence a 98 percent confidence interval is given by

$$\bar{x} - \bar{y} - (2.821)s_p\sqrt{\frac{1}{m} + \frac{1}{n}} < \mu_1 - \mu_2 < \bar{x} - \bar{y} + (2.821)s_p\sqrt{\frac{1}{m} + \frac{1}{n}}$$

Substitution produces the upper and lower limits of the confidence interval

$$(43 - 45) \pm 2.821(5.517)\sqrt{\frac{1}{5} + \frac{1}{6}}.$$

After simplifying, we get

$$-11.424 < \mu_1 - \mu_2 < 7.424. \quad \blacksquare$$

SECTION 8-2   EXERCISES

If two independent random samples are picked from two populations, both having the same variance σ^2, find a pooled estimate of σ^2 in Exercises 1–6.

1. $\sum_{i=1}^{8} (x_i - \bar{x})^2 = 8.47$ and $\sum_{i=1}^{6} (y_i - \bar{y})^2 = 6.84$

2. $s_1^2 = 10$ with $m = 16$ and $s_2^2 = 14$ with $n = 11$

3. $\sum_{i=1}^{10} (x_i - \bar{x})^2 = 0.189$ and $\sum_{i=1}^{7} (y_i - \bar{y})^2 = 0.168$

4. $\sum_{i=1}^{5} (x_i - \bar{x})^2 = 26.24$ and $s_2^2 = 8.62$ with $n = 6$

5. $x_1 = 2.2$, $x_2 = 3.2$, $x_3 = 4.1$, $x_4 = 1.8$, $x_5 = 2.4$; and $y_1 = 3.5$, $y_2 = 4.1$, $y_3 = 3.0$, $y_4 = 3.7$.

6. $x_1 = 16.1$, $x_2 = 18.3$, $x_3 = 14.6$, $x_4 = 13.6$; and $y_1 = 11.8$, $y_2 = 9.6$, $y_3 = 8.2$, $y_4 = 9.2$, $y_5 = 10.8$.

7. Assume that we have two populations, both normally distributed with a common variance σ^2 that is unknown. If $\bar{x} = 10.2$, $\bar{y} = 8.5$, $m = 10$, $n = 12$, $s_1^2 = 2.8$, and $s_2^2 = 3.2$, assuming the samples are independent, find the following:

   (a) a pooled estimate, s_p

   (b) the degrees of freedom

   (c) a 90 percent confidence interval for $\mu_1 - \mu_2$.

In Exercises 8–11, assume that the two populations are normally distributed with the same, but unknown, variance.

8. The mean length of 14 trout caught in Clear Lake was 10.5 inches with s_1 of 2.1 inches, and the mean length of 11 trout caught in Blue Lake was 9.6 inches with s_2 of 1.8 inches. Construct a 90 percent confidence interval for the difference in the true mean lengths of trout in the two lakes.

9. A random sample of 10 light bulbs manufactured by Company A had a mean life of 1850 hours and a standard deviation s_1 of 130 hours. Also, a random sample of 12 light bulbs manufactured by Company B had a mean life of 1940 hours with a standard deviation s_2 of 140 hours. Construct a 95 percent confidence interval for the difference in the true means.

10. In ten half-hour morning programs, the mean time devoted to commercials was 6.8 minutes with s_1 of 1 minute. In twelve half-hour evening programs, the mean time was 5.6 minutes with s_2 of 1.3 minutes. Estimate the difference in the true mean times devoted to commercials during the morning and the evening half-hour programs, using a 90 percent confidence interval.

11. A psychologist measured the reaction times (in seconds) of 8 individuals who were not given any stimulant and 6 individuals who were given an alcoholic stimulant. The figures are given below:

Reaction time without stimulant	3.0	2.0	1.0	2.5	1.5	4.0	1.0	2.0
Reaction time with stimulant	5.0	4.0	3.0	4.5	2.0	2.5		

Determine a 95 percent confidence interval for the difference in true mean times for the two populations.

8-3 CONFIDENCE INTERVAL FOR THE DIFFERENCE OF POPULATION PROPORTIONS

Let us consider the following example, where we wish to compare the proportion of women favoring the death penalty with the proportion of men. The collection of all women represents one population. Let us denote the proportion favoring the death penalty in this population by p_1. The collection of all men represents the other population, where the proportion is p_2. The values of p_1 and p_2 are not known to us and it is our intention to construct a confidence interval for $p_1 - p_2$. The first step is to pick two random samples, one from each population, taking care that the two samples are independent.

Suppose we interview m women. Let X denote the number among them who favor the death penalty. Then from what we learned earlier, X/m is a good point estimator of p_1. Similarly, if we interview n males and Y denotes the number among them favoring the death penalty, then Y/n will be a point estimator of p_2. It would then seem natural that we use $X/m - Y/n$ as a point estimator of $p_1 - p_2$. For example, if among 100 women interviewed, 30 are in favor, and among 200 men, 72 are in favor, then a point estimate of $p_1 - p_2$ is $30/100 - 72/200 = -0.06$.

Next, let us construct a confidence interval for $p_1 - p_2$. First we must determine the sampling distribution of its estimator, namely, $X/m - Y/n$. At this point we will have to make some assumptions. Let us assume that both n and m are large (both greater than 30) and that neither p_1 nor p_2 is near 0 or 1.

The distribution of $\dfrac{X}{m} - \dfrac{Y}{n}$ is approximately a normal distribution with mean $p_1 - p_2$ and standard deviation

$$\sqrt{\frac{p_1(1 - p_1)}{m} + \frac{p_2(1 - p_2)}{n}}.$$

if n and m are large (both greater than 30) and neither p_1 nor p_2 is near 0 or 1.

Hence, using the argument adopted in Section 7-5 while setting a confidence interval for the proportion p of a single population, an approximate $(1 - \alpha)100$ percent confidence interval for $p_1 - p_2$ would be given by the endpoints

$$\left(\frac{x}{m} - \frac{y}{n}\right) \pm z_{\alpha/2} \sqrt{\frac{p_1(1 - p_1)}{m} + \frac{p_2(1 - p_2)}{n}}.$$

As we see, the left and right extremities involve p_1 and p_2. But we do not know these quantities! To estimate these quantities is precisely our goal! However, it turns out that not much error results if we use x/m and y/n, respectively, in their place. Thus we have the following consequence:

> ### LARGE SAMPLES CONFIDENCE INTERVAL FOR $p_1 - p_2$
>
> The upper and lower limits of a $(1 - \alpha)100$ percent confidence interval for $p_1 - p_2$ are approximately given by
>
> $$\left(\frac{x}{m} - \frac{y}{n}\right) \pm z_{\alpha/2} \sqrt{\frac{\frac{x}{m}\left(1 - \frac{x}{m}\right)}{m} + \frac{\frac{y}{n}\left(1 - \frac{y}{n}\right)}{n}}$$
>
> where we assume that both m and n are large (greater than 30). Here x/m and y/n are the sample proportions of the attribute in the two populations.

EXAMPLE 1   In a survey conducted on a college campus, 48 out of 150 professors were smokers and 114 out of 300 students were smokers.

(a) Obtain a point estimate of $p_1 - p_2$, the difference in the proportions, where p_1 is the proportion of smokers among all professors and p_2 the proportion among all students.

(b) Obtain an approximate 95 percent confidence interval for $p_1 - p_2$.

SOLUTION   (a) We are given that $x = 48$, $m = 150$, $y = 114$, and $n = 300$. Therefore,

$$\frac{x}{m} = \frac{48}{150} = 0.32, \quad \text{and} \quad \frac{y}{n} = \frac{114}{300} = 0.38.$$

A point estimate of $p_1 - p_2$ is $\frac{x}{m} - \frac{y}{n} = 0.32 - 0.38 = -0.06$.

(b) Since $\alpha = 0.05$, $z_{\alpha/2} = 1.96$. First let us compute

$$\sqrt{\frac{\frac{x}{m}\left(1 - \frac{x}{m}\right)}{m} + \frac{\frac{y}{n}\left(1 - \frac{y}{n}\right)}{n}} = \sqrt{\frac{(0.32)(0.68)}{150} + \frac{(0.38)(0.62)}{300}}$$

$$= 0.0473.$$

Since $\frac{x}{m} - \frac{y}{n} = -0.06$, a 95 percent confidence interval is given by

$$-0.06 - 1.96(0.0473) < p_1 - p_2 < -0.06 + 1.96(0.0473).$$

Simplifying, we find

$$-0.153 < p_1 - p_2 < 0.033. \quad \blacksquare$$

SECTION 8-3   EXERCISES

Suppose two independent random samples are picked from two populations where we are interested in comparing proportions of an attribute in them. In Exercises 1–3, determine a confidence interval for $p_1 - p_2$ at the indicated level of confidence.

1. $x = 42$, $m = 110$; and $y = 25$, $n = 60$; 95 percent

2. $x = 17$, $m = 48$; and $y = 36$, $n = 112$; 98 percent

3. $x = 35$, $m = 86$; and $y = 62$, $n = 156$; 90 percent.

4. Of 150 Democrats interviewed, 90 favor a certain proposition, and of 120 Republicans interviewed, 80 favor it. Set an 80 percent confidence interval on the difference in the proportions of Democrats and Republicans in the population who favor the proposition.

5. Of 200 college students interviewed, 150 favored strict environmental controls, whereas of 300 college graduates interviewed, 195 favored them. Estimate the difference in the true proportions among college students and college graduates who favor strict environmental controls.

6. When a salesman contacted 100 women about a utility product, he sold the product to 43 of them. When he contacted 150 men, he sold the product to 72 of them. Set a 95 percent confidence interval on the difference in the true proportions among males and females who will buy the product if contacted.

7. A survey was conducted to compare the proportions of males and females who favor government assistance for child care. It was found that among 64 males interviewed, 40 favored assistance, and among 100 females, 70 favored assistance. Construct a 90 percent confidence interval on the difference in true porportions among all males and females who favor government assistance for child care.

8. The records of the Internal Revenue Service showed that of 200 business executives picked at random from its files, 46 had defaulted on tax payments, and of 100 white-collar workers, also picked at random, 18 had defaulted. Set an 85 percent confidence interval for the "true proportion of defaulting business executives *minus* the true proportion of defaulting white-collar workers."

9. When a chemist tested Dynamite spray insecticide on 250 insects, 160 of them were killed. When she tested Quick spray on 200 insects, 120 were killed. Estimate the difference in true proportions killed by the two insecticides using a 95 percent confidence interval.

10. In a laboratory test, of the 200 electric bulbs of Brand A, 80 lasted beyond 800 hours, and of the 160 bulbs of Brand B, 48 lasted beyond 800 hours. Determine 90 percent confidence limits for the difference in population proportions of the bulbs that last beyond 800 hours for the two brands.

11. A retailer ordered 100 cathode tubes from Company A and 80 tubes from Company B. Upon inspection of the merchandise, he found that 18 tubes from Company A were defective and 16 from Company B were defective. Find an 88 percent confidence interval for the difference in the true proportions of defective tubes from the two companies.

KEY TERMS AND EXPRESSIONS

difference of means
difference of proportions
pooled estimate of variance

KEY FORMULAS

Summary of Confidence Intervals (two populations)

Nature of the population	Parameter on which confidence interval is set	Procedure	Limits of confidence interval with confidence coefficient $1 - \alpha$
Quantitative data; variances σ_1^2 and σ_2^2 are *known;* both populations are normally distributed.	$\mu_1 - \mu_2$, the difference of the population means	Draw samples of sizes m and n from the two populations, get the respective means $\bar{x}$ and $\bar{y}$, and find the value of $\bar{x} - \bar{y}$, the estimate of $\mu_1 - \mu_2$.	$(\bar{x} - \bar{y}) \pm z_{\alpha/2}\sqrt{\dfrac{\sigma_1^2}{m} + \dfrac{\sigma_2^2}{n}}$
Quantitative data; σ_1^2 and σ_2^2 are *not* known but assumed *equal;* both populations are normally distributed.	$\mu_1 - \mu_2$, the difference of the population means	Draw samples of sizes m and n from the two populations; compute $\bar{x} - \bar{y}$ and $$s_p^2 = \frac{(m-1)s_1^2 + (n-1)s_2^2}{m+n-2}.$$	$(\bar{x} - \bar{y}) \pm t_{m+n-2,\alpha/2} s_p \sqrt{\dfrac{1}{m} + \dfrac{1}{n}}$
Quantitative data; σ_1^2 and σ_2^2 are *not* known and *not* assumed equal; populations may *not* be normal; sample sizes m and n are large.	$\mu_1 - \mu_2$, the difference of the population means	Draw samples of sizes m and n compute $\bar{x} - \bar{y}$; compute s_1^2 and s_2^2.	$\bar{x} - \bar{y} \pm z_{\alpha/2}\sqrt{\dfrac{s_1^2}{m} + \dfrac{s_2^2}{n}}$ Confidence interval is approximate.
Qualitative data; p_1 is the proportion in Population 1 and p_2 in Population 2; sample sizes m and n are assumed large.	$p_1 - p_2$, the difference of the population proportions	Draw two samples, one from each population, of sizes m and n; find x/m and y/n, the estimates of p_1 and p_2 respectively; then get $x/m - y/n$.	$\left(\dfrac{x}{m} - \dfrac{y}{n}\right) \pm z_{\alpha/2}\sqrt{\dfrac{\dfrac{x}{m}\left(1 - \dfrac{x}{m}\right)}{m} + \dfrac{\dfrac{y}{n}\left(1 - \dfrac{y}{n}\right)}{n}}$ Confidence interval is based on the central limit theorem, hence is approximate.

CHAPTER 8 TEST

In Exercises 1–5, find a confidence interval for the indicated difference of parameters at the specified level of confidence. Mention the assumptions that would be considered necessary in each case.

1. 98 percent for $\mu_1 - \mu_2$ if $\bar{x} = 6.8$, $m = 6$, $s_1^2 = 2.7$; and $\bar{y} = 4.5$, $n = 10$, $s_2^2 = 3.9$

2. 95 percent for $\mu_2 - \mu_1$ if $\bar{x} = 1.8$, $m = 10$, $\sigma_1^2 = 1.69$; and $\bar{y} = -3.0$, $n = 16$, $\sigma_2^2 = 2.32$

3. 98 percent for $p_1 - p_2$ if $x = 28$, $m = 72$; and $y = 40$, $n = 90$

4. 92 percent for $p_2 - p_1$ if $x = 42$, $m = 100$; and $y = 53$, $n = 125$

5. 90 percent for $\mu_1 - \mu_2$ if $\bar{x} = -3.2$, $m = 48$, $s_1^2 = 7.6$; and $\bar{y} = -6.4$, $n = 60$, $s_2^2 = 5.9$.

6. Eight fertilized eggs were incubated for 14 days and six fertilized eggs were incubated for 18 days. When a determination was made of the amount (in milligrams) of total brain actin, the following figures were obtained.

Incubated 14 days	1.2	1.4	1.5	1.2	1.4	1.7	1.5	1.7
Incubated 18 days	1.3	1.8	1.9	2.0	1.5	2.2		

Assuming a normal distribution of total brain actin in each case, find a 95 percent confidence interval for $\mu_{14} - \mu_{18}$, where the subscripts stand for number of days of incubation.

7. The following information was collected on the amount of insulin release when pancreatic tissues of some animals were treated in a laboratory with two drugs, Drug A and Drug B. Seven tissues were treated with Drug A and eight with Drug B.

	Sample mean	Sample variance
Drug A	5.24	0.813
Drug B	7.06	1.572

Assuming that the amount of insulin release is normally distributed with the same variance for both drugs, set a 90 percent confidence interval for $\mu_A - \mu_B$.

8. The table below gives calorific values of dry organic matter ashed with and without benzoic acid.

Ashing with benzoic acid (cal/gm)	Ashing without benzoic acid (cal/gm)
4729	4327
5416	5233
5318	4383
4626	4675
4781	4428
5508	
4962	

Assume that calorific values are normally distributed in both situations. Suppose μ_B and μ_N are, respectively, true mean calorific values when ashed with and ashed without benzoic acid. Find a 98 percent confidence interval for $\mu_B - \mu_N$.

9. Out of 400 Europeans interviewed, 123 were of blood type O, and out of 200 Japanese, 57 were of that blood type. Set a 90 percent confidence interval for the difference in true proportions of O blood types for the two groups.

10. Five blue-collar workers and six white-collar workers, interviewed regarding the amount of time they spent watching television during a weekend, gave the following figures (in hours):

Blue-collar workers	10	9	14	4	8	
White-collar workers	3	7	8	6	10	8

Set a 90 percent confidence interval for the difference in the true mean times for the two types of workers. Assume that time spent watching television is normally distributed for the two groups with the same variance.

9

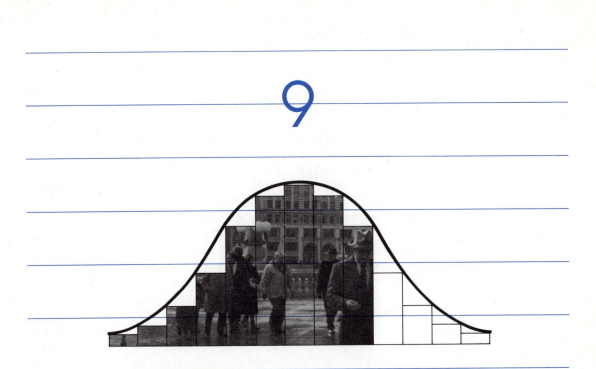

TESTS OF HYPOTHESES

9-1 The Jargon of Statistical Testing

9-2 The Mechanics of Carrying Out Tests of Hypotheses

9-3 Tests of Hypotheses (A Single Population)

9-4 Tests of Hypotheses (Two Populations)

INTRODUCTION

In Chapters 7 and 8 we considered one aspect of statistical inference, the estimation of population parameters. We shall now turn our attention to another aspect, called *hypothesis testing*. The main objective here is to formulate rules that lead to decisions culminating in acceptance or rejection of statements about the population parameters.

On a cloudy day, a man looks out the window and says, "It is going to rain today." What that man has done is make a statement forecasting the weather for that day. We shall say that he has made a hypothesis regarding the weather. He is now faced with the following decision problem. Should he or should he not take his raincoat? Whatever his action, it is going to result in precisely one of the following four situations: He takes his raincoat and it rains (a wise decision); he does not take his raincoat and it does not rain (again, a wise decision); he takes his raincoat and it does not rain (a wrong decision); and, finally, he does not take his raincoat and it rains (a wrong decision). The last two situations are not desirable in that the actions result in erroneous decisions. Ideally, what this man would like is to have his raincoat with him if it does rain and not to have it with him if it does not rain. But, alas, he will not know if it will rain when the day begins, and he must make his decision before the day begins. Before starting from home, he will have to take into account such factors as how overcast the sky is, his past experience, and so on, and balance the consequences of erroneous decisions. If he is highly susceptible to catching pneumonia, he might prefer to err on the side of taking the raincoat with him even if it does not rain that day. On the other hand, if he is physically fit and prone to misplace his raincoat, he might prefer to leave the raincoat home.

It is ideas of this nature that crop up in the context of statistical hypothesis testing. Fortunately our intuition will guide us well in our discussion. ▬▬

9-1 THE JARGON OF STATISTICAL TESTING

WHAT IS A HYPOTHESIS?

A **statistical hypothesis** is a statement, assertion, or claim about the nature of a population. It is a basic object of any experimental inquiry.

Numerous examples can be cited where a person hypothesizes about a situation. Anyone who has watched commercial television cannot fail to be aware of the constant barrage of claims. Brand *X* detergent will wash white clothes sparkling white; with a supergasoline your car will get more miles to the gallon than before; a new radial tire will give more than 40,000 miles of trouble-free driving; and so on and so on. In the framework of statistical investigation, very often a hypothesis takes the form of stipulating the values of the unknown parameters of the population being studied. For example:

(a) A coin is unbiased. If $P(\text{head}) = p$, the hypothesis is that $p = \frac{1}{2}$.

(b) A tire manufacturer claims that the new radial tire produced by his company will give more than 40,000 miles of trouble-free driving. If μ represents the mean number of miles, the hypothesis that the manufacturer has in mind is $\mu > 40,000$.

(c) A social scientist asserts that students with an urban background perform better than those with a rural background. Suppose the measure used is the percent of students who graduate from high school. If p_U and p_R represent, respectively, the population proportions of students with urban and rural backgrounds who graduate, then the hypothesis is that $p_U > p_R$.

(d) An engineer claims that the machine manufactures bolts within specifications. If she uses the standard deviation of the diameter of the bolts as a measure, the hypothesis may be that $\sigma < 0.1$ inch.

The only sure way of finding the truth or falsity of a hypothesis is by examining the entire population. Since this is not always feasible, instead we might examine a sample for the purpose of drawing conclusions.

A hypothesis that is being tested for the purpose of possible rejection is called a **null hypothesis.** It is a common convention to designate this hypothesis as H_0.

If, for example, the null hypothesis is that the coin is a fair coin, letting $P(\text{head}) = p$, we will write

$$H_0: p = \frac{1}{2}.$$

As another example, the null hypothesis that the mean life of bulbs produced by a machine is at least 700 hours will be written as

$$H_0: \mu \geq 700.$$

Also, in our discussion of the topic, we will always state the null hypothesis in such a way that it contains the equality sign. Thus we will *never* have a null hypothesis stating that the mean life of bulbs is *greater* than 700, that is, we will *not* state $H_0: \mu > 700$.

The hypothesis against which the null hypothesis is tested is called the **alternative hypothesis.** We shall denote it as H_A. This is the hypothesis that is accepted when the null hypothesis is rejected.

The values of the parameter(s) stated under the alternative hypothesis are outside the region stated under the null hypothesis. That is, there is no overlap between the set of parameter values stipulated under the null hypothesis and those stipulated under the alternative hypothesis. For example, against the null hypothesis $H_0: p = \frac{1}{2}$ stated above, we might have as an alternative hypothesis $H_A: p \neq \frac{1}{2}$, or $H_A: p = \frac{3}{4}$, or $H_A: p > \frac{2}{3}$. However, against $H_0: p = \frac{1}{2}$ we will never have $H_A: p > \frac{1}{4}$.

CHOICE OF H_0 AND H_A

In a given problem, which hypothesis constitutes the null hypothesis and which one the alternative hypothesis is an important question and one that has to do with the logic of a statistical test procedure. In formulating tests of hypotheses, the essential idea is that of proof by contradiction. For example, suppose a teacher strongly suspects that students coming from families at a higher economic level perform better than students at the lower level. This is what the teacher believes she can establish and hence the onus of proof is on her. Under H_0 we shall say that this is not the case and state

H_0: Students at a higher economic level do *not* perform better.

Against this null hypothesis we shall give the alternative hypothesis as

H_A: Students at a higher economic level do perform better.

It is for this reason that the alternative hypothesis is sometimes referred to as the research hypothesis. The null hypothesis asserts, in essence, that there is no merit to the researcher's claim. If the sample evidence is such that it leads us to reject H_0, then we shall, of course, accept H_A.

At the heart of setting the two hypotheses this way is the question, "Do students coming from families at a higher economic level perform better than students at the lower level?" Thus, in the final analysis, the null hypothesis and the alternative hypothesis are determined by the question posed in the statement of the problem. Here are some additional examples:

(a) A businessman wonders: "Are the sales this year going to be better than last year?" Here we write

H_0: The sales are *not* going to be better than last year.

H_A: The sales will be better than last year.

(b) An agronomist wants to know: "Does the new breeding method increase grain production?" We write

H_0: The new method does *not* increase grain production.

H_A: The new method does increase grain production.

(c) A pollster wonders: "Is the proportion p_1 of Republicans in New Hampshire greater than the proportion p_2 of Republicans in California?" In this case

H_0: $p_1 \leq p_2$, that is, p_1 is *not* greater than p_2.

H_A: $p_1 > p_2$.

In each of the examples given above, the null hypothesis denies the claim posed in the question, explaining in part the usage of the term *null*. Hence it makes good sense to pose the question appropriately, depending upon what it is that the researcher is interested in establishing.

> **FORMULATION OF H_0 AND H_A**
>
> When we wish to establish a statement about the population with evidence obtained from the sample, the negation of the statement is what we take as the null hypothesis H_0. The statement itself constitutes the alternative hypothesis H_A.

A *test of a statistical hypothesis* is a rule or procedure that leads to a decision to accept or reject the hypothesis when the experimental sample values are obtained. This rule, formulated before drawing the sample, is often referred to as a **decision rule.**

TYPE I AND TYPE II ERRORS

In any hypothesis-testing problem, since we take action based on incomplete information, there is a built-in danger of an erroneous decision. A statistical test procedure based on sample data will lead to precisely one of the following four situations. Two of these situations will entail correct decisions and the other two, incorrect decisions.

1. H_0 is true and H_0 is accepted—a correct decision.
2. H_0 is true and H_0 is rejected—an incorrect decision.
3. H_0 is false and H_0 is accepted—an incorrect decision.
4. H_0 is false and H_0 is rejected—a correct decision.

Rejection of the null hypothesis when in fact it is true is called a **Type I error** or a *rejection error*. The probability of committing this error is denoted by the Greek letter α (alpha) and is referred to as the **level of significance** of the test.

Acceptance of H_0 when it is false is called a **Type II error** or an *acceptance error*. The probability of making this error is denoted by the Greek letter β (beta).

The four possibilities mentioned above are summarized in Table 9-1.

TABLE 9-1

Four possibilities based on the decision taken and the truth or falsity of H_0

	True state of nature (unknown)	
Test procedure conclusion	H_0 *is true.*	H_0 *is false* (H_A *is true*).
Accept H_0.	Correct decision; probability is $1 - \alpha$.	Incorrect decision; Type II error; probability is β.
Reject H_0 (accept H_A).	Incorrect decision; Type I error; probability is α.	Correct decision; probability is $1 - \beta$.

Ideally we would like to have both α and β very low. In fact, if it were possible, we would eliminate both these errors and set their probabilities equal to zero. However, once the sample size is agreed upon, there is no way to exercise simultaneous control over both errors. The only way to accomplish this simultaneous reduction is to increase the sample size and, if we desire to have both α and β equal to zero, to explore the entire population.

To understand the basic approach to hypothesis testing, we might recall the familiar presumption under our judicial system, "the accused is innocent until proven guilty beyond a reasonable doubt." Is the accused guilty? That is the question. We state the null hypothesis as

H_0: The accused is not guilty.

The alternative hypothesis is

H_A: The accused is guilty.

It is up to the prosecution to provide evidence to destroy the null hypothesis. If the prosecution is unable to provide such evidence, the accused goes free. If the null hypothesis is refuted, we accept the alternative hypothesis and declare that the accused is guilty. Bear in mind that if the accused goes free, it does not mean that the accused is indeed innocent. It simply means that there just was not enough evidence to find the accused guilty. Nor, if the accused is convicted, does it mean that the accused did indeed commit the crime. It simply means that the evidence was so overwhelming that it is highly improbable that the accused is innocent. Only the accused knows the truth.

In this context, suppose the accused is innocent, in fact, but is found guilty. Then a Type I error has been made because the null hypothesis has been rejected erroneously. Thus the probability of convicting the innocent would be α, and we would like to keep this value rather low. On the other hand, if a guilty person is declared not guilty, a Type II error has been made with probability β.

In approaching the problem of testing a statistical hypothesis, our attitude will be to assume initially that the null hypothesis H_0 is correct. It will be up to the experimental data to provide evidence, beyond reasonable doubt, which will refute this notion. We will then reject H_0 and opt for H_A. Otherwise the status quo prevails in that we have no reason to believe otherwise. The evidence from the experimental data should be extremely strong for us to go along with the hypothesis H_A. When we reject the null hypothesis, we have not proved that it is false, for no statistical test can give 100 percent assurance of anything. However, if we reject H_0 with a small α, then we are able to assert that H_0 is false and H_A is true *beyond a reasonable doubt*. Thus, in any test procedure, it makes good sense to let α be small.

A further understanding of the roles of the null and alternative hypotheses will be developed in Exercises 1–5 in Section 9-1 Exercises which follow.

SECTION 9-1 EXERCISES

1. Suppose a drug company has developed a new serum that could help prevent a disease. The question is whether the serum is effective or not. The course of action open is whether to market the serum or not to market it.

   (a) What are the two errors that could be committed?

   (b) As far as the drug company is concerned, which error is more serious?

   (c) As far as the Federal Drug Administration (which protects consumer interests) is concerned, which error is more serious?

2. Referring to Exercise 1, in deciding whether to market the serum or not, suppose the company does not want to miss the opportunity of marketing an effective drug. Explain why the company would state the hypotheses as

   H_0: The drug is effective.

   H_A: The drug is ineffective.

3. Referring to Exercise 1, if the Federal Drug Administration is interested in testing the claim of the drug comapny, state how the null and the alternative hypotheses should be formulated.

4. Various claims have been made over the past few years to the effect that smoking leads to lung cancer. We realize fully well that the consumer protection agencies are not out to destroy the tobacco industry. Also it is not the purpose of the tobacco industry to promote callously a product that would lead to cancer.

   (a) From the viewpoint of the tobacco industry which error is more serious?

   (b) From the viewpoint of the consumer protection agencies which error is more serious?

   (c) If an executive of some tobacco company feels that the onus of proof is on the consumer protection agencies, how would she formulate the null hypothesis and the alternative hypothesis?

5. Mr. and Mrs. Diaz have inherited some money and plan to invest it in a business venture. It is possible that the business will flourish. By the same token, it is quite possible that it might fail. Mr. and Mrs. Diaz formulate the null and alternative hypotheses as follows:

   H_0: The business venture will not flourish.

   H_A: The business venture will flourish.

   State the Type I and Type II errors in this context. From the viewpoint of Mr. And Mrs. Diaz, which error do you suppose would be more serious?

6. Suppose you are a cigarette smoker who wants to minimize your risk of cancer. For some reason, you feel that if the mean nicotine content is less than 1 mg, it is safe for you to smoke that brand. If a new cigarette is introduced on the market and you wish to decide by statistical testing whether it is safe for you to switch to this brand, how would you formulate the null and the alternative hypotheses? (Remember you are not too eager to switch to a new brand.)

7. Suppose the probability of Type I error is 0.08. State whether the following statements are true or false:

(a) The probability of rejecting the null hypothesis is 0.08.

(b) The probability of rejecting the null hypothesis when it is true is 0.08.

(c) The probability of Type II error is 0.92.

(d) The probability of Type II error is a number between 0 and 1, both inclusive.

(e) The probability of Type II error will depend on the alternative hypothesis.

(f) The probability of accepting the null hypothesis is 0.92.

(g) The probability of accepting the null hypothesis when it is true is 0.92.

8. For each of the following, state what action would constitute a Type I error and what action would constitute a Type II error. (In each case take H_A to be the negation of the statement under H_0.)

(a) H_0: The sulphur dioxide content in the atmosphere is at a dangerous level.

(b) H_0: The new process produces a better product than the old process.

(c) H_0: Most Americans favor gun control.

(d) H_0: The mean velocity of light in vacuum is 299,795 km/sec.

(e) H_0: Eighty percent of doctors recommend aspirin for headaches.

(f) H_0: A geographic basin with a certain type of rock formation is oil-bearing.

(g) H_0: The life expectancy of people living in a certain geographic region is 52 years.

(h) H_0: Supersonic transport will affect the ozone level in the stratosphere.

(i) H_0: The old factory should be scrapped and a new factory should be installed.

(j) H_0: Nuclear power does not pose hazards to the public.

(k) H_0: The diet will help reduce weight.

9. An industrial engineer tested two types of steel wires to test H_0: The mean strengths of the two types of steel wires are the same, against H_A: The mean strengths are different. It is revealed subsequently that in coming to her conclusion, the engineer committed a Type II error. What is the actual state of affairs and what was the decision taken?

10. In Exercise 9, suppose in the long run the engineer does not mind if 10 percent of the time she decides erroneously that the two types of wires are different when in fact they are not. What type of error are we alluding to? What is its approximate probability?

11. An educational testing service wanted to compare the mean IQ of university students in a certain geographic region with that of the rest of the nation. The null hypothesis is given as follows:

   H_0: The mean IQ of the region is less than or equal to the mean for the rest of the nation.

After carrying out the statistical analysis, the following conclusion is reached: At the 5 percent level of significance, the mean IQ of the region is higher. Interpret the meaning of the conclusion and the stated level of significance.

12. Widespread skepticism resulted from the report in June 1985 that the body which had been exhumed from a graveyard in Embu, Brazil was that of Joseph Mengele, the infamous Nazi doctor. After a team of international forensic experts examined the skeletal remains, Dr. Lowell Levine, a forensic specialist from the U.S. Justice Department, announced the experts' unanimous conclusion: "The skeleton is that of Joseph Mengele within a *reasonable scientific certainty*." Comment on this announcement in light of the likely null hypothesis, alternative hypothesis, and the level of significance.

9-2 THE MECHANICS OF CARRYING OUT TESTS OF HYPOTHESES

Having been introduced to terms used in statistical testing, we shall now work out the details of test procedures concerning the population mean when σ^2 is known. Though assuming that σ^2 is known may seem unrealistic in actual situations, it initiates our study of the topic in the correct direction. We shall address this aspect later. Our principal concern will be to formulate decision rules based on the nature of the alternative hypotheses and the levels of the two types of errors. The ideas developed here will guide us in cases that will be studied in Sections 9-3 and 9-4.

Let us consider the following example: Suppose a factory manufactures electric bulbs. An engineer employed by the factory believes that she has found a new manufacturing technique that will increase the mean life of the bulb, which is presently believed to be 450 hours. Let us assume that the length of time that a bulb lasts is normally distributed and the standard deviation is 60 hours and has not changed under the engineer's invention.

To put the problem formally in a statistical framework, we have two hypotheses. The engineer is making a claim, so let her prove it; it is the research hypothesis. Our attitude is that the engineer should provide reasonable evidence that will result in our rejecting the hypothesis that there is no improvement. Hence we set up the hypotheses as

Null hypothesis H_0: $\mu = 450$, that is, the new technique *does not* result in improvement.

Alternative hypothesis H_A: $\mu > 450$, that is, the new technique *does* result in improvement.

To test the engineer's claim, we will take a sample of bulbs from the lot of bulbs produced using the new technique and, as might be expected, discuss the merits of her claims on the basis of the sample mean $\bar{x}$. Accordingly, suppose we pick a sample of 9 bulbs produced using the new technique and obtain the mean life span of these bulbs.

FIGURE 9-1

Critical value
and rejection
region

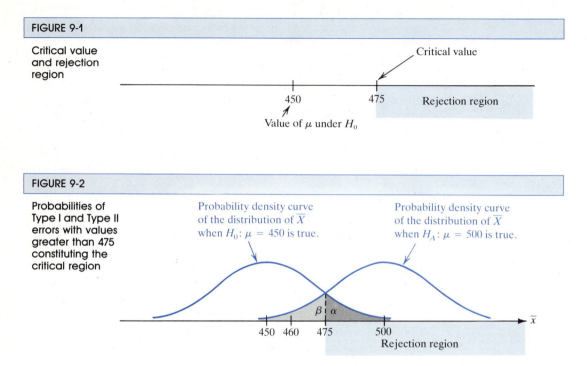

FIGURE 9-2

Probabilities of
Type I and Type II
errors with values
greater than 475
constituting the
critical region

Now a low value of the sample mean would, of course, dispose us in favor of H_0. In fact, the lower the value of $\bar{x}$, the stronger would be our conviction in favor of H_0. On the other hand, a large value of $\bar{x}$ would cause us to lean in favor of the alternative hypothesis H_A. The higher the value of $\bar{x}$, the more we would favor H_A.

Suppose we decide, rather arbitrarily at this stage, to choose a cutoff point of the sample mean at 475 and adopt the following decision rule:

Decision rule: Reject H_0 if the sample mean $\bar{x}$ is greater than 475 hours.

We shall call 475 a **critical value.** The values to the right of 475 will constitute what is called a **rejection region,** also called a **critical region** (see Figure 9-1). If, when we actually pick a sample, for instance, it yields a mean of 470 hours, then our conclusion will be that the engineer has not provided evidence to back her claim. If, on the other hand, the sample mean turns out to be 490 hours, we will reject H_0 and accept the engineer's claim.

TYPE I AND TYPE II ERRORS

We have agreed that if the sample mean yields a value greater than 475, we will reject H_0 and grant the engineer's claim. However, due to sampling fluctuations, it is very possible that the bulbs are indeed from a population where the mean life is

450, but that our sample yields a mean greater than 475. If we grant the engineer's claim (that is, reject H_0) when she is wrong (that is, when H_0 is true), we will be committing the Type I error. Since the test is based on the value of $\bar{x}$, the probability of this error will depend on the distribution of $\bar{X}$.

The sampling distribution of $\bar{X}$ is normal because the population is normally distributed. Its mean is 450 if H_0: $\mu = 450$ is true and its standard deviation is $\sigma/\sqrt{n} = 60/\sqrt{9} = 20$. The distribution is shown by the curve on the left in Figure 9-2.

The alternative hypothesis contemplates a wide range of possible values (greater than 450) for the parameter μ. Suppose for definiteness, the alternative hypothesis is specifically H_A: $\mu = 500$. The distribution of $\bar{X}$ in this case is normal with mean 500 and standard deviation $\sigma/\sqrt{n} = 20$. This distribution is shown by the curve on the right in Figure 9-2.

Type I Error

The probability of a Type I error α is the level of significance of the test. So we want to find

The probability of rejecting H_0 when H_0 is true.

That is, we want

Probability $\bar{X} > 475$ when $\mu = 450$

or, equivalently, we want

Probability $\bar{X} > 475$ when the distribution of $\bar{X}$ is given by the curve on the left in Figure 9-2.

This is simply the shaded area to the *right* of 475. Since the standard deviation of $\bar{X}$ is 20, conversion to the z-scale shows that the shaded area to the right of 475 is equal to the area to the right of $(475 - 450)/20$, or 1.25, and from Appendix II, Table A-4 is equal to 0.11 (approximately). Hence α, the level of significance, is 0.11. This means that in a long series of identical experiments in which 9 bulbs are tested from the production, in approximately 11 out of 100 experiments the engineer's claim will be granted when the engineer is wrong in her claim. That is the risk we are running with the decision procedure we have adopted.

Type II Error

Next let us find β, the probability of not granting the engineer's claim when she is correct and μ is in fact 500. (We are assuming that 500 is the value of μ stipulated under H_A.) According to our decision rule, we do not grant the engineer's claim if $\bar{x}$ is less than 475. Thus we want to find

Probability $\bar{X} < 475$ when $\mu = 500$.

Equivalently we want

Probability $\overline{X} < 475$ when the distribution of $\overline{X}$ is given by the curve on the right in Figure 9-2.

This is the shaded area to the left of 475. After converting to the z-scale, this is equal to the area to the left of $(475 - 500)/20$, or -1.25, and is approximately equal to 0.11.

Interplay Between α and β

Suppose the company is well established and conservative in its decisions and feels that granting the engineer's claim when she is wrong is fraught with much more serious consequences for the company than not granting her claim when she is right. In other words, instead of having a decision rule with $\alpha = \beta = 0.11$, the company wants a decision rule for which α is smaller. An easy way to accomplish this is by moving the critical point to the right as we have done in Figure 9-3. We have picked 490 as the new critical point. The area shaded to the right of this critical point is now reduced but the area shaded to the left has increased.

Thus α has gone down but β has gone up. In a like manner, if we were to move the critical point to the left, β would go down but then α would go up. This shows that we cannot control both errors simultaneously once the sample size is determined.

In any decision making process, what chances of errors an experimenter (whether business person, student, or scientist) is prepard to tolerate is basically up to the individual. However, as we mentioned in the previous section, it is most desirable to exercise control over α and keep it small. The value of α chosen is usually between 0.01 and 0.1, the most common value being 0.05.

FORMULATION OF THE DECISON RULE

Continuing with our example, suppose we wish to carry out a test of

$H_0: \mu = 450$ against $H_A: \mu > 450$

at the level of significance α. How do we formulate a decision rule with this significance level in mind?

From our earlier discussion, in view of the nature of the alternative hypothesis, the decision rule will be

Reject H_0 if $\overline{x} > c$

(that is, reject H_0 for large values of $\overline{x}$) where the critical value c is to be determined in such a way that the probability of rejecting H_0 when H_0 is true must be α. In other words, we want

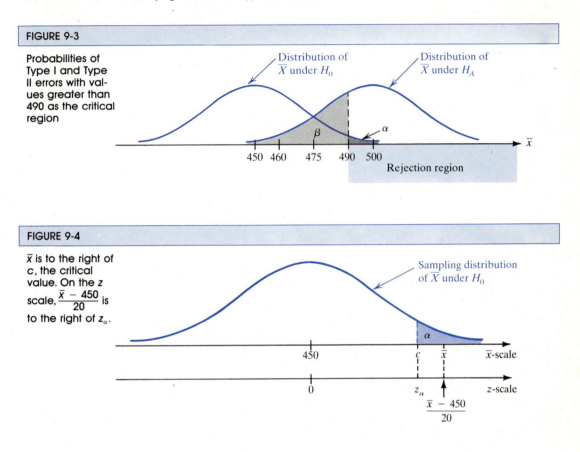

FIGURE 9-3

Probabilities of
Type I and Type
II errors with val-
ues greater than
490 as the critical
region

Distribution of
$\overline{X}$ under H_0

Distribution of
$\overline{X}$ under H_A

β

α

450 460   475   490 500

Rejection region

FIGURE 9-4

$\bar{x}$ is to the right of
c, the critical
value. On the z
scale, $\dfrac{\bar{x} - 450}{20}$ is
to the right of z_α.

Sampling distribution
of $\overline{X}$ under H_0

α

450   c $\bar{x}$ $\bar{x}$-scale

0   z_α z-scale

$\dfrac{\bar{x} - 450}{20}$

$$P(\overline{X} > c \text{ when } \mu = 450) = \alpha.$$

This simply says that c is the value such that to its right there is an area α under the
normal curve with mean 450 and standard deviation 20 as shown in Figure 9-4. But
we know that if we convert to the z-scale, by definition, it is to the right of z_α that
there is an area equal to α. Thus in terms of z-scores, the decision rule takes the form

Reject H_0 if $\dfrac{\bar{x} - 450}{20} > z_\alpha$.

In general, if we have

$$H_0\colon \mu = \mu_0 \qquad \text{against} \qquad H_A\colon \mu > \mu_0$$

and $\bar{x}$ is the mean based on n observations from a normal population with variance
σ^2, then the decision rule at the level of significance α would be

value of μ stipulated under H_0

Reject H_0 if $\dfrac{\overline{X} - \mu_0}{\sigma/\sqrt{n}} > z_\alpha$.

This procedure using standard scores of $\overline{X}$ obviates the need to compute c. (If your curiosity gets the better of you, the formula for c is

$$c = \mu_0 + z_\alpha \frac{\sigma}{\sqrt{n}}$$

which is obtained by setting

$$z_\alpha = \frac{c - \mu_0}{\sigma/\sqrt{n}}$$

and solving for c.)

A statistic such as

$$\frac{\overline{X} - \mu_0}{\sigma/\sqrt{n}}$$

used in formulating the decision rule is called a **test statistic.**

The set of values of the test statistic that lead to the rejection of the null hypothesis is called the **rejection region** or the **critical region.**

The set of values of the test statistic that are not in the critical region constitutes the **acceptance region.**

A value of the test statistic that separates the acceptance and rejection regions is called a **critical value.**

A **test of a statistical hypothesis** is completely determined by providing the relevant test statistic and the rejection region.

ONE-TAILED AND TWO-TAILED TESTS

The nature of the critical region for a statistical test procedure depends on the alternative hypothesis. In the following discussion we shall consider three cases of the alternative hypothesis, (1) H_A: $\mu > \mu_0$, (2) H_A: $\mu < \mu_0$, and (3) H_A: $\mu \neq \mu_0$, where μ_0 is a given specific value (for instance, 450 in our example).

1. **A right-tailed test.** In discussing the engineer's claim, we have considered the principle of testing the null hypothesis

   $$H_0: \mu = \mu_0$$

   against the alternative hypothesis

   $$H_A: \mu > \mu_0.$$

   The decision rule we formulated there at the level of significance α was

   $$\text{Reject } H_0 \text{ if } \frac{\overline{x} - \mu_0}{\sigma/\sqrt{n}} > z_\alpha.$$

FIGURE 9-5

A critical region
that is in the left
tail of the distri-
bution.

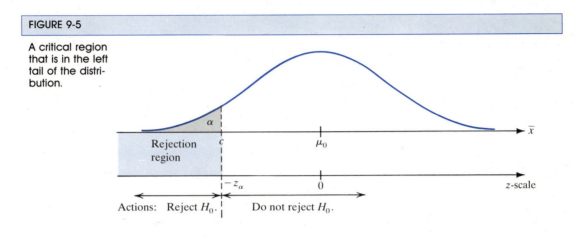

It is the one-sided nature of the alternative hypothesis (greater than, $>$) that prompts the rejection of H_0 if the value of the statistic falls in the right tail of its distribution. The test is therefore called a **one-tailed test,** specifically, a **right-tailed test.**

2. **A left-tailed test.** Suppose the null and alternative hypotheses are given as

$$H_0: \mu = \mu_0$$
$$H_A: \mu < \mu_0.$$

Once again, the alternative hypothesis is one-sided (less than, $<$). Guided by common sense, we would prefer to reject H_0 for smaller values of $\bar{x}$, leading to the rejection of H_0 if the value falls in the left tail of the distribution of $\overline{X}$ as indicated in Figure 9-5. This gives a one-tailed test that is specifically a **left-tailed test.**

Following the argument of the right-tailed test, we can easily see that, in terms of standard scale, the decision rule would be

$$\text{Reject } H_0 \text{ if } \frac{\bar{x} - \mu_0}{\sigma/\sqrt{n}} < -z_\alpha.$$

3. **A two-tailed test.** Finally, a test leads to a two-tailed test if the alternative hypothesis is two-sided. Consider the following example. Suppose a machine is adjusted to manufacture bolts to the specification of a 1-inch diameter, and we state the null hypothesis and the alternative hypothesis as

$$H_0: \mu = 1$$
$$H_A: \mu \neq 1.$$

If the sample mean of the diameters was too far off on either side of 1, we would favor rejecting H_0. In other words, if the value of $\bar{x}$ falls in either tail of the distribution of $\overline{X}$, we will reject H_0.

FIGURE 9-6

A two-tailed
critical region
for testing
$H_0: \mu = \mu_0$
against
$H_A: \mu \neq \mu_0$
with level of
significance α.

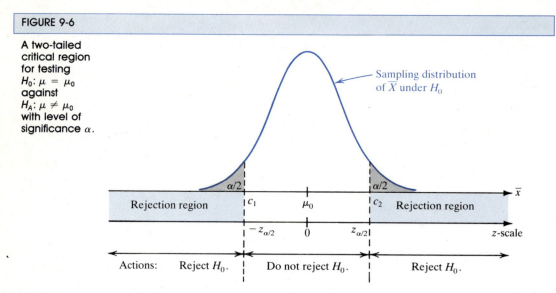

If the test is carried out at the level of significance α, we distribute α equally between the two tails as in Figure 9-6 and so have two critical values c_1 and c_2. In terms of standardized scores the decision rule is then given by

Reject H_0 if $\dfrac{\bar{x} - \mu_0}{\sigma/\sqrt{n}}$ is less than $-z_{\alpha/2}$ or greater than $z_{\alpha/2}$.

In general, a **two-tailed test** is one where the rejection region is located in the two tails of the distribution and results when the alternative hypothesis is two-sided.

EXAMPLE 1   Suppose a population is normally distributed with $\sigma^2 = 36$.

(a) Based on a sample of 16, describe a suitable test procedure at the 10 percent level of significance to test $H_0: \mu = 50$ against $H_A: \mu < 50$.

(b) If a sample of 16 yields a mean of 48.5, what decision would be appropriate at the 10 percent level of significance?

SOLUTION   (a) Since the alternative hypothesis is one-sided (less than), we have a one-tailed (left-tailed) test. We are given that $\mu_0 = 50$. (It is the value stipulated under H_0.) Also, $z_\alpha = z_{0.1} = 1.28$, $\sigma = \sqrt{36} = 6$, and $n = 16$. Therefore

$$\frac{\bar{x} - \mu_0}{\sigma/\sqrt{n}} = \frac{\bar{x} - 50}{6/\sqrt{16}} = \frac{\bar{x} - 50}{1.5}.$$

Since the test is left-tailed, the decision rule is

Reject H_0 if $\dfrac{\bar{x} - 50}{1.5} < -1.28$.

FIGURE 9-7

Areas give *P*-values, the probabilities of getting extreme values when H_0 is true.

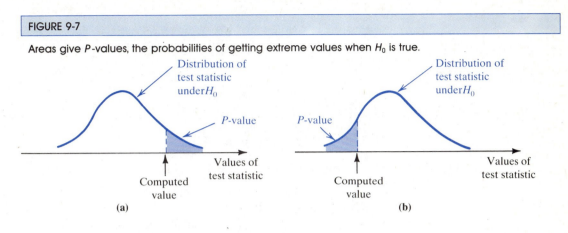

(b) The sample mean is $\bar{x} = 48.5$. From Part (a) the computed value of the test statistic is

$$\frac{\bar{x} - 50}{1.5} = \frac{48.5 - 50}{1.5} = -1.$$

Since this value is not less than -1.28, we do not reject H_0.   ▬▬

We shall see more of this approach in Sections 9-3 and 9-4.

SIGNIFICANCE PROBABILITY (*P*-VALUE)

The approach we have adopted so far involves picking a specific level of significance α at the outset and then, on the basis of the value of the test statistic, rejecting H_0 or not rejecting it at that level. An alternate way which is widely used in journals and technical reports is to provide the probability of obtaining a value of the test statistic as significant as, or even more significant than, the one computed from the data, if indeed H_0 is true. This probability is just the area in the appropriate tail(s) of the distribution of the test statistic when H_0 is true and is called *P*-value or the significance probability of the observed value (see Figure 9-7). When calculating a *P*-value, it is important to keep the alternative hypothesis H_A in perspective since the nature of H_A (less than, greater than, not equal to) determines the tail(s) in which the values of the test statistic are significant. The smaller the *P*-value, the stronger the justification for rejection of H_0. For example, if the *P*-value was reported as 0.003, we would feel very confident in rejecting H_0.

There is an advantage to this method of reporting results by providing the *P*-values. We are free to pick our own level of significance in arriving at a decision. Suppose, for instance, that the *P*-value in a certain situation is 0.083. Then, at the 5 percent level of significance, we would not reject H_0; however, we would reject it at the 10 percent level. Thus, if the calculated *P*-value is less than or equal to the stipulated level of significance of α, then we reject H_0. On the other hand, if the calculated *P*-value is greater than the stipulated α, then we do not reject H_0.

> The P-value associated with test of a hypothesis is the smallest α for which the observed data would call for rejection of the null hypothesis in favor of the alternative.

EXAMPLE 2   Consider the engineer's claim where we tested

$$H_0\colon \mu = 450 \qquad \text{against} \qquad H_A\colon \mu > 450.$$

Suppose a sample of 9 bulbs is found to have a mean life of 492 hours. Calculate the P-value.

SOLUTION   We already know that $\sigma/\sqrt{n} = 20$. In view of H_A, the P-value is the probability of getting a value greater than 492 for the sample mean when H_0 is true, that is, when $\mu = 450$. After converting to the z-scale, this is the area to the right of $(492 - 450)/20$, or 2.10, under the standard normal curve (see Figure 9-8). From Table A-4 we get this probability as 0.0179.

   If we were to carry out the test at a level of significance where α was greater than 0.0179, say $\alpha = 0.05$, then we would reject H_0. However, we would not reject H_0 for an α value less than 0.0179, say $\alpha = 0.01$.   ▬

EFFECT OF THE SAMPLE SIZE

It can be shown that for testing

$$H_0\colon \mu = 450 \qquad \text{against} \qquad H_A\colon \mu = 500$$

at the 5 percent level of significance and with a sample of size 9, the critical value on the $\bar{x}$ scale is 482.5. The resultant value of β turns out to be 0.192 (the area in the left tail of the standard normal distribution, to the left of $(482.5 - 500)/20$, or -0.875). We have discussed the interplay between α and β and are aware that once the sample size is fixed, there is no way of decreasing β below 0.192 without

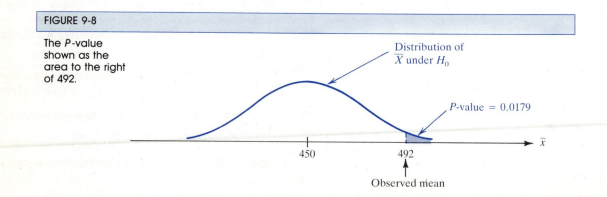

FIGURE 9-8

The P-value shown as the area to the right of 492.

Distribution of $\bar{X}$ under H_0

P-value = 0.0179

450

492

Observed mean

$\bar{x}$

interfering with α and increasing its value beyond 0.05, and vice versa. The only way to reduce both α and β simultaneously is by increasing the sample size. For instance, instead of carrying out the test with a sample of 9, let us pick a sample of 25. The distribution of $\overline{X}$ has less spread now, since the standard deviation of its distribution is $60/\sqrt{25}$, or 12. The distributions of $\overline{X}$ under H_0 and H_A are given in Figure 9-9 for $n = 9$ and $n = 25$.

The reader can see by comparing the respective shaded regions in Figure 9-9 that *both* α and β are reduced when the sample size is increased from 9 to 25. Actual computations will show that α is reduced from 0.05 to 0.0032 and β from 0.192 to 0.0764. Thus a sure way to improve the validity of a test is by taking a larger sample.

A CONNECTION BETWEEN HYPOTHESIS TESTING AND CONFIDENCE INTERVALS

Suppose we test

$$H_0: \mu = \mu_0$$

against

$$H_A: \mu \neq \mu_0.$$

(Notice that the alternative hypothesis is two-sided.) As we have seen, the region where we would not reject H_0 at the α level of significance is given by

$$-z_{\alpha/2} < \frac{\overline{x} - \mu_0}{\sigma/\sqrt{n}} < z_{\alpha/2}.$$

We would reject H_0 outside this region. Now, resorting to some algebraic manipu-

FIGURE 9-9

Effect on Type I and Type II errors when sample size is increased from 9 to 25

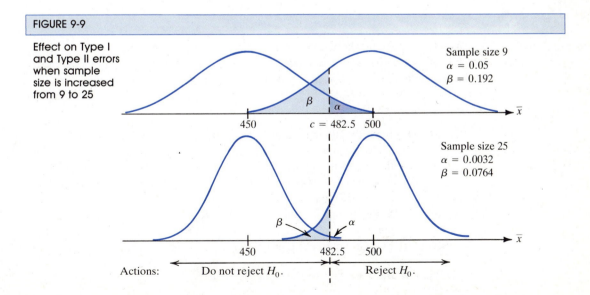

lations, we can say equivalently that we would not reject H_0 if

$$\bar{x} - z_{\alpha/2} \frac{\sigma}{\sqrt{n}} < \mu_0 < \bar{x} + z_{\alpha/2} \frac{\sigma}{\sqrt{n}}.$$

We recognize at once that the endpoints of the interval above are those we found previously for a $(1 - \alpha)100$ percent confidence interval for μ. *Thus we do not reject H_0 at the α level of significance if μ_0 stipulated under it is inside the $(1 - \alpha)100$ percent confidence interval for μ and reject it if μ_0 is outside the confidence interval.*

Thus suppose we have (56.54, 59.86) as a 90 percent confidence interval for μ as in Example 2 in Section 7-2. Based on the data of that example, we would not reject any of the null hypotheses

$$H_0: \mu = 57.3 \qquad \text{against} \qquad H_A: \mu \neq 57.3$$
$$H_0: \mu = 58.5 \qquad \text{against} \qquad H_A: \mu \neq 58.5$$
$$H_0: \mu = 59 \qquad \text{against} \qquad H_A: \mu \neq 59$$

at the 10 percent level since in each case μ_0 is inside the confidence interval (56.54, 59.86). However we would reject

$$H_0: \mu = 55 \qquad \text{against} \qquad H_A: \mu \neq 55$$

at the 10 percent level since the value stipulated under H_0 falls outside the 90 percent confidence interval.

Having considered in detail the factors that enter our considerations when a statistical test is devised, we shall now outline the general procedure in the following steps. We plan to follow these steps scrupulously in the next two sections, even at the risk of seeming repetitious. We add a comment before giving the steps.

STEPS TO BE FOLLOWED FOR TESTING A HYPOTHESIS

Comment: When the alternative hypothesis H_A is stated, for example, as $H_A: \mu < 40$, the values of μ that are not included in H_A are $\mu \geq 40$ and one would usually state the null hypothesis as $H_0: \mu \geq 40$. For our purpose of testing hypothesis, it is relevant that we state H_0 as $H_0: \mu = 40$, giving a specific value which is nearest to those given under H_A.

Step 1   State the alternative hypothesis H_A. (This is the research hypothesis in the investigation and specifies a range of possible values for the parameter that is being tested. It is important in deciding whether the test is one-tailed or two-tailed. Rejection of H_0 leads to the acceptance of H_A.)

Step 2*   State the null hypothesis. Under H_0 give a specific value of the parameter.

Step 3   Pick an appropriate test statistic.

---

*The order of Steps 1 and 2 can be reversed if one so prefers.

Step 4   Stipulate the level of significance α, the probability of rejecting H_0 wrongly. The critical point(s) are determined by the value of α. Together with Step 1, formulate the decision rule. That is, determine the values of the test statistic that will lead to the rejection of H_0 (the critical region). If the value of α is not stipulated, then provide the P-value in Step 6 below.

Step 5   Take a random sample and compute the value of the test statistic.

Step 6   The final step consists of making the decision in light of the decision rule formulated in Step 4. It is important to translate the conclusions into nonstatistical language for the benefit of the uninitiated.

SECTION 9-2   EXERCISES

1. For testing a null hypothesis H_0 against an alternative hypothesis H_A, suppose the critical region is as described in Figure 9-10. Copy the curves and shade appropriate regions under them to represent the probability of Type I error and the probability of Type II error.

2. In Figure 9-11 are given critical regions for testing H_0 against H_A. In each case shade appropriate regions under the curves to represent the probabilities of Type I and Type II errors. In which of the cases would you say that the choice of the critical region was not prudently made?

3. Peggy is given a jar containing 5 chips and told that either the chips are marked 1, 3, 5, 7, and 9 or they are marked 2, 4, 6, 8, and 10. She is allowed to pick one chip at random. If Peggy sets the null hypothesis as H_0: The chips are marked 1, 3, 5, 7, and 9, what is the most appropriate decision rule that she can devise? Find the corresponding probabilities of Type I and Type II errors.

FIGURE 9-10

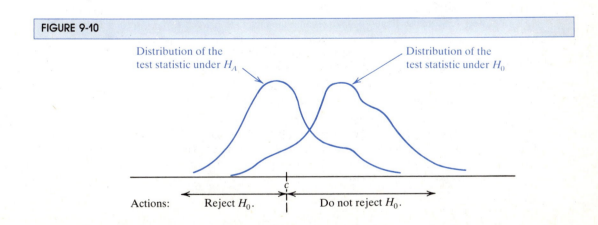

Distribution of the test statistic under H_A

Distribution of the test statistic under H_0

Actions: Reject H_0. c Do not reject H_0.

FIGURE 9-11

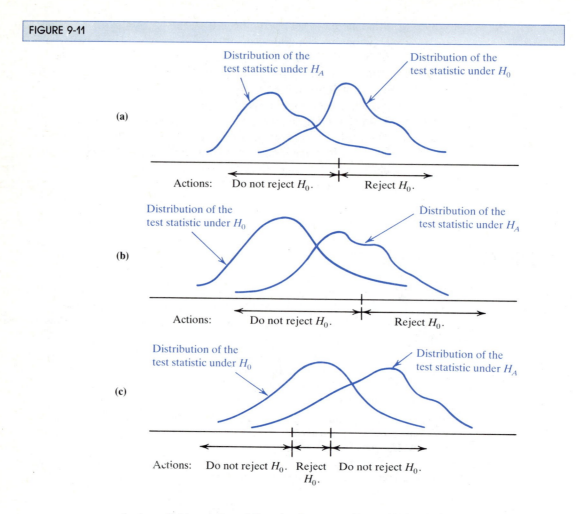

(a) Distribution of the test statistic under H_A

Distribution of the test statistic under H_0

Actions: Do not reject H_0. Reject H_0.

(b) Distribution of the test statistic under H_0

Distribution of the test statistic under H_A

Actions: Do not reject H_0. Reject H_0.

(c) Distribution of the test statistic under H_0

Distribution of the test statistic under H_A

Actions: Do not reject H_0. Reject H_0. Do not reject H_0.

4. As a slight variation of Exercise 3, suppose Peggy is given a jar containing 5 chips and told that either the chips are marked 1, 2, 3, 4, and 5 or they are marked 5, 6, 7, 8, and 9. Suppose H_0 states that the chips are marked 1, 2, 3, 4, and 5. Peggy is allowed to pick one chip at random and has to decide according to the following decision rule: If the observed chip is one of the chips 5, 6, 7, 8, or 9, then reject H_0. Find the probabilities of Type I and Type II errors.

5. In Exercise 4 devise a decision rule such that the probability of Type I error is zero and the probability of Type II error is a minimum.

6. A population is known to be normally distributed with variance 9. However, there is a dispute as to whether the mean μ is 50 or 60. Suppose you have to decide by picking one observation whether H_0: $\mu = 50$ is true. The following decision rule is adopted: *Reject H_0 if the observed value is greater than 56.* Find the probabilities of Type I and Type II errors.

7. Suppose you cannot reject a null hypothesis at the 5 percent level of significance. Would you reject it at the 3 percent level? Would you reject it at any level less than 5 percent? Explain.

8. Suppose we know that a population is normally distributed with the variance σ^2 equal to 16, but we have doubts as to whether the mean is 40 or greater than 40. Suppose we wish to test

$$H_0: \mu = 40 \qquad \text{against} \qquad H_A: \mu > 40.$$

What decision would you take at the 10 percent level of significance if a sample of 25 yielded a mean equal to 42?

9. Suppose we want to test

$$H_0: \mu = 100 \qquad \text{against} \qquad H_A: \mu \ne 100$$

for a population that is normally distributed with a variance of 9. What decision would you take at the 10 percent level of significance if a sample of 16 yields a mean of 97?

10. A population is known to be normally distributed with variance 17.64 but unknown mean μ. Suppose we wish to test

$$H_0: \mu = 100 \qquad \text{against} \qquad H_A: \mu < 100.$$

(a) What decision would you take at the 4 percent level of significance if a sample of 9 yields a mean of 97.6?

(b) In Part (a) determine the P-value.

11. Suppose we have a normal population with $\sigma = 4$ and wish to test

$$H_0: \mu = 20 \qquad \text{against} \qquad H_A: \mu < 20$$

A sample of 9 items yields a mean $\bar{x} = 18.6$.

(a) Compute the value of an appropriate test statistic.

(b) Determine the P-value.

12. Determine P-values in the following cases:

(a) $H_0: \mu = 12.5 \qquad \text{against} \qquad H_A: \mu > 12.5 \qquad$ if $\bar{x} = 14.8$, $\sigma = 6$, and $n = 25$.

(b) $H_0: \mu = 100 \qquad \text{against} \qquad H_A: \mu \ne 100 \qquad$ if $\bar{x} = 82.8$, $\sigma = 24$, and $n = 15$.

13. Mr. Neuman is running for election in a certain district. Suppose p represents the proportion of voters who favor him. The following hypotheses are set up:

$$H_0: p = 0.6 \qquad \text{against} \qquad H_A: p = 0.4.$$

It is agreed that a sample of 8 individuals picked at random will be interviewed and if 6 or more people declare in favor of Mr. Neuman, then H_0 will be accepted. Find the probability of (a) Type I error and (b) Type II error.

[*Hint:* Use the binomial tables, Table A-2.]

14. A stockbroker claims to be able to tell correctly whether a stock will go up or not during a trading session. She asserts that she is correct 80 percent of the time. Of course, if she is guessing, her probability of being correct is 0.5. The question is "Does the stockbroker have the ability that she claims to have?"

   (a) Set up the null hypothesis and the alternative hypothesis.

   (b) What decisions would constitute Type I and Type II errors?

15. In Exercise 14, suppose in order to test the claim of the broker, she is shown 12 unrelated stocks and it is agreed that if the broker is correct nine times or more, her claim will be granted. Determine α and β.

     [*Hint:* Use the binomial tables, Table A-2.]

9-3 TESTS OF HYPOTHESES (A SINGLE POPULATION)

THE POPULATION MEAN WHEN THE POPULATION VARIANCE IS KNOWN

In the preceding section, we considered tests of a population mean when the population variance was known. A basic assumption about the population in such a case is that it is normally distributed. In the absence of a normally distributed population, we will require that the sample size be large, that is, greater than 30. The level of significance is then approximately 100α percent.

As we have seen, the relevant statistic in this case is

$$\frac{\overline{X} - \mu_0}{\sigma/\sqrt{n}}.$$

A summary of the test criteria developed in the last section to test $H_0: \mu = \mu_0$ against the three forms of alternative hypotheses is given in Table 9-2.

TABLE 9-2	
Procedure to test $H_0: \mu = \mu_0$ when σ^2 is known.	
Alternative hypothesis	*The decision rule is to reject H_0 if the computed value is*
1. $\mu > \mu_0$	greater than z_α
2. $\mu < \mu_0$	less than $-z_\alpha$
3. $\mu \neq \mu_0$	less than $-z_{\alpha/2}$ or greater than $z_{\alpha/2}$

   Let us solve some examples following the steps outlined at the end of the last section.

EXAMPLE 1   After taking a refresher course, a salesman found that his sales (in dollars) on 9 random days were

     1280, 1250, 990, 1100, 880, 1300, 1100, 950, 1050.

Has the refresher course had the desired effect, in that his mean sale is now more than 1000 dollars? Assume $\sigma = 100$, and the probability of erroneously saying that the refresher course is beneficial should not exceed 0.01. Also, assume that sales are normally distributed.

SOLUTION

The approach is outlined below:

1. H_0: $\mu = 1000$ (actually $\mu \leqslant 1000$)
2. H_A: $\mu > 1000$
3. The appropriate test statistic is $\dfrac{\overline{X} - 1000}{\sigma/\sqrt{n}}$, since $\mu_0 = 1000$.
4. The alternative hypothesis is one-sided (right-sided). Also, $\alpha = 0.01$ so that $z_\alpha = 2.33$. Therefore, the decision rule is *Reject H_0 if the computed value of the test statistic is greater than 2.33*.
5. Since $\sigma = 100$, $n = 9$, and $\bar{x} = 1100$, the computed value of the test statistic is

$$\frac{\bar{x} - 1000}{\sigma/\sqrt{n}} = \frac{1100 - 1000}{100/3} = 3 \text{ (see Figure 9-12).}$$

FIGURE 9-12

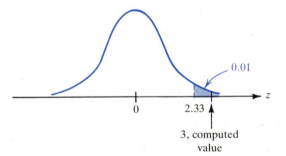

6. Since the computed value is greater than 2.33, we reject H_0. Incidentally, since the computed value is 3, from Table A-4 the P-value is 0.0013.

Conclusion: At the 1 percent level of significance, there is evidence indicating that the refresher course was beneficial to the salesman. It is indeed very unlikely (probability less than 0.01) that we would get a sample with such an extreme value of $\overline{X}$ if the course was not beneficial. ▪

EXAMPLE 2

An IQ test was administered to 9 students and their mean IQ was found to be 95. Assuming the population variance is 144, is it true that the mean IQ in the population is less than 100? Use $\alpha = 0.15$, and assume that IQ is normally distributed.

SOLUTION   In view of the question posed, the null hypothesis will stipulate that the mean IQ is not less than 100. Thus

1. H_0: $\mu = 100$ (actually $\mu \geq 100$)
2. H_A: $\mu < 100$
3. The test statistic is $\dfrac{\overline{X} - 100}{\sigma/\sqrt{n}}$.
4. $\alpha = 0.15$ so that $z_\alpha = 1.04$. Since the alternative hypothesis is one-sided (on the left), the decision rule is *Reject H_0 if the computed value of the test statistic is less than -1.04.*
5. $\overline{x} = 95$, $\sigma = \sqrt{144} = 12$, and $n = 9$. Therefore, the computed value is

$$\frac{\overline{x} - 100}{\sigma/\sqrt{n}} = \frac{95 - 100}{12/3} = -1.25 \text{ (see Figure 9-13)}.$$

FIGURE 9-13

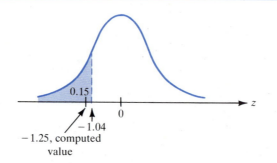

6. Since the computed value is less than -1.04, we reject H_0 (*P*-value 0.1056).

Conclusion: The evidence indicates that the mean IQ is less than 100 at the 15 percent level of significance. ■

EXAMPLE 3   A machine can be adjusted so that when under control, the mean amount of sugar filled in a bag is 5 pounds. From past experience, the standard deviation of the amount filled is known to be 0.15 pound. To check if the machine is under control, a random sample of 16 bags was weighed and the mean weight was found to be 5.1 pounds. At the 5 percent level of significance, is the adjustment out of control? (Assume a normal distribution of the amount of sugar filled in a bag.)

SOLUTION   The adjustment is out of control if it overfills or underfills. Hence the alternative hypothesis is two-sided.

1. $H_0: \mu = 5$
2. $H_A: \mu \neq 5$
3. The test statistic is $\dfrac{\overline{X} - 5}{\sigma/\sqrt{n}}$, since $\mu_0 = 5$.
4. The alternative hypothesis is two-sided. Therefore, we have a two-tailed test. Since $\alpha = 0.05$, $z_{\alpha/2} = z_{0.025} = 1.96$. The decision rule is *Reject H_0 if the computed value is less than -1.96 or greater than 1.96.*
5. We have $\overline{x} = 5.1$, $\sigma = 0.15$, and $n = 16$. Therefore, the computed value of the test statistic is

$$\frac{\overline{x} - 5}{\sigma/\sqrt{n}} = \frac{5.1 - 5}{0.15/\sqrt{16}} = 2.67 \text{ (see Figure 9-14)}.$$

FIGURE 9-14

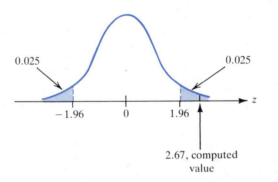

6. Since the computed value is greater than 1.96, we reject H_0 (P-value 2(0.0038), or 0.0076).

Conclusion: There is a strong evidence indicating that the adjustment is out of control.

THE POPULATION MEAN WHEN THE POPULATION VARIANCE IS UNKNOWN

In tests of hypotheses regarding the population mean, in more realistic situations, the population variance is unknown. We wish to test the following null hypothesis:

$$H_0: \mu = \mu_0$$

In the case where σ was known, we used the test statistic

$$\frac{\overline{X} - \mu_0}{\sigma/\sqrt{n}}.$$

Since σ is not known, we will use its estimator S. Hence the appropriate test statistic is

$$T = \frac{\overline{X} - \mu_0}{S/\sqrt{n}}.$$

At this point we need the added assumption that *the population is normally distributed,* especially if n is small. Since, under this assumption, the statistic T has Student's t distribution with $n - 1$ degrees of freedom, we get the decision rules given in Table 9-3, depending upon the particular alternative hypothesis.

TABLE 9-3	
Procedure to test H_0: $\mu = \mu_0$ when σ^2 is not known.	
Alternative hypothesis	*The decision rule is to reject H_0 if the computed value of T is*
1. $\mu > \mu_0$	greater than $t_{n-1,\alpha}$
2. $\mu < \mu_0$	less than $-t_{n-1,\alpha}$
3. $\mu \neq \mu_0$	less than $-t_{n-1,\alpha/2}$ or greater than $t_{n-1,\alpha/2}$

If σ is unknown but n is large (that is, $n > 30$), then, even if the population is not normally distributed, the test statistic

$$\frac{\overline{X} - \mu_0}{S/\sqrt{n}}$$

will have, approximately, a standard normal distribution. So we can use the test procedure described in Table 9-2 with the only difference that we have S in place of σ in the test statistic. Of course we have to bear in mind that the level of significance will be, approximately, 100α percent.

EXAMPLE 4   A car salesman claims that a particular model of car would give a mean mileage of greater than 20 miles per gallon. To test the claim, a field experiment was conducted where 10 cars were each run on one gallon of gasoline. The results (in miles) were:

    23,   18,   22,   19,   19,   22,   18,   18,   24,   22.

Is the salesman's claim justified? Use $\alpha = 0.05$, and assume a normal distribution for mileage per gallon.

SOLUTION   1. H_0: $\mu = 20$
2. H_A: $\mu > 20$
3. The standard deviation of the distribution is not known; we will have to estimate it from the sample. Hence the test statistic is $\dfrac{\overline{X} - 20}{S/\sqrt{n}}$.
4. Here $\alpha = 0.05$ and, since $n = 10$, we have $10 - 1$, or 9 degrees of freedom. Hence, from the table for Student's t distribution (Table A-5), we find

$t_{9,0.05} = 1.833$. The decision rule is *Reject H_0 if the computed value of the test statistic is greater than 1.833.*

5. It can be easily verified from the data that

$$\bar{x} = 20.5 \quad \text{and} \quad s = \sqrt{\frac{\sum\limits_{i=1}^{10}(x_i - \bar{x})^2}{n-1}} = 2.32.$$

Hence the computed value is

$$\frac{\bar{x}-20}{s/\sqrt{n}} = \frac{20.5-20}{2.32/\sqrt{10}} = 0.681 \text{ (see Figure 9-15).}$$

FIGURE 9-15

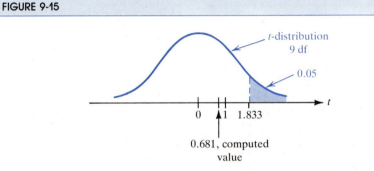

6. We do not reject H_0, since the computed value is less than 1.833 (P-value greater than 0.1).

Conclusion: The evidence at hand does not corroborate the salesman's claim at the 5 percent level of significance.   ▬▬▬

EXAMPLE 5   An auto dealer believes that his new model will give mean trouble-free service of at least 12,000 miles. In a simulated test with 4 cars, the following numbers of trouble-free miles were obtained:

11,000,   12,000,   11,800,   11,200.

Test the null hypothesis that the mean mileage is at least 12,000 miles against the alternative hypothesis that it is not. Use $\alpha = 0.05$. (Assume a normal distribution.)

SOLUTION   1. $H_0\colon \mu = 12,000$ (actually $\mu \geqslant 12,000$)
2. $H_A\colon \mu < 12,000$
3. The test statistic is $\dfrac{\bar{X} - 12,000}{S/\sqrt{n}}$.
4. $\alpha = 0.05$ and the number of degrees of freedom is $4 - 1 = 3$. Therefore, the decision rule is *Reject H_0 if the computed value of the test statistic is less than* $-t_{3,0.05} = -2.353.$

5. Routine computations give $\bar{x} = 11,500$ and $s = 476.095$. Hence the computed value of the test statistic is

$$\frac{\bar{x} - 12,000}{s/\sqrt{n}} = \frac{11,500 - 12,000}{476.095/\sqrt{4}} = -2.10 \text{ (see Figure 9-16).}$$

FIGURE 9-16

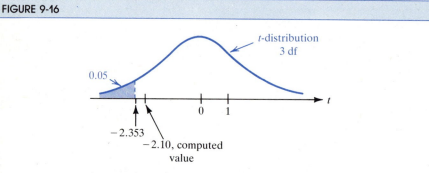

6. Since the computed value (-2.10) is greater than -2.353, we do not reject H_0 (P-value greater than 0.05).

Conclusion: There is no conclusive evidence at the 5 percent level of significance to refute the dealer's claim. ▮

EXAMPLE 6   A machine can be adjusted so that when under control, the mean amount of sugar filled in a bag is 5 pounds. To check if the machine is under control, six bags were picked at random and their weights (in pounds) were found to be

$$5.3, \quad 5.2, \quad 4.8, \quad 5.2, \quad 4.8, \quad 5.3.$$

At the 5 percent level of significance, is it true that the machine is not under control? (Assume a normal distribution for the weight of a bag.)

SOLUTION σ is not known and so will have to be estimated from the sample.

1. H_0: $\mu = 5$
2. H_A: $\mu \neq 5$
3. The test statistic is $\dfrac{\bar{X} - 5}{S/\sqrt{n}}$.
4. We have a two-tailed test. Since $\alpha = 0.05$ and $n = 6$, $t_{n-1,\alpha/2} = t_{5,0.025} = 2.571$. Therefore, the decision rule is *Reject H_0 if the computed value is less than -2.571 or greater than 2.571.*
5. $\bar{x} = \dfrac{30.6}{6} = 5.1$, and $s^2 = \dfrac{0.28}{5} = 0.056$.

   Hence

   $$\frac{\bar{x} - 5}{s/\sqrt{n}} = \frac{5.1 - 5}{\sqrt{0.056}/\sqrt{6}} = 1.04 \text{ (see Figure 9-17).}$$

FIGURE 9-17

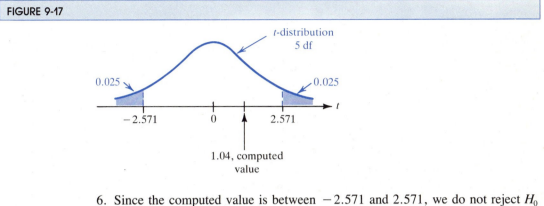

6. Since the computed value is between -2.571 and 2.571, we do not reject H_0 (P-value greater than $2(0.1)$, or 0.2).

Conclusion: There is no evidence to indicate that the machine is not under control. ■

THE POPULATION VARIANCE

While testing hypotheses concerning the population mean, we used the population variance when it was available and estimated it when it was not available. So we are already aware of instances where there is need for some information about the population variance σ^2. Knowledge about σ^2 is also important in many practical problems, such as in quality control, where the magnitude of the variance will be very crucial in determining whether the process is under control. For example, suppose a machine packs a mean amount of 5 pounds per bag exactly as the specifications require. In spite of this, we would regard the machine not under control if the standard deviation should turn out to be 1 pound. It is therefore easy to see that very often the variance of the population will be the main object of the researcher's interest and study.

We are interested in testing a null hypothesis that postulates that the population variance has a specified value σ_0^2 against various alternative hypotheses. In other words, we want to test

$$H_0: \sigma^2 = \sigma_0^2.$$

For testing hypotheses about μ, as a natural first step, we initiated our discussion by using the sample mean $\bar{x}$. This eventually led to our using the test statistic

$$\frac{\overline{X} - \mu_0}{\sigma/\sqrt{n}}.$$

For testing hypotheses concerning the population variance σ^2, what would be more natural than starting with its sample counterpart, namely, the sample variance,

$$s^2 = \frac{\sum_{i=1}^{n}(x_i - \bar{x})^2}{n-1}?$$

It turns out that this leads to the test statistic

$$\frac{(n-1)S^2}{\sigma_0^2}$$

on the basis of which decision rules are formulated.

At this time we will have to know the distribution of $(n-1)S^2/\sigma_0^2$. In this connection, we shall assume that *the population from which the sample is picked has a normal distribution.* Now we know from Chapter 7 that if the population is normally distributed with variance σ^2, then the distribution of $(n-1)S^2/\sigma^2$ is chi-square with $n-1$ degrees of freedom, where n represents the sample size. Therefore, *if H_0 is true*, that is, if the population variance is σ_0^2, then the distribution of $(n-1)S^2/\sigma_0^2$ will be chi-square with $n-1$ degrees of freedom.

In Table 9-4, we give test procedures depending upon whether the alternative hypothesis is right-sided, left-sided, or two-sided. (Recall that $\chi_{n-1,\alpha}^2$ is the table value of a chi-square distribution with $n-1$ degrees of freedom such that the area to its right is α.)

TABLE 9-4
Procedure to test H_0: $\sigma = \sigma_0$

Alternative hypothesis	*The decision rule is to reject H_0 if the computed value of* $\dfrac{(n-1)S^2}{\sigma_0^2}$ *is*
1. $\sigma > \sigma_0$	greater than $\chi_{n-1,\alpha}^2$
2. $\sigma < \sigma_0$	less than $\chi_{n-1,1-\alpha}^2$
3. $\sigma \neq \sigma_0$	less than $\chi_{n-1,1-\alpha/2}^2$ or greater than $\chi_{n-1,\alpha/2}^2$

EXAMPLE 7   The following measurements (in ounces) were obtained on the weights of six laboratory mice picked at random: 12, 8, 7, 12, 14, and 13. At the 5 percent level of significance, is the population variance of the weights of such mice greater than 2.25?

SOLUTION
1. H_0: $\sigma^2 = 2.25$
2. H_A: $\sigma^2 > 2.25$
3. The test statistic is $\dfrac{(n-1)S^2}{2.25}$.
4. The alternative hypothesis is right-sided. Since $n = 6$, the chi-square has $6 - 1$ or 5 degrees of freedom. From Table A-6, $\chi_{5,0.05}^2 = 11.07$. Therefore, the decision rule is *Reject H_0 if the computed value of the test statistic exceeds 11.07.*
5. It can be verified that $\bar{x} = 11$ and $(n-1)s^2 = \sum_{i=1}^{n}(x_i - \bar{x})^2 = 40$. Therefore, the computed value of the test statistic is

$$\frac{(n-1)s^2}{2.25} = \frac{40}{2.25} = 17.8 \text{ (see Figure 9-18)}.$$

FIGURE 9-18

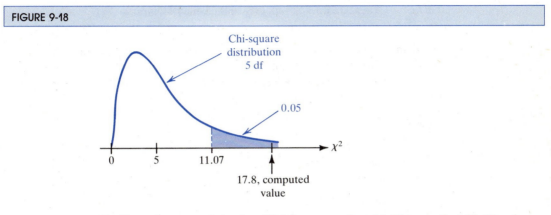

6. Since the computed value 17.8 is greater than 11.07, we reject H_0 (P-value less than 0.005).

Conclusion: At the 5 percent level, there is evidence indicating that the population variance exceeds 2.25.   ▬

EXAMPLE 8   A salesman claims that the standard deviation of the diameter of the bolts manufactured by his company is less than 0.1 inch. Based on a sample of eleven bolts, the sample standard deviation was found to be 0.07. At the 5 percent level of significance, is the salesman's claim valid?

SOLUTION   1. H_0: $\sigma^2 = (0.1)^2 = 0.01$
2. H_A: $\sigma^2 < 0.01$
3. The test statistic is $\dfrac{(n-1)S^2}{0.01}$.
4. Since $n = 11$, there are 10 degrees of freedom. The test is a left-tailed test and $\chi^2_{10,0.95} = 3.94$. Therefore, the decision rule is *Reject H_0 if the computed value of the test statistic is less than 3.94*.
5. $s = 0.07$. Hence the computed value of the test statistic is

$$\frac{(n-1)s^2}{0.01} = \frac{10(0.07)^2}{0.01} = 4.9 \text{ (see Figure 9-19)}.$$

FIGURE 9-19

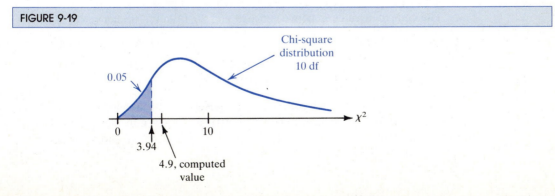

6. Since the computed value is greater than 3.94, we do not reject H_0 (P-value greater than 0.05).

Conclusion: The data do not support the salesman's claim. ▰▰

EXAMPLE 9   A random sample of 15 bulbs produced by a machine was picked, and each bulb was tested for the number of hours it lasted before it blew. If $s = 13$ hours, at the 5 percent level of significance, is the population standard deviation different from 10 hours?

SOLUTION   1. $H_0: \sigma = 10$
2. $H_A: \sigma \neq 10$
3. The test statistic is $\dfrac{(n-1)S^2}{10^2}$.
4. The alternative hypothesis is two-sided so that the test is two-tailed. Since $n = 15$ and $\alpha = 0.05$,
$$\chi^2_{n-1,\alpha/2} = \chi^2_{14,0.025} = 26.119 \qquad \text{and} \qquad \chi^2_{n-1,1-\alpha/2} = \chi^2_{14,0.975} = 5.629.$$
The decision rule is *Reject H_0 if the computed value of the test statistic is less than 5.629 or greater than 26.119.*
5. Since $s = 13$, the computed value of the test statistic is
$$\frac{(n-1)s^2}{100} = \frac{(14)(169)}{100} = 23.66 \text{ (see Figure 9-20).}$$

FIGURE 9-20

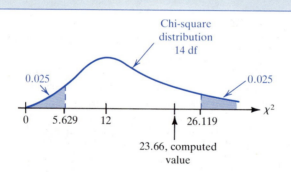

6. The computed value is between 5.629 and 26.119. We do not reject H_0.

Conclusion: There is insufficient evidence to warrant that $\sigma \neq 10$. ▰▰

THE POPULATION PROPORTION FOR QUALITATIVE DATA

So far in this chapter we have concerned ourselves with treating data where the observed variable can be measured on a numerical scale. We shall now treat the case of a qualitative variable, where the data are recorded as tall-short, black-green, defective-nondefective, and so forth. Our objective will be to test hypotheses regarding the proportion p of a certain attribute in the population. We shall specifically consider the problem of testing the null hypothesis $H_0: p = p_0$, where p_0 is a number between 0 and 1, against various alternative hypotheses. For example, we might be interested in the proportion of defective items produced by a machine and wish to test

$$H_0: p = 0.2 \qquad \text{against} \qquad H_A: p > 0.2$$

or $\qquad H_0: p = 0.2 \qquad \text{against} \qquad H_A: p < 0.2$

or $\qquad H_0: p = 0.2 \qquad \text{against} \qquad H_A: p \neq 0.2$

To carry out a test of a hypothesis regarding the population proportion, we begin by picking a sample of independent observations and using the sample proportion as the statistic on which the test is based. Now if p is the proportion in the population, then, as we know, the sample proportion X/n has a sampling distribution with mean p and standard deviation $\sqrt{p(1-p)/n}$. Furthermore, if the sample is large in such a way that both np and $n(1-p)$ are greater than 5, the shape of the distribution of X/n is approximately normal. Consequently, *under the null hypothesis,* which postulates that the population proportion is p_0, X/n has a distribution that is approximately normal with mean p_0 and standard deviation $\sqrt{p_0(1-p_0)/n}$, provided np_0 and $n(1-p_0)$ are both greater than 5.

We now have a situation analogous to the one where we tested hypotheses regarding the population mean when σ^2 was known. The role of $\overline{X}$ is played by X/n, that of μ_0 by p_0, and that of $\sigma/\sqrt{n}$ by $\sqrt{p_0(1-p_0)/n}$. We list in Table 9-5 the three cases based on the nature of the alternative hypothesis.

It is worth keeping in mind that the present test procedure is based on the application of the central limit theorem. Hence the significance level is *approximately* 100α percent. This should be understood even when it is not mentioned in a specific problem.

TABLE 9-5

Procedure to test $H_0: p = p_0$ when sample size is large so that $np_0 \geqslant 5$ and $n(1-p_0) \geqslant 5$

Alternative hypothesis	*The decision rule is to reject H_0 if the computed value of* $\dfrac{(X/n) - p_0}{\sqrt{\dfrac{p_0(1-p_0)}{n}}}$ *is*
1. $p > p_0$	greater than z_α
2. $p < p_0$	less than $-z_\alpha$
3. $p \neq p_0$	less than $-z_{\alpha/2}$ or greater than $z_{\alpha/2}$

EXAMPLE 10   When a coin was tossed 200 times, it showed heads 120 times. Is the coin biased in favor of heads? Carry out the test at the 5 percent level of significance.

SOLUTION

Let p = probability of getting a head on a toss. The question posed in the problem is ''Is $p > 0.5$?'' Under null hypothesis, we state that this is not the case.

1. $H_0: p = 0.5$ (actually $p \leqslant 0.5$)
2. $H_A: p > 0.5$
3. Since $p_0 = 0.5$, the test statistic is $\dfrac{(X/n) - 0.5}{\sqrt{(0.5)(0.5)/n}}$.
4. $\alpha = 0.05$, therefore the decision rule is *Reject H_0 if the computed value is greater than $z_{0.05} = 1.645$.*
5. Since $x = 120$ and $n = 200$, the computed value of the statistic is

$$\frac{(x/n) - 0.5}{\sqrt{(0.5)(0.5)/n}} = \frac{\dfrac{120}{200} - 0.5}{\sqrt{0.25/200}} = 2.83 \text{ (see Figure 9-21)}.$$

FIGURE 9-21

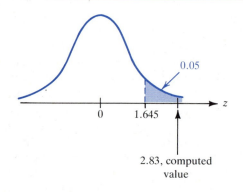

6. Since the computed value 2.83 is greater than 1.645, we reject H_0 (P-value 0.0023).

Conclusion: At the 5 percent level of significance, there is strong evidence that the coin is biased in favor of heads. ▬

EXAMPLE 11

A machine is known to produce 30 percent defective tubes. After repairing the machine, it was found that it produced 22 defective tubes in the first run of 100. Is it true that after the repairs the proportion of defective tubes is reduced? Use $\alpha = 0.01$.

SOLUTION

Let p denote the proportion of defective tubes.

1. $H_0: p = 0.3$
2. $H_A: p < 0.3$

3. The test statistic is $\dfrac{(X/n) - 0.3}{\sqrt{(0.3)(0.7)/n}}$.

4. Since $\alpha = 0.01$, the decision rule is *Reject H_0 if the computed value is less than* $-z_{0.01} = -2.33$.

5. Since $x = 22$ and $n = 100$, the computed value is

$$\frac{(x/n) - 0.3}{\sqrt{(0.3)(0.7)/n}} = \frac{0.22 - 0.3}{\sqrt{0.21/100}} = -1.746 \text{ (see Figure 9-22)}.$$

FIGURE 9-22

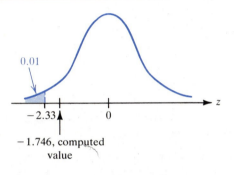

6. Since the computed value -1.746 is greater than -2.33, we do not reject H_0 (P-value 0.0401).

Conclusion: There is no evidence (at the 1 percent level) that the proportion of defective tubes produced by the machine is reduced.   ▬

EXAMPLE 12 The proportion of Americans who traveled abroad last year was 20 percent. To find the attitude of people on foreign travel this year, 100 people were interviewed. Of these, 15 said they would travel and the remaining 85 said they would not. Has the attitude changed from last year? Use $\alpha = 0.10$.

SOLUTION Let p represent the proportion of those who would travel.

1. H_0: $p = 0.2$
2. H_A: $p \neq 0.2$
3. Since $p_0 = 0.2$, the test statistic is $\dfrac{(X/n) - 0.2}{\sqrt{(0.2)(0.8)/n}}$.
4. Since $\alpha = 0.10$, $\alpha/2 = 0.05$, and $z_{\alpha/2} = 1.645$, the decision rule is *Reject H_0 if the computed value is less than* -1.645 *or greater than* 1.645.
5. Since $x = 15$ and $n = 100$,

$$\frac{(x/n) - 0.2}{\sqrt{(0.2)(0.8)/n}} = \frac{0.15 - 0.2}{\sqrt{0.16/100}} = -1.25 \text{ (see Figure 9-23)}.$$

FIGURE 9-23

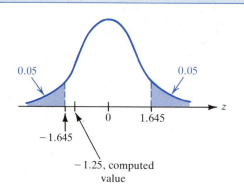

6. The computed value -1.25 is between -1.645 and 1.645, and so we do not reject H_0 (P-value 2(0.1056), or 0.2112).

Conclusion: At the 10 percent level of significance, there is no evidence to indicate that the attitude this year is different from that of last year. ▬▬

SECTION 9-3   EXERCISES

In Exercises 1–6, determine appropriate test statistics and critical regions that you would use in carrying out tests of the hypotheses. In each case provide the decision rule. You may assume that the populations are normally distributed.

1. H_0: $\mu = 10.2$ against H_A: $\mu > 10.2$ when $\sigma = 1.3$, $n = 16$, $\alpha = 0.05$

2. H_0: $\mu = 110$ against H_A: $\mu < 110$ when $\sigma = 4.32$, $n = 25$, $\alpha = 0.025$

3. H_0: $\mu = 32.8$ against H_A: $\mu < 32.8$ when $\sigma = 6.8$, $n = 14$, $\alpha = 0.01$

4. H_0: $\mu = 17.8$ against H_A: $\mu > 17.8$ when $\sigma = 2.8$, $n = 20$, $\alpha = 0.02$

5. H_0: $\mu = -3.3$ against H_A: $\mu \neq -3.3$ when $\sigma = 4.8$, $n = 20$,
   $\alpha = 0.02$

6. H_0: $\mu = 11.6$ against H_A: $\mu \neq 11.6$ when $\sigma = 3.2$, $n = 12$, $\alpha = 0.1$

7. The time (in minutes) taken by a biological cell to divide into two cells has a normal distribution. From past experience, the population standard deviation σ can be assumed to be 3.5 minutes. When 16 cells were observed, the mean time taken by them to divide into two was 31.6 minutes. At the 5 percent level of significance, test

   (a) H_0: $\mu = 30$ against H_A: $\mu \neq 30$

   (b) H_0: $\mu = 30$ against H_A: $\mu > 30$.

8. A breeder of crabs claims that he can breed crabs yielding a mean weight of greater than 28 ounces. Suppose the standard deviation σ is known to be 3 ounces. A random sample of 16 crabs had a mean weight of 29.2 ounces. At the 5 percent level of significance, is the breeder's claim justified?

9. The label on a can of pineapple slices states that the mean carbohydrate content per serving of canned pineapple is over 50 grams. It may be assumed that the standard deviation of the carbohydrate content σ is 4 grams. A random sample of 25 servings has a mean carbohydrate content of 52.3 grams. Is the company correct in its claim? Use $\alpha = 0.05$.

10. From his long-standing experience, a farmer believes that the mean yield of grain per plot on his farm is 150 bushels. When a new seed introduced on the market was tried on 16 randomly picked experimental plots, the mean yield was 158 bushels. Suppose the yield per plot can be assumed to be normally distributed with a standard deviation of yield σ of 20 bushels. Is the new seed significantly better? Use $\alpha = 0.02$.

11. The contention of the factory management is that the mean weekly idle time per machine is over 5 hours. The foreman does not agree with this. When a random sample of 25 machines was observed, the mean weekly idle time was found to be 5.8 hours. From past experience, it is safe to assume that the idle time of a machine during a week is normally distributed with a standard deviation of 2 hours. At the 3 percent level of significance, is the contention of the management justified?

In Exercises 12–17, determine appropriate test statistics and critical regions for carrying out the tests of hypotheses. In each case, provide the decision rule. You may assume that the populations are normally distributed with unknown variance σ^2.

12. H_0: $\mu = 10.2$ against H_A: $\mu > 10.2$ when $s = 1.3$, $n = 16$, $\alpha = 0.05$

13. H_0: $\mu = 110$ against H_A: $\mu < 110$ when $s = 4.32$, $n = 25$, $\alpha = 0.025$

14. H_0: $\mu = 32.8$ against H_A: $\mu < 32.8$ when $s = 6.8$, $n = 14$, $\alpha = 0.01$

15. H_0: $\mu = 17.8$ against H_A: $\mu > 17.8$ when $s = 2.8$, $n = 20$, $\alpha = 0.025$

16. H_0: $\mu = -3.3$ against H_A: $\mu \neq -3.3$ when $s = 4.8$, $n = 20$,
    $\alpha = 0.02$

17. H_0: $\mu = 11.6$ against H_A: $\mu \neq 11.6$ when $s = 3.2$, $n = 12$, $\alpha = 0.1$

18. A random sample of size $n = 16$ is drawn from a population having a normal distribution. The sample mean and the sample variance are given, respectively, as $\bar{x} = 23.8$ and $s^2 = 10.24$. At the 5 percent level of significance, test

    (a) H_0: $\mu = 25$ against H_A: $\mu \neq 25$

    (b) H_0: $\mu = 25$ against H_A: $\mu < 25$.

19. When a skier took ten runs on the downhill course, it was found that the mean time was 12.3 minutes with $s = 1.2$ minutes. Is the true mean time of the skier down the course less than 13 minutes? Use $\alpha = 0.025$.

20. A sample of eight workers in a clothing-manufacturing company gave the following figures for the amount of time (in minutes) needed to join a collar to a shirt:

         10,   12,   13,   9,   8,   14,   10,   11.

    At the 5 percent level of significance, is the true mean time to join a collar more than 10 minutes?

21. A manufacturer of automobile tires claims that the mean number of trouble-free miles

given by a new tire introduced by his company is more than 36,000. When 16 randomly picked tires were tested, the mean number of miles was 38,900 with the sample standard deviation s equal to 8200 miles. At the 5 percent level of significance, is the manufacturer's claim justified?

22. The following figures refer to the amount of coffee (in ounces) filled by a machine in six randomly picked jars;

$$15.7, \qquad 15.9, \qquad 16.3, \qquad 16.2, \qquad 15.7, \qquad 15.9.$$

Is the true mean amount of coffee filled in a jar less than 16 ounces? Use $\alpha = 0.05$.

In Exercises 23–26, determine appropriate test statistics and critical regions for carrying out the tests of hypotheses. In each case provide the decision rule.

[*Hint:* Notice n is large in each case.]

23. $H_0: \mu = 9.8$ against $H_A: \mu < 9.8$ when $s = 1.6$, $n = 46$, $\alpha = 0.05$

24. $H_0: \mu = 82.0$ against $H_A: \mu \neq 82.0$ when $s = 6.5$, $n = 64$, $\alpha = 0.02$

25. $H_0: \mu = -3.0$ against $H_A: \mu > -3.0$ when $s = 2.4$, $n = 80$, $\alpha = 0.1$

26. $H_0: \mu = 0$ against $H_A: \mu \neq 0$ when $s = 1$, $n = 50$, $\alpha = 0.1$

27. In a survey in which 100 randomly picked middle-class families were interviewed, it was found that their mean medical expenses during a year were $770 with $s = \$120$. Are the true mean medical expenses during the year significantly greater than $750? Use $\alpha = 0.025$.

28. A biologist claims that the mean cholesterol content of chicken eggs can be reduced by giving a special diet to the chickens. When 36 eggs obtained in this way were tested, it was found that the mean cholesterol content was 245 mg with a sample standard deviation equal to 20 mg. Is the true mean cholesterol content significantly less than 250 mg? Use $\alpha = 0.05$

In Exercises 29–34, determine appropriate test statistics and critical regions for carrying out the tests of hypotheses. In each case provide the decision rule. You may assume that the populations are normally distributed.

29. $H_0: \sigma^2 = 8.2$ against $H_A: \sigma^2 \neq 8.2$ when $n = 16$, $\alpha = 0.05$

30. $H_0: \sigma^2 = 0.36$ against $H_A: \sigma^2 < 0.36$ when $n = 7$, $\alpha = 0.025$

31. $H_0: \sigma = 6.8$ against $H_A: \sigma > 6.8$ when $n = 14$, $\alpha = 0.05$

32. $H_0: \sigma^2 = 10$ against $H_A: \sigma^2 \neq 10$ when $n = 18$, $\alpha = 0.1$

33. $H_0: \sigma^2 = 0.78$ against $H_A: \sigma^2 < 0.78$ when $n = 8$, $\alpha = 0.01$

34. $H_0: \sigma^2 = 9$ against $H_A: \sigma^2 > 9$ when $n = 11$, $\alpha = 0.05$

35. A random sample of size $n = 11$ drawn from a normal population gave $s^2 = 2.45$. At the 5 percent level of significance, test

(a) $H_0: \sigma^2 = 7$ against $H_A: \sigma^2 \neq 7$

(b) $H_0: \sigma^2 = 7$ against $H_A: \sigma^2 < 7$

36. An archaeologist found that the mean cranial width of 21 skulls was 5.3 inches with the sample standard deviation equal to 0.55 inch. Is the true standard deviation of cranial width significantly greater than 0.5? Use $\alpha = 0.025$.

37. A grade school teacher has obtained the following figures for the time (in minutes) that ten randomly picked students in her class took to complete a task:

    15,   10,   12,   8,   11,   10,   10,   13,   12,   9.

    At the 5 percent level of significance, is the true variance of the time different from 2.4 minutes squared?

38. A coffee-dispensing machine is regulated to dispense 6 ounces of coffee per cup. The machine will need to be adjusted if the true variance of the amount per cup is greater than 1.1 ounces squared. When a sample of 16 random cups was inspected, it was found that s^2 was 1.95. At the 5 percent level of significance, does the machine need adjustment?

In Exercises 39–44 determine appropriate test statistics and critical regions for carrying out the tests of hypotheses. In each case provide the critical region.

39. $H_0: p = 0.4$ against $H_A: p \neq 0.4$ when $n = 60$, $\alpha = 0.05$

40. $H_0: p = 0.2$ against $H_A: p < 0.2$ when $n = 80$, $\alpha = 0.025$

41. $H_0: p = 0.7$ against $H_A: p > 0.7$ when $n = 36$, $\alpha = 0.05$

42. $H_0: p = 0.5$ against $H_A: p \neq 0.5$ when $n = 64$, $\alpha = 0.1$

43. $H_0: p = 0.8$ against $H_A: p < 0.8$ when $n = 120$, $\alpha = 0.01$

44. $H_0: p = 0.3$ against $H_A: p > 0.3$ when $n = 96$, $\alpha = 0.05$

45. A drug company claims that more than 80 percent of the people given a vaccine will develop immunity to a disease. Of 100 randomly picked people who were given the vaccine, 88 developed immunity. On the basis of this evidence, is the claim of the drug company valid? Use $\alpha = 0.03$.

46. In a sample of 100 cathode tubes inspected, 14 were found to be defective. If the true proportion of defectives is significantly higher than the 8 percent that the company considers tolerable, repairs in the machine are in order. At the 5 percent level of significance, does the machine need repairs?

47. It has been claimed that the majority of Americans are in favor of the proposed Star Wars strategic defense. In a survey in which 400 people were interviewed, it was found that 224 were in favor of the Star Wars defense. At the 5 percent level of significance, is the claim correct?

48. A wire service flashed the following news item: "The credibility of Congress is down to less than 20 percent." In a survey in which 400 random individuals were interviewed, 70 thought the performance of Congress was satisfactory. Is the wire service correct in its news flash? Use $\alpha = 0.10$.

49. Should a national lottery be instituted? A random sample of 400 adults gave the following results: 144 in favor, 256 opposed. Is it true that less than 40 percent of the population favor a national lottery? Use $\alpha = 0.05$.

9-4 TESTS OF HYPOTHESES (TWO POPULATIONS)

In this section we shall test hypotheses concerning the difference of means of two populations and the difference of proportions of an attribute in two populations. The setup in each case regarding the nature of the sampled populations (notations included) is as defined in the corresponding counterpart in Chapter 8.

DIFFERENCE IN POPULATION MEANS WHEN THE VARIANCES ARE KNOWN

The null hypothesis under test is

$$H_0: \mu_1 = \mu_2, \text{ that is } \mu_1 - \mu_2 = 0$$

and the test statistic appropriate for the purpose is

$$\frac{\overline{X} - \overline{Y}}{\sqrt{\dfrac{\sigma_1^2}{m} + \dfrac{\sigma_2^2}{n}}}.$$

The decision rules for various forms of alternative hypothesis are given in Table 9-6.

TABLE 9-6
Procedure to test $H_0: \mu_1 = \mu_2$ when the populations are normally distributed and variances are known

Alternative hypothesis	*The decision rule is to reject H_0 if the computed value is*
1. $\mu_1 > \mu_2$	greater than z_α
2. $\mu_1 < \mu_2$	less than $-z_\alpha$
3. $\mu_1 \neq \mu_2$	less than $-z_{\alpha/2}$ or greater than $z_{\alpha/2}$

The assumption of normal distribution of populations is not crucial if sample sizes are large, that is, as a general rule, m and n both greater than 30. Actually, in this case, even the population variances need not be known. We can use the sample variances s_1^2 and s_2^2 in place of the respective population variances σ_1^2 and σ_2^2. The test procedure is the same as given in Table 9-6 except that we use the test statistic

$$\frac{\overline{X} - \overline{Y}}{\sqrt{\dfrac{s_1^2}{m} + \dfrac{s_2^2}{n}}}.$$

EXAMPLE 1   For a sample of 15 adult Europeans picked at random, the mean weight $\overline{x}$ was 154 pounds, whereas for a sample of 18 adults in the United States, the mean weight $\overline{y}$ was 162 pounds. From past surveys it is known that the variance of weight in Europe σ_1^2 is 100 and in the United States σ_2^2 is 169. Is it true that there is significant difference between mean weights in the two places? Use $\alpha = 0.05$. (Assume that the weights are normally distributed.)

SOLUTION

1. $H_0: \mu_1 = \mu_2$, that is, $\mu_1 - \mu_2 = 0$
2. $H_A: \mu_1 \neq \mu_2$
3. The test statistic is $\dfrac{\overline{X} - \overline{Y}}{\sqrt{\dfrac{\sigma_1^2}{m} + \dfrac{\sigma_2^2}{n}}}$.
4. Since $\alpha = 0.05$, $z_{\alpha/2} = 1.96$. The decision rule is *Reject H_0 if the computed value falls outside the interval -1.96 to 1.96.*
5. $\overline{x} = 154$, $\overline{y} = 162$, $\sigma_1^2 = 100$, $\sigma_2^2 = 169$, $m = 15$, and $n = 18$. Therefore, the computed value is

$$\frac{\overline{x} - \overline{y}}{\sqrt{\dfrac{\sigma_1^2}{m} + \dfrac{\sigma_2^2}{n}}} = \frac{154 - 162}{\sqrt{\dfrac{100}{15} + \dfrac{169}{18}}} = \frac{-8}{\sqrt{16.056}} = -2 \text{ (see Figure 9-24)}.$$

FIGURE 9-24

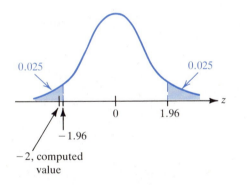

6. Since the computed value is less than -1.96, we reject H_0 (P-value 0.0456).

Conclusion: There is a significant difference between the mean weights in the two places, with the data indicating that the mean weight of adults in the United States is more than that of adults in Europe. ■

EXAMPLE 2

In order to compare two brands of cigarettes, Brand A and Brand B, for their tar content, a sample of 60 was inspected from Brand A and a sample of 40 from Brand B. The results of the tests are summarized as follows:

Brand A	$\overline{x} = 15.4$	$s_1^2 = 3$
Brand B	$\overline{y} = 16.8$	$s_2^2 = 4$

At the 5 percent level of significance, do the two brands differ in their mean tar content?

SOLUTION   Since the two samples are large, we can use s_1^2 in place of σ_1^2 and s_2^2 in place of σ_2^2. Let $\mu_{\text{Brand }A}$ represent the mean for Brand A, with a similar interpretation for $\mu_{\text{Brand }B}$.

1. H_0: $\mu_{\text{Brand }A} = \mu_{\text{Brand }B}$
2. H_A: $\mu_{\text{Brand }A} \neq \mu_{\text{Brand }B}$
3. The test statistic is $\dfrac{\overline{X} - \overline{Y}}{\sqrt{\dfrac{S_1^2}{m} + \dfrac{S_2^2}{n}}}$.
4. Since $\alpha = 0.05$, $z_{\alpha/2} = 1.96$. The decision rule is *Reject H_0 if the computed value is less than -1.96 or greater than 1.96.*
5. The computed value of the statistic is

$$\frac{\overline{x} - \overline{y}}{\sqrt{\dfrac{s_1^2}{m} + \dfrac{s_2^2}{n}}} = \frac{15.4 - 16.8}{\sqrt{\dfrac{3}{60} + \dfrac{4}{40}}} = \frac{-1.4}{\sqrt{\dfrac{3}{20}}} = -3.62. \text{ (See Figure 9-25.)}$$

FIGURE 9-25

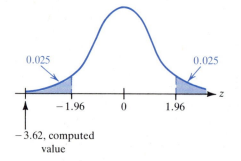

6. The computed value -3.62 is less than -1.96. We therefore reject H_0 (P-value less than 0.002).

Conclusion: There is a strong reason to conclude that the two brands of cigarettes differ in their tar content. ▬

DIFFERENCE IN POPULATION MEANS WHEN THE VARIANCES ARE UNKNOWN BUT ARE ASSUMED EQUAL

The following test procedure is particularly suited for the case when *small* independent samples are drawn from normally distributed populations both having the *same variance*.

We are interested in testing the null hypothesis

$$H_0: \mu_1 = \mu_2.$$

When the variances were known, we used the statistic

$$\frac{\overline{X} - \overline{Y}}{\sqrt{\dfrac{\sigma_1^2}{m} + \dfrac{\sigma_2^2}{n}}}.$$

But we are given that the variances are equal. So suppose $\sigma_1^2 = \sigma_2^2$ and let σ^2 represent the common value. The above test statistic then reduces to

$$\frac{\overline{X} - \overline{Y}}{\sigma\sqrt{\dfrac{1}{m} + \dfrac{1}{n}}}.$$

Since σ is not known, we shall use its pooled estimate s_p, where

$$s_p^2 = \frac{(m-1)s_1^2 + (n-1)s_2^2}{m+n-2}.$$

Therefore, the test statistic appropriate for carrying out the test of H_0 is

$$\frac{\overline{X} - \overline{Y}}{S_p\sqrt{\dfrac{1}{m} + \dfrac{1}{n}}}.$$

When H_0 is true, this statistic has Student's t distribution with $m + n - 2$ degrees of freedom. The test procedures for the various forms of the alternative hypothesis are given in Table 9-7.

TABLE 9-7	
Procedure to test H_0: $\mu_1 = \mu_2$ when the variances are unknown but are assumed equal	
Alternative hypothesis	*The decision rule is to reject H_0 if the computed value is*
1. $\mu_1 > \mu_2$	greater than $t_{m+n-2,\alpha}$
2. $\mu_1 < \mu_2$	less than $-t_{m+n-2,\alpha}$
3. $\mu_1 \neq \mu_2$	less than $-t_{m+n-2,\alpha/2}$ *or* greater than $t_{m+n-2,\alpha/2}$

EXAMPLE 3   A nitrogen fertilizer was used on 10 plots and the mean yield per plot $\overline{x}$ was found to be 82.5 bushels with s_1 equal to 10 bushels. On the other hand, 15 plots treated with phosphate fertilizer gave a mean yield $\overline{y}$ of 90.5 bushels per plot with s_2 equal to 20 bushels. At the 5 percent level of significance, are the two fertilizers significantly different?

SOLUTION   1. H_0: $\mu_1 = \mu_2$
2. H_A: $\mu_1 \neq \mu_2$

3. The test statistic is $\dfrac{\bar{X} - \bar{Y}}{S_p\sqrt{\dfrac{1}{m} + \dfrac{1}{n}}}$.

4. $m = 10$ and $n = 15$. Therefore, the number of degrees of freedom is $m + n - 2 = 23$. Since $\alpha = 0.05$, $t_{23,\alpha/2} = t_{23,0.025} = 2.069$. The decision rule is *Reject H_0 if the computed value is less than -2.069 or greater than 2.069.*

5. $\bar{x} = 82.5$, $\bar{y} = 90.5$; $s_1 = 10$, $s_2 = 20$; $m = 10$, $n = 15$. Therefore,

$$s_p^2 = \frac{(m - 1)s_1^2 + (n - 1)s_2^2}{m + n - 2} = \frac{9(10)^2 + 14(20)^2}{10 + 15 - 2} = 282.6$$

so that $s_p = \sqrt{282.6} = 16.8$. The computed value of the test statistic is

$$\frac{\bar{x} - \bar{y}}{s_p\sqrt{\dfrac{1}{m} + \dfrac{1}{n}}} = \frac{82.5 - 90.5}{16.8\sqrt{\dfrac{1}{10} + \dfrac{1}{15}}} = \frac{-8}{6.9} = -1.16. \text{ (See Figure 9-26.)}$$

FIGURE 9-26

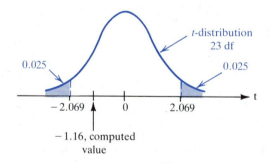

6. Since the computed value -1.16 is between -2.069 and 2.069, we do not reject H_0 (P-value greater than 0.05).

Conclusion: The data do not support the contention that the fertilizers are significantly different. ▬▬

DIFFERENCE IN POPULATION MEANS USING THE PAIRED *t* TEST

The method of a **paired *t* test** reduces the problem of comparing the means of two populations to that of a one-sample *t* test. Suppose we wish to compare two treatments which we will call Treatment 1 and Treatment 2. The two treatments could be two stimuli, two diets, two fertilizers, or two methods of teaching, and so on. The idea involves taking experimental units in pairs so that within each pair, the units are as homogeneous as possible thereby eliminating the influence of extraneous factors. Treatment 1 is then assigned to one of the members of the pair at random and Treatment 2 to the other member.

For instance, if we wish to compare two teaching methods we might pick, for example, 10 pairs of students so that within each pair both students are of the same socioeconomic background, same age, same IQ. (Pairs of twins might turn out to be ideal.) We then assign one student in a pair to be taught by one method and the other student by the second method.

A common method of obtaining matched pairs is to match an experimental unit with itself leading to what is popularly referred to as a "before–after" study.

The data can be recorded as ordered pairs (x, y), where x is the response to Treatment 1 and y is the response to Treatment 2. If there are n pairs which are independent of each other, we get a set of n independent pairs of values (x_1, y_1), $(x_2, y_2), \ldots , (x_n, y_n)$. When data are collected in this way, obviously the two samples of x values and y values are not independent. (If one member of a pair performs well (poorly) on a test, then we might reasonably expect the other one to do well (poorly) also.) Our previous test for comparing means was based on the strict assumption of independence of samples and so cannot be applied here. Therefore we devise a different test which is based on the difference of responses within pairs.

Let us denote the difference of the pair (x_i, y_i) as $d_i = x_i - y_i$. The n differences $d_1, d_2, \ldots , d_n$ now constitute a sample of n independent observations from a population of "within-pair differences D" which we shall assume to be normally distributed. If μ_1 is the mean for Treatment 1 and μ_2 that for Treatment 2, then it can be shown that μ_D, the population mean of paired differences, is equal to $\mu_1 - \mu_2$. We are interested in testing $H_0: \mu_1 = \mu_2$, which is the same thing as testing $\mu_D = 0$.

At this stage we have all the ingredients appropriate for applying the one-sample t-test. We have a sample of n independent observations $d_1, d_2, \ldots , d_n$ from a normal population whose variance is unknown, and we wish to test the hypothesis about its mean μ_D. We use the one-sample t test with $n - 1$ degrees of freedom to test $H_0: \mu_D = 0$ against the various alternative hypotheses using the test statistic whose value is

$$\frac{\bar{d} - 0}{s_d / \sqrt{n}}, \qquad \text{that is, } \frac{\bar{d} \sqrt{n}}{s_d}$$

where

$$\bar{d} = \frac{\sum\limits_{i=1}^{n} d_i}{n}, \qquad \text{and} \qquad s_d^2 = \frac{1}{n-1} \sum\limits_{i=1}^{n} (d_i - \bar{d})^2.$$

We have the decision rule given in Table 9-8.

TABLE 9-8	
Procedure for paired difference test of a hypothesis $H_0: \mu_1 = \mu_2$	
Alternative hypothesis	*The decision rule is Reject H_0 if the computed value of $\dfrac{\bar{d}\sqrt{n}}{s_d}$ is*
$\mu_D > 0$, that is, $\mu_1 > \mu_2$	greater than $t_{n-1,\alpha}$
$\mu_D < 0$, that is, $\mu_1 < \mu_2$	less than $-t_{n-1,\alpha}$
$\mu_D \neq 0$, that is, $\mu_1 \neq \mu_2$	less than $-t_{n-1,\alpha/2}$ or greater than $t_{n-1,\alpha/2}$

EXAMPLE 4   A dietician claims that a certain diet will reduce the dieter's weight. The following data record the weights (in pounds) of eight people, before and after the diet and the differences of these weights. At the 5 percent level of significance does the diet reduce weight?

Person	Before the diet x_i	After the diet y_i	Difference $d_i = x_i - y_i$
1	175	170	5
2	168	169	−1
3	140	133	7
4	130	132	−2
5	143	150	−7
6	128	116	12
7	138	130	8
8	165	163	2

SOLUTION
1. $H_0: \mu_D = 0$, that is, weight is not reduced.
2. $H_A: \mu_D > 0$, that is, weight is reduced.
3. The test statistic is $\dfrac{\bar{d}\sqrt{n}}{s_d}$.
4. $\alpha = 0.05$ and, since $n = 8$, there are 7 degrees of freedom. Since $t_{7,0.05} = 1.895$, the decision rule is *Reject H_0 if the computed value of the test statistic is greater than 1.895.*
5. From column 4 of the table, $\sum_{i=1}^{8} d_i = 24$, and $\sum_{i=1}^{8} d_i^2 = 340$. Therefore
   $$\bar{d} = 24/8 = 3$$
   and
   $$s_d^2 = \frac{1}{7}\left(340 - \frac{24^2}{8}\right) = 38.286$$
   so that $s_d = 6.1875$.
   Hence the computed value of the test statistic is
   $$\frac{\bar{d}\sqrt{n}}{s_d} = \frac{3\sqrt{8}}{6.1875} = 1.371 \text{ (see Figure 9-27).}$$

FIGURE 9-27

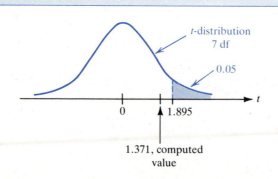

6. Since the computed value is less than 1.895 we do not reject H_0 (P-value greater than 0.1).

Conclusion: The claim of the dietician is not corroborated by the data.   ▬▬

DIFFERENCE IN POPULATION PROPORTIONS

Very often we are interested in deciding whether the observed difference between two sample proportions is due to sampling error or due to the fact that the proportions in the parent populations from which the samples are taken are inherently different. As an example, let us consider the following experiment, where Vaccine A is given to 200 and Vaccine B to 300 randomly picked people.

	Vaccine A	*Vaccine B*
Infected	80	105
Not infected	120	195
Total	200	300

Of the 200 people who were given Vaccine A, 80 were infected, that is, 40 percent were infected. On the other hand, with Vaccine B, only 105 out of 300 were infected, that is, 35 percent. The difference in the sample proportions is $0.40 - 0.35$, or 0.05. We might ask whether this difference is large enough to suggest that the two vaccines are different in their effectiveness; or is it just possible that the difference is due to sampling error, in which case we should not read much into it? In short, if p_1 and p_2 represent proportions of those infected in the two populations, we want to test the null hypothesis

$$H_0: p_1 = p_2, \text{ that is } p_1 - p_2 = 0.$$

If H_0 is true and $p_1 = p_2$, let us denote this common value as p. Suppose among the m people given Vaccine A, X are infected, and among the n people given Vaccine B, Y are infected. We can then obtain a pooled estimate of p using the estimator

$$\hat{p} = \frac{X + Y}{m + n}.$$

Writing p_e to represent the estimate, in our example we get

$$p_e = \frac{80 + 105}{500} = 0.37.$$

It can be shown that the appropriate test statistic in this case is

$$\frac{\dfrac{X}{m} - \dfrac{Y}{n}}{\sqrt{\hat{p}(1 - \hat{p})\left(\dfrac{1}{m} + \dfrac{1}{n}\right)}}.$$

Since $p_e = 0.37$, the value of the test statistic in our present case is

$$\frac{0.40 - 0.35}{\sqrt{0.37(1 - 0.37)\left(\dfrac{1}{200} + \dfrac{1}{300}\right)}} = \frac{0.05}{\sqrt{0.001942}} = 1.134.$$

If m and n are large (both greater than 30), then the test procedures for various forms of the alternative hypothesis are given in Table 9-9.

TABLE 9-9	
Procedure to test $H_0\!:\!p_1 = p_2$ when samples are large	
Alternative hypothesis	*The decision rule is to reject H_0 if the computed value of the test statistic is*
1. $p_1 > p_2$	greater than z_α
2. $p_1 < p_2$	less than $-z_\alpha$
3. $p_1 \neq p_2$	less than $-z_{\alpha/2}$ or greater than $z_{\alpha/2}$

Since the tests are based on the application of the central limit theorem, the level of significance is *approximately* 100α percent.

EXAMPLE 5   A jar containing 120 flies was sprayed with an insecticide of Brand A and it was found that 95 of the flies were killed. When another jar containing 145 flies of the same type was sprayed with Brand B, 124 flies were killed. At the 2 percent level of significance, do the two brands differ in their effectiveness?

SOLUTION   Notice that the data is qualitative in nature.

1. $H_0\!: p_1 = p_2$, that is $p_1 - p_2 = 0$
2. $H_A\!: p_1 \neq p_2$

3. The test statistic is $\dfrac{\dfrac{X}{m} - \dfrac{Y}{n}}{\sqrt{\hat{p}(1 - \hat{p})\left(\dfrac{1}{m} + \dfrac{1}{n}\right)}}.$

4. Since $\alpha = 0.02$, $z_{\alpha/2} = z_{0.01} = 2.33$. Hence the decision rule is *Reject H_0 if the computed value is less than -2.33 or greater than 2.33.*
5. Since $x = 95$, $y = 124$, $m = 120$, and $n = 145$,

$$p_e = \frac{x + y}{m + n} = \frac{95 + 124}{120 + 145} = 0.826.$$

Therefore, the computed value of the test statistic is

$$\frac{\dfrac{95}{120} - \dfrac{124}{145}}{\sqrt{0.826(1 - 0.826)\left(\dfrac{1}{120} + \dfrac{1}{145}\right)}} = \frac{0.792 - 0.855}{\sqrt{(0.144)(0.0152)}}$$

$$= -1.35 \text{ (see Figure 9-28).}$$

FIGURE 9-28

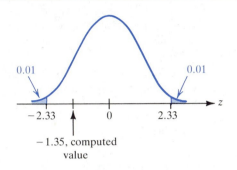

6. The computed value -1.35 is between -2.33 and 2.33. There is no reason to discard H_0 (P-value 0.177).

Conclusion: There is no evidence to indicate that one spray is significantly better than the other (at the 2 percent level). ▬▬

SECTION 9-4   EXERCISES

1. Suppose there are two normally distributed populations. One population has variance σ_1^2 of 10, and when a sample of size 20 was picked from it, the mean $\bar{x}$ was 8. The second population has variance σ_2^2 of 12, and a sample of 16 yielded a mean $\bar{y}$ of 7.2. Assuming that the samples are independent, at the 5 percent level of significance, test the null hypothesis H_0: $\mu_1 = \mu_2$ against the following alternative hypotheses:

    (a) $\mu_1 \neq \mu_2$ (b) $\mu_1 > \mu_2$.

2. A placement exam in mathematics was given to 15 students who had the "new" math background and to 10 students who had the "old" math background. The mean score of the "new" math students was 88 points and that of the "old" math students was 82 points. Suppose it may be assumed that the variances of the score for new math and old math are known and are, respectively, $\sigma_1^2 = 20$ and $\sigma_2^2 = 12$. At the 5 percent level of significance, do the true mean scores for the two groups differ significantly? Assume that scores in the two populations are normally distributed.

3. Suppose there are two populations. When a sample of 100 items was picked from one population, it was found that $\bar{x}$ is 14 and s_1^2 is 1.2. A sample of 200 items drawn independently from the other population yielded $\bar{y} = 14.4$ with $s_2^2 = 1.4$. At the 10 percent level of significance, test the null hypothesis H_0: $\mu_1 = \mu_2$ against the following alternative hypotheses:

    (a) $\mu_1 \neq \mu_2$ (b) $\mu_1 < \mu_2$.

4. A company has two branches, one on the East Coast and the other on the West Coast. The mean daily sales of the company on the East Coast, when observed on 100 days, were found to be $32,000 with s_1 of $2400. For the company on the West Coast, when observed on 200 days, the mean daily sales were $36,000 with s_2 of $2700. At the 5 percent level of significance, do the two branches differ in their mean daily sales?

5. In a test to compare the performance of two models of cars, Thunderball and Fireball, 128 cars of each model were driven on the same terrain with ten gallons of gasoline in each car. The mean number of miles for Thunderball was 240 miles with $s_1 = 10$ miles; the mean number of miles for Fireball was 248 miles with $s_2 = 16$ miles. At the 2 percent level of significance, are the two models of cars significantly different?

6. When 100 raw commercial apples of Variety A and 150 of Variety B were assayed for their iron content (in milligrams), the following information was obtained:

	Sample mean (in milligrams)	Sample standard deviation s
Variety A	0.32	0.08
Variety B	0.29	0.05

At the 5 percent level of significance do the two varieties of apples differ in their iron content?

7. A rancher tried Feed A on 256 cattle and Feed B on 144 cattle. The mean weight of cattle given Feed A was found to be 1350 pounds with s_1 of 180 pounds. On the other hand, the mean weight of the cattle given Feed B was found to be 1430 pounds with s_2 of 210 pounds. At the 5 percent level of significance, are the two feeds different?

8. Assume that we have two populations, both normally distributed with a common variance σ^2 that is unknown. If $\bar{x} = 10.2$, $\bar{y} = 8.5$, $m = 10$, $n = 12$, $s_1^2 = 2.8$, and $s_2^2 = 3.2$, assuming the samples are independent, test the hypothesis $H_0: \mu_1 = \mu_2$ against the following alternative hypotheses at the 5 percent level:

(a) $\mu_1 \neq \mu_2$

(b) $\mu_1 > \mu_2$.

In Exercises 9 through 15, assume that the two populations are normally distributed with the same, but unknown, variance.

9. The mean length of 14 trout caught in Clear Lake was 10.5 inches with s_1 of 2.1 inches, and the mean length of 11 trout caught in Blue Lake was 9.6 inches with s_2 of 1.8 inches. Is the true mean length of trout in Clear Lake significantly greater than the true mean length of trout in Blue Lake? Use $\alpha = 0.1$.

10. A random sample of 10 light bulbs manufactured by Company A had a mean life of 1850 hours and a sample standard deviation of 130 hours. Also, a random sample of 12 light bulbs manufactured by Company B had a mean life of 1940 hours with a sample standard deviation of 140 hours. Is the true mean life of bulbs manufactured by Company B significantly different from the true mean life of bulbs manufactured by Company A? Use $\alpha = 0.05$.

11. In ten half-hour morning programs, the mean time devoted to commercials was 6.4 minutes with s_1 of 1 minute. In twelve half-hour evening programs, the mean time

was 5.6 minutes with s_2 of 1.3 minutes. At the 10 percent level of significance, is the true mean time devoted to commercials in the evening significantly less than the true mean time devoted to commercials in the morning?

12. A sociologist wants to compare the dexterity of children living in a well-to-do neighborhood with that of children living in a poor neighborhood. A random sample of 5 children is selected from the well-to-do neighborhood and 6 children from the poor neighborhood, and the time that each child takes to assemble a certain toy is recorded. The following figures give the times:

Well-to-do	6	8	7	6	10	
Poor	4	10	9	7	11	8

At the 5 percent level of significance, are children from the well-to-do neighborhood more adept at assembling the toy than those in the poor neighborhood?

13. A psychologist measured reaction times (in seconds) of 8 individuals who were not given any stimulant and 6 individuals who were given an alcoholic stimulant. The figures are given below:

Reaction time without stimulant	3.0	2.0	1.0	2.5	1.5	4.0	1.0	2.0
Reaction time with stimulant	5.0	4.0	3.0	4.5	2.0	2.5		

At the 5 percent level of significance is the true mean reaction time increased significantly by alcoholic stimulant?

14. The figures below refer to the amounts of carbohydrates (in grams) per serving of two varieties of canned peaches:

Variety A	32	32	27	38	29		
Variety B	27	28	33	30	26	29	28

Based on the above sample data, at the 5 percent level of significance, are the two varieties significantly different in their true mean carbohydrate content?

15. Ten voters were picked at random from those who voted in favor of a certain proposition, and twelve voters from those who voted against it. The following figures give their ages:

In favor	28	33	27	31	29	25	50	30	25	41		
Against	31	43	49	32	40	41	48	30	29	39	42	36

At the 5 percent level of significance, is the true mean age of those voting against the proposition significantly different from the true mean age of those voting for it?

16. Suppose you are interested in comparing means of two populations by using paired differences. The following data are available:

Pair	1	2	3	4	5
x	17	25	36	16	14
y	19	18	30	21	10

(a) Find the differences d_i, that is, $x_i - y_i$, and compute $\bar{d}$ and s_d.

(b) Suppose you wish to test

$$H_0: \mu_D = 0 \text{ against } H_A: \mu_D > 0.$$

Determine the critical region at the 5 percent level.

(c) Compute the value of the test statistic.

(d) What conclusion do you arrive at using the 5 percent level of significance?

17. Suppose you want to compare means of two populations using paired observations. If $\bar{d} = -2.85$, $s_d = 3.72$, and $n = 15$, test the hypothesis $H_0: \mu_1 = \mu_2$ against the following alternative hypotheses using $\alpha = 0.05$:

(a) $\mu_1 \neq \mu_2$ (that is, $\mu_D \neq 0$)

(b) $\mu_1 < \mu_2$ (that is, $\mu_D < 0$).

18. A dealer of a certain Japanese car claims that it is easier to change spark plugs on a 4-cylinder model that he sells than on a comparable U.S. model. To test the claim, eight mechanics of proven ability were assigned, each to change spark plugs on the two models. Their times (in minutes) are recorded below:

Mechanic	Japanese model	U.S. model
Tom	15	19
Adam	21	24
Susan	22	19
Henry	13	16
Wong	18	16
Chavez	21	23
Bob	20	19
Wendel	25	26

(a) Are the two samples independent? Why?

(b) At the 1 percent level of significance, is the dealer's claim valid?

19. Scores of nine students before and after a special coaching program are given below.

Student	Before coaching	After coaching
1	91	82
2	78	80
3	47	62
4	37	49
5	64	55
6	54	73
7	43	59
8	33	58
9	53	43

At the 5 percent level of significance, does the special coaching benefit students?

20. Is it safe for an individual to drink two beers and drive? To test whether this amount of beer impairs driving ability, the department of motor vehicles conducted a simulated experiment with seven volunteers each of whom drove through an obstacle course first, without consuming any alcohol and then, after drinking two beers. Driv-

ing ability was measured by recording the total reaction time (in seconds) of a driver in avoiding the various obstacles on the course. The data are presented below.

Driver	1	2	3	4	5	6	7
Without alcohol	8.82	6.62	9.84	10.81	9.20	11.67	4.42
With alcohol	10.61	8.18	12.24	13.45	8.88	10.88	5.38

At the 5 percent level of significance, is it true that consumption of two beers impairs driving ability?

The *confidence interval for* μ_D *using paired differences* is given by

$$\bar{d} - t_{n-1, \alpha/2} \frac{s_d}{\sqrt{n}} < \mu_D < \bar{d} + t_{n-1, \alpha/2} \frac{s_d}{\sqrt{n}}$$

21. Determine a 90 percent confidence interval for μ_D in Exercise 18.

22. Determine a 95 percent confidence interval for μ_D in Exercise 19.

23. Determine a 95 percent confidence interval for μ_D in Exercise 20.

24. Suppose we wish to compare proportions of an attribute in two populations. When a sample of 80 was picked from one population it was found that $x = 36$ had the particular attribute, whereas a sample of 60 picked from the other population showed that $y = 18$ individuals had the attribute. Assuming that the two samples are independent, at the 2 percent level of significance test the hypothesis $H_0: p_1 = p_2$ against the following alternative hypotheses:

(a) $H_A: p_1 \neq p_2$

(b) $H_A: p_1 > p_2$

25. Of the 150 Democrats interviewed, 90 favor a proposition, and of the 120 Republicans interviewed, 80 favor it. Is the true proportion of individuals in favor of the proposition significantly larger among Republicans than among Democrats? Use $\alpha = 0.01$.

26. A survey was conducted to compare the proportions of males and females who favor government assistance for child care. It was found that among 64 males interviewed, 40 favored assistance, and among 100 females, 70 favored assistance. At the 5 percent level of significance, are the true proportions among males and females in the population who favor government assistance for child care significantly different?

27. Two hundred students in a class are divided into two groups of 100. One of the groups was given a vaccine against the common cold. The other group was not given any vaccine. It was found that in the group that was given the vaccine, 45 got colds. In the other group, 54 got colds. At the 2 percent level of significance is the vaccine effective against the common cold?

28. In the course of his campaign, a presidential candidate made a "foot-in-the-mouth" statement. When a random sample of 400 were interviewed to see if this had affected the candidate's standing, 212 said that they favored the candidate. However, during the week before this unfortunate episode, when a random sample of 600 had been interviewed, 348 had voted in favor. At the 5 percent level of significance, did the statement hurt the candidate?

29. Of the 200 college students interviewed, 150 favored strict environmental controls, whereas, of the 300 college graduates interviewed, 195 favored them. At the 5 percent level of significance, do college students tend to be more in favor of environmental controls than college graduates?

KEY FORMULAS

Summary of Tests of Hypotheses (Two Populations)

Nature of the population	Parameter	Null hypothesis	Alternative hypothesis	Test statistic	Decision rule: Reject H_0 if the computed value is
Quantitative data; variances σ_1^2 and σ_2^2 are known; both populations normally distributed.	$\mu_1 - \mu_2$	$\mu_1 = \mu_2$	1. $\mu_1 > \mu_2$ 2. $\mu_1 < \mu_2$ 3. $\mu_1 \neq \mu_2$	$\dfrac{\bar{X} - \bar{Y}}{\sqrt{\dfrac{\sigma_1^2}{m} + \dfrac{\sigma_2^2}{n}}}$	1. greater than z_α 2. less than $-z_\alpha$ 3. less than $-z_{\alpha/2}$ or greater than $z_{\alpha/2}$
Quantitative data; σ_1, σ_2 unknown; populations not necessarily normal; sample sizes are large.	$\mu_1 - \mu_2$	$\mu_1 = \mu_2$	1. $\mu_1 > \mu_2$ 2. $\mu_1 < \mu_2$ 3. $\mu_1 \neq \mu_2$	$\dfrac{\bar{X} - \bar{Y}}{\sqrt{\dfrac{S_1^2}{m} + \dfrac{S_2^2}{n}}}$	1. greater than z_α 2. less than $-z_\alpha$ 3. less than $-z_{\alpha/2}$ or greater than $z_{\alpha/2}$
Quantitative data; σ_1^2 and σ_2^2 are not known but are assumed to be equal; both populations are normally distributed.	$\mu_1 - \mu_2$	$\mu_1 = \mu_2$	1. $\mu_1 > \mu_2$ 2. $\mu_1 < \mu_2$ 3. $\mu_1 \neq \mu_2$	$\dfrac{\bar{X} - \bar{Y}}{S_p \sqrt{\dfrac{1}{m} + \dfrac{1}{n}}}$ where $S_p^2 = \dfrac{(m-1)S_1^2 + (n-1)S_2^2}{m+n-2}$	1. greater than $t_{m+n-2,\alpha}$ 2. less than $-t_{m+n-2,\alpha}$ 3. less than $-t_{m+n-2,\alpha/2}$ or greater than $t_{m+n-2,\alpha/2}$
Paired observations.	$\mu_1 - \mu_2$ $(= \mu_D)$	$\mu_D = 0$ i.e., $\mu_1 = \mu_2$	1. $\mu_D > 0$ 2. $\mu_D < 0$ 3. $\mu_D \neq 0$	$\dfrac{\bar{d}\sqrt{n}}{s_d}$ where $\bar{d} = \sum\limits_{i=1}^{n} d_i / n$ and $s_d^2 = \dfrac{1}{n-1}\sum\limits_{i=1}^{n}(d_i - \bar{d})^2$	1. greater than $t_{n-1,\alpha}$ 2. less than $-t_{n-1,\alpha}$ 3. less than $-t_{n-1,\alpha/2}$ or greater than $t_{n-1,\alpha/2}$
Qualitative data; binomial case.	$p_1 - p_2$	$p_1 = p_2$	1. $p_1 > p_2$ 2. $p_1 < p_2$ 3. $p_1 \neq p_2$	$\dfrac{\dfrac{X}{m} - \dfrac{Y}{n}}{\sqrt{\hat{p}(1-\hat{p})\left(\dfrac{1}{m} + \dfrac{1}{n}\right)}}$ where $\hat{p} = \dfrac{X+Y}{m+n}$ Both m and n are assumed large.	1. greater than z_α 2. less than $-z_\alpha$ 3. less than $-z_{\alpha/2}$ or greater than $z_{\alpha/2}$ The level of significance is approximately 100α percent.

Summary of Tests of Hypotheses (Single Population)

Nature of the population	Param- eter	Null hypoth- esis	Alternative hypothesis	Test statistic	Decision rule: Reject H_0 if the computed value is
Quantitative data; variance σ^2 is known; population normally distributed.	μ	$\mu = \mu_0$	1. $\mu > \mu_0$ 2. $\mu < \mu_0$ 3. $\mu \neq \mu_0$	$\dfrac{\overline{X} - \mu_0}{\sigma/\sqrt{n}}$	1. greater than z_α 2. less than $-z_\alpha$ 3. less than $-z_{\alpha/2}$ or greater than $z_{\alpha/2}$
Quantitative data; variance σ^2 is *not* known; population normally distributed.	μ	$\mu = \mu_0$	1. $\mu > \mu_0$ 2. $\mu < \mu_0$ 3. $\mu \neq \mu_0$	$\dfrac{\overline{X} - \mu_0}{S/\sqrt{n}}$	1. greater than $t_{n-1,\alpha}$ 2. less than $-t_{n-1,\alpha}$ 3. less than $-t_{n-1,\alpha/2}$ or greater than $t_{n-1,\alpha/2}$
Quantitative data; variance σ^2 unknown; population not necessarily normal; sample large.	μ	$\mu = \mu_0$	1. $\mu > \mu_0$ 2. $\mu < \mu_0$ 3. $\mu \neq \mu_0$	$\dfrac{\overline{X} - \mu_0}{S/\sqrt{n}}$	1. greater than z_α 2. less than $-z_\alpha$ 3. less than $-z_{\alpha/2}$ or greater than $z_{\alpha/2}$ The level of significance is approximately 100α percent.
Quantitative data; population normally distributed.	σ^2	$\sigma = \sigma_0$	1. $\sigma > \sigma_0$ 2. $\sigma < \sigma_0$ 3. $\sigma \neq \sigma_0$	$\dfrac{(n-1)S^2}{\sigma_0^2}$	1. greater than $\chi^2_{n-1,\alpha}$ 2. less than $\chi^2_{n-1,1-\alpha}$ 3. less than $\chi^2_{n-1,1-\alpha/2}$ or greater than $\chi^2_{n-1,\alpha/2}$
Qualitative data; binomial case.	p	$p = p_0$	1. $p > p_0$ 2. $p < p_0$ 3. $p \neq p_0$	$\dfrac{(X/n) - p_0}{\sqrt{\dfrac{p_0(1-p_0)}{n}}}$ Here n is assumed large.	1. greater than z_α 2. less than $-z_\alpha$ 3. less than $-z_{\alpha/2}$ or greater than $z_{\alpha/2}$ The level of significance is approximately 100α percent.

KEY TERMS AND EXPRESSIONS

statistical hypothesis
null hypothesis
alternative hypothesis
decision rule
Type I error
Type II error
level of significance
test statistic
rejection region

critical region
acceptance region
critical value
test of a statistical hypothesis
one-tailed test (right, left)
two-tailed test
P-value (significance probability)
paired t test

CHAPTER 9   TEST

1. Define the following terms: null hypothesis; alternative hypothesis; Type I error; Type II error; level of significance; critical region.

2. In carrying out the test of a hypothesis, why is it so important to state the alternative hypothesis?

3. Suppose the null hypothesis states: Vitamin C is effective in controlling colds. State what action will constitute a Type I error and what action will constitute a Type II error.

4. Suppose we have $H_0: \sigma^2 = 25$ against $H_A: \sigma^2 = 9$. State which of the following statements are true:

   (a) $\dfrac{(n-1)S^2}{25}$ has a chi-square distribution.

   (b) $\dfrac{(n-1)S^2}{9}$ has a chi-square distribution.

   (c) $\dfrac{(n-1)S^2}{9}$ has a chi-square distribution if H_0 is false.

   (d) $\dfrac{(n-1)S^2}{25}$ has a chi-square distribution if H_0 is true.

5. Suppose we have $H_0: p = 0.4$ against $H_A: p = 0.3$. A sample of 100 is picked. Find the approximate distribution of $X/100$, the sample proportion,

   (a) when H_0 is true   (b) when H_A is true.

6. The following figures refer to weights (in ounces) of 10 one-pound cans of peaches distributed by a company:

   16.3,   15.7,   16.8,   16.5,   15.5,   16.0,   16.2,   15.8,   15.4,   16.8.

   At the 5 percent level of significance,

   (a) is the true mean of all the cans distributed by the company different from 16 ounces?

   (b) is the true variance of all the cans distributed by the company different from 0.16 ounce squared?

7. Thirty experimental launches are made with a rocket-launching system. The outcomes are given below. (S is recorded if successful and U if unsuccessful.)

S	S	U	U	S	S	S	U	S	S
U	S	S	S	S	U	U	S	S	U
S	S	S	S	U	S	U	S	S	S.

   At the 5 percent level of significance is the true proportion of successful launches among all possible launches different from 0.8?

8. A study was conducted to compare two brands of batteries. When ten batteries of Brand A were tested, it was found that the mean life was 1200 days with s_1 of 120 days. On the other hand, eight batteries of Brand B yielded a mean life of 1300 days with s_2 of 150 days. At the 5 percent level of significance, are the two brands different? Assume that battery life is normally distributed for both brands.

9. When a salesman contacted 100 women about a utility product, he sold the product to 43 of them. On the other hand, when he contacted 150 men, he sold the product to 78 of them. At the 10 percent level of significance, is the true proportion among men who buy the product when contacted different from the true proportion among women?

10. A precision tool company needs plates of uniform thickness. A manufacturing company claims that it can supply plates of thickness with $\sigma < 0.1$ inch. When a random sample of 30 plates was examined, it was found that $s = 0.095$. At the 5 percent level of significance, is the claim of the manufacturing company valid?

11. Sprouts grow from dormant buds around root collars and along sides of many stumps of cut redwood trees. An investigation was carried out to determine if sprouts on smaller redwood stumps (diameter less than 60 inches) grow differently from those on larger stumps (diameter greater than 60 inches). Suppose the data below gives the summary regarding height growth at the end of five years for unthinned stumps.*

Stump diameter	Sample mean (in feet)	Sample standard deviation (in feet)	Sample size
Less than 60 inches	12.4	1.38	12
Greater than 60 inches	11.5	1.62	15

At the 5 percent level of significance do the true mean height growths for the two classifications of stump diameters differ significantly?

*Kenneth N. Boe, Research note PSW-290, USDA Forest Service, 1974.

10

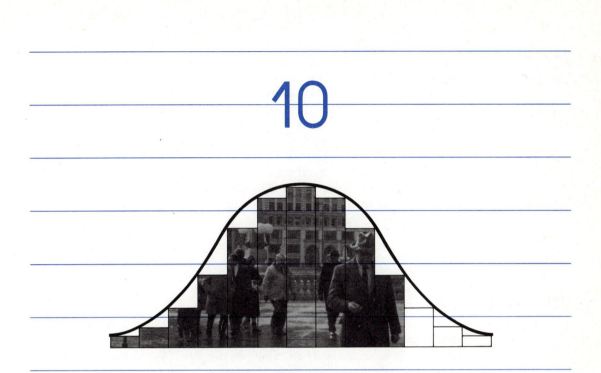

GOODNESS OF FIT
AND CONTINGENCY TABLES

INTRODUCTION

In the preceding chapter, we considered tests of hypotheses regarding the mean of a population and also comparison of the means of two populations. In this connection, the variables measured were quantitative in nature, where very often we had to assume that the populations were normally distributed. We also discussed tests of hypotheses about a population proportion and comparison of proportions of an attribute in two populations. The variables in these cases were essentially qualitative. We now extend these ideas, with the qualitative case to be covered in this chapter and the quantitative case in Chapter 12, Analysis of Variance. ■

10-1 A TEST OF GOODNESS OF FIT

In the discussion of tests of hypotheses about a population proportion, the items that were inspected were classified into one of two categories. For instance, a coin could land heads or tails, a person could be a smoker or a nonsmoker, an item could be defective or nondefective, and so on. The independent choice of n items from such a population leads to the binomial distribution, as we have seen. An obvious generalization is when the population can be separated into more than two mutually exclusive categories. For example, a coin could land heads, tails, or on edge; a die could land showing up any one of six faces; a person might be a Democrat, a Republican, or neither; a person might be an A, B, O, or AB blood type, and so on. If n independent observations are made from such a population, we get a generalized concept of the binomial distribution called the *multinomial distribution*.

The preceding chapter equips us to test the following null hypothesis:

H_0: The proportion of Democrats in the United States is 0.60 (of course, this implies that the proportion of non-Democrats is 0.40).

Now, as a generalization, this section will, for instance, enable us to test a null hypothesis of the following type:

H_0: In the United States, the proportion of Democrats is 0.55, the proportion of Republicans is 0.35, and the proportion of other voters belonging to neither party is 0.10.

To test the hypothesis above, suppose we interview 1000 people picked at random. On the basis of the stipulated null hypothesis, we would *expect* 550 Democrats, 350 Republicans, and 100 others. If we actually *observe* 568 Democrats, 342 Republicans, and 90 others in this sample, we might be willing to go along with the null hypothesis. On the other hand, if the sample yields 460 Democrats, 400 Republicans, and 140 others, we would be reluctant to accept H_0. Thus in the final analysis the statistical test will have to be based on how good a fit or closeness there is between the observed numbers and the numbers that one would expect from the hypothesized distribution. A test of this type which determines whether the sample data are in conformity with the hypothesized distribution is called a **test of goodness of fit,** since it literally tests how good the fit is.

For the purpose of indicating the procedure, let us consider the following example. Suppose we want to test whether a die is a fair die. Let p_1 represent the probability of getting a 1, p_2 that of getting a 2, and so on. Then the null hypothesis H_0 can be stated as follows:

$$H_0: \quad p_1 = \tfrac{1}{6}, p_2 = \tfrac{1}{6}, p_3 = \tfrac{1}{6}, p_4 = \tfrac{1}{6}, p_5 = \tfrac{1}{6}, p_6 = \tfrac{1}{6}.$$

This hypothesis will be tested against the alternative hypothesis H_A, stated as

$H_A:$ The die is not fair.

To test the hypothesis, suppose we roll the die 120 times and obtain the data given in Table 10-1. If the die is a fair die, we would *expect* each face to show up $120(\tfrac{1}{6})$, or 20, times. Any departure from this could be simply due to sampling error or the fact that the die is indeed a loaded die. A radical departure from the expected frequency of 20 would definitely make the die suspect in our judgment and the discrepancy hardly attributable to chance. We are thus faced with two questions. First, how do we measure this departure? Second, for how large a departure would we consider rejecting H_0? In short, what test criterion do we use? The test criterion is provided by a statistic **X** whose value for any sample is given as a number χ^2 defined by

TABLE 10-1						
Observed frequencies in 120 tosses of a die						
Number on the die	1	2	3	4	5	6
Frequency	17	25	17	23	15	23

$$\chi^2 = \sum_{i=1}^{6} \frac{(O_i - E_i)^2}{E_i}$$

where O_i represents the **observed frequency** of the face marked i on the die and E_i the corresponding **expected frequency** obtained by assuming that the null hypothesis is true. The idea is that for each face on the die, we find [(observed $-$ expected)^2/expected] and then add these quantities for all the faces.

Notice that the observed frequencies of the outcomes will change from sample to sample. That is, if we were to roll our die another 120 times, we would probably get a different set of observed frequencies from those given in Table 10-1. Thus **X** is a random variable which provides a measure of departure of the observed frequencies, since $(O - E)^2$ will be small if the observed and expected frequencies are close, and it will be large if they are far apart.

The procedure for computing the χ^2 value for a given sample is shown in Table 10-2. The total in column 6 gives the χ^2 value as equal to 4.30.

TABLE 10-2					
Computation of the χ^2 value for 120 tosses of a die					
Outcome	Observed frequency O	Expected frequency E	$(O - E)$	$(O - E)^2$	$(O - E)^2/E$
1	17	20	−3	9	0.45
2	25	20	5	25	1.25
3	17	20	−3	9	0.45
4	23	20	3	9	0.45
5	15	20	−5	25	1.25
6	23	20	3	9	0.45
Total	120		0		4.30

As to how the number 4.30 reflects in determining whether the die is fair or not, we should know the distribution of **X**. It turns out that *if none of the expected frequencies is less than 5, then the distribution of* **X** *can be approximately represented by a chi-square distribution*. The next question is, how many degrees of freedom does the chi-square distribution have? The number of degrees of freedom associated with the chi-square is $6 - 1$, or 5. The rationale is that although we are filling 6 cells, or entries in the table, in reality we have freedom to fill only 5 cells. The number in the sixth cell is determined automatically due to the constraint that the total should be 120.

Intuition tells us that we should reject the null hypothesis that the die is fair if the computed χ^2 value is large. In other words, we have a one-tailed test with the critical region consisting of the right tail of the chi-square distribution with 5 degrees of freedom, as indicated in Figure 10-1.

At the 5 percent level of significance, for 5 degrees of freedom the table value of chi-square from Table A-6 is 11.07. Since the computed value of 4.30 does not exceed 11.07, we do not reject the null hypothesis. Thus, at the 5 percent level of significance, there is no reason to believe that the die is loaded.

EXAMPLE 1   It is believed that the proportions of people with A, B, O, and AB blood types in the population are, respectively, 0.4, 0.2, 0.3, and 0.1. When 400 randomly picked people were examined, the observed numbers of each type were 148, 96, 106, and 50. At the 5 percent level of significance, test the hypothesis that these data bear out the stated belief.

FIGURE 10-1

Probability den-
sity curve for the
chi-square distri-
bution with 5 df
and the decision
rule

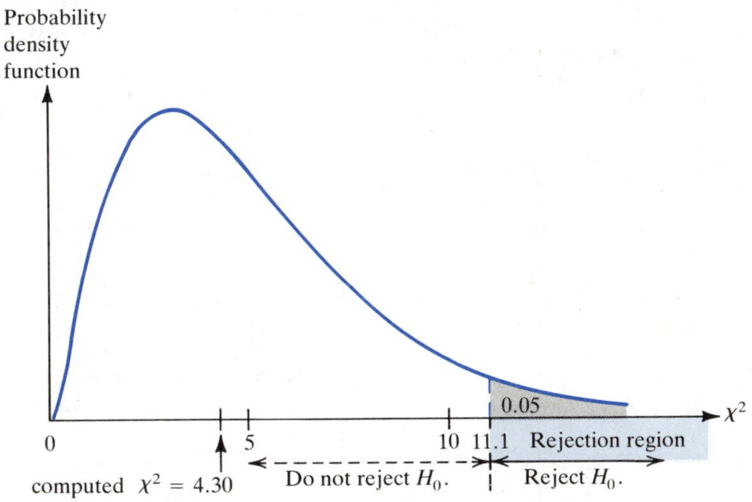

Probability
density
function

SOLUTION

Let p_A be the probability that a person has type A blood, and so on. Then

$$H_0: \quad p_A = 0.4,\ p_B = 0.2,\ p_O = 0.3,\ p_{AB} = 0.1.$$

We will state the alternative hypothesis H_A as

$H_A:$ The population proportions of blood types are not as stated in H_0.

Our first goal should be to find the expected frequency in each cell. Since
$n = 400$ and $p_A = 0.4$, the expected number of people of type A blood is
$400(0.4)$, or 160. Other expected frequencies are obtained similarly. The computations for finding the χ^2 value are presented in Table 10–3.

TABLE 10-3

**Observed and expected frequencies of blood
types and computation of the χ^2 value**

Blood type	Observed frequency O	Expected frequency E	$(O - E)$	$(O - E)^2$	$(O - E)^2/E$
A	148	160	−12	144	0.90
B	96	80	16	256	3.20
O	106	120	−14	196	1.63
AB	50	40	10	100	2.50
Total	400	400	0		8.23

Since there are four cells, the chi-square distribution has $4 - 1$, or 3, degrees of freedom. From Table A-6, $\chi^2_{3, 0.05} = 7.815$. Since the computed value of 8.23 is larger than 7.815, we reject the null hypothesis. Thus there is strong evidence that the belief stated above regarding the distribution of blood types is not correct. ■

EXAMPLE 2   According to the Mendelian law of segregation in genetics, when two doubly heterozygous walnut comb fowls are crossed, the offsprings consist of four types of fowls, namely, walnut, rose, pea, and single, with frequencies expected in the ratio 9:3:3:1. Among 320 offsprings, it was found that there were 160 walnut, 70 rose, 65 pea, and 25 single. Based on the data, are the probabilities for each type not in keeping with the Mendelian law?

SOLUTION   Let p_{walnut} represent the probability that the offspring is walnut, and so on. Since the frequencies are expected in the ratio 9:3:3:1, the null hypothesis is given as

$$H_0: \quad p_{walnut} = \tfrac{9}{16}, p_{rose} = \tfrac{3}{16}, p_{pea} = \tfrac{3}{16}, p_{single} = \tfrac{1}{16}.$$

The alternative hypothesis H_A is

$H_A:$ The probabilities are not as given by the Mendelian law.

If the null hypothesis is true, then we expect $320(\tfrac{9}{16})$, or 180, walnut types. Other expected frequencies can be obtained similarly. The computation of the χ^2 value is presented in Table 10-4.

TABLE 10–4

Observed and expected frequencies of comb types and computation of the χ^2 value

Comb type	Observed frequency O	Expected frequency E	$(O - E)^2$	$(O - E)^2/E$
Walnut	160	180	400	2.22
Rose	70	60	100	1.67
Pea	65	60	25	0.42
Single	25	20	25	1.25
Total	320	320		5.56

Since there are four comb types, the chi-square distribution has 3 degrees of freedom. The table value of $\chi^2_{3, 0.05}$ is 7.815. The computed value 5.56 is less than 7.815, and so we do not reject H_0. Thus there is no reason to suspect that there is a disagreement with the Mendelian ratio 9:3:3:1. ■

The essence of the previous examples can be summarized as follows:

1. The population is divided into k categories (or classes), $c_1, c_2, \ldots, c_k$.

2. The null hypothesis stipulates that the probability that an individual belongs to category c_1 is p_1, that it belongs to category c_2 is p_2, and so on.

3. To test this hypothesis, a random sample of n individuals is picked. The observed frequencies of the categories are recorded as $O_1, O_2, \ldots, O_k$.

4. If the null hypothesis is true, then the expected frequencies $E_1, E_2, \ldots E_k$ are obtained as follows:

$$E_1 = np_1, \ E_2 = np_2, \ldots, E_k = np_k.$$

5. The departure of the observed frequencies from those expected is measured by means of a statistic $\mathbf{X}$ whose value χ^2 is given by

$$\chi^2 = \frac{(O_1 - E_1)^2}{E_1} + \frac{(O_2 - E_2)^2}{E_2} + \cdots + \frac{(O_k - E_k)^2}{E_k}$$

The computations leading to the evaluation of the statistic above are displayed in Table 10-5.

TABLE 10-5

Observed and expected frequencies and computation of the χ^2 value

Category	Observed frequency O	Expected frequency E	$(O - E)^2/E$
c_1	O_1	$E_1 = np_1$	$(O_1 - E_1)^2/E_1$
c_2	O_2	$E_2 = np_2$	$(O_2 - E_2)^2/E_2$
$\vdots$	$\vdots$	$\vdots$	$\vdots$
c_k	O_k	$E_k = np_k$	$(O_k - E_k)^2/E_k$
Total	n	n	$\sum_{i=1}^{k} (O_i - E_i)^2/E_i$

6. *If none of the expected frequencies is less than 5, the distribution of* $\mathbf{X}$ *can be approximated very closely by a chi-square distribution.* Since there are k categories, *the number of degrees of freedom associated with the chi-square is $k - 1$.*

7. If the null hypothesis is true, then the departure of the observed frequencies from the expected frequencies will be small. In that case, the χ^2 value will be small. However, if there is a large discrepancy between the observed and expected frequencies, then we shall tend to question the null hypothesis. Such a large discrepancy will be reflected in making the χ^2 value large. The critical region for a given level of significance will therefore consist of the right tail of the chi-square dis-

FIGURE 10-2

Probability density curve for chi-square distribution with $k - 1$ df and the decision rule

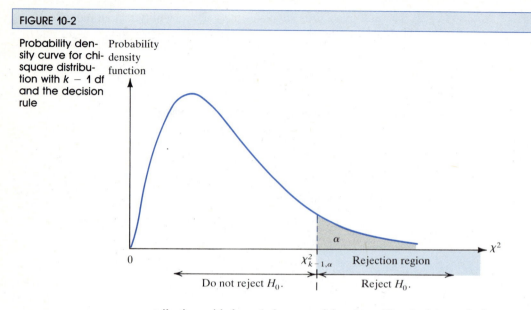

tribution with $k - 1$ degrees of freedom. The decision rule is:

Reject H_0 if the computed χ^2 value is greater than the table value $\chi^2_{k-1,\alpha}$. (See Figure 10.2.)

It should be emphasized that the distribution of the statistic **X** employed here is only approximately chi-square. It should not be used if one or more of the expected frequencies is less than 5.

SECTION 10-1   EXERCISES

1. In one progeny consisting of 172 seedlings, a botanist found that there were 38 yellow and 134 green seedlings. At the 5 percent level of significance, test the hypothesis that the segregation ratio of green seedlings to yellow is 3:1. Use the following tests:

   (a) the chi-square test developed in this section

   (b) a test concerning population proportion given in Section 9-3

2. According to genetic theory, a cross between a homozygous fruit fly and a heterozygous fruit fly is expected to yield homozygous and heterozygous offspring in the ratio 1:1. In an experimental cross, there were 27 homozygous offspring in a progeny of 68. At the 5 percent level, test the null hypothesis that the true ratio is 1:1.

3. Suppose according to a survey conducted in 1960 the *probability* distribution of the age of an adult over twenty years of age is as given in the first two columns of the

following table. In 1983, when a sample of 1000 adults over twenty years of age was interviewed, the frequency distribution given in column 3 was obtained. Test the null hypothesis that the probability distribution in 1983 is the same as that in 1960. Use $\alpha = 0.05$.

Age group	Probability distribution	Observed sample frequencies in 1983
20–29	0.25	270
30–39	0.23	202
40–49	0.20	180
50–59	0.15	160
60–69	0.08	88
70 and over	0.09	100

4. The number of accidents in a factory during the months of a year are given below:

Jan	Feb	Mar	Apr	May	June	Jul	Aug	Sep	Oct	Nov	Dec
25	28	24	18	17	27	9	18	22	14	12	26

   Using a 5 percent level of significance, test the null hypothesis that the number of accidents does not depend on the month of the year. [*Hint:* If a person has an accident, then under the null hypothesis the probability of that accident taking place in any month is $1/12$.]

5. The Mendelian law of segregation in genetics states that when certain types of peas are crossed, the following four varieties, round and yellow (RY), round and green (RG), wrinkled and yellow (WY), and wrinkled and green (WG), are obtained in the ratio $9:3:3:1$. In a greenhouse experiment, an agriculturist found that there were 652 RY, 200 RG, 185 WY, and 83 WG. Using a 5 percent level of significance, test the null hypothesis that these observed frequencies are consistent with those expected in theory.

6. A physicist theorizes that the probabilities that a radioactive substance emits 0, 1, 2, 3, or 4 particles of a certain kind during a one-hour period are, respectively, 0.05, 0.3, 0.3, 0.25, and 0.1. A record of 200 one-hour periods gave the following distribution of the number of particles emitted:

Number of particles	0	1	2	3	4
Number of times	16	68	51	46	19

   At the 5 percent level of significance, test the null hypothesis that there is no significant departure from what the physicist theorizes.

7. Suppose the probability that a newborn baby is a boy is $1/2$.

   (a) In a family with four children, find the probabilities of 0, 1, 2, 3, or 4 sons.

   (b) If 160 families, each with four children, are interviewed, find the expected number of families with 0, 1, 2, 3, and 4 sons.

   (c) When 160 families, each with four children, were interviewed, the following information was obtained:

Number of sons	0	1	2	3	4
Number of families	12	35	53	44	16

At the 5 percent level of significance, test the null hypothesis that the probability that a child is a son is $1/2$.

8. A machine is supposed to mix three types of candy—caramel, butterscotch, and coconut-chocolate—in the proportion $4:3:2$. A sample of 270 pieces of candy was found to contain 135 caramel, 70 butterscotch, and 65 coconut-chocolate. At the 5 percent level of significance, test the null hypothesis that the machine is mixing the candy in the proportion $4:3:2$.

9. Of the 500 cars observed on a certain freeway, 160 were GM, 100 Ford, 80 Chrysler, 25 American, and 135 foreign imports. Using a 5 percent level of significance, test the hypothesis that the expected frequencies of GM, Ford, Chrysler, American, and foreign imports on that freeway are in the ratio $30:15:10:5:20$.

10. The distribution of the number of clients that an accountant had on each day during 400 randomly picked days is as follows:

Number of clients	0	1	2	3	4	5	6
Number of days	50	112	125	60	28	16	9

At the 5 percent level of significance, is there departure from the theoretical distribution given below?

Number of clients	0	1	2	3	4	5	6
Probability	0.10	0.27	0.27	0.18	0.09	0.05	0.04

10-2 A TEST OF INDEPENDENCE

The problem that we shall consider next has no previous counterpart. It differs from the earlier situations where we conspicuously observed only one characteristic of any individual. For example, in classifying an individual as A, B, O, or AB blood type, we observed the characteristic "blood type"; in classifying a person as smoker or nonsmoker, we observed the "smoking habit." However, there are times when we might be interested in observing more than one variable on each individual to find if there exists a relationship between these variables. As an example, for each person we might observe both blood type and eye color and investigate if these characteristics are related in any way. In short, our goal is a **test of independence,** specifically, to find whether two observed attributes of members of a population are independent.

As a first step, we pick a sample of size n and classify the data in a two-way table on the basis of the two variables. Such a table for determining whether the distribution according to one variable is contingent on the distribution of the other is appropriately called a **contingency table.** A table with r rows and c columns is an $r \times c$ contingency table, read r by c contingency table.

Going back to our example of blood type and eye color, let us investigate if there is evidence indicating a relationship between these two variables. For this purpose, suppose we collect information on 400 randomly picked people and classify them according to their blood type and eye color. The data recording the observed frequencies is contained in Table 10-6, which is a 2 × 4 contingency table since there are 2 rows and 4 columns.

TABLE 10-6

A 2 × 4 contingency table giving the observed frequencies

Eye color	Blood type				
	A	B	O	AB	Total
Blue	95	40	80	25	240
Brown	65	50	40	5	160
Totals	160	90	120	30	400

There are 8 cells corresponding to the 2 rows and 4 columns. The row and column totals are referred to as the *marginal totals,* since they appear in the margin. The null hypothesis and the alternative hypothesis can be stated as follows:

H_0: The blood type of a person and the eye color of that person are independent variables.

H_A: The two attributes are not independent.

To test the *null* hypothesis, our objective will be the same as before, namely, to see how big a discrepancy there is between the observed frequencies and the frequencies that we would expect assuming the null hypothesis is true, that is, assuming the two attributes are independent of each other. Hence we obtain the expected frequencies in the 8 cells under the assumption of independence.

To indicate the method for obtaining the expected frequencies, as an illustration, let us find the number of people that we expect to have type A blood and blue eyes. Consider two events E and F defined as follows:

E: A person has blue eyes.

F: A person has type A blood.

There are 240 people with blue eyes in a sample of 400. Therefore, we can use 240/400 as a *point estimate* of the probability, $P(E)$. Similarly, since there are 160 people of type A blood among 400 people, a point estimate of $P(F)$ is 160/400. If, as per H_0, blood type and eye color are independent variables, then we must have $P(E \text{ and } F) = P(E) \cdot P(F)$. (See Chapter 3, page 118.) Consequently, we can obtain a point estimate of the probability that a person has blue eyes *and* is of type A blood as $\left(\frac{240}{400}\right)\left(\frac{160}{400}\right)$. Since there are 400 people in all in the sample, the *expected* frequency of people with blue eyes and type A blood is

$$400\left(\frac{240}{400}\right)\left(\frac{160}{400}\right) = \frac{(240)(160)}{400}.$$

A close examination will show that this is simply equal to

$$\frac{(\text{Total of Row 1})(\text{Total of Column 1})}{n}$$

where n is the total number in the sample.

In general, it is true that the expected frequency in any cell is obtained by multiplying the corresponding marginal row and column totals and then dividing this product by n. That is, the expected frequency in the cell that represents the intersection of row i and column j, which we shall denote by E_{ij}, is given by

$$E_{ij} = \frac{(\text{Total of row } i)(\text{Total of column } j)}{n}.$$

Thus, the expected number of people with brown eyes and type O blood is $\frac{(160)(120)}{400}$ or 48, that of people with brown eyes and type AB blood is $\frac{(160)(30)}{400}$ or 12, and so on. All the expected frequencies are given in Table 10-7.

TABLE 10-7					
Expected frequencies assuming eye color and blood type are independent variables					
Eye color	*Blood type*				
	A	*B*	*O*	*AB*	*Totals*
Blue	96	54	72	18	240
Brown	64	36	48	12	160
Totals	160	90	120	30	400

Remember that the expected frequencies are obtained under the assumption that eye color and blood type are independent variables. If there is a large discrepancy between these and the corresponding observed frequencies, it will cast doubt on the validity of the null hypothesis. Thus once again we use a statistic **X** whose value χ^2 for any sample is given by

$$\chi^2 = \sum \frac{(O - E)^2}{E}$$

where the symbol Σ means that we are summing over all the cells. In each cell we divide the square of the difference between the observed and expected frequencies by the expected frequency of that cell and then add these quantities over all the cells. These computations are shown in Table 10-8.

TABLE 10-8

Observed frequencies, expected frequencies assuming eye color and blood type are independent, and the χ^2 value

Eye color	Blood type	Observed frequency O	Expected frequency E	$(O - E)$	$(O - E)^2$	$(O - E)^2/E$
Blue	A	95	96	-1	1	0.01
Blue	B	40	54	-14	196	3.63
Blue	O	80	72	8	64	0.89
Blue	AB	25	18	7	49	2.72
Brown	A	65	64	1	1	0.02
Brown	B	50	36	14	196	5.44
Brown	O	40	48	-8	64	1.33
Brown	AB	5	12	-7	49	4.08
Totals		400	400	0		18.12

If n is large so that each of the expected frequencies is at least 5, then the statistic $\mathbf{X}$ has a distribution that is close to a chi-square distribution, and for the particular problem we are discussing, with 3 degrees of freedom. We can explain why there are 3 degrees of freedom as follows: There are 8 cells to be filled. Toward this goal we had to estimate the probabilities of three blood types. (The estimate of the probability of the fourth blood type is determined automatically, since the sum of all the four estimated probabilities must be 1.) Also, we had to estimate the probability of one of the two eye colors, say, the probability that the eye color is blue. (The estimate of the probability that the eye color is brown is determined automatically since the sum of the two probabilities should be 1.) Thus we had to estimate four parameters. Also, a further constraint is that the total of all the expected frequencies has to be equal to n, the sample size. The degrees of freedom are given by

$$\binom{\text{degrees of}}{\text{freedom}} = \binom{\text{number}}{\text{of cells}} - \binom{\text{number of}}{\substack{\text{estimated} \\ \text{parameters}}} - \binom{\text{number of}}{\text{constraints}}$$

$$= 8 - 4 - 1 = 3$$

Notice that we can write the 3 degrees of freedom as $(2 - 1)(4 - 1)$. In general, *if there are r rows and c columns in the contingency table, then the corresponding chi-square distribution has $(r - 1)(c - 1)$ degrees of freedom.*

The critical region consists of the values in the right tail of the chi-square distribution. The decision rule for a level of significance α is:

Reject H_0 if the computed χ^2 value is greater than the table value, $\chi^2_{(r-1)(c-1),\alpha}$.

To test the hypothesis of independence for the data in our example, let us take $\alpha = 0.05$. There are 3 degrees of freedom, and from the chi-square table, $\chi^2_{3,0.05} = 7.815$. Since the computed value of 18.12 is larger than the table value of 7.815, we reject the null hypothesis. The data does not support the hypothesis of independence on which the calculations are based. There is reason to believe that eye color and blood type are related.

To see what our conclusion means, let us examine column 5 in Table 10-8, which gives the difference $O - E$. There is an indication that blue eyes and type O blood, blue eyes and type AB blood, and brown eyes and type B blood associate together more often, and that blue eyes and type B blood, brown eyes and type O blood, and brown eyes and type AB blood associate together less often than what one might expect if eye color and blood type were independent.

EXAMPLE 1

In a certain community, 360 randomly picked people were classified according to their age group and political leaning. The data is presented in Table 10-9.

TABLE 10-9

Observed frequencies giving the distribution of political leaning and age group of people in a community

Political leaning	Age group			Total
	20–35	36–50	Over 50	
Conservative	10	40	10	60
Moderate	80	85	45	210
Liberal	30	25	35	90
Total	120	150	90	360

Test the hypothesis that a person's age and political leaning are not related (use $\alpha = 0.05$).

SOLUTION

1. The null and alternative hypotheses are given as follows:

H_0: A person's age and political leaning are independent.

H_A: The two variables are not independent.

2. The expected frequency in each cell is obtained by multiplying the corresponding row and column totals and then dividing the product by 360, the total number observed. These frequencies are given in Table 10-10.

TABLE 10-10

Expected frequencies assuming political leaning and age level are independent

| Political leaning | Age group | | | Total |
	20–35	36–50	Over 50	
Conservative	20	25.0	15.0	60
Moderate	70	87.5	52.5	210
Liberal	30	37.5	22.5	90
Total	120	150	90	360

3. The value χ^2 of the test statistic is $\chi^2 = \sum \dfrac{(O - E)^2}{E}$. The computations are given in Table 10-11, where we obtain the computed χ^2 value as 29.35.

TABLE 10-11

Observed frequencies, expected frequencies assuming independence, and computation of the χ^2 value

Political leaning	Age group	Observed frequency O	Expected frequency E	$(O - E)^2$	$(O - E)^2/E$
Conservative	20–35	10	20.0	100	5.00
Conservative	36–50	40	25.0	225	9.00
Conservative	over 50	10	15.0	25	1.67
Moderate	20–35	80	70.0	100	1.43
Moderate	36–50	85	87.5	6.25	0.07
Moderate	over 50	45	52.5	56.25	1.07
Liberal	20–35	30	30.0	–	——
Liberal	36–50	25	37.5	156.25	4.17
Liberal	over 50	35	22.5	156.25	6.94
	Total	360	360		29.35

4. The contingency table has 3 rows and 3 columns. Hence the chi-square distribution has $(3 - 1)(3 - 1)$, or 4, degrees of freedom. For $\alpha = 0.05$, the table value of chi-square is $\chi^2_{4,0.05} = 9.488$.

5. The computed value, 29.35, is greater than the table value, 9.488, and so we reject H_0.

6. The magnitude of the computed value indicates rather emphatically that age group and political leaning are strongly related. ▬

Note that if we do conclude that two attributes are dependent, it does not imply any cause and effect relationship. For instance, if we conclude that eye color and blood type are related, it should not be construed to mean that eye color causes blood type, or vice versa.

As a final comment, since the distribution of X is approximately chi-square, we emphasize once again that n should be large, and the criterion for the test of independence of two attributes is not recommended if any expected frequency is less than 5.

SECTION 10-2 EXERCISES

1. A survey was conducted to investigate whether alcohol drinking and smoking are related. The following information was compiled for 600 individuals.

	Smoker	Nonsmoker
Drinker	193	165
Nondrinker	89	153

Using a 2.5 percent level of significance, test the null hypothesis that alcohol drinking and smoking are not related.

2. A store inquired of 500 of its customers whether they were satisfied with the service or not. The response was classified according to the sex of the customer and is contained in the table below:

	Male	Female
Satisfied	140	231
Not satisfied	39	90

At the 5 percent level of significance, test the null hypothesis that whether a customer is satisfied or not does not depend on the sex of the customer.

3. In a survey 600 people were classified with respect to hypertension and heart ailment and the following data were obtained.

Hypertension	Heart condition	
	Ailment	No ailment
Constant	51	89
Occasional	72	280
Never	19	89

At the 1 percent level of significance, test the null hypothesis that hypertension and heart ailment are not related.

4. The table below gives the distribution of students according to the type of music they prefer and their IQ:

Music preferred	IQ		
	High	*Medium*	*Low*
Classical	45	32	13
Semiclassical	58	62	30
Rock	87	126	87

Using a 5 percent level of significance, test the null hypothesis that music taste and IQ level are independent attributes.

5. The office of the dean has compiled the following information on 500 students regarding their grade point averages and the way they ranked their instructor:

Rating	Grade point average			
	Above 4.0	*3.0–3.9*	*2.0–2.9*	*Below 2.0*
Outstanding	40	50	55	20
Good	25	35	45	20
Average	10	25	50	36
Poor	10	15	40	24

Test the null hypothesis that a student's grade point average and evaluation of the instructor are not related. Use a 5 percent level of significance.

6. The data given below refer to achievement in high school and family economic level of 400 students:

Achievement	Family economic level		
	High	*Medium*	*Poor*
High	15	80	20
Average	40	100	50
Low	25	30	40

At the 5 percent level of significance, is there a relationship between achievement in high school and family economic level?

7. The following table gives the distribution of weight and blood pressure of 300 subjects:

Weight	Blood pressure		
	High	*Normal*	*Low*
Overweight	40	34	18
Normal	36	77	27
Underweight	16	33	19

At the 5 percent level of significance, are weight and blood pressure related?

8. The following figures give information regarding eye color and a certain eye trait of 300 individuals:

Trait	Eye color			
	Blue	*Brown*	*Green*	*Other*
Present	50	69	45	34
Not present	25	24	30	23

At the 5 percent level of significance, is there any association between eye color and the eye trait?

10-3 A TEST OF HOMOGENEITY

In Chapter 9 we saw how to test two populations for equality of proportions of some characteristic. With that background we are in a position, for example, to test whether the proportion of Democrats in California is the same as the proportion of Democrats in New York if we are given the data in Table 10-12 where 200 people were interviewed in New York and 300 in California.

TABLE 10-12		
Party affiliation	**State**	
	New York	*California*
Democrat	122	168
Non-Democrat	78	132
Total	200	300

We might generalize this procedure in several directions. For instance, we might want to compare the proportions of Democrats in four states such as New York, California, Indiana, and Florida. In other words, we might want to compare the proportions of a characteristic in more than two populations. Another generalization might be if we considered three states, for example, New York, California, and Indiana, to test whether, in these three states, the proportions of Republicans are the same, whether the proportions of Democrats are the same, and whether the proportions of voters belonging to neither party are the same. In short, what we are interested in is whether the three states are *homogeneous* with respect to the party affiliations of their residents. A test that deals with problems of this type is called a **test of homogeneity.**

Let us consider the last case. As has always been our practice, we shall pick a sample. Since each of the states has to be represented, we pick a certain number of individuals from each state. Suppose we interview 200 people in California, 200 in New York, and 100 in Indiana, and the survey produces the results given in Table 10-13.

TABLE 10-13

Observed frequencies of Democrats, Republicans, and others in California, New York, and Indiana in the samples

| Party | State | | | Total |
	California	New York	Indiana	
Democrats	95	105	40	240
Republicans	80	60	50	190
Others	25	35	10	70
Total	200	200	100	500

Notice that when we considered independence of attributes, we picked a random sample of a given size and then *both* the column and row totals were random numbers determined by chance. In our present discussion, the number of individuals that are to be interviewed in each state is predetermined; that is, the column totals are fixed.

In Table 10-14, suppose p_{11}, p_{12}, and so on, represent *theoretical probabilities* of the cells.

TABLE 10-14

Theoretical probabilities of the cells

| Party | State | | |
	California	New York	Indiana
Democrats	p_{11}	p_{12}	p_{13}
Republicans	p_{21}	p_{22}	p_{23}
Others	p_{31}	p_{32}	p_{33}

In testing for homogeneity, we are actually setting the following null hypothesis:

$$H_0: \begin{cases} p_{11} = p_{12} = p_{13} & \text{The proportions of Democrats are equal.} \\ p_{21} = p_{22} = p_{23} & \text{The proportions of Republicans are equal.} \\ p_{31} = p_{32} = p_{33} & \text{The proportions of others are equal.} \end{cases}$$

The alternative hypothesis will, of course, be

H_A: There is no homogeneity.

Let us find the expected number of Democrats in each of the states, assuming $p_{11} = p_{12} = p_{13}$.

Since there are 240 Democrats in the sample of 500, an *estimate* of the probability that a person is a Democrat is 240/500. (If there is homogeneity, as H_0 stipulates, we are entitled to obtain an estimate by pooling for all the states.) Now there are 200 people representing the state of California. Therefore, the expected number of Democrats from this state is $200\left(\frac{240}{500}\right)$. The same reasoning gives $200\left(\frac{240}{500}\right)$ and $100\left(\frac{240}{500}\right)$ as the expected number of Democrats from New York and Indiana, respectively. Actually, once we have found the expected number of Democrats for any two states, the expected number of Democrats in the third state is determined automatically in view of the fact that the total number of Democrats in the sample is 240. A careful observation shows that *the expected frequency in any cell is obtained by multiplying together the corresponding row and column totals and then dividing the product by n*, the same as when we carried out the test for independence of variables. The expected numbers of Republicans and other voters who are neither Republican nor Democrat in the three states are computed in a similar way. These results are presented in Table 10-15. Notice, for instance, that once we have determined that there are 96 Democrats and 76 Republicans in California, the number of other voters in that state has to be 28, since there are a total of 200 Californians in the sample.

TABLE 10-15

Expected frequencies assuming homogeneity

Party	State			
	California	New York	Indiana	Total
Democrats	96	96	48	240
Republicans	76	76	38	190
Others	28	28	14	70
Total	200	200	100	500

Once again, the measure of departure from homogeneity is provided by a statistic **X** whose value for any sample is given by

$$\chi^2 = \sum \frac{(O - E)^2}{E}.$$

The distribution of the statistic is approximately chi-square with $(r - 1)(c - 1)$ degrees of freedom, where r represents the number of rows and c the number of columns. The approximation is satisfactory if none of the expected frequencies is less than 5.

In our particular example, there are $(3 - 1)(3 - 1)$, or 4, degrees of freedom, and the rejection region consists of the right tail of the distribution. Using a 5 percent level of significance, we would reject H_0 if the computed value is greater than the table value, $\chi^2_{4, 0.05} = 9.488$. The computations are shown in Table 10-16.

The computed value, 12.77, of the statistic is greater than the table value of 9.488, and so we reject H_0. If we examine column 5, which gives the difference

$O - E$, there are too few Republicans in New York and too many Republicans in Indiana for compatibility with the hypothesis of homogeneity.

TABLE 10-16

Computation of the χ^2 value for test of homogeneity

State	Party affiliation	Observed frequency O	Expected frequency E	O − E	(O − E)²	(O − E²)E
California	Democrat	95	96	−1	1	0.01
California	Republican	80	76	4	16	0.21
California	Others	25	28	−3	9	0.32
New York	Democrat	105	96	9	81	0.84
New York	Republican	60	76	−16	256	3.37
New York	Others	35	28	7	49	1.75
Indiana	Democrat	40	48	−8	64	1.33
Indiana	Republican	50	38	12	144	3.80
Indiana	Others	10	14	−4	16	1.14
	Total	500	500	0		12.77

In summary, the approach for the test of homogeneity is the same as for the test of independence of variables, although the conclusions to be drawn are of a slightly different nature.

EXAMPLE 1

In order to investigate whether the distribution of the blood types in Europe is the same as in the United States, information was collected on 200 randomly picked people in Europe and 300 people in the United States. From the data summarized in Table 10-17, is it true that the distributions of blood types in Europe and the United States are significantly different?

TABLE 10-17

Observed frequencies of the distribution of blood types in Europe and the United States

Blood type	Location		
	Europe	United States	Total
A	95	125	220
B	50	70	120
O	45	90	135
AB	10	15	25
Total	200	300	500

SOLUTION   1. The null and alternative hypotheses are given as follows:

H_0:   The distribution of blood types in Europe is the same as in the United States.

H_A: There is no homogeneity.

2. The expected frequencies are obtained by multiplying appropriate row and column totals together and then dividing the product by $n = 500$. The computations to find the χ^2 value are given in Table 10-18.

TABLE 10-18
Computation of the χ^2 value

Location	Blood type	Observed frequency O	Expected frequency E	$O - E$	$(O - E)^2$	$(O - E)^2/E$
Europe	A	95	88	7	49	0.55
Europe	B	50	48	2	4	0.08
Europe	O	45	54	-9	81	1.50
Europe	AB	10	10	—	—	—
U.S.	A	125	132	-7	49	0.37
U.S.	B	70	72	-2	4	0.06
U.S.	O	90	81	9	81	1.00
U.S.	AB	15	15	—	—	—
	Total	500	500	—		3.56

3. The chi-square distribution has $(4 - 1)(2 - 1)$, or 3, degrees of freedom.
4. Using a 5 percent level of significance, the decision rule is to reject H_0 if the computed value is greater than $\chi^2_{3,0.05} = 7.815$.
5. Since the computed value 3.56 is not greater than the table value of 7.815, we do not reject H_0.

Conclusion: There is no reason to believe that the distribution of blood types in Europe is any different from the distribution in the United States.   ■

SECTION 10-3   EXERCISES

1. In a laboratory experiment involving two crosses of fruit flies, the following data were obtained giving the number of progeny with curly wings for the two crosses:

	Cross 1	Cross 2
Number of progeny with curly wings	63	40
Number of progeny	180	144

At the 5 percent level, is there a significant difference in the true proportions of curly-winged progeny for the two crosses? Use

(a) the chi-square test developed in this section
(b) a test for comparing two population proportions given in Section 9-4.

2. The following data give information regarding 80 patients, 50 of whom were administered a certain drug and 30 of whom were administered a placebo:

	Drug	Placebo
Cured	35	17
Not cured	15	13

Do the data indicate that the true proportion cured by the drug is different from that using placebo? Use $\alpha = 0.05$.

3. Three launching pads are each fitted with a different device. The following data give the number of successful launches along with the total number of launches from each pad:

	Pad 1	Pad 2	Pad 3
Number of attempts	50	80	60
Number of successful attempts	30	52	33

At the 5 percent level of significance, is there a significant difference in the true proportions of successful launches from the three pads?

4. In a certain area, 200 Democrats, 150 Republicans, and 100 voters who belong to neither party were asked if they supported the welfare programs. The following figures give their response:

	Support	Do not support
Democrats	133	67
Republicans	82	68
Other voters	59	41

At the 5 percent level of significance, test the null hypothesis that the support is the same among the three groups.

5. To find the attitude of people on the issue of school busing, 100 people in each of the four regions, the East, the West, the Midwest, and the South, were interviewed. The following sample data were obtained:

	East	West	Midwest	South
In favor	42	44	30	36

Using a 2.5 percent level of significance, test the null hypothesis that the sentiment in favor of busing is the same in the four regions.

6. Four jars, each containing 100 insects, were sprayed with four different insecticides. The number of insects killed were 46, 70, 60, and 52. At the 5 percent level of significance, test the null hypothesis that there is no significant difference between the insecticides.

7. Professor Kieval feels convinced that the traditional approach is best suited to teaching a beginning course in calculus. He feels that the avant-garde approach does not provide any insight and, certainly, does not equip a student with the basic skills. To test the validity of this assertion, a section of 60 students was taught using the traditional approach and another section of 80 students was taught using the avant-garde approach. At the end of the quarter, the same test (insight-oriented) was given to the two sections and the performance of the students graded as excellent, satisfactory, or poor. The following table presents the results:

	Traditional	Avant-garde
Excellent	31	28
Satisfactory	21	27
Poor	8	25

At the 5 percent level of significance, would you say that Professor Kieval is justified in his assertion?

8. In a course in elementary statistics, there were 40 chemistry majors, 50 forestry majors, and 60 fisheries majors. The figures given below refer to the number of students who passed the course:

	Chemistry	Forestry	Fisheries
Number passed	34	35	44

At the 5 percent level, test the null hypothesis that the true proportions of students passing are the same for the three majors.

9. Three pain relievers, Extrarelief, Superrelief, and Mightyrelief, were each tried on 100 patients. The following figures refer to the extent of relief (recorded as excellent, moderate, or poor):

	Extrarelief	Superrelief	Mightyrelief
Excellent	42	51	55
Moderate	38	27	17
Poor	20	22	28

At the 5 percent level of significance, test the null hypothesis that the three pain relievers are basically the same in their effectiveness.

10. The following table gives the marital status of 200 blue-collar workers, 300 white-collar workers, and 150 university professors:

	Blue-collar workers	White-collar workers	University professors
Married	105	147	90
Single (never married)	45	66	18
Divorced	50	87	42

At the 5 percent level of significance, test the null hypothesis that the true distribution of the marital status among the three groups is the same.

11. A single-dose study to compare three drugs was conducted over a four-hour period. The experiment involved 180 patients who were all suffering from one of the following conditions: sprain, strain, dislocation, fracture, or postsurgical pain. Each drug was tested on 60 patients allotted to it at random. The figures below give the number of patients reporting effective pain relief and those requesting additional medication.

	Drug A	Drug B	Drug C
Relief	46	33	40
No relief	14	27	20

At the one percent level of significance, are the true proportions of patients getting relief different for the three drugs?

KEY TERMS AND EXPRESSIONS

test of goodness of fit
observed frequency
expected frequency
test of independence
contingency tables
test of homogeneity

KEY FORMULAS

For goodness of fit, expected frequency E_i is

$$E_i = np_i$$

where n is the sample size and p_i is the probability of category c_i hypothesized under H_0.

For tests of independence and homogeneity, in the cell which represents the intersection of row i and column j, expected frequency $E_{i,j}$ is

$$E_{i,j} = \frac{(\text{Total of row } i)(\text{Total of column } j)}{n}$$

where n is the total number in the sample.

Test criterion for goodness of fit, independence, and homogeneity

$$\chi^2 = \sum \frac{(O - E)^2}{E}$$

The sum is found over all the cells where O represents the observed frequency and E, the expected frequency.

CHAPTER 10 TEST

1. What null hypothesis do we stipulate in chi-square tests of goodness of fit? What is the alternative hypothesis?

2. What type of null hypothesis do we make when we carry out chi-square tests of independence of classification? What is the alternative hypothesis?

3. Eighty children were asked to pick one of the four numbers 1, 2, 3, 4, whichever came to their minds. The following figures give the distribution of the numbers picked:

Number picked:	1	2	3	4
Number of children	25	14	15	26

Use a 5 percent level of significance to test the hypothesis that the numbers were picked at random.

4. The table below gives the distribution of eye color and hair color of 200 individuals picked at random:

	Hair color	
Eye color	Blonde	Brunette
Blue	32	40
Green	40	24
Brown	10	54

At the 5 percent level, are eye color and hair color related?

5. According to the Hardy-Weinberg law in genetics, if the proportion of A alleles in the population is p and that of a alleles is q (that is, $1 - p$), and the population is *mating at random,* then the proportions of AA, Aa, and aa genotypes in the progeny will be in the ratios $p^2:2pq:q^2$. Out of 200 individuals from a random mating population, it was found that there were 26 AA genotypes, 90 Aa genotypes, and 84 aa genotypes. Test the null hypothesis that the proportion of A alleles in the parent population is 0.4.

6. A survey was conducted in Chicago, Denver, and St. Louis. When 200 people were interviewed in each of these cities as to whether they favored mandatory car-safety regulations, the following response was obtained:

	Chicago	Denver	St. Louis
Favor	60	86	70
Oppose	105	76	83
No opinion	35	38	47

At the 5 percent level, test the three cities for the homogeneity of attitude toward mandatory car-safety regulations.

11

LINEAR REGRESSION
AND LINEAR CORRELATION

INTRODUCTION

In the most part thus far, we have restricted our study to discussing statistical methods analyzing a single variable. Such a discussion about a single variable is commonly known as the **univariate** case. However, often in dealing with problems in social sciences, business, biology, or economics, we are interested in determining whether a discernible relationship exists between two or more variables. It will be recalled that in Chapter 10 we considered only briefly a situation when two response variables are observed on each member. The variables considered were qualitative, and using contingency tables, we went on to discuss whether they were related. In this chapter we now turn our attention to the case where two *numerical* variables are measured, obtaining a pair of measurements for each member. This is called the **bivariate** case. For example, an educator might be interested in finding out a relationship, if any, between the number of hours a student studies and his or her score on a scholastic test; a marketing specialist might wish to find a relationship between the amount spent on advertising and the volume of sales; a social scientist might be interested in knowing whether the income of a family and the amount spent on entertainment are related in some way; and so on.

In cases of this kind, many questions occur. In the first place, is there a reason to suspect that a mathematical relationship exists between the variables? If there is a relation, how can we describe it effectively? An important reason for establishing a relation between two variables is for the purpose of making predictions. How reliable are such predictions?

We shall treat the simplest case, in which the relation is linear. That is, the association can be expressed graphically by means of a straight line relation on the coordinate system. Specifically, we shall discuss two concepts that are extensively used in scientific research, namely, **linear correlation** and **linear regression.*** It should be realized that in investigating an association between variables, these two concepts, though interrelated, serve different purposes. The role of linear correlation is to find out if there exists a *linear* relation between the variables. If a linear relation exists, the correlation coefficient will indicate how close to a straight line such a relation is. This closeness is often referred to as the strength of their association. The regression theory, on the other hand, goes on to describe the specific linear relation in quantitative terms, assuming that there does exist a linear relation. Once the equation describing the straight-line relation is established, we may use it for making predictions about one variable from the knowledge of the other. The variable that is predicted is called the *dependent* variable (also called the *output* variable or *response* variable). The variable used for making the prediction is called the *independent* variable (also

---

*It seems that the term *regression* was first used by Francis Galton to describe laws of human inheritance.

called the *input* variable). If, for example, the IQ is used as a predictor of performance on a scholastic test, then IQ is the independent variable and performance on the test, the dependent variable. ▪▬▬

11-1 COMPUTATION OF A LINEAR CORRELATION COEFFICIENT AND FITTING OF A REGRESSION LINE

WHAT IS A LINEAR RELATION?

As mentioned in the introduction, a linear relation between two variables is one that can be expressed graphically by means of a straight line. How do we express a straight-line relation by means of an equation? Let us consider the following examples.

Suppose a taxi driver charges 50¢ per mile. If x represents the number of miles traveled and y the total amount charged (in dollars), then we would write

$$y = 0.5x.$$

Now suppose the driver says that the initial charge is $2 just for turning on the meter and that each mile costs 50¢. Then we would write

$$y = 2 + 0.5x.$$

For different numbers of miles traveled x, we can compute the amount charged y as shown in Table 11-1. When the points that represent the pairs of values (x, y) computed in this

TABLE 11-1

Number of miles traveled and amount charged.

Number of miles x	Amount charged (in dollars) y
0	2.0
1	2.5
2	3.0
3	3.5
4	4.0

way are plotted on the coordinate system as in Figure 11-1, we see that they all lie on a straight line that rises from left to right.

FIGURE 11-1

A straight line ris-
ing from left to
right.

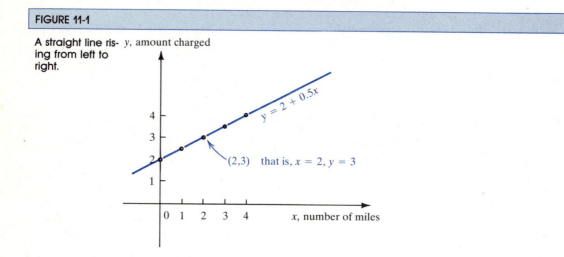

As another example, suppose a person with $40 in his pocket hires a taxi and is charged at a rate of 50¢ a mile. If x represents the number of miles traveled and y the amount that the passenger has left after traveling x miles, then clearly

$$y = 40 - 0.5x.$$

The values of y for different values of x are given in Table 11-2 on page 381. As can be seen from Figure 11-2, all the points (x, y) lie on a straight line that falls from left to right.

In general, a straight-line relation between two variables x and y can be expressed mathematically by means of an equation as

$$y = a + bx$$

where a and b are the constants that determine the nature of the particular straight line.

FIGURE 11-2

A straight line
falling from left
to right.

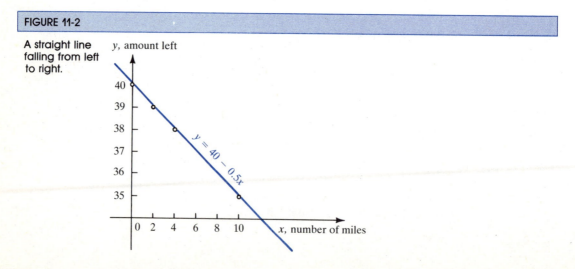

TABLE 11-2

Number of miles traveled and amount left.

Number of miles x	Amount left y
0	40
2	39
4	38
10	35

If b, the coefficient of x, is positive, the straight line rises from left to right as in Figure 11-1, and if b is negative, it falls from left to right as in Figure 11-2.

When $x = 0$, we get the y value equal to a. We call a the **y-intercept.**

To attach a meaning to the constant b, consider a unit increase in x. That is, x changes to $x + 1$. Correspondingly, suppose y changes to y_1. We then have

$$y = a + bx$$

and

$$y_1 = a + b(x + 1).$$

Taking the difference, we get

$$y_1 - y = b$$

Thus b represents the change in y for a *unit* increase in x and is called the **slope** of the line. The y-intercept and the slope of a typical straight line are shown in Figure 11-3.

FIGURE 11-3

The y-intercepts a and slopes b for two straight lines $y = 1.5 + 2.5x$ and $y = 6.5 - 1.5x$.

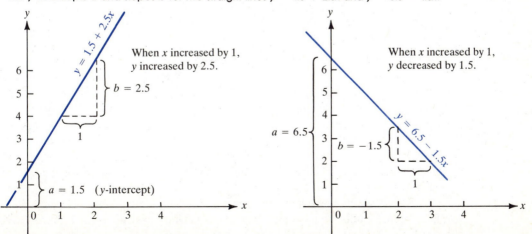

Thus, we summarize as follows:

EQUATION OF A STRAIGHT LINE

The **equation of a straight line** may be written in the form

$$y = a + bx$$

where the constant a represents the **y-intercept** and the constant b represents the **slope** which is the change in y for a unit increase in x. If b is positive, then y increases as x increases; if b is negative, then y decreases as x increases; and if $b = 0$, then y is constant and $y = a$.

THE LINEAR CORRELATION COEFFICIENT

Let us consider the data given in Table 11-3 which represents the IQ and grade point averages of ten students.

TABLE 11-3
IQ and grade point averages of ten students.

IQ x	Grade point average y
100	3.0
120	3.8
110	3.1
105	2.9
85	2.6
95	2.9
130	3.6
100	2.8
105	3.1
90	2.4

Designating IQ by x and grade point average by y, we have ten ordered pairs (x, y), namely, (100, 3.0), (120, 3.8), . . . , (90, 2.4). We call these pairs *ordered pairs* with the understanding that the first component will refer to the value of x and the second component to the value of y. In treating problems involving bivariate data, as a preliminary step, it is a good idea to plot the data as points in the xy-coordinate system. The pictorial representation thus obtained is called a **scatter diagram.** When the points are plotted, a pattern may evolve suggesting a particular type of relation such as linear, curvilinear, and so on. The scatter diagram for the data in Table 11-3 is shown in Figure 11-4.

We see from the scatter diagram that students with higher IQ's appear to obtain higher grade point averages, and students with lower IQ's tend to have lower

FIGURE 11-4

Scatter diagram for IQ and grade point average data given in Table 11-3.

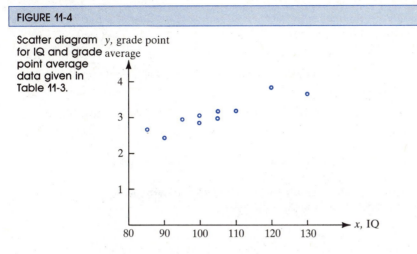

grade point averages. Since high values of x tend to be associated with high values of y, and, conversely, since low values of x tend to be associated with low values of y, we shall say that our data show a *positive relationship* between IQ and grade point average. On the other hand, if it should turn out that high values of x are associated with low values of y, and low values of x are associated with high values of y, we shall say that a *negative* or an *inverse relationship* exists. An example of a negative relationship is that of the price of a used car and the number of years it has been used.

From the scatter diagram in Figure 11-4, we see that the points are scattered in such a way that, more or less, a straight-line pattern is indicated. The measure characterizing the strength of this relationship, namely, closeness to a straight line, is provided by the **coefficient of linear correlation,** also called the **correlation coefficient.** It is commonly designated by the letter r, and if (x_1, y_1), (x_2, y_2), . . . , (x_n, y_n) are n pairs of values, it is defined by the following formula:

$$r = \frac{\sum_{i=1}^{n} (x_i - \bar{x})(y_i - \bar{y})}{\sqrt{\sum_{i=1}^{n} (x_i - \bar{x})^2 \cdot \sum_{i=1}^{n} (y_i - \bar{y})^2}}.$$

This coefficient is also called the *Pearson product-moment correlation coefficient.* For computational purposes, a more convenient version is given by the following:

COEFFICIENT OF LINEAR CORRELATION

$$r = \frac{n \sum_{i=1}^{n} x_i y_i - \sum_{i=1}^{n} x_i \cdot \sum_{i=1}^{n} y_i}{\sqrt{\left[n \sum_{i=1}^{n} x_i^2 - \left(\sum_{i=1}^{n} x_i \right)^2 \right] \left[n \sum_{i=1}^{n} y_i^2 - \left(\sum_{i=1}^{n} y_i \right)^2 \right]}}$$

Thus in order to compute r, we need $\sum_{i=1}^{n} x_i$, $\sum_{i=1}^{n} y_i$, $\sum_{i=1}^{n} x_i^2$, $\sum_{i=1}^{n} y_i^2$, $\sum_{i=1}^{n} x_i y_i$, and n, the number of paired observations.

EXAMPLE 1   Obtain the coefficient of linear correlation between IQ and grade point average based on the data of Table 11.3.

SOLUTION   The data is reproduced in the first two columns of Table 11-4 below and the relevant computations are carried out in columns 3, 4, and 5.

TABLE 11-4

Computations for obtaining correlation coefficient between IQ and grade point average based on the data of Table 11-3.

	x	y	x^2	y^2	xy
	100	3.0	10,000	9.00	300.0
	120	3.8	14,400	14.44	456.0
	110	3.1	12,100	9.61	341.0
	105	2.9	11,025	8.41	304.5
	85	2.6	7225	6.76	221.0
	95	2.9	9025	8.41	275.5
	130	3.6	16,900	12.96	468.0
	100	2.8	10,000	7.84	280.0
	105	3.1	11,025	9.61	325.5
	90	2.4	8100	5.76	216.0
Total	1040	30.2	109,800	92.80	3187.5

Since $\sum_{i=1}^{n} x_i = 1040$, $\sum_{i=1}^{n} y_i = 30.2$, $\sum_{i=1}^{n} x_i^2 = 109{,}800$, $\sum_{i=1}^{n} y_i^2 = 92.8$, and $\sum_{i=1}^{n} x_i y_i = 3187.5$, substituting in the formula we get

$$r = \frac{10(3187.5) - (1040)(30.2)}{\sqrt{[10(109{,}800) - (1040)^2][10(92.80) - (30.2)^2]}}$$

$$= \frac{467}{\sqrt{(16{,}400)(15.96)}}$$

$$= 0.913. \quad \blacksquare$$

How does r measure the strength of a linear relationship? What does a value of $r = 0.913$ in Example 1 mean? We shall consider these questions later when we consider correlation analysis in Section 11-3. But in the meantime, we shall list some of the properties of r.

Properties of the Correlation Coefficient

1. The coefficient of correlation always yields a value between -1 and 1, inclusive. That is $-1 \le r \le 1$.
2. A value of $r = 1$ means that there is a perfect positive correlation and that in the scatter diagram all the points lie on a *straight line* rising upward to the right, as shown in Figure 11-5(a).
3. A value of $r = -1$ means that there is a perfect negative correlation and that all the points lie on a *straight line* falling downward to the right, as in Figure 11-5(b). Any value of r in the vicinity of $+1$ or -1, such as 0.95 or -0.9, indicates that the points are scattered closely around a straight line.
4. A value of r near zero means either of the following:
   a. No clear relationship exists between the two variables, as, for example, is indicated by the scatter diagram in Figure 11-5(c). There is no definite pattern of dependence between x and y and the relation could be shown by a *horizontal line*.
   b. The variables are related, but *not* by a linear relation. For instance, for Figure 11-5(d), the value of r would be very near zero. However, there is obviously a definite relation that is *curvilinear*.
5. A positive value of r indicates that the *linear trend* is upward from left to right, as in Figures 11-5(a) and (e). A negative value of r implies that the *linear trend* is downward from left to right, as in Figures 11-5(b) and (f).

FITTING A REGRESSION LINE TO DATA

From the magnitude of r we are able to decide whether two variables are linearly related and also to determine the closeness of this relationship. However, a correlation coefficient, even when it indicates a rather strong linear relationship, does not tell us anything about the way the two variables are linearly related, since r by itself gives us neither the slope nor the y-intercept. Hence the mere knowledge of r is inadequate for prediction purposes. What we need now is the theory of regression. We tacitly assume that there is a straight-line relation and proceed to determine the constants that describe it. In regression problems it is conventional to denote the independent variable as x and to plot its values along the horizontal axis. The dependent variable is denoted by y, and its values are plotted along the vertical axis.

FIGURE 11-5

Some possible scatter diagrams.

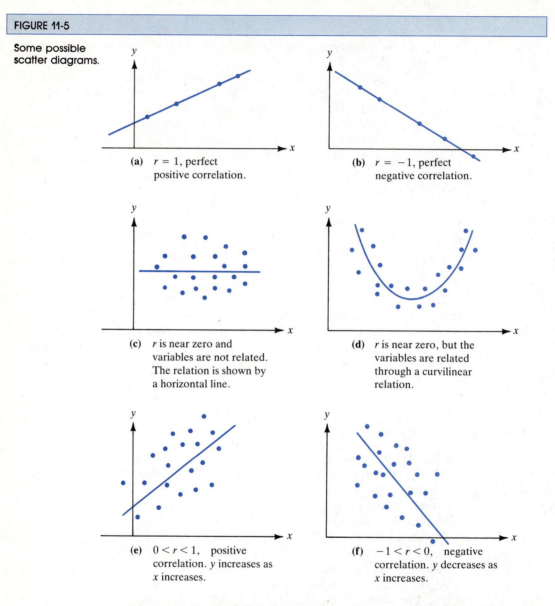

(a) $r = 1$, perfect positive correlation.

(b) $r = -1$, perfect negative correlation.

(c) r is near zero and variables are not related. The relation is shown by a horizontal line.

(d) r is near zero, but the variables are related through a curvilinear relation.

(e) $0 < r < 1$, positive correlation. y increases as x increases.

(f) $-1 < r < 0$, negative correlation. y decreases as x increases.

Let us consider the following example, where we wish to predict the yield of grain (bushels per acre) as a linear function of the amount of fertilizer applied (pounds per acre).* Since yield is the predicted variable, it is the dependent variable. The amount of fertilizer applied is the variable used for the purpose of making a prediction; hence it is the independent variable.

---

*This example will be referred to later in this section and in Section 11-2.

Table 11-5 gives the observed yields of grain for corresponding applications of fertilizer. These pairs of values are plotted in Figure 11-6 to obtain the scatter

TABLE 11-5	
Yield of grain and the amount of fertilizer applied.	
Amount of fertilizer x	*Yield* y
30	43
40	45
50	54
60	53
70	56
80	63

diagram. Clearly a linear relation is indicated. We may therefore want to fit a straight line to bring out this linear relation. If the points were exactly on a straight line, there would be no problem about drawing the line. Since the points are not on a straight line, theoretically the number of lines that can be drawn between points is unlimited. Thus given a set of data, our problem concerns drawing the particular straight line that best reflects the linear trend indicated by the points. In other words, the problem is simply that of determining appropriate constants a and b associated with the straight line.

The principle involved in obtaining "the line of best fit" is called the *method of least squares* and was developed by Adrien-Marie Legendre (1752–1833). We shall now explain the central idea behind this principle.

FIGURE 11-6

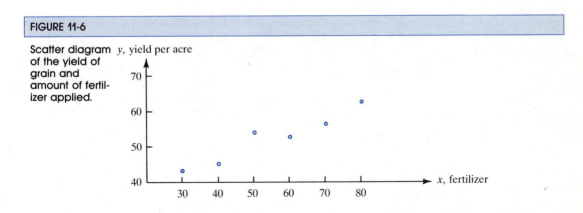

Scatter diagram of the yield of grain and amount of fertilizer applied.

Suppose $\hat{y} = a + bx$ ($\hat{y}$ is read *y hat*) is the line that best fits the data consisting of the n pairs of values (x_1, y_1), (x_2, y_2), . . . , (x_n, y_n). We shall call this the **prediction equation.** For example, $\hat{y}_i = a + bx_i$, represents the predicted value of the dependent variable when the value of the independent variable is x_i. Of course, y_i is the corresponding observed y-value in the data. The quantities $y_1 - \hat{y}_1$, $y_2 - \hat{y}_2$, . . . , $y_n - \hat{y}_n$, that is, the quantities $y_1 - (a + bx_1)$, $y_2 - (a + bx_2)$, . . . , $y_n - (a + bx_n)$ give deviations of the observed y values from the corresponding predicted values and each is called the **residual** or **error,** denoted e. This information is summarized in Table 11-6 where we let $e_i = y_i - \hat{y}_i = y_i - (a + bx_i)$.

TABLE 11-6

The observed x- and y-values together with the predicted y-values and the corresponding residuals.

	y-values		*Residual*
x	*Observed*	*Predicted*	*(or error)*
x_1	y_1	$\hat{y}_1 = a + bx_1$	$y_1 - \hat{y}_1 = e_1$
x_2	y_2	$\hat{y}_2 = a + bx_2$	$y_2 - \hat{y}_2 = e_2$
.	.	.	.
.	.	.	.
.	.	.	.
x_i	y_i	$\hat{y}_i = a + bx_i$	$y_i - \hat{y}_i = e_i$
.	.	.	.
.	.	.	.
.	.	.	.
x_n	y_n	$\hat{y}_n = a + bx_n$	$y_n - \hat{y}_n = e_n$

FIGURE 11-7

Deviations of the observed y values from the predicted y values $\hat{y} = a + bx$.

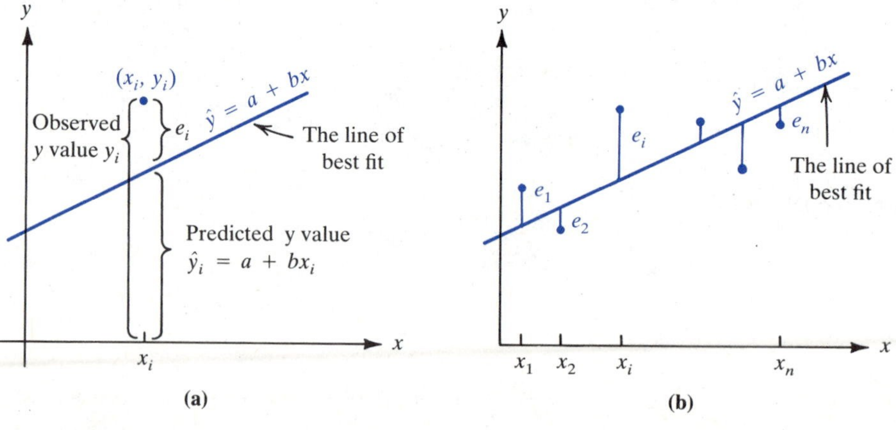

Graphically, e_i which is $y_i - (a + bx_i)$ denotes the vertical distance (plus or minus) from the observed value y_i to the line of best fit, as shown in Figures 11-7(a) and (b).

The principle of the **method of least squares** is stipulated as follows:

METHOD OF LEAST SQUARES

Of all the possible straight lines that can be drawn on a scatter diagram, choose as the **line of best fit** the one for which the sum of the squares of the deviations of the observed y-values from the predicted y-values is a minimum. That is, determine the constants a and b in such a way that

$$\sum_{i=1}^{n} e_i^2 = \sum_{i=1}^{n} [y_i - (a + bx_i)]^2$$

is a minimum.

The straight line obtained using this criterion is called the **least squares regression line.** The values of a and b that determine the line of best fit can be obtained from the following formulas. (These formulas are derived using methods of calculus.)

THE LEAST SQUARES FORMULAS:

The *slope* of the line of best fit is given by

$$b = \frac{n \sum_{i=1}^{n} x_i y_i - \sum_{i=1}^{n} x_i \cdot \sum_{i=1}^{n} y_i}{n \sum_{i=1}^{n} x_i^2 - \left(\sum_{i=1}^{n} x_i \right)^2}$$

and its *y-intercept* by

$$a = \bar{y} - b\bar{x}$$

where n is the number of pairs of observations in the data, $\bar{y}$ is the mean of the observed y-values, and $\bar{x}$ is the mean of the x-values.

The slope b given by the method of least squares is referred to as the **sample regression coefficient.** The specific line

$$\hat{y} = a + bx$$

obtained in this way is called the line of **regression of y on x.**

EXAMPLE 2   For the data on fertilizer application and yield of grain considered earlier, find the regression of yield of grain on the amount of fertilizer applied. The data are given in the first two columns of Table 11-7 below.

SOLUTION   First we should compute a and b. The quantities that are needed for this purpose are $\sum_{i=1}^{n} x_i$, $\sum_{i=1}^{n} y_i$, $\sum_{i=1}^{n} x_i^2$, $\sum_{i=1}^{n} x_i y_i$, and n, the number of paired observations. These computations are displayed in columns 3, 4, and 5 of Table 11-7. (Many handheld calculators are equipped to provide all these quantities by pressing a few keys.)

TABLE 11-7

Computations for finding the regression of yield of grain on the amount of fertilizer applied.

Fertilizer x	Yield y	x^2	xy	y^2
30	43	900	1290	1849
40	45	1600	1800	2025
50	54	2500	2700	2916
60	53	3600	3180	2809
70	56	4900	3920	3136
80	63	6400	5040	3969
Total 330	314	19,900	17,930	16,704

Since $n = 6$, $\sum_{i=1}^{6} x_i = 330$, $\sum_{i=1}^{6} y_i = 314$, $\sum_{i=1}^{6} x_i^2 = 19,900$, $\sum_{i=1}^{6} x_i y_i = 17,930$, we obtain the following value for b:

$$b = \frac{n \sum_{i=1}^{n} x_i y_i - \sum_{i=1}^{n} x_i \cdot \sum_{i=1}^{n} y_i}{n \sum_{i=1}^{n} x_i^2 - \left(\sum_{i=1}^{n} x_i\right)^2}$$

$$= \frac{6(17,930) - (330)(314)}{6(19,900) - (330)^2}$$

$$= 0.377$$

Since $\bar{x} = \dfrac{330}{6} = 55$ and $\bar{y} = \dfrac{314}{6} = 52.33$, we get

$$a = \bar{y} - b\bar{x}$$

$$= 52.33 - (0.377)(55)$$

$$= 31.593$$

Finally, the line of regression of y on x (that is, the prediction equation) is given by

$$\hat{y} = 31.593 + (0.377)x$$

predicted   fixed   change in yield   pounds of
yield   yield   per one pound   fertilizer
                           increase in   applied
                           fertilizer

The value of $b = 0.377$ signifies that for an increase of one pound per acre of fertilizer, there corresponds an increase of 0.377 bushels of grain. ▬

Once the equation of the regression line is obtained, it is a simple matter to draw it. We need only two points in the plane to draw a straight line. Thus we pick any two values of x and for these values obtain corresponding $\hat{y}$ values using the equation. In Example 2, for instance,

for $x = 50$, $\hat{y} = 31.593 + 0.377(50) = 50.443$

and

for $x = 60$, $\hat{y} = 31.593 + 0.377(60) = 54.213$

Joining the two points (50, 50.443) and (60, 54.213), we get the line in Figure 11-8.

It is interesting to note from the formula for the y-intercept that $\bar{y} = a + b\bar{x}$. It therefore follows that *the regression line always passes through the point* $(\bar{x}, \bar{y})$.

FIGURE 11-8

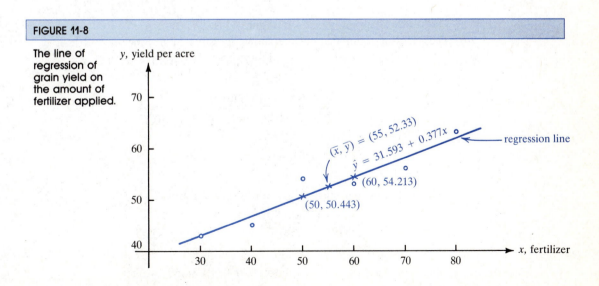

The line of regression of grain yield on the amount of fertilizer applied.

EXAMPLE 3　　The scores x of ten students on the midterm exam and their scores y on the final exam are given in the first two columns of Table 11-8.

TABLE 11-8				
Scores of ten students on the midterm and final exams and other computations.				
Scores on midterm x	Scores on final y	x^2	y^2	xy
58	55	3364	3025	3190
64	75	4096	5625	4800
80	82	6400	6724	6560
74	63	5476	3969	4662
40	82	1600	6724	3280
70	83	4900	6889	5810
72	78	5184	6084	5616
26	50	676	2500	1300
80	65	6400	4225	5200
74	57	5476	3249	4218
Total　638	690	43,572	49,014	44,636

(a) Plot the scatter diagram.

(b) Compute the coefficient of correlation.

(c) Obtain the regression of the scores in the final exam on the scores in the midterm.

(d) Draw the least squares regression line.

SOLUTION　　(a) The scatter diagram is given in Figure 11-9.

(b) The necessary computations are shown in Table 11-8. Since $n = 10$, $\sum_{i=1}^{10} x_i = 638$, $\sum_{i=1}^{10} y_i = 690$, $\sum_{i=1}^{10} x_i^2 = 43,572$, $\sum_{i=1}^{10} y_i^2 = 49,014$, and $\sum_{i=1}^{10} x_i y_i = 44,636$, from the formula for r we get

$$r = \frac{10(44,636) - (638)(690)}{\sqrt{[10(43,572) - (638)^2][10(49,014) - (690)^2]}}$$

$$= \frac{6140}{\sqrt{(28,676)(14,040)}}$$

$$= 0.306.$$

(c) The regression coefficient b is given from the formula as

$$b = \frac{10(44,636) - (638)(690)}{[10(43,572) - (638)^2]}$$

$$= 0.214.$$

Since $\bar{x} = 638/10 = 63.8$, and $\bar{y} = 690/10 = 69$, we get

FIGURE 11-9

Scatter diagram and the line of regression of final scores on the midterm scores.

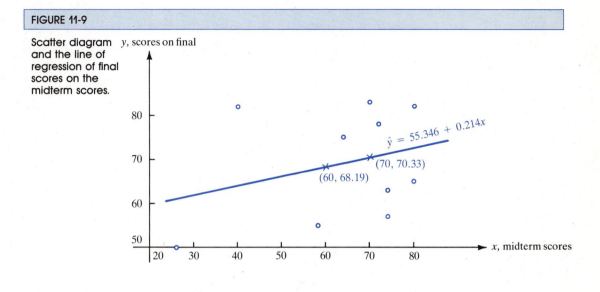

$$a = \bar{y} - b\bar{x}$$
$$= 69 - (0.214)(63.8)$$
$$= 55.346.$$

Hence the line of regression of final scores on the midterm scores is

$$\hat{y} = 55.346 + 0.214x.$$

(d) To draw the least squares regression line we pick any two convenient values of x and find the corresponding $\hat{y}$ values.

For $x = 60$, $\hat{y} = 55.346 + (0.214)(60) = 68.19$.
For $x = 70$, $\hat{y} = 55.346 + (0.214)(70) = 70.33$.

Joining the two points (60, 68.19) and (70, 70.33), we get the line in Figure 11-9. (We could have used $(\bar{x}, \bar{y})$, or (63.8, 69), as one of the points.)

SECTION 11-1   EXERCISES

1. Suppose two variables x and y are related by the linear relation $2y = 3 - 4x$. Complete the following table:

x	3	0	-2	–	–
y	–	–	–	23	-7

2. Draw the lines given by the following equations.

   (a) $y = 5 - 3x$

(b) $2y - 4x = 15$

(c) $20x + 8y - 4 = 0$

3. State whether each relationship between x and y described below is a positive relationship or an inverse relationship:

(a) $y = -5 + 4x$

(b) $9y + 51x = 60$

(c) $3y - 8x + 5 = 0$

4. Without plotting the graphs, find the y-intercepts and the slopes of the lines given below.

(a) $y = -10 - 15x$

(b) $2y = -5 + 4x$

(c) $3y + 17x = 14$

(d) $8x + 3y + 1 = 0$

5. A linear relation between x and y is described by the equation $4y = 6 - 3x$. Find

(a) the change in y for a unit increase in x

(b) the change in y for two units of increase in x.

6. What does the value of $r = 0$ mean?

7. What is the value of r if all the points in a scatter diagram lie

(a) on a straight line sloping upward?

(b) on a straight line sloping downward?

8. Of the two correlation coefficients, $r = 0.3$ and $r = -0.8$, based on the same number of pairs of observations, which would you say suggests a stronger linear relation?

9. The following are the heights x (in inches) and weights y (in pounds) of seven athletes:

Heights x	69	66	68	73	71	74	71
Weights y	163	153	185	186	157	220	190

(a) Find $\sum_{i=1}^{7} x_i$, $\sum_{i=1}^{7} y_i$, $\sum_{i=1}^{7} x_i^2$, $\sum_{i=1}^{7} y_i^2$, $\sum_{i=1}^{7} x_i y_i$

(b) Compute r, the coefficient of correlation between heights and weights.

10. The scores on a scholastic test of ten randomly picked students and the number of hours they devoted preparing for it are given below:

Number of hours x	18	27	20	10	30	24	32	27	12	16
Scores y	68	82	77	90	78	72	94	88	60	70

Compute the coefficient of correlation r.

11. The following data give the average daily temperature on fifteen days and the quantity of soft drinks (in thousands of gallons) sold by a company on each of those days:

Temperature	70	75	80	90	93	98	72	75	75	80	90	95	98	91	98
Quantity	30	28	40	52	57	54	27	38	32	46	49	51	62	48	58

(a) Plot the scatter diagram.

(b) Compute the coefficient of correlation.

12. Compute the coefficient of correlation between the annual income (in thousands of dollars) and the amount of life insurance (in thousands of dollars) of eight families of the same size:

Annual income	10	13	15	18	21	24	27	30
Amount of insurance	15	12	25	20	25	30	35	32

13. This is the same situation as in Exercise 12, but now the annual income is doubled and the insurance amount is tripled. That is,

Annual income	20	26	30	36	42	48	54	60
Amount of insurance	45	36	75	60	75	90	105	96

Compare the value of the coefficient of correlation obtained here with that obtained in Exercise 12. Can you venture a guess, in general, as to what happens to the coefficient of correlation when the x values are multiplied by a constant and the y values are multiplied by another constant?

14. This is the same situation as in Exercise 12, but now the annual income is increased by $5000 and the amount of insurance is increased by $12,000 for each family. Find the coefficient of correlation and compare it with the answers in Exercises 12 and 13. Make a general comment.

15. In a research problem where two variables x and y were measured on each of 40 items, the following data were obtained:

$$\sum_{i=1}^{40} x_i = 293, \ \sum_{i=1}^{40} y_i = 357, \ \sum_{i=1}^{40} x_i^2 = 2685, \ \sum_{i=1}^{40} x_i y_i = 3667$$

(a) Find the slope and the y-intercept of the line of best fit.

(b) Give the least squares regression line, $\hat{y} = a + bx$.

16. In a regression problem, the following information is available:

$$n = 50, \ \sum_{i=1}^{50} x_i = 170, \ \sum_{i=1}^{50} y_i = -282, \ \sum_{i=1}^{50} x_i^2 = 4350, \ \sum_{i=1}^{50} y_i^2 = 37,610,$$
$$\sum_{i=1}^{50} x_i y_i = -12,590$$

(a) Find a and b and obtain the least square regression line $\hat{y} = a + bx$.

(b) Draw the regression line.

17. The following pairs of values (x, y) are given:

x	7	6	5	3	9	5	2	10
y	5	1	-1	-5	3	4	0	7

(a) Plot the scatter diagram.

(b) Find the regression equation, $\hat{y} = a + bx$.

(c) Draw the regression line.

18. Consider the following bivariate data set.

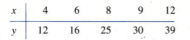

x	4	6	8	9	12
y	12	16	25	30	39

(a) Find the equation of the regression line $\hat{y} = a + bx$.

(b) Using this regression line what value $\hat{y}$ would you predict if $x = 7.2$?

19. Suppose the fitted linear regression of y, the score on a test, and x, the number of hours a student studies, is given by the relation

$$\hat{y} = 55.28 + 4.1x.$$

Predict a score for a student who studies

(a) 1 hour   (b) 4 hours.

20. The scores on the final tests in mathematics, physics, and English for eight randomly picked students are given below:

Mathematics	50	58	67	70	75	82	86	92
Physics	62	54	63	78	81	78	88	90
English	79	82	70	74	67	62	64	56

(a) Find the regression of scores in physics on scores in mathematics.

(b) Find the regression of scores in English on scores in mathematics.

(c) Plot the scatter diagram for the physics-mathematics scores and draw the fitted regression line.

(d) Do the same as in (c) for the English-mathematics scores.

21. Using the data in Exercise 20,

(a) find the correlation coefficient between scores in mathematics and scores in physics

(b) find the correlation coefficient between scores in mathematics and scores in English

(c) comment on the signs of the correlation coefficients above in light of the signs of the corresponding regression coefficients in Exercise 20.

22. The following data refer to the amount spent (in thousands of dollars) on advertising and the volume of sales (in millions of dollars) from 1976 to 1983.

Year	1976	1977	1978	1979	1980	1981	1982	1983
Advertising	30	35	50	40	60	80	60	75
Sales	0.8	1.1	1.3	1.8	1.2	2.2	3.1	1.8

(a) Find the regression of the volume of sales during a year on the amount spent on advertising during that year.

(b) Find the regression of the volume of sales on the amount spent on advertising during the previous year. That is, find the regression for the following data:

x	30	35	50	40	60	80	60
y	1.1	1.3	1.8	1.2	2.2	3.1	1.8

(c) Comment on your findings in (a) and (b).

23. For the data in Exercise 11, obtain the least squares regression line, $\hat{y} = a + bx$, where x represents the average temperature during a day and y the quantity of soft drinks. Estimate the quantity of soft drinks sold if the average temperature on a day is 85 degrees.

24. Based on your experience, guess in each of the following cases if there is a positive correlation, a negative correlation, or no correlation:

(a) the price of a commodity and the demand for it

(b) the supply of a certain food item and its price

(c) the number of unemployed and the volume of exports

(d) the height and weight of a person

(e) the collar size of an individual and the amount of milk that person consumes

(f) the number of years a machine is in use and the number of yearly repairs

(g) the price of gold per ounce (in dollars) and the exchange rate of the dollar

(h) the resale value of a machine and the number of years it is in use

(i) the waist size of a person and the person's bank balance

(j) the number of automobile accidents per year and the age of the driver (up to the age of 60 years).

25. A highway agency interested in finding a relation between the highway speed of a car and mileage per gallon conducted the following experiment. Three test runs were made with a car of a certain make at each of the five speeds (measured to the nearest 5 miles) 45, 50, 55, 60, 65 over a distance of 100 miles, and the mileage per gallon was noted for each run. The following data were obtained:

Speed (mph)	45	50	55	60	65
Mileage per gallon	24.0	27.0	28.0	26.0	23.5
	23.0	26.5	29.0	26.5	23.5
	24.5	26.0	27.5	26.0	24.0

(a) draw the scatter diagram

(b) draw a free-hand curve that might best bring out the relation between speed and mileage per gallon

(c) estimate what speed will give the best mileage per gallon for a car of the above make

26. For the data in Exercise 25, compute the coefficient of correlation between speed and mileage per gallon. What explanation can you provide for a value of r so close to zero?

11-2 REGRESSION ANALYSIS

THE LINEAR MODEL

In the preceding section, using the method of least squares, we computed the straight line that best represents a set of *n* pairs of observations. The *n* pairs, of course, constitute a sample drawn from a much larger population. We therefore call that specific regression line the sample regression line.

Our primary interest in the sample regression line is to investigate the corresponding counterpart in the population. What we have in our mind is a **linear model** in the population, that is, we feel that a straight line best depicts the relation between *x* and *y* in the population. This is not to say that we envision a precise relation of the type

$$Y = A + Bx.$$

(While we used lowercase letters *a* and *b* to represent the *y*-intercept and the slope, respectively, of the sample regression line, we shall use uppercase letters *A* and *B* for the corresponding analogues in the population.) If this was the case, then any given value of *x* would always *determine* exactly the same *y*-value and we would have what is called a **deterministic model.** Of course, when we consider the fertilizer application–grain yield type of problems, we realize that this does not happen. The same amount of fertilizer applied to different lots of the same size can give different amounts of grain yield. To explain this phenomenon, we need a model which illustrates the overall linear trend in the population that we suspect and also takes into account the random error. Thus we think of a response which has a basic linear component added to a random error component so that

response = (linear component) + (random error component).

A model of this type is called a **linear statistical model.** If *Y* represents the response variable, then our model stipulates that

$$Y = (A + Bx) + \text{(random error component)}$$

where *A* and *B* are population constants, that is, parameters. The assumptions that we make regarding the random error component are that, for any *x*, its mean is 0, its variance is σ^2, and its distribution is normal. This, in turn, implies that, for any *x*, the distribution of *Y* is normal with mean $A + Bx$ and variance σ^2.

To be more specific let us now consider the example of the preceding section where we obtained the regression of the amount of yield of grain on the amount of fertilizer applied. The levels of fertilizer are under the experimenter's control. Therefore, we shall assume that the levels of fertilizer are fixed in that they are not decided by chance. It is the yield *Y* of grain that is a random variable, since one acre of land will not give the same yield as another acre, even with the same amount of fertilizer.

Now let us think about the totality of all the acres to which 50 pounds of fertilizer might be applied. The yields of these plots will constitute a subpopulation having a certain distribution. In keeping with our linear statistical model, we shall assume that the distribution is normal and has a mean equal to $A + 50B$ and a variance of σ^2. In a similar way, we can think of subpopulations of yields for different amounts of fertilizer applications. For any level x of the fertilizer, we assume that the subpopulation of yields is normally distributed with the *same* variance σ^2 and a mean yield equal to $A + Bx$. In short, we assume that the means of the subpopulations fall on a straight line with the y-intercept equal to A and the slope equal to B. If we denote by $\mu_{Y|x}$ the mean yield of the subpopulation when x pounds of fertilizer are applied, then

$$\mu_{Y|x} = A + Bx.$$

Last, we assume that the yield of one plot is not influenced by the yield of another plot. That is, yields of plots are independent.

Summarizing, we now state formally the basic assumptions underlying a simple linear regression model:

Assumptions

1. The values of the independent variable x are fixed. They are non-random.
2. For each x, Y is normally distributed with mean $A + Bx$, where A and B are constants. That is, $\mu_{Y|x} = A + Bx$. We shall call the line given by $\mu_{Y|x} = A + Bx$ the **population regression line** or the **true regression line**. (In Figure 11-10 we sketch two straight lines, one repre-

FIGURE 11-10

Sample regression line for the fertilizer–grain yield data and the population regression line.

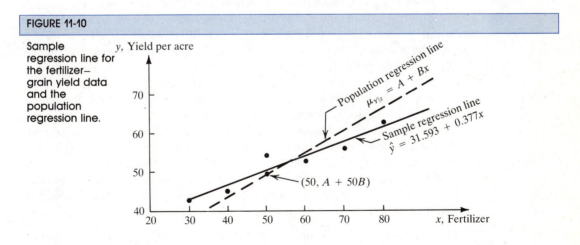

FIGURE 11-11

A three-dimensional display showing the distribution of Y for given values of x, the true regression line, and the sample regression line.

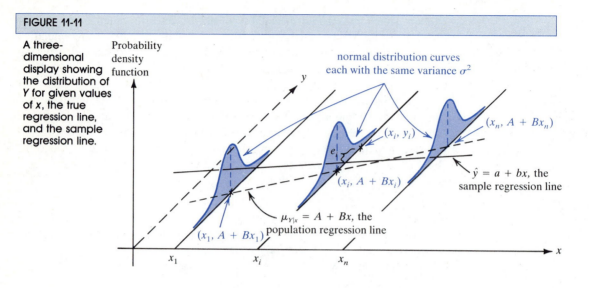

senting the sample regression line based on the given data and the other, the population regression line generally unknown but drawn here for the sake of illustration.)

3. For each x, the variance of the distribution of Y is the same, say σ^2. The implication of this is that the sizes of the normal curves are the same for different values of x, as in Figure 11-11.

4. The random variables representing the response variable are independently distributed.

By and large, the population regression line will not be known to the experimenter. In fact, it will be the experimenter's goal to investigate it. Also we should be aware that this line is a fixed straight line and will not be the same as the sample regression line which changes from sample to sample, just as we recognize, for instance, that the population mean μ is a fixed constant and will not be the same as the sample mean $\bar{x}$ which also changes from sample to sample. Now recall at this point that when we considered the univariate case, we first discussed the sample mean $\bar{x}$ and the sample variance s^2 because we were interested in the corresponding parameters μ and σ^2. Subsequently we developed methods to set confidence intervals and to test hypotheses regarding these parameters. In like manner, having considered the sample regression line in the preceding section, we now propose to investigate the population insofar as the linear trend is depicted in the population.

In view of the setup of the linear regression model, the statistical problem of regression analysis involves the following aspects:

1. Obtain estimates of the two important parameters A and B of the population regression line. From a practical standpoint even more useful are inferences regarding $\mu_{Y|x_0} = A + Bx_0$, the mean value of Y for a given

x value. For example, we might be interested in the true mean amount of yield Y of grain when x_0 pounds of fertilizer are applied per acre. Also, of paramount importance is the estimation of σ^2 which, as you will soon find out, is involved in the variances of the various estimators.

2. Construct confidence intervals for the parameters above.
3. Finally, test hypotheses regarding the parameters.

It can be shown that the least squares regression line $\hat{y} = a + bx$, which was fitted to the sample data in Section 11-1, provides an unbiased estimate for the population regression equation. Also, a and b provide unbiased estimates of A and B, respectively. In fact, writing $\hat{A}, \hat{B}, \hat{Y}_0 (= \hat{A} + \hat{B}x_0)$ for the point estimators of $A, B, \mu_{Y|x_0}$, it can be shown that all these estimators are normally distributed and that their expected values and variances are given as follows:

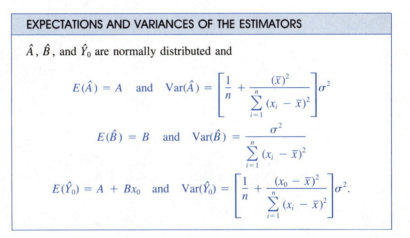

EXPECTATIONS AND VARIANCES OF THE ESTIMATORS

$\hat{A}, \hat{B}$, and $\hat{Y}_0$ are normally distributed and

$$E(\hat{A}) = A \quad \text{and} \quad \mathrm{Var}(\hat{A}) = \left[\frac{1}{n} + \frac{(\bar{x})^2}{\sum_{i=1}^{n}(x_i - \bar{x})^2}\right]\sigma^2$$

$$E(\hat{B}) = B \quad \text{and} \quad \mathrm{Var}(\hat{B}) = \frac{\sigma^2}{\sum_{i=1}^{n}(x_i - \bar{x})^2}$$

$$E(\hat{Y}_0) = A + Bx_0 \quad \text{and} \quad \mathrm{Var}(\hat{Y}_0) = \left[\frac{1}{n} + \frac{(x_0 - \bar{x})^2}{\sum_{i=1}^{n}(x_i - \bar{x})^2}\right]\sigma^2.$$

Since $\hat{A}, \hat{B}$, and $\hat{Y}_0$ are estimators, they represent random variables. Their values, namely, a, b, and $\hat{y}_0$, computed from a given sample will provide estimates of A, B, and $\mu_{Y|x_0}$, respectively.

VARIANCE ESTIMATES

As can be seen from the formulas above for the variances of the estimators $\hat{A}, \hat{B}$, and $\hat{Y}_0$, they all involve the parameter σ^2 which, for the most part, will not be known to the experimenter. So before we can discuss the inferential aspects, we should obtain an estimate of σ^2. We shall use this estimate, in turn, to estimate the variances of $\hat{A}, \hat{B}$, and $\hat{Y}_0$.

One of the assumptions of the linear regression model is that for any x, the distribution of Y has mean $A + Bx$ and the *same* variance σ^2, independent of x. Because of this, it can be shown that an unbiased estimate of σ^2, denoted by s_e^2, is given by the following formula:

ESTIMATE OF σ^2

In a linear regression model involving n pairs of observations, an unbiased estimate of σ^2 is provided by

$$s_e^2 = \frac{\sum_{i=1}^{n} (y_i - \hat{y}_i)^2}{n - 2}.$$

The reason for dividing by $n - 2$ instead of by n is that to obtain $\hat{y}_i$, which is used in $\sum_{i=1}^{n} (y_i - \hat{y}_i)^2$, we had to estimate two parameters A and B. Thus we have lost two degrees of freedom.

EXAMPLE 1   Find the estimate of σ^2 for the data on fertilizer application and yield of grain of the preceding section (see the first two columns of Table 11-9 below).

SOLUTION   It will be recalled that in this case in Example 2 of the preceding section, we obtained the sample regression line as $\hat{y} = 31.593 + 0.377x$. Thus the computations to obtain s_e^2 could be carried out as shown in Table 11-9.

TABLE 11-9

Computations for finding estimate of σ^2 for the grain yield–amount of fertilizer data in Table 11-5.

Amount of fertilizer x	Yield y	$\hat{y} = 31.593 + 0.377x$	$(y - \hat{y})^2$
30	43	42.903	0.010
40	45	46.673	2.800
50	54	50.443	12.652
60	53	54.213	1.472
70	56	57.983	3.932
80	63	61.753	1.555
Total			22.421

From column 4 of the table, $\sum_{i=1}^{n} (y_i - \hat{y}_i)^2 = 22.421$. Hence, applying the formula, the estimate of σ^2 is given as

$$s_e^2 = \frac{\sum_{i=1}^{n} (y_i - \hat{y}_i)^2}{n - 2}$$

$$= \frac{22.421}{6 - 2}$$

$$= 5.61. \quad \blacksquare$$

In order to be able to use the above formula for computing s_e^2, we have to obtain the predicted value $\hat{y}_i$ for each x_i. It is clearly unsuitable if there are a large number of observations. The following alternate formula is more manageable from the computational point of view:

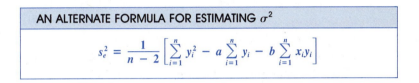

AN ALTERNATE FORMULA FOR ESTIMATING σ^2

$$s_e^2 = \frac{1}{n - 2} \left[\sum_{i=1}^{n} y_i^2 - a \sum_{i=1}^{n} y_i - b \sum_{i=1}^{n} x_i y_i \right]$$

There are $n - 2$ degrees of freedom associated with s_e^2 since we used the n observations to compute two quantities, namely a and b, which are involved in the formula for s_e^2 (see the similar rationale given on page 57 in the context of sample variance).

Caution: In using the alternate version, compute a and b to enough decimal places to compensate for the magnitudes of $\sum_{i=1}^{n} y_i$ and $\sum_{i=1}^{n} x_i y_i$. You must realize that the products $a \sum_{i=1}^{n} y_i$, $b \sum_{i=1}^{n} x_i y_i$ are involved in the formula.

Let us now compute s_e^2 for the data in Example 1 using the alternate formula. In Section 11-1 (page 390) we have already computed all of the quantities involved in the formula but will give them below with a and b expressed to more decimal places. They are

$$n = 6, \ a = 31.5905, \ b = 0.377143, \ \sum_{i=1}^{n} y_i = 314,$$

$$\sum_{i=1}^{n} y_i^2 = 16{,}704, \ \sum_{i=1}^{n} x_i y_i = 17{,}930.$$

Therefore

$$s_e^2 = \frac{1}{4}[16{,}704 - (31.5905)(314) - (0.377143)(17{,}930)]$$

$$= 5.61.$$

Let us denote the estimates of the variances of $\hat{A}$, $\hat{B}$, and $\hat{Y}_0$ by s_a^2, s_b^2, and $s_{\hat{y}_0}^2$, respectively. They are obtained directly from the corresponding expressions for the variances of the estimators (given on page 401) simply by replacing σ^2 by s_e^2. Therefore we have the following formulas for the various estimates:

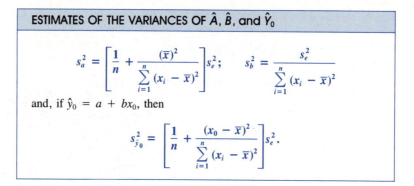

ESTIMATES OF THE VARIANCES OF $\hat{A}$, $\hat{B}$, and $\hat{Y}_0$

$$s_a^2 = \left[\frac{1}{n} + \frac{(\bar{x})^2}{\sum_{i=1}^{n}(x_i - \bar{x})^2}\right]s_e^2; \qquad s_b^2 = \frac{s_e^2}{\sum_{i=1}^{n}(x_i - \bar{x})^2}$$

and, if $\hat{y}_0 = a + bx_0$, then

$$s_{\hat{y}_0}^2 = \left[\frac{1}{n} + \frac{(x_0 - \bar{x})^2}{\sum_{i=1}^{n}(x_i - \bar{x})^2}\right]s_e^2.$$

EXAMPLE 2   In Example 1 we have already found a point estimate of σ^2 as $s_e^2 = 5.61$ for the fertilizer application–yield of grain data. Now compute the variance estimates s_a^2, s_b^2, and $s_{\hat{y}}^2$ when $x = 45$ lb.

SOLUTION   Since $n = 6$, $\bar{x} = \dfrac{330}{6} = 55$, $\sum_{i=1}^{n} x_i = 330$ and $\sum_{i=1}^{n} x_i^2 = 19{,}900$, we get

$$\sum_{i=1}^{n}(x_i - \bar{x})^2 = \sum_{i=1}^{n} x_i^2 - \frac{\left(\sum_{i=1}^{n} x_i\right)^2}{n}$$

$$= 19{,}900 - \frac{(330)^2}{6} = 1750.$$

Therefore,

$$s_a^2 = \left[\frac{1}{n} + \frac{(\bar{x})^2}{\sum_{i=1}^{n}(x_i - \bar{x})^2}\right]s_e^2$$

$$= \left[\frac{1}{6} + \frac{55^2}{1750}\right](5.61)$$

$$= 10.6323$$

and

$$s_b^2 = \frac{s_e^2}{\sum_{i=1}^{n}(x_i - \bar{x})^2}$$

$$= \frac{5.61}{1750}$$

$$= 0.00321.$$

Since $x_0 = 45$, the estimate of the variance of the estimated yield is

$$s_{\hat{y}}^2 = \left[\frac{1}{n} + \frac{(x_0 - \bar{x})^2}{\sum\limits_{i=1}^{n} (x_i - \bar{x})^2} \right] s_e^2$$

$$= \left[\frac{1}{6} + \frac{(45 - 55)^2}{1750} \right](5.61)$$

$$= 1.2556.$$

Taking square roots, we get $s_a = 3.261$, $s_b = 0.0566$. Also $s_{\hat{y}} = 1.121$, where the yield is estimated for 45 pounds of fertilizer application. ▆▆▆

CONFIDENCE INTERVALS AND PREDICTION INTERVAL

Our first objective is to contruct confidence intervals for the parameters A, B, and $\mu_{Y|x_0}$. We shall then discuss the prediction interval.

CONFIDENCE INTERVALS FOR A, B, $\mu_{Y|x_0}$

1. A $(1 - \alpha)100$ percent confidence interval for A is given by

$$a - t_{n-2,\alpha/2}\, s_a < A < a + t_{n-2,\alpha/2}\, s_a .$$

Or equivalently the limits of this interval are

$$a \pm t_{n-2,\alpha/2}\, s_e \sqrt{ \frac{1}{n} + \frac{(\bar{x})^2}{\sum\limits_{i=1}^{n} (x_i - \bar{x})^2} }.$$

2. A $(1 - \alpha)100$ percent confidence interval for B is given by

$$b - t_{n-2,\alpha/2}\, s_b < B < b + t_{n-2,\alpha/2}\, s_b .$$

Or equivalently the limits of the interval are

$$b \pm t_{n-2,\alpha/2}\, \frac{s_e}{\sqrt{\sum\limits_{i=1}^{n} (x_i - \bar{x})^2}}.$$

3. A $(1 - \alpha)100$ percent confidence interval for $\mu_{Y|x_0}$ $(= A + Bx_0)$, the mean of Y corresponding to a value x_0 of the independent variable, is given by

$$(a + bx_0) - t_{n-2,\alpha/2}\, s_{\hat{y}_0} < A + Bx_0 < (a + bx_0) + t_{n-2,\alpha/2}\, s_{\hat{y}_0}.$$

Or equivalently the limits of the interval are

$$(a + bx_0) \pm t_{n-2,\alpha/2}\, s_e \sqrt{ \frac{1}{n} + \frac{(x_0 - \bar{x})^2}{\sum\limits_{i=1}^{n} (x_i - \bar{x})^2} }.$$

Confidence Intervals

The upper and lower limits of the confidence intervals can be obtained using the following rule which was established in Chapter 7 for the case when the estimator is normally distributed but its variance must be estimated. The limits are given by

$$
\begin{pmatrix} \text{Estimate of} \\ \text{the parameter} \end{pmatrix} \pm \begin{pmatrix} \text{Appropriate value} \\ \text{from Student's} \\ t \text{ distribution} \end{pmatrix} \cdot \begin{pmatrix} \text{Estimated} \\ \text{standard} \\ \text{deviation} \end{pmatrix}.
$$

In the present case, the t distribution has $n - 2$ degrees of freedom because there are $n - 2$ degrees of freedom associated with s_e^2. Hence we have confidence intervals for A, B, and $\mu_{Y|x_0}$ as shown at the bottom of page 405.

EXAMPLE 3   Construct 95 percent confidence intervals for the parameters A, B, and $\mu_{Y|45}$ involved in our fertilizer–yield of grain problem discussed earlier.

SOLUTION   In Example 2 of the preceding section, we found $a = 31.593$ and $b = 0.377$. Also, in Example 2 of this section we found $s_a = 3.261$, $s_b = 0.0566$, and $s_{\hat{y}_0} = 1.121$ when $x_0 = 45$. Now, since $n = 6$, there are 4 degrees of freedom and, since $\alpha = 0.05$, we have $t_{n-2,\alpha/2} = t_{4,0.025} = 2.776$. Therefore, 95 percent confidence intervals are obtained as follows:

1. Since $a = 31.593$ and $s_a = 3.261$, a 95 percent confidence interval for A is

$$a - t_{4,0.025}\, s_a < A < a + t_{4,0.025}\, s_a$$
$$31.593 - (2.776)(3.261) < A < 31.593 + (2.776)(3.261)$$
$$22.540 < A < 40.646.$$

2. Since $b = 0.377$ and $s_b = 0.0566$, a 95 percent confidence interval for B is

$$b - t_{4,0.025}\, s_b < B < b + t_{4,0.025}\, s_b$$
$$0.377 - (2.776)(0.0566) < B < 0.377 + (2.776)(0.0566)$$
$$0.220 < B < 0.534.$$

3. As a first step toward constructing a confidence interval for the mean yield when 45 pounds of fertilizer are applied (that is, for $\mu_{Y|45}$), we compute the estimated yield when $x_0 = 45$. This is given by

$$\hat{y}_0 = a + bx_0 = 31.593 + 0.377(45) = 48.558.$$

Now, a 95 percent confidence interval for $\mu_{Y|45}$ is

$$(a + bx_0) - t_{4,0.025}\, s_{\hat{y}_0} < \mu_{Y|45} < (a + bx_0) + t_{4,0.025}\, s_{\hat{y}_0}.$$

Since $s_{\hat{y}_0} = 1.121$ and $a + bx_0 = 48.558$, we get

$$48.558 - (2.776)(1.121) < \mu_{Y|45} < 48.558 + (2.776)(1.121)$$
$$45.446 < \mu_{Y|45} < 51.670. \quad \blacksquare$$

Prediction Intervals

As we know well, the expression *confidence interval* is used in the context of estimating intervals for population parameters. Very often it happens that of considerable importance to the experimenter is the problem of predicting a y-value for a given value of x, for example, x_0. We realize that we can get different y-values for the same x_0. Thus we are interested in estimating an interval for a value of a random variable Y when $x = x_0$. For instance, we might be interested in estimating an interval for the amount of grain yield when x_0 pounds of fertilizer are applied per acre (as opposed to the *mean* amount of grain yield $\mu_{Y|x_0}$ which is a specific constant). The expression *prediction interval* is commonly used in the context of setting an interval for the value of a random variable.

As before, suppose that corresponding to the value x_0 of the independent variable the value from the sample regression line is $\hat{y}_0$, so that $\hat{y}_0 = a + bx_0$. Then it can be shown that the prediction interval is given as follows:

PREDICTION INTERVAL FOR y-VALUE WHEN $x = x_0$

The limits for a $(1 - \alpha)100$ percent prediction interval for a single y-value corresponding to a value x_0 of the independent variable are given by

$$(a + bx_0) \pm t_{n-2,\alpha/2}\, s_e \sqrt{1 + \frac{1}{n} + \frac{(x_0 - \bar{x})^2}{\sum\limits_{i=1}^{n}(x_i - \bar{x})^2}}.$$

EXAMPLE 4   Obtain a 95 percent prediction interval for yield y_0 of grain when 45 pounds of fertilizer are applied with reference to the fertilizer–yield of grain problem we have been discussing.

SOLUTION   We know that $x_0 = 45$, and we have seen earlier that $n = 6$, $\bar{x} = 55$, $\sum\limits_{i=1}^{n}(x_i - \bar{x})^2 = 1750$, $s_e = \sqrt{5.61} = 2.368$, and $\hat{y}_0 = 48.558$. Also $t_{n-2,\alpha/2} = t_{4,0.025} = 2.776$. Hence, after substituting in the formula, the limits of a 95 percent confidence interval for the amount of yield of grain y_0 are given by

$$48.558 \pm (2.776)(2.368)\sqrt{1 + \frac{1}{6} + \frac{(45 - 55)^2}{1750}}.$$

That is, the interval is

$$48.558 - 7.272 < y_0 < 48.558 + 7.272$$
$$41.286 < y_0 < 55.830. \quad \blacksquare$$

TEST OF HYPOTHESIS REGARDING B

Since the main function of the regression theory is to estimate the mean value of the dependent variable from the knowledge of the independent variable, a question of considerable importance is whether the independent variable is at all relevant for the prediction purpose. This would lead to our testing $H_0: B = 0$. (Notice that if $B = 0$, we would get $\mu_{Y|x} = A + 0x$, that is, $\mu_{Y|x} = A$, implying that the mean value of Y for any x is not dependent on x.)

The problem of a general nature is to test

$$H_0: B = B_0$$

where B_0 is some given number. For example, in the fertilizer–grain yield example we might wish to test the null hypothesis that increase in grain yield for an increase of one pound of fertilizer application is 1.5 bushels. In this case $B_0 = 1.5$ and we would be testing $H_0: B = 1.5$.

The alternative hypothesis could be one-sided or two-sided. The reader will recall that this is what will decide the nature of the test as to whether it is one-tailed or two-tailed. The decision rule is based on Student's t distribution with $n - 2$ degrees of freedom and depends on the value of $\dfrac{b - B_0}{s_b}$ and is carried out as shown in Table 11-10.

TABLE 11-10
Procedure to test $H_0: B = B_0$.

Alternative hypothesis	*The decision rule is to reject H_0 if the computed value of $\dfrac{b - B_0}{s_b}$ is*
$B > B_0$	greater than $t_{n-2,\alpha}$
$B < B_0$	less than $-t_{n-2,\alpha}$
$B \neq B_0$	less than $-t_{n-2,\alpha/2}$ or greater than $t_{n-2,\alpha/2}$

EXAMPLE 5 For the fertilizer–yield of grain problem, at the 5 percent level of significance test the null hypothesis

$$H_0: B = 0$$

against the two-sided alternative hypothesis

$$H_A: B \neq 0.$$

SOLUTION At the 5 percent level of significance, since $t_{n-2,\alpha/2} = t_{4,0.025} = 2.776$, the decision rule is *Reject H_0 if the computed value $\dfrac{b - 0}{s_b}$, or $\dfrac{b}{s_b}$, is less than -2.776 or greater than 2.776* (see Fig. 11-12). Since $b = 0.377$ and $s_b = 0.0566$,

$$\frac{b}{s_b} = \frac{0.377}{0.0566} = 6.66.$$

FIGURE 11-12

t distribution with 4 degrees of freedom and the critical region.

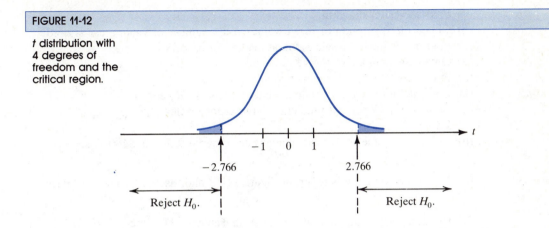

Because the computed value is greater than 2.776, we reject the hypothesis that $B = 0$ and conclude that the yield of grain depends on the level of fertilizer application. ■

For a given set of bivariate data, problems of statistical inference involve lengthy computations. At the very least, one needs the estimates a, b, and s_e^2. The following example illustrates, step by step, the general procedure involved.

EXAMPLE 6 Table 11-11 gives the resale values of a certain model of car (in hundreds of dollars) and the number of years each car has been in use (along with other necessary computations).

TABLE 11-11

Resale value of used cars, and related computations.

Number of years x	Resale value (in hundreds of dollars) y	x^2	xy	y^2
1	30	1	30	900
1	32	1	32	1024
2	25	4	50	625
2	27	4	54	729
2	26	4	52	676
4	16	16	64	256
4	18	16	72	324
4	16	16	64	256
4	20	16	80	400
6	10	36	60	100
6	8	36	48	64
6	11	36	66	121
Total 42	**239**	**186**	**672**	**5475**

(a) Plot the scatter diagram.

(b) Obtain the regression of resale value on the number of years a car is used.

(c) Find the estimated resale value at the end of 5 years.

(d) Obtain an estimate of the variance σ^2.

(e) Find s_a, s_b, and $s_{\hat{y}_0}$ for $x = 5$ years.

(f) Construct 95 percent confidence intervals for A, B, and $\mu_{Y|5}$.

(g) Construct a 95 percent prediction interval for resale value y_0 when $x = 5$ years.

(h) Test the hypothesis that the annual rate of depreciation is $400. Use $\alpha = 0.05$.

SOLUTION

(a) The scatter diagram is given in Figure 11-13 and, as can be seen, a linear trend is evident.

(b) The necessary computations are given in Table 11-11. Since $n = 12$,
$$\sum_{i=1}^{n} x_i = 42, \ \sum_{i=1}^{n} y_i = 239, \ \sum_{i=1}^{n} x_i^2 = 186, \ \sum_{i=1}^{n} x_i y_i = 672, \ \sum_{i=1}^{n} y_i^2 = 5475$$
we get

$$b = \frac{n \sum_{i=1}^{n} x_i y_i - \sum_{i=1}^{n} x_i \sum_{i=1}^{n} y_i}{n \sum_{i=1}^{n} x_i^2 - \left(\sum_{i=1}^{n} x_i\right)^2}$$

FIGURE 11-13

The scatter diagram, and the line of regression of resale value of a car on the number of years in use.

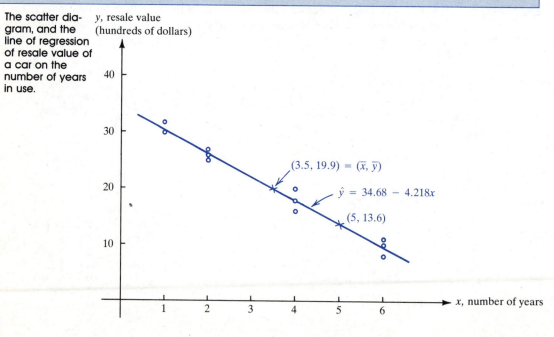

$$= \frac{12(672) - (42)(239)}{12(186) - (42)^2}$$

$$= -4.218$$

and

$$a = \bar{y} - b\bar{x}$$

$$= \frac{239}{12} - (-4.218)\left(\frac{42}{12}\right)$$

$$= 34.68.$$

Hence, the regression of resale value on the number of years is given by

$$\hat{y} = 34.68 - 4.218x.$$

The regression line is drawn in Figure 11-13. From the regression equation, since $b = -4.218$, the depreciation of a car per year is \$421.80.

(c) The estimated resale value at the end of 5 years is

$$\hat{y}_0 = 34.68 - (4.218)(5)$$

$$= 13.59$$

That is, the estimated resale value after 5 years is \$1359.

(d) We have

$$s_e^2 = \frac{1}{n-2}\left[\sum_{i=1}^{n} y_i^2 - a\sum_{i=1}^{n} y_i - b\sum_{i=1}^{n} x_i y_i\right]$$

$$= \frac{1}{12-2}[5475 - (34.68)(239) - (-4.218)(672)]$$

$$= 2.098.$$

(e) Since $\sum_{i=1}^{n}(x_i - \bar{x})^2 = \sum_{i=1}^{n} x_i^2 - \dfrac{\left(\sum_{i=1}^{n} x_i\right)^2}{n} = 186 - \dfrac{42^2}{12} = 39$, and $\bar{x} = \dfrac{42}{12}$

$= 3.5$, we get

$$s_a^2 = \left[\frac{1}{n} + \frac{(\bar{x})^2}{\sum_{i=1}^{n}(x_i - \bar{x})^2}\right]s_e^2$$

$$= \left[\frac{1}{12} + \frac{(3.5)^2}{39}\right](2.098) = 0.8338$$

$$s_b^2 = \frac{s_e^2}{\sum_{i=1}^{n}(x_i - \bar{x})^2}$$

$$= \frac{2.098}{39} = 0.0538$$

and, for $x_0 = 5$ years,

$$s_{\hat{y}_0}^2 = \left[\frac{1}{n} + \frac{(x_0 - \bar{x})^2}{\sum_{i=1}^{n}(x_i - \bar{x})^2} \right] s_e^2$$

$$= \left[\frac{1}{12} + \frac{(5 - 3.5)^2}{39} \right](2.098) = 0.2959.$$

Hence, $s_a = 0.913$, $s_b = 0.232$, and $s_{\hat{y}_0} = 0.544$.

(f) At this stage we need to make some assumptions. We assume that for any given number of years x a car has been in use, its resale value is normally distributed. Also, we assume that the variance of the distribution of resale value is the same for any x. Since $n = 12$, the t distribution has 10 degrees of freedom and $t_{n-2,\alpha/2} = t_{10,0.025} = 2.228$.

(i) $a = 34.68$ and $s_a = 0.913$. Therefore, a 95 percent confidence interval for A is

$$a - t_{n-2,\alpha/2}\, s_a < A < a + t_{n-2,\alpha/2}\, s_a$$
$$34.68 - 2.228(0.913) < A < 34.68 + 2.228(0.913)$$
$$32.646 < A < 36.714.$$

(ii) $b = -4.218$ and $s_b = 0.232$. Therefore, a 95 percent confidence interval for B is

$$b - t_{n-2,\alpha/2}\, s_b < B < b + t_{n-2,\alpha/2}\, s_b$$
$$-4.218 - 2.228(0.232) < B < -4.218 + 2.228(0.232)$$
$$-4.735 < B < -3.701.$$

(iii) From Part (c), $\hat{y}_0 = 13.59$. Also $s_{\hat{y}_0} = 0.544$. Therefore, a 95 percent confidence interval for $\mu_{Y|5}$ is

$$\hat{y}_0 - t_{n-2,\alpha/2}\, s_{\hat{y}_0} < \mu_{Y|5} < \hat{y}_0 + t_{n-2,\alpha/2}\, s_{\hat{y}_0}$$
$$13.59 - 2.228(0.544) < \mu_{Y|5} < 13.59 + 2.228(0.544)$$
$$12.378 < \mu_{Y|5} < 14.802.$$

(g) We have $n = 12$, $x_0 = 5$, $\bar{x} = 3.5$, $\sum_{i=1}^{n}(x_i - \bar{x})^2 = 39$, $s_e = \sqrt{2.098} = 1.448$, $\hat{y}_0 = 13.59$, and $t_{n-2,\alpha/2} = 2.228$. Consequently, substituting in the formula

$$\hat{y}_0 \pm t_{n-2,\alpha/2}\, s_e \sqrt{1 + \frac{1}{n} + \frac{(x_0 - \bar{x})^2}{\sum_{i=1}^{n}(x_i - \bar{x})^2}}$$

we find the limits of a 95 percent prediction interval for y_0 when $x = 5$ years are

$$13.59 \pm (2.228)(1.448)\sqrt{1 + \frac{1}{12} + \frac{(5 - 3.5)^2}{39}}.$$

This can be simplified to yield the interval

$$10.144 < y_0 < 17.036.$$

FIGURE 11-14

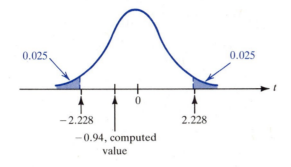

0.025

0.025

t

-2.228 0 2.228

-0.94, computed
value

(h) We want to test the null hypothesis that the annual rate of depreciation is
$400. In other words, we want to test $H_0: B = -4$.

Suppose the alternative hypothesis is given as $H_A: B \neq -4$. Since $b = -4.218$ and $s_b = 0.232$, the computed value of the test statistic is

$$\frac{b - B_0}{s_b} = \frac{-4.218 - (-4)}{0.232}$$

$$= -0.94. \text{ (See Figure 11-14.)}$$

Now $t_{10, 0.025} = 2.228$, and since the computed value is between -2.228 and
2.228, we do not reject the null hypothesis.   ▪▪▪

SECTION 11-2   EXERCISES

In the following exercises, in the context of setting confidence intervals and testing hypotheses, for each x assume that the response variable Y is normally distributed with the same variance.

1. The following table gives x, the age (in months) of a certain breed of calf in the tropics, and y, the mean length of its horns (in inches):

Age x	12	16	16	24	24	30	36	36
Mean length y	2.0	3.0	2.5	4.0	4.5	5.0	6.0	5.5

(a) Find n, $\sum_{i=1}^{n} x_i$, $\sum_{i=1}^{n} y_i$, $\sum_{i=1}^{n} x_i^2$, $\sum_{i=1}^{n} y_i^2$, and $\sum_{i=1}^{n} x_i y_i$.

(b) Find a and b, and obtain the least squares regression line $\hat{y} = a + bx$.

(c) Construct a table giving the values $\hat{y}$ for the different values of x, and obtain
$$\sum_{i=1}^{n} (y_i - \hat{y}_i)^2.$$

(d) Find s_e^2, an unbiased estimate of σ^2, using the result in Part (c).

(e) Alternatively, obtain s_e^2 by using the formula

$$s_e^2 = \frac{1}{n-2}\left[\sum_{i=1}^{n} y_i^2 - a\sum_{i=1}^{n} y_i - b\sum_{i=1}^{n} x_i y_i\right].$$

2. The following pairs of values (x, y) are obtained in a laboratory experiment:

x	2	4	6	8	12	16
y	3.8	18.4	27.2	45.8	62.3	100.3

(a) Find n, $\sum_{i=1}^{n} x_i$, $\sum_{i=1}^{n} y_i$, $\sum_{i=1}^{n} x_i^2$, $\sum_{i=1}^{n} y_i^2$, $\sum_{i=1}^{n} x_i y_i$.

(b) Find a and b.

(c) Find s_e^2.

3. Consider the following bivariate data:

x	6	8	10	11	13
y	10	14	23	28	37

Find an unbiased estimate of σ^2.

4. In Exercise 1, obtain

(a) s_a^2

(b) a 90 percent confidence interval for A.

5. In Exercise 1, obtain

(a) s_b^2

(b) a 90 percent confidence interval for B.

6. In Exercise 1,

(a) estimate the mean length of the horns of a calf of 20 months

(b) set a 90 percent confidence interval for $\mu_{Y|20}$.

7. From a sample of 12 paired observations (x, y), the following information was obtained:

$$\sum_{i=1}^{12} x_i = 18, \ \sum_{i=1}^{12} x_i^2 = 38, \text{ and } s_e^2 = 1.21.$$

Find   (a) $\sum_{i=1}^{12} (x_i - \bar{x})^2$ (b) s_a^2 and s_b^2.

8. The following data are given on two variables x (independent) and y (dependent) measured on 20 experimental units:

$$\sum_{i=1}^{20} x_i = 92, \ \sum_{i=1}^{20} y_i = 183, \ \sum_{i=1}^{20} x_i^2 = 460,$$

$$\sum_{i=1}^{20} y_i^2 = 2184, \ \sum_{i=1}^{20} x_i y_i = 717$$

(a) Find the least squares regression equation.

(b) Obtain the variance estimates s_e^2, s_a^2, and s_b^2.

(c) Set 95 percent confidence intervals on A and B.

9. In Exercise 22 of Section 11-1, at the 5 percent level of significance, test the null hypothesis $B = 0$, where

(a) B is the regression coefficient of the volume of sales during a year on the amount spent on advertising during that year

(b) B is the regression coefficient of the volume of sales on the amount spent on advertising during the previous year.

10. The following data refer to the stopping distance y (in feet) of an automobile traveling at speed x (in miles per hour) in a simulated experiment:

Speed x	25	25	30	35	45	45	55	55	65	65
Stopping distance y	63	56	84	107	153	164	204	220	285	303

(a) Plot the scatter diagram. Do you think that a linear model would be appropriate?

(b) Fit a least squares regression line to the data.

(c) Estimate the stopping distance for a car traveling at 50 mph.

(d) Obtain an estimate of σ^2.

(e) Find s_a^2 and s_b^2.

(f) Test the hypothesis that $B = 5$.

(g) Set a 95 percent confidence interval for $\mu_{Y|50}$, the mean stopping distance for a car traveling at 50 mph.

11. The following data refer to the hardness (in Rockwell units) of an alloy when 8 samples were treated at 500°C for different periods of time (in hours):

Time	6	9	9	12	12	15	18	21
Hardness	50	53	54	60	57	62	67	71

(a) Plot the scatter diagram.

(b) Obtain the regression of hardness on the time.

(c) Find the estimated hardness at the end of 10 hours.

(d) Obtain an estimate of the variance σ^2.

(e) Find s_a, s_b, and $s_{\hat{y}}$ for $x = 10$ hours.

(f) Construct 95 percent confidence intervals for A, B, and $\mu_{Y|10}$.

(g) Test the null hypothesis that the hourly rate of increase of hardness is 1.5 Rockwell units against a suitable alternative hypothesis. (Use $\alpha = 0.10$.)

12. The following figures refer to the temperature x (in degrees Celsius) and the amount y (in grams) of a chemical substance extracted from one pound of a mineral soil:

Temperature x	210	250	270	290	310	340
Amount y	2.1	5.8	8.1	10.8	12.8	14.3

(a) Find the regression of the amount of the chemical substance extracted on the temperature.

(b) Find the estimated yield at 300°C.

(c) Obtain an estimate of σ^2.

(d) Find s_a, s_b, and $s_{\hat{y}}$ when $\hat{y}$ is the estimated yield at 300°C.

(e) Construct 95 percent confidence intervals for A, B, and $\mu_{Y|300}$.

(f) Construct a 95 percent prediction interval for the amount of chemical substance extracted when the temperature is 300°C.

(g) Test the null hypothesis that per degree Celsius increase in temperature, the yield of the extract is increased by 0.1 gram against a suitable alternative hypothesis. (Use $\alpha = 0.05$.)

13. The figures below refer to concentrations of standard manganese solutions and filter photometer readings:

Concentration of standard solution, mg Mn/250 ml x	Photometer reading y
0.5	57
1.0	110
2.0	205
3.0	298
5.0	492

(a) Set up a simple regression of photometer readings on the concentration of solution and determine the constants in the regression equation.

(b) At the 5 percent level of significance, test the null hypothesis that per milligram increase in concentration per 250 milliliters of solution, the increase in photometer readings is 100.

14. A botanist has obtained the following information regarding y, the heights of seedlings (in inches), and x, the number of weeks after planting:

Number of weeks x	2	3	4	5	6	8
Height y (in inches)	2.5	4.8	8.0	10.5	13.0	18.0

(a) Find the regression of the seedling height on the number of weeks after planting.

(b) Set a 95 percent confidence interval on B.

(c) Test the null hypothesis that $B = 2.4$.

(d) Set a 95 percent confidence interval for $\mu_{Y|5}$, the mean seedling height at the end of 5 weeks.

(e) Construct a 95 percent prediction interval for the height of seedling at the end of 5 weeks.

15. The figures below give the resting potential (in millivolts) of frog voluntary muscle fibers and body temperature (in degrees Celsius). Two readings of the resting potential were obtained at each of the temperatures:

Temperature x	5	10	15	20	25	30
Resting	61	63	64	68	69	72
potential y	62	65	66	68	70	72

Fit the regression equation of resting potential on temperature and test the hypothesis that $B = 0.4$ at the 5 percent level of significance.

11-3 CORRELATION ANALYSIS

COEFFICIENT OF CORRELATION AS A MEASURE OF LINEAR ASSOCIATION

We shall now explain why the coefficient of correlation is taken as a measure of the strength of linear association. If the values of y were considered without any reference to their dependence on some variable x, we would find the total variation in the y values in the usual way as $\sum_{i=1}^{n} (y_i - \bar{y})^2$. This is often called the total sum of squares and is commonly denoted by **SST.** Thus

$$\textbf{SST} = \sum_{i=1}^{n} (y_i - \bar{y})^2.$$

As can be seen from Figure 11-15, the deviation of y_i from $\bar{y}$ consists of two parts: one that represents the deviation of the observed value from the regression line, namely, $y_i - \hat{y}_i$, and the other that represents the deviation of the predicted value from the mean $\bar{y}$, namely, $\hat{y}_i - \bar{y}$. Algebraically, this is expressed by the relation

$$y_i - \bar{y} = (y_i - \hat{y}_i) + (\hat{y}_i - \bar{y}).$$

FIGURE 11-15

The deviation $y_i - \bar{y}$ expressed as the sum of the deviations $y_i - \hat{y}_i$ and $\hat{y}_i - \bar{y}$.

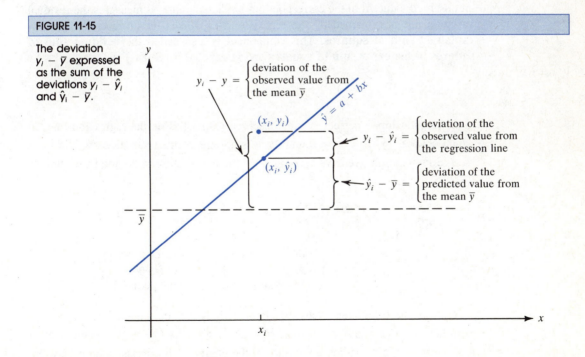

Using these deviations, one could compute $\sum_{i=1}^{n} (y_i - \bar{y})^2$, $\sum_{i=1}^{n} (y_i - \hat{y}_i)^2$, and $\sum_{i=1}^{n} (\hat{y}_i - \bar{y})^2$. A remarkable algebraic fact is that

$$\sum_{i=1}^{n} (y_i - \bar{y})^2 = \sum_{i=1}^{n} (y_i - \hat{y}_i)^2 + \sum_{i=1}^{n} (\hat{y}_i - \bar{y})^2.$$

This relation shows that the total sum of squares is made up of two components, $\sum_{i=1}^{n} (y_i - \hat{y}_i)^2$ and $\sum_{i=1}^{n} (\hat{y}_i - \bar{y})^2$.

The component $\sum_{i=1}^{n} (\hat{y}_i - \bar{y})^2$ is called the **explained sum of squares,** meaning that it explains or accounts for the portion of the total variation through the linear relationship of y with x. If the linear relationship is perfect, all the observed values will be equal to the predicted values; that is, $y_i = \hat{y}_i$. Consequently $\sum_{i=1}^{n} (y_i - \hat{y}_i)^2 = 0$ and we get $\sum_{i=1}^{n} (y_i - \bar{y})^2 = \sum_{i=1}^{n} (\hat{y}_i - \bar{y})^2$, showing that *all* of the total variation is explained by means of the linear relationship. The explained sum of squares is also called the **regression sum of squares** and is denoted by **SSR.** Thus

$$\mathbf{SSR} = \sum_{i=1}^{n} (\hat{y}_i - \bar{y})^2.$$

The other component, $\sum_{i=1}^{n} (y_i - \hat{y}_i)^2$, is the sum of squares of the deviations from the linear regression relation since, for each i, $y_i - \hat{y}_i$ denotes a deviation of an observed value from the corresponding predicted value. It measures the residual variation left unexplained by the regression line and is therefore called the **unexplained sum of squares.** This component is also called the **residual sum of squares** or the **error sum of squares** and is denoted by **SSE.** Thus

$$\mathbf{SSE} = \sum_{i=1}^{n} (y_i - \hat{y}_i)^2.$$

If the relationship is perfect, all the observed values fall on the regression line, in which case $\sum_{i=1}^{n} (y_i - \hat{y}_i)^2 = 0$ and there is no unexplained variation.

Summarizing, we have accomplished the partitioning of the total variation as follows:

$$\sum_{i=1}^{n} (y_i - \bar{y})^2 \quad = \quad \sum_{i=1}^{n} (\hat{y}_i - \bar{y})^2 \quad + \quad \sum_{i=1}^{n} (y_i - \hat{y}_i)^2$$

$$\begin{pmatrix} \text{SST, total sum} \\ \text{of squares} \end{pmatrix} = \begin{pmatrix} \text{SSR, regression} \\ \text{sum of squares} \\ \text{(explained)} \end{pmatrix} + \begin{pmatrix} \text{SSE, error sum} \\ \text{of squares} \\ \text{(unexplained)} \end{pmatrix}$$

From the foregoing discussion, it is obvious that the quantity $\sum_{i=1}^{n} (\hat{y}_i - \bar{y})^2$ expressed as a fraction of the total sum of squares $\sum_{i=1}^{n} (y_i - \bar{y})^2$ would play a prominent role in providing a measure of the degree of linear association between x and y. This fraction, namely,

$$\frac{\sum\limits_{i=1}^{n} (\hat{y}_i - \bar{y})^2}{\sum\limits_{i=1}^{n} (y_i - \bar{y})^2} = \frac{\text{explained sum of squares}}{\text{total sum of squares}}$$

is called the **coefficient of determination** and represents the fraction of the total variation that can be ascribed to the linear relation. Through some algebraic manipulations, it can be shown that this quantity is precisely the square of the Pearson product–moment correlation coefficient r that we studied in Section 11-1. Thus

$$r^2 = \frac{\sum\limits_{i=1}^{n} (\hat{y}_i - \bar{y})^2}{\sum\limits_{i=1}^{n} (y_i - \bar{y})^2} = \text{coefficient of determination.}$$

Notice that the explained variation cannot exceed the total variation. Hence $r^2 \leq 1$. Also r^2 is a ratio of two positive quantities. Hence r^2 is greater than or equal to zero. Thus, $0 \leq r^2 \leq 1$, or, equivalently, $-1 \leq r \leq 1$.

As we remarked earlier, if all the observed values fall on the regression line, then $\sum\limits_{i=1}^{n} (y_i - \bar{y})^2 = \sum\limits_{i=1}^{n} (\hat{y}_i - \bar{y})^2$, in which case we will have $r^2 = 1$, that is, $r = 1$ or $r = -1$. On the other hand, $r^2 = 1$ will mean that $\sum\limits_{i=1}^{n} (y_i - \hat{y}_i)^2 = 0$, showing that all the observed values fall on the straight line. (Why?) Hence $r = +1$ or -1 means that there is a perfect linear relation among the observed pairs in the sample. If the observed values are close to the fitted line, the ratio r^2 will be close to 1. It will deteriorate and get closer to zero as the scatter gets larger.

Remark: The reader is now aware of two methods for computing the coefficient of correlation, one that was given in Section 11-1 and the other, just introduced, as the square root of the coefficient of determination. It should be mentioned that for computational purposes, the formula introduced in Section 11-1 is preferred. Should one choose to compute r by using the relation

$$r = \pm \sqrt{\frac{\sum\limits_{i=1}^{n} (\hat{y}_i - \bar{y})^2}{\sum\limits_{i=1}^{n} (y_i - \bar{y})^2}}$$

how does one decide whether r is positive or negative? It turns out that the sign of r is always the same as that of the regression coefficient b.

EXAMPLE 1 For the data in Table 11-4 of Section 11-1, we have already obtained the coefficient of correlation between IQ and grade point average as equal to 0.913. Now alternatively, find it as the square root of the coefficient of determination.

SOLUTION In order to obtain the coefficient of determination, we must compute the various components of variation. In Table 11-4 we computed the following quantities:

$$n = 10, \sum_{i=1}^{10} x_i = 1040, \sum_{i=1}^{10} y_i = 30.2, \sum_{i=1}^{10} x_i^2 = 109{,}800$$

$$\sum_{i=1}^{10} y_i^2 = 92.80, \sum_{i=1}^{10} x_i y_i = 3187.5$$

Therefore:

(a) The regression coefficient is

$$b = \frac{10(3187.5) - (1040)(30.2)}{10(109{,}800) - (1040)^2}$$

$$= 0.0285.$$

(b) Since $\bar{y} = 3.02$ and $\bar{x} = 104$, the y-intercept is

$$a = \bar{y} - b\bar{x}$$

$$= 3.02 - (0.0285)(104)$$

$$= 0.056.$$

(c) The total sum of squares is

$$\sum_{i=1}^{10} (y_i - \bar{y})^2 = \sum_{i=1}^{10} y_i^2 - \frac{\left(\sum_{i=1}^{10} y_i\right)^2}{n}$$

$$= 92.80 - \frac{(30.2)^2}{10}$$

$$= 1.596.$$

(d) Using the formula $\sum_{i=1}^{10} (y_i - \hat{y}_i)^2 = \sum_{i=1}^{10} y_i^2 - a \sum_{i=1}^{10} y_i - b \sum_{i=1}^{10} x_i y_i$, we find the unexplained sum of squares is

$$\sum_{i=1}^{10} (y_i - \hat{y}_i)^2 = 92.80 - 0.056(30.2) - (0.0285)(3187.5)$$

$$= 0.265.$$

From (c) and (d), the explained sum of squares $\sum_{i=1}^{10} (\hat{y}_i - \bar{y})^2$ is given by

$$\sum_{i=1}^{10} (\hat{y}_i - \bar{y})^2 = \sum_{i=1}^{10} (y_i - \bar{y})^2 - \sum_{i=1}^{10} (y_i - \hat{y}_i)^2$$

$$= 1.596 - 0.265$$

$$= 1.331.$$

A summary of the components of variation is given in Table 11-12.

TABLE 11-12

Components of variation.

Component	Sum of squares
Explained (regression)	1.331
Unexplained (error)	0.265
Total	1.596

The coefficient of determination is therefore given as

$$r^2 = \frac{\text{explained variation}}{\text{total variation}}$$

$$= \frac{1.331}{1.596}$$

$$= 0.834.$$

Since b is positive, we take the positive square root of 0.834 and obtain $r = \sqrt{0.834} = 0.913$, the same value we had obtained rather easily earlier. The coefficient of determination of 0.834 signifies that 83.4 percent of the variation in the grade point averages can be explained through the linear regression relation. ▬▬▬

EXAMPLE 2   For the data in Example 3 in Section 11-1, find

(a) the different components of variation
(b) the coefficient of determination
(c) the coefficient of correlation.

SOLUTION   (a) We have the following information from Example 3, Section 11-1:

$$n = 10, \sum_{i=1}^{10} y_i = 690, \sum_{i=1}^{10} y_i^2 = 49{,}014, \sum_{i=1}^{10} x_i y_i = 44{,}636,$$

$$a = 55.346, b = 0.214$$

Hence the total sum of squares is

$$\sum_{i=1}^{10} (y_i - \bar{y})^2 = \sum_{i=1}^{10} y_i^2 - \frac{\left(\sum_{i=1}^{10} y_i\right)^2}{10}$$

$$= 49{,}014 - \frac{(690)^2}{10}$$

$$= 1404.$$

and the unexplained sum of squares is

$$\sum_{i=1}^{10} (y_i - \hat{y}_i)^2 = \sum_{i=1}^{10} y_i^2 - a \sum_{i=1}^{10} y_i - b \sum_{i=1}^{10} x_i y_i$$
$$= 49{,}014 - (55.346)(690) - (0.214)(44{,}636)$$
$$= 1273.16.$$

Thus we get Table 11-13, giving the components of variation. The explained component in the table is obtained by subtracting the unexplained component from the total variation.

TABLE 11-13

Components of variation.

Component	Sum of squares	Ratio, explained/total
Explained (regression)	130.84	0.09319
Unexplained (error)	1273.16	
Total	1404.00	

(b) The coefficient of determination is given in column 3 of Table 11-13. The small value of 0.09319 tells us that only 9.3 percent of the variation in the final scores can be explained through linear regression on the midterm scores.

(c) Since b is positive, the coefficient of correlation is given as the positive square root of the coefficient of determination and hence is equal to $\sqrt{0.09319} = 0.306$, the value we obtained in Example 3 in Section 11-1. ▪

TEST OF HYPOTHESIS REGARDING POPULATION CORRELATION COEFFICIENT

We have discussed the meaning and significance of the correlation coefficient computed from a sample of n pairs of observations. To be more specific, we should refer to r as the *sample* correlation coefficient. Different samples of size n drawn from the population will give rise to different values of r. Thus, conceptually we can think of the distribution of r. (Notice our transgression in using r to denote both a random variable and its value. We take the liberty in yielding to the customary usage of the letter r and thus avoid introducing new symbols.) The parameter that provides a measure of association between two variables in the population analogous to the way r does in the sample is called the **population correlation coefficient** and is denoted by the Greek letter ρ (rho).

Suppose we obtain a certain value of r from a given set of data. What does it suggest regarding ρ? We shall consider only the simple case where the null hypothesis in which we are interested is

$$H_0: \rho = 0$$

meaning that there is no linear relationship between the two variables in the population. We test this against the alternative hypothesis

$$H_A: \rho \neq 0.$$

It is at this point that we need to make some assumptions regarding the distributions of the two variables. Whereas in regression analysis one of the variables is nonrandom and the other is random, in correlation analysis we require that *both* the variables be random variables. Their joint distribution is given by what is called a *bivariate normal distribution*. In such a case, $\rho = 0$ *does* imply that the two variables are independent.

THE TEST STATISTIC

The test statistic to carry out the test of the null hypothesis $H_0: \rho = 0$ is

$$r\sqrt{\frac{n - 2}{1 - r^2}}.$$

If H_0 is true, then this statistic has the Student's t distribution with $n - 2$ degrees of freedom.

In view of the alternative hypothesis we have a two-tailed test and the decision rule is *Reject H_0 if the computed value of the test statistic is less than $-t_{n-2, \alpha/2}$ or greater than $t_{n-2, \alpha/2}$.*

EXAMPLE 3 Consider the following data from a bivariate normal distribution:

x	6	13	18	24	30
y	12	16	14	23	20

Test $H_0: \rho = 0$ against $H_A: \rho \neq 0$ at the 5 percent level.

SOLUTION First we compute r using the formula in Section 11-1. Since $n = 5$, $\sum\limits_{i=1}^{n} x_i = 91$, $\sum\limits_{i=1}^{n} x_i^2 = 2005$, $\sum\limits_{i=1}^{n} y_i = 85$, $\sum\limits_{i=1}^{n} y_i^2 = 1525$, and $\sum\limits_{i=1}^{n} x_i y_i = 1684$ we get

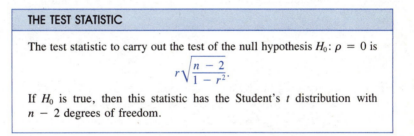

$$r = \frac{5(1684) - (91)(85)}{\sqrt{[5(2005) - (91)^2][5(1525) - (85)^2]}} = 0.82.$$

Then, the computed value of the test statistic is

$$r\sqrt{\frac{n - 2}{1 - r^2}} = (0.82)\sqrt{\frac{5 - 2}{1 - (0.82)^2}} = 2.481.$$

At the 5 percent level of significance $t_{n-2,\alpha/2} = t_{3,0.025} = 3.182$. The hypothesis that $\rho = 0$ cannot be rejected and so there is no evidence that the variables are dependent (in view of the bivariate normal distribution assumption). ▬▬

Remarks: (1) It can be shown that testing $H_0: \rho = 0$ is tantamount to testing $H_0: B = 0$. So, in essence, we have dealt with this problem in Section 11-2. A statistical test is available to test $H_0: \rho = \rho_0$ where ρ_0 is not equal to zero. But to carry out this test one has to perform what is called a *logarithmic transformation* and is not presented here.

(2) On first glance it may seem from Example 3 that a value of $r = 0.82$ is rather high and should be statistically significant. A closer scrutiny reveals that about 67.2 percent of the total variation can be ascribed to the linearity relation. This fraction in itself may not be negligible, but coupled with the fact that the sample size is very small, it explains why the test in Example 3 showed the result not significant. If we were to obtain the same r-value (that is, 0.82) from a sample of, for example, 12 pairs of observations, then the result would be highly significant since the computed value would be $\dfrac{0.82\sqrt{10}}{\sqrt{1 - 0.82^2}}$, or 4.53, and $t_{n-2,0.025} = t_{10,0.025} = 2.228$.

Thus once again we see how the sample size plays a dominant role in a decision-making process.

EXAMPLE 4

Concerning the IQ and grade point average data given in Example 1, we obtained $r = 0.913$ based on a sample with $n = 10$. Test the null hypothesis $H_0: \rho = 0$ against $H_A: \rho \neq 0$ at the 5-percent level of significance.

SOLUTION

The computed value of the test statistic is

$$r\sqrt{\frac{n-2}{1-r^2}} = (0.913)\sqrt{\frac{10-2}{1-(0.913)^2}}$$

$$= 6.33.$$

At the 5 percent level of significance, $t_{n-2,\alpha/2} = t_{8,0.025} = 2.306$. We reject H_0, since the computed value does not fall in the interval -2.306 to 2.306, and we conclude that IQ and grade point average are linearly related. ▬▬

EXAMPLE 5

In Example 3 in section 11-1 we considered the data giving scores of ten students on their midterm and final exams. The value of r was computed to be 0.306. Test $H_0: \rho = 0$ against $H_A: \rho \neq 0$ at the 5-percent level of significance.

SOLUTION

To test the hypothesis $H_0: \rho = 0$, we assume that scores on the midterm exam and the final exam have a bivariate normal distribution and we compute the value of the statistic

$$r\sqrt{\frac{n-2}{1-r^2}}.$$

The computed value is

$$(0.306)\sqrt{\frac{10-2}{1-(0.306)^2}} = 0.909.$$

The null hypothesis is tested against the alternative hypothesis $H_A: \rho \neq 0$. At the 5 percent level of significance, $t_{n-2,\alpha/2} = t_{8,0.025} = 2.306$. We cannot reject H_0, since the computed value does not fall in the critical region. There is no evidence to indicate that the relation between the final scores and the midterm scores is linear.   ■

SECTION 11-3   EXERCISES

In the following exercises, for the purpose of testing hypotheses regarding ρ, assume that the two variables have a bivariate normal distribution.

1. The scores on a scholastic test and the number of hours that 10 randomly picked students devoted preparing for it are given below:

Number of hours x	18	27	20	10	30	24	32	27	12	16
Scores y	68	82	77	90	78	72	94	88	60	70

   Determine

   (a) the slope b
   (b) the y-intercept, $a = \bar{y} - b\bar{x}$
   (c) the total sum of squares $\sum_{i=1}^{n} (y_i - \bar{y})^2$
   (d) the unexplained sum of squares using the formula

   $$\sum_{i=1}^{n} (y_i - \hat{y}_i)^2 = \sum_{i=1}^{n} y_i^2 - a\sum_{i=1}^{n} y_i - b\sum_{i=1}^{n} x_i y_i$$

   (e) a table showing the partitioning of the total sum of squares
   (f) the coefficient of determination.

2. If the value of r is given as 0.8 and the unexplained component of variation is given as 50.3, find the other components of variation.

3. Forty-nine percent of the total variation in y is explained by the linear regression of y on x. What is the coefficient of correlation?

4. If the coefficient of correlation between x and y is -0.6, what percent of the total variation is explained by the straight-line relation? If the total variation is 33.2, find the unexplained component of variation.

5. The following information is available regarding the annual yield of wheat per acre y and the annual rainfall x:

$$n = 9, \sum_{i=1}^{9} (x_i - \bar{x})^2 = 91.5, \sum_{i=1}^{9} (y_i - \bar{y})^2 = 2066.9,$$

$$\sum_{i=1}^{9} (x_i - \bar{x})(y_i - \bar{y}) = 310$$

(a) Compute r.

(b) Compute the coefficient of determination.

(c) What does the coefficient of determination signify insofar as the variation in wheat yield is concerned?

6. The figures given below pertain to the annual disposable income x and consumption expenditure y of 5 families:

Disposable income x (in thousands of dollars)	8.5	11.6	14.6	20.6	22.8
Consumption expenditure y (in thousands of dollars)	7.6	13.2	12.8	16.9	18.6

(a) Find the total sum of squares of y's.

(b) Find the explained and unexplained components of variation.

(c) Compute the coefficient of determination r^2.

(d) What does r^2 measure?

In Exercises 7–11, at the 5 percent level of significance, test the null hypothesis $\rho = 0$ against the alternative hypothesis $\rho \neq 0$.

7. $r = -0.5$, $n = 66$

8. $r = -0.28$, $n = 51$

9. $r = -0.5$, $n = 11$

10. $r = 0.3$, $n = 198$

11. The coefficient of determination is 0.49 and $n = 38$.

12. An experiment is conducted to determine the effect of pressure in a certain brand of tires on the number of miles the tire lasts. The following data are obtained:

Pressure x	27	27.5	28	28.5
Miles y	12,928	14,036	14,700	14,919

Pressure x	29	29.5	30
Miles y	14,694	14,024	12,910

(a) Plot a scatter diagram.

(b) Obtain the coefficient of correlation.

(c) Comment on the magnitude of ρ in view of the scatter diagram.

13. The following are heights and weights of 7 athletes:

Height x (in inches)	69	66	68	73	71	74	71
Weight y (in pounds)	163	153	185	186	157	220	190

(a) Follow Parts a–f in Exercise 1. (Construct a table showing the partitioning of the total sum of squares of weights.)

(b) Find the coefficient of correlation.

(c) Test the hypothesis that $\rho = 0$ at the 5 percent level.

14. The following data refer to y, the level of cholesterol (in milligrams per hundred milliliters) of 10 males, and x, their average daily intake of saturated fat (in grams).

Fat intake x	28	32	35	40	42	48	52	56	58	60
Cholesterol y	170	145	155	160	210	180	240	190	220	200

(a) Find the regression of the level of cholesterol on the intake of saturated fat.
(b) Find the components of variation explained and unexplained by the linear regression.
(c) Find the coefficient of determination. Interpret the value obtained in light of the components of variation.
(d) Find the sample coefficient of correlation r.
(e) Test $H_0: \rho = 0$ at the 5-percent level of significance.

15. A survey of 66 families was conducted to find the relationship between x, the annual family income, and y, the amount spent on entertainment. It was found that the total variation was 409,600 and the unexplained component of variation was 77,824.

(a) What percent of the variation in the amount spent on entertainment is accounted for by the linear regression relation?
(b) Compute the coefficient of correlation r. Comment fully.
(c) Test the hypothesis that ρ, the coefficient of correlation for all potential families, is zero.

16. The data below refer to the atomic radius (in nanometers) and ionization potential (in electron volts) for some second period elements of the periodic table of elements:

Element	Li	Be	B	C	N	O	F
Atomic radius	0.123	0.089	0.080	0.077	0.074	0.074	0.072
Ionization potential	5.4	9.3	8.3	11.3	14.5	13.6	17.4

Find the coefficient of correlation between atomic radius and ionization potential and test the null hypothesis $\rho = 0$ at the 5-percent level of significance.

CONCLUDING REMARKS

Given any n pairs of observations, we can always compute the regression line and the linear correlation coefficient. It is when the inferential nature of the investigation is involved that we have to make some assumptions regarding the distributions of X and Y. For regression analysis the values of the independent variable are fixed, and for each value of x the subpopulation of y values has a normal distribution. For correlation analysis, X and Y are both random variables.

Regression technique is valuable for the purpose of making prediction. But the prediction should be limited to the range of the values of the independent variable that led to obtaining the prediction equation in the first place. As an example, if a regression equation for predicting the yield of grain is fitted by considering levels of fertilizer ranging from 30 to 80 pounds per acre, it would be foolhardy to use the equation to predict the yield for 100 pounds of fertilizer. Crops are known to die with excessive use of fertilizer! And yet, in some situations, such as in economic forecasts, some amount of extrapolation beyond the range is almost a necessity. For instance, if a prediction equation is set up on the basis of the economic activity from 1970 to 1985, an economist may want to forecast what is in store in later years, say in 1986 and 1987. Whenever this is done, one should exercise utmost caution and be fully aware of other related factors that could invalidate the conclusions.

Finally, it should be clearly understood that a relationship between variables does not mean that there is a cause-and-effect relation. Even though we use terms such as dependent and independent variables, there is no implication that one variable is the cause of the other. There may exist a causal relation and there may not. For instance, in considering the relationship between the amount of fertilizer applied and the amount of grain produced, we clearly have a causal relation in mind. But consider the following situation: During the 1950s production of cars increased every year in the United States and so did the population of Latin America. A mathematical relation could perhaps be established between the number of cars produced in the U.S. and the population of Latin America during that period. However, it would be ridiculous to conclude that more cars in the U.S. caused there to be more people in Latin America or vice versa. We must be careful in deciding whether a relation is causal or simply mathematical.

KEY TERMS AND EXPRESSIONS

univariate case

bivariate case

linear correlation

linear regression

y-intercept

slope

equation of a straight line

scatter diagram

coefficient of linear correlation

sample correlation coefficient r

population correlation
  coefficient ρ

prediction equation

residual or error

method of least squares

line of best fit

least squares regression line

sample regression coefficient

regression of y on x

linear model

deterministic model

linear statistical model

population, or true, regression line

prediction interval

total sum of squares

explained, or regression, sum of squares

unexplained, residual, or error,
  sum of squares

coefficient of determination

KEY FORMULAS

correlation coefficient

$$r = \frac{n\sum_{i=1}^{n} x_i y_i - \sum_{i=1}^{n} x_i \cdot \sum_{i=1}^{n} y_i}{\sqrt{\left[n\sum_{i=1}^{n} x_i^2 - \left(\sum_{i=1}^{n} x_i\right)^2\right]\left[n\sum_{i=1}^{n} y_i^2 - \left(\sum_{i=1}^{n} y_i\right)^2\right]}}$$

line of best fit $\hat{y} = a + bx$
where the slope b is

slope

$$b = \frac{n\sum_{i=1}^{n} x_i y_i - \sum_{i=1}^{n} x_i \cdot \sum_{i=1}^{n} y_i}{n\sum_{i=1}^{n} x_i^2 - \left(\sum_{i=1}^{n} x_i\right)^2}$$

y-intercept $a = \bar{y} - b\bar{x}$

variance estimates

$$s_e^2 = \frac{1}{n-2}\left[\sum_{i=1}^{n} y_i^2 - a\sum_{i=1}^{n} y_i - b\sum_{i=1}^{n} x_i y_i\right]$$

$$s_a^2 = \left[\frac{1}{n} + \frac{(\bar{x})^2}{\sum_{i=1}^{n}(x_i - \bar{x})^2}\right] s_e^2$$

$$s_b^2 = \frac{s_e^2}{\sum_{i=1}^{n}(x_i - \bar{x})^2}$$

$$s_{\hat{y}_0}^2 = \left[\frac{1}{n} + \frac{(x_0 - \bar{x})^2}{\sum_{i=1}^{n}(x_i - \bar{x})^2}\right] s_e^2$$

confidence intervals

$a - t_{n-2,\alpha/2}\, s_a < A < a + t_{n-2,\alpha/2}\, s_a$

$b - t_{n-2,\alpha/2}\, s_b < B < b + t_{n-2,\alpha/2}\, s_b$

$(a + bx_0) - t_{n-2,\alpha/2}\, s_{\hat{y}_0} < \mu_{Y|x_0} < (a + bx_0) + t_{n-2,\alpha/2}\, s_{\hat{y}_0}$

prediction interval limits

$$(a + bx_0) \pm t_{n-2,\alpha/2}\, s_e \sqrt{1 + \frac{1}{n} + \frac{(x_0 - \bar{x})^2}{\sum_{i=1}^{n}(x_i - \bar{x})^2}}$$

Procedure used to test $H_0: B = B_0$

Alternative hypothesis	The decision rule is to reject H_0 if the computed value of $\dfrac{b - B_0}{s_b}$ is
$B > B_0$	greater than $t_{n-2,\alpha}$
$B < B_0$	less than $-t_{n-2,\alpha}$
$B \neq B_0$	less than $-t_{n-2,\alpha/2}$ or greater than $t_{n-2,\alpha/2}$

total sum of squares
$$SST = \sum_{i=1}^{n} (y_i - \bar{y})^2$$

regression sum of squares
$$SSR = \sum_{i=1}^{n} (\hat{y}_i - \bar{y})^2$$

error sum of squares
$$SSE = \sum_{i=1}^{n} (y_i - \hat{y}_i)^2$$

coefficient of determination
$$r^2 = \frac{SSR}{SST}$$

CHAPTER 11 TEST

1. Describe the roles of linear correlation and linear regression in statistical analysis of bivariate data.

2. An electronics company has found the following relation between sales y (in millions of dollars) and advertising x (in thousands of dollars):

   $$y = 20.18 + 0.42x$$

   when x is between 60 and 80. What does the coefficient 0.42 of x mean?

3. List the basic properties of the coefficient of linear correlation.

4. In a research problem, two variables x and y were measured on each of 7 items. The data are presented below.

x	0.15	0.20	0.20	0.35	0.45	0.55	0.55
y	18	20	19	28	42	50	58

   (a) Plot the scatter diagram.
   (b) Find the regression of y on x.
   (c) Find the coefficient of correlation.

5. A research article reported findings for which the linear regression equation of y on x, and the correlation coefficient r were given as follows:

   $$y = 4.009 - 0.026x; \quad r = 0.329.$$

   This information is inconsistent. Explain why.

6. The following data give the total length (in millimeters) of a certain species of fish (*Epiplatys bifasciatus*) inhabiting a swampy region and its head length (in millimeters) based on a sample of 8 fish.

head length	3.8	4.1	4.0	4.3	4.8	4.6	5.1	6.0
total length	20	25	27	30	34	38	40	45

(a) Find the regression of head length on the total length.
(b) Find the coefficient of correlation.

7. A study has asserted that based on 78 observations, the regression of age y (in years) on diameter at breast height (dbh) x (in inches), and correlation were found to be

$$y = 103.95 + 5.27x; r = 0.81.[1]$$

(a) What fraction of the total variation in age can be explained through linear regression on dbh?
(b) At the 5 percent level of significance test the null hypothesis $H_0: \rho = 0$ against $H_A: \rho \neq 0$.
(c) Estimate the age of a tree when dbh is 30 inches.

8. The following measurements were obtained (in feet) for the length and the girth of 10 full grown sea cows. Find the coefficient of correlation between girth and length and, at the 5 percent level, test the hypothesis that the population coefficient of correlation is zero.

Length x	10.5	12.2	11.5	9.6	13.3	12.0	10.7	11.0	12.6	12.9
Girth y	7.6	8.3	8.1	7.2	8.7	8.4	7.3	7.9	8.2	8.5

9. The following data give the increase in weight (in grams) of male rats during a one-month period when they were fed different doses of vitamin A (in milligrams):

Dose	0.3	0.6	0.9	1.2	1.5
Increase in weight	33 29	36 34	40 37	41	45

There were two rats at each of the doses 0.3, 0.6, and 0.9, and one rat at each of the doses 1.2 and 1.5. Estimate B, the population regression coefficient of weight increase on vitamin A dosage, and test the null hypothesis that it is equal to 12. (Use $\alpha = 0.05$.)

10. The volume of a tree is commonly regarded as a function of many variables such as height, basal area, age, and so on. The following data give net 10-year volume stand growth y (in cubic feet) for trees larger than 4.5 inches at breast height and their basal area per acre x (in square feet) where stand basal area is a measure expressing the degree of land occupation by trees[2]:

Volume stand growth	610	1,034	1,753	2,387	2,972	3,522	4,046	4,549
Basal area per acre	50	100	200	300	400	500	600	700

1. Kenneth N. Boe, "Growth and Mortality After Regeneration Cuttings in Old-growth Redwood," USDA Forest Service Research Paper PSW-104.

2. James Lindquist and Marshal N. Palley, "Prediction of Stand Growth of Young Redwood," California Agriculture Experimental Station Bulletin 831.

(a) Obtain the regression of volume stand growth on the basal height.

(b) Find a 90 percent confidence interval for B.

(c) Test the null hypothesis $H_0: B = 5.5$ against the alternative hypothesis $H_A: B \neq 5.5$. Use $\alpha = 0.10$.

(d) Find a 90 percent confidence interval for $\mu_{Y|400}$.

(e) Find a 90 percent prediction interval for volume stand growth when basal area is 400 square feet.

11. Unlike most organisms, crustacean animals such as crabs are enclosed in a rigid exterior skeleton. Through a process called molting, the entire exoskeleton is shed. Only then, before the formation of the new rigid shell, can the animal grow. The following data give the specific molt increment y (in millimeters) and the premolt crab size x (in millimeters) for Dungeness crab found off the coast of Northern California[3]:

Premolt size (mm)	126.2	131.0	131.7	132.8	136.2	136.8	138.3	138.3	140.0	143.9
Molt Increment (mm)	17.1	16.5	15.1	15.9	14.8	15.1	14.4	12.8	11.9	11.7

(a) Plot the scatter diagram.

(b) Find the regression of molt increment on premolt size.

(c) Find the coefficient of correlation.

(d) Test $H_0: \rho = 0$ against $H_A: \rho \neq 0$. (Use $\alpha = 0.05$.)

---

3. Nancy Diamond, Master's Thesis, Humboldt State University, Arcata, California.

12

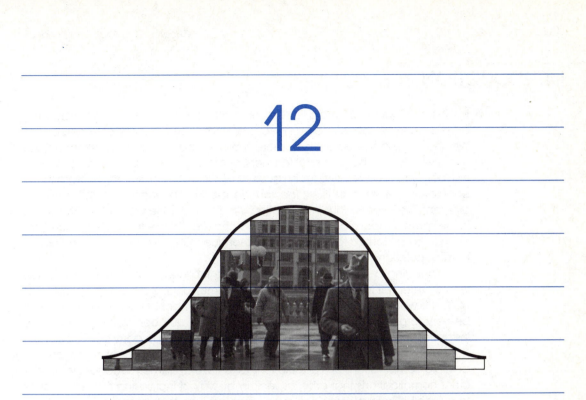

ANALYSIS OF VARIANCE

INTRODUCTION

In Chapter 9 we described a test for comparing two population means by using a *t* test when the population variances are not known but are assumed equal. We shall now generalize this to more than two populations. For example, we might wish to compare several fertilizers, or manufacturing processes, or teaching methods, or advertising techniques. As usual, we will take samples from these populations. More than likely the sample means obtained from them will be different. The intent of our investigation is to find out if the observed differences in the sample means can be ascribed to chance. Or, are these differences too large to be attributed to chance, but rather due to the fact that the samples are from populations that are not alike?

In testing a hypothesis regarding more than two population means, we would prefer not to compare them two at a time, running a battery of *t* tests. First of all, this would be cumbersome, because the number of *t* tests involved would be very large, even for comparing a moderate number of populations. For instance, to compare eight populations we would have to carry out 28 *t* tests. (A student familiar with counting techniques can determine that there are 28 possible combinations of 8 populations taken two at a time.) Secondly, there are other complications that arise, especially by way of increasing the Type 1 error to a prohibitive level. We need a different test, one that, at the very least, is an extension of the *t* test used for testing two means. Such an extension is provided by the method of **analysis of variance,** called briefly ANOVA, which gives a single test for comparing several means. True to its name, it analyzes variances but in the process tests for equality of means.

The method of analysis of variance owes its beginning to Sir Ronald A. Fisher (1890–1962) and received its early impetus in applications to agricultural research. It is such a powerful statistical tool that it has since found applications in nearly every branch of scientific research, including economics, psychology, marketing research, and industrial engineering, to mention just a few. We shall consider here only the simplest case, that of *one-way classification,* where the observations are classified into groups on the basis of a single criterion of classification. ▬▬

12-1 ONE-WAY CLASSIFICATION—EQUAL SAMPLE SIZES

THE TEST CRITERION AND THE RATIONALE FOR IT

Let us consider the following example, in which a car rental company is interested in comparing three brands of gasoline designated Brand 1, Brand 2, and Brand 3. Suppose it is agreed to make four runs with each brand, obtaining the mileage per gallon for each run as the response variable. Certainly we do not expect the runs to be made for different brands with different-sized engines. As a matter of fact, it might be desirable to make all the runs using one car, with Brand 1 tested at random, say, on runs 2, 5, 9, and 11, Brand 2 on runs 1, 6, 8, and 12, and Brand 3 on runs 3, 4, 7, and 10. We randomize in this way to minimize any bias that might result as the tune-up of the car deteriorates. Also, it is to be hoped that, as far as possible, the runs will be made on similar terrains. Such considerations as these form the essence of a well-designed experiment. As much as possible, we reduce the influence of the factors that might impair the results of the experiment. All other factors that might contribute to variability and cannot be eliminated are then lumped together and designated as *the experimental error*.

Suppose the results of the twelve runs are as presented in Table 12-1.

TABLE 12-1		
Miles per gallon on 12 runs, four with each brand.		
Brand 1	*Brand 2*	*Brand 3*
15	19	22
19	17	17
14	16	19
16	20	18
Mean mileage per gallon 16	18	19

From the table we see that individual readings vary, as do the means of the brands. Are these observed differences in the brand means due to chance fluctuations or are they due to the fact that the brands are basically different? In other words, if μ_1, μ_2, and μ_3 represent the true mean mileage per gallon for Brands 1, 2, and 3, respectively, the null hypothesis that we are interested in testing is

$$H_0: \mu_1 = \mu_2 = \mu_3.$$

In what follows we shall denote this common value of the population means as μ. The null hypothesis will be tested against the alternative hypothesis

H_A: Not all the brands have the same mean mileage per gallon.

Thus the alternative hypothesis entertains possibilities where all the population means are different from each other, or two of the means are equal but different from the third mean.

At this point we shall make some assumptions that are basic to applying the analysis of variance technique. These assumptions are the following:

1. The *response variable* (miles per gallon in our illustration) is normally distributed.

2. The variance of the distribution is the same for each brand. This assumption implies that although different brands of gasoline may have different mean mileages per gallon, the spread of the distribution for each of them is the same.
3. The samples are drawn in such a way that the results of one sample do not influence and are not influenced by those of others. In other words, the samples are independent.

As you will recall, these assumptions are the same as those made in Chapter 9 for testing equality of two means using the t test.

Our common sense would indicate that if the sample means differed substantially from each other, we would be inclined to doubt H_0 and regard the brands as being different. On the other hand, if the sample means differed very little from each other, we would be inclined to support the validity of H_0. However, we cannot rely entirely on this kind of argument as the following discussion will show when we consider the two sets of data in Tables 12-2 and 12-3 below where the means for respective brands in the two tables are equal.

TABLE 12-2		
Miles per gallon where the response for each brand is erratic.		
Brand 1	*Brand 2*	*Brand 3*
13	14	19
15	19	15
19	21	24
17	18	18
Mean mileage per gallon 16	18	19

The data in Table 12-2 would support the contention under H_0 that the population means are equal. The performance of the brands is so erratic and the overlap is so complete that we might regard the three samples corresponding to the three brands as one large sample drawn from one population as shown in Figure 12-1.

The observed differences among the brand means could be ascribed to experimental or chance variations. In this case, it would stand to reason to regard the observed values of 16, 18, and 19 for brand means simply as three point estimates of the common mean μ.

On the other hand, a different picture emerges from the data in Table 12-3. The variability within brands is so small that the situation shown in Figure 12-2 seems more plausible. It would support the contention under the alternative hypothesis that the differences in the sample means are due to real differences among the brands. (Later, in Example 1, we shall actually carry out the test of significance.)

FIGURE 12-1

The data in Table 12-2 which can be conceived of as coming from one single population.

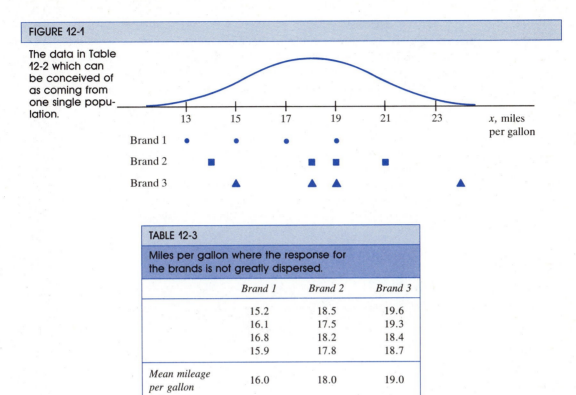

TABLE 12-3

Miles per gallon where the response for the brands is not greatly dispersed.		
Brand 1	*Brand 2*	*Brand 3*
15.2	18.5	19.6
16.1	17.5	19.3
16.8	18.2	18.4
15.9	17.8	18.7
Mean mileage per gallon 16.0	18.0	19.0

From the data in Tables 12-2 and 12-3, even though the observed sample means are exactly the same for respective brands, we are led to different conclusions. This situation arises because in the decision procedure it is appropriate to take into account not only the extent to which the sample means differ from each other but also variability within the samples. This is why we call this procedure "analysis of variance."

FIGURE 12-2

The data in Table 12-3, which can be conceived of as coming from three distinct populations.

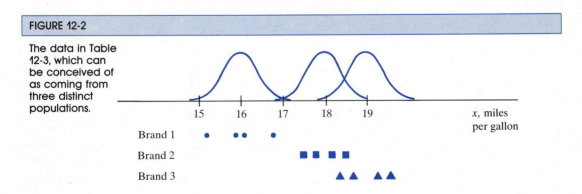

Having determined the importance of the variability between the sample means together with the variability within the samples, exactly what criterion do we use that will take into account both these variabilities? Also, how do we provide a rationale for it? To accomplish this, we shall describe a procedure that involves obtaining two unbiased estimates of σ^2 and comparing them.

Let us consider a general situation where we have k populations which we shall refer to as treatments and where, for the sake of simplicity, we shall assume that each sample consists of the same number n of observations (we shall generalize this later in Section 12-2). We might present the results as in Table 12-4, where, using double subscripts, we represent the outcomes as $x_{i,j}$, with the first subscript i designating the treatment and the second subscript j denoting the observation number within that treatment. For example, $x_{2,4}$ represents the fourth outcome in the second treatment. Thus with reference to Table 12-1, we would have $x_{2,4} = 20$, $x_{3,1} = 22$, $x_{1,4} = 16$, and so on.

TABLE 12-4		
General response with n observations per treatment.		
Treatment 1	*Treatment 2* . . .	Treatment k
$x_{1,1}$	$x_{2,1}$	$x_{k,1}$
$x_{1,2}$	$x_{2,2}$	$x_{k,2}$
.	.	.
.	.	.
.	.	.
$x_{1,n}$	$x_{2,n}$	$x_{k,n}$
Mean $\quad \bar{x}_1 = \dfrac{\sum_{j=1}^{n} x_{1,j}}{n}$	$\bar{x}_2 = \dfrac{\sum_{j=1}^{n} x_{2,j}}{n}$ . . .	$\bar{x}_k = \dfrac{\sum_{j=1}^{n} x_{k,j}}{n}$

The **grand mean,** designated $\bar{\bar{x}}$ and read *x double bar,* is obtained by adding observations in all the samples and dividing the total by kn, the total number of observations. Since we are assuming that the sample sizes are equal, the grand mean is equal to the mean of the sample means. That is

$$\bar{\bar{x}} = \frac{\bar{x}_1 + \bar{x}_2 + \cdots + \bar{x}_k}{k}.$$

The Within-samples Sum of Squares

Basic to our approach is the assumption that the parent populations have the same variance σ^2 and that they are normally distributed. (If for some reason these assumptions are not satisfied, one should not apply the ANOVA technique for testing equality of means.*) Since $x_{1,1}, x_{1,2}, \ldots, x_{1,n}$ is a sample,

$$s_1^2 = \frac{\sum\limits_{j=1}^{n} (x_{1,j} - \bar{x}_1)^2}{n - 1}.$$

is an unbiased estimate of the variance of the population from which it was picked. Also,

$$s_2^2 = \frac{\sum\limits_{j=1}^{n} (x_{2,j} - \bar{x}_2)^2}{n - 1}, \ldots, s_k^2 = \frac{\sum\limits_{j=1}^{n} (x_{k,j} - \bar{x}_k)^2}{n - 1}$$

are unbiased estimates of the variances of their respective parent populations. Now since all the populations are assumed to have the same variance σ^2, we can justifiably obtain an improved estimate of σ^2 by pooling whatever information $s_1^2, s_2^2, \ldots, s_k^2$ can contribute toward its knowledge. Such a pooled unbiased estimate is given as

$$\frac{\sum\limits_{j=1}^{n} (x_{1,j} - \bar{x}_1)^2 + \sum\limits_{j=1}^{n} (x_{2,j} - \bar{x}_2)^2 + \cdots + \sum\limits_{j=1}^{n} (x_{k,j} - \bar{x}_k)^2}{(n - 1) + (n - 1) + \cdots + (n - 1)}$$

(This is analogous to what we did in Section 9-4 when comparing means of two populations using the t test.) Using double summation, this pooled estimate s_p^2 can be written more compactly as

$$s_p^2 = \frac{\sum\limits_{i=1}^{k} \sum\limits_{j=1}^{n} (x_{i,j} - \bar{x}_i)^2}{k(n - 1)}.$$

In statistical literature the term appearing in the numerator of s_p^2 is called the **error sum of squares,** and is commonly denoted as **SSE.** It is also called the **within-samples sum of squares,** since it is obtained by pooling variability *within* the samples. The number of degrees of freedom associated with SSE is equal to the sum of the degrees of freedom associated with $s_1^2, s_2^2, \ldots, s_k^2$ and is equal to $k(n - 1)$. Thus

$$SSE = \sum\limits_{i=1}^{k} \sum\limits_{j=1}^{n} (x_{i,j} - \bar{x}_i)^2.$$

The quantity $SSE/[k(n - 1)]$, which is an estimate of σ^2, is called the *mean error sum of squares* and is denoted by MSE. *This estimate is an unbiased estimate of σ^2 irrespective of whether H_0 is valid or not.*

For the data in Table 12-1 we see that

---

*In some situations transformations exist to bring the data into line with the assumptions.

Brand 1: $\sum_{j=1}^{n} (x_{1,j} - \bar{x}_1)^2 = (15 - 16)^2 + (19 - 16)^2 + (14 - 16)^2 + (16 - 16)^2$
$$= 14$$

Brand 2: $\sum_{j=1}^{n} (x_{2,j} - \bar{x}_2)^2 = (19 - 18)^2 + (17 - 18)^2 + (16 - 18)^2 + (20 - 18)^2$
$$= 10$$

and

Brand 3: $\sum_{j=1}^{n} (x_{3,j} - \bar{x}_3)^2 = (22 - 19)^2 + (17 - 19)^2 + (19 - 19)^2 + (18 - 19)^2$
$$= 14.$$

Consequently, the within sum of squares obtained by pooling variability within samples is

$$SSE = \sum_{i=1}^{3} \sum_{j=1}^{4} (x_{i,j} - \bar{x}_i)^2 = 14 + 10 + 14 = 38.$$

The pooled estimate of σ^2 is now obtained as

$$SSE/[k(n - 1)] = 38/[3(4 - 1)] = 4.222.$$

The Between Means Sum of Squares

The null hypothesis H_0 that we are interested in testing is

$$H_0: \mu_1 = \mu_2 = \cdots = \mu_k.$$

If the null hypothesis is true, which implies that all the populations have the same mean μ, we are in a position to obtain another estimate of σ^2. To H_0 being true, add our basic assumption that the populations have the same variance. The means $\bar{x}_1, \bar{x}_2, \ldots, \bar{x}_k$ then represent a sample of k means drawn from one single population with mean μ and variance σ^2. Using the formula for s^2 in Chapter 2, if we compute the quantity

$$\frac{(\bar{x}_1 - \bar{\bar{x}})^2 + (\bar{x}_2 - \bar{\bar{x}})^2 + \cdots + (\bar{x}_k - \bar{\bar{x}})^2}{k - 1} = \frac{\sum_{i=1}^{k} (\bar{x}_i - \bar{\bar{x}})^2}{k - 1}$$

and denote it as $s_{\bar{x}}^2$, we realize that $s_{\bar{x}}^2$ is simply an unbiased estimate of the variance of the distribution of the sample means. But the variance of the distribution of $\bar{X}$ is σ^2/n, as will be recalled from Chapter 6. It follows then that $s_{\bar{x}}^2$ is an unbiased estimate of σ^2/n so that $ns_{\bar{x}}^2$ is an unbiased estimate of σ^2. (We accept this statement intuitively.) That is,

$$\frac{n \sum_{i=1}^{k} (\bar{x}_i - \bar{\bar{x}})^2}{k - 1}$$

is another unbiased estimate of σ^2 if the null hypothesis is true. The term in the numerator, $n \sum_{i=1}^{k} (\bar{x}_i - \bar{\bar{x}})^2$, is called the **between means sum of squares** since it is based on the variation between the means of the samples. The between means sum of squares is also referred to as the *sum of squares for treatments*. It is commonly denoted as SSB and has $k - 1$ degrees of freedom. (We will have $3 - 1$, or 2, degrees of freedom in our illustration.) Thus

$$SSB = n \sum_{i=1}^{k} (\bar{x}_i - \bar{\bar{x}})^2.$$

The quantity $SSB/(k - 1)$ is a new estimate of σ^2. It is called the *mean between sum of squares* and is denoted as MSB.

For the data in Table 12–1, the grand mean is given by

$$\bar{\bar{x}} = \frac{16 + 18 + 19}{3}$$

$$= 17.67.$$

Therefore,

$$SSB = n[(\bar{x}_1 - \bar{\bar{x}})^2 + (\bar{x}_2 - \bar{\bar{x}})^2 + (\bar{x}_3 - \bar{\bar{x}})^2]$$
$$= 4[16 - 17.67)^2 + (18 - 17.67)^2 + (19 - 17.67)^2]$$
$$= 18.667.$$

Hence, $SSB/(k - 1) = 18.667/2 = 9.333$ is an estimate of σ^2 based on the assumption that H_0 is true, that is, based on the assumption that the three population means are equal.

The Test of Hypothesis

We have obtained two estimates of σ^2, namely, 4.222 and 9.333. The estimate 4.222 is obtained assuming equal population variances. It should be emphasized that the second estimate 9.333 is contingent on the assumption that the population variances are equal and, more importantly, on the truth of the null hypothesis that the population means are all equal. If the null hypothesis is true, then the two estimates $SSE/[k(n - 1)]$ and $SSB/(k - 1)$ are valid unbiased estimates of σ^2 and should be of the same order of magnitude. For the purpose of comparison, we use the following ratio, called the **variance ratio:**

$$f = \frac{ns_{\bar{x}}^2}{s_p^2} = \frac{SSB/(k - 1)}{SSE/[k(n - 1)]}$$

We might ascribe a slight departure of the above ratio from 1 to chance fluctuations. However, if the ratio is substantially larger than 1, it will cast doubts on the validity of H_0. For, if H_0 is not true, then $ns_{\bar{x}}^2$ tends to overestimate σ^2, because it now incorporates not only the experimental error but also real differences in the population means. The question that we are faced with now is "How much larger than 1 must the variance ratio be to cause us to reject H_0?" This will, of course, depend on the distribution of the variance ratio. The corresponding statistic, which we denote as **F**, has a distribution called the F distribution which we now consider.

As a broad theoretical statement, a statistic **F** which is the ratio of two independent chi-square variables, each divided by the corresponding degrees of freedom, has an F distribution. If the chi-square in the numerator has ν_1 degrees of freedom and the one in the denominator has ν_2 degrees of freedom, then the F distribution is said to have ν_1 and ν_2 degrees of freedom. It is important that the number of degrees of freedom be given in this specific order.

Since the F distributions depend on two parameters, namely, the degrees of freedom associated with the numerator and those associated with the denominator, they constitute what is called a two-parameter family of distributions. Because it is a ratio of two positive terms, **F** can assume only positive values, and the curve describing the distribution is located to the right of the vertical axis at zero. As can be seen from Figure 12-3, the distribution is skewed to the right, but the skewness disappears rapidly as the numbers of degrees of freedom increase. In these respects it behaves very much like the chi-square distribution.

Appendix II, Table A-7 gives certain probabilities and corresponding f values for different F distributions specified by the parameters v_1 and v_2. Since an F distribution depends on two parameters, we are restricted even more than in the case of the t distribution and the chi-square distribution in providing complete tables. In Table A-7, the top row in each table is assigned to the set of degrees of freedom ν_1, the number associated with the numerator. The left-hand column is assigned to the other set of degrees of freedom, ν_2. Each table then presents an f value (in the body of the table) for a specific probability α contained in the right tail of the distribution. An f value for an F distribution with ν_1 and ν_2 degrees of freedom leaving an area

---

FIGURE 12-3

Probability density curves for F distributions with different pairs of degrees of freedom.

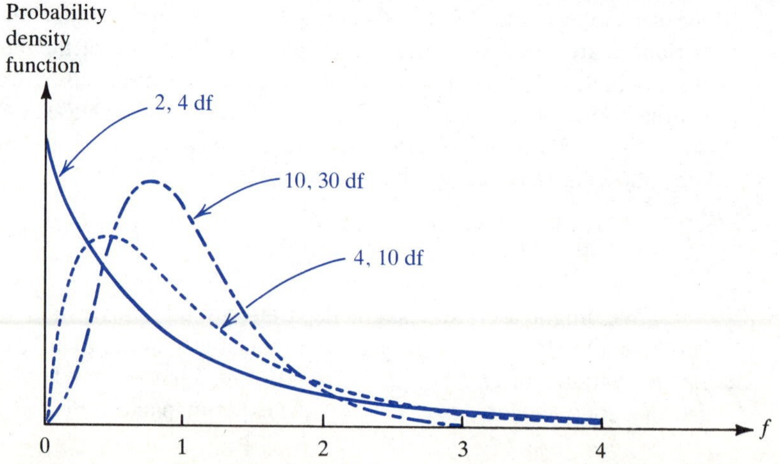

FIGURE 12-4

An *f* value leaving an area α in the right tail.

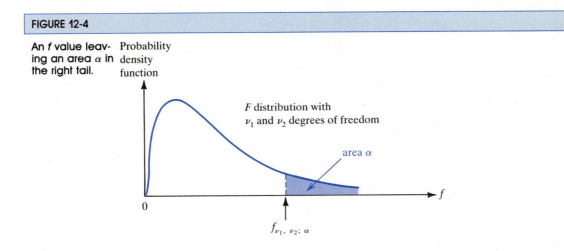

α in the right tail of the distribution is commonly denoted as $f_{\nu_1,\nu_2;\alpha}$ and is called an upper percentage value of the *F* distribution. This is shown in Figure 12-4 above. There will be as many tables of *f* values as there are α levels for which one might wish to provide tables. For most statistical purposes, the α levels of interest are $\alpha = 0.1, 0.05, 0.025,$ and 0.01. For $\nu_1 = 4$ and $\nu_2 = 10$, we give below the upper percentage points $f_{4,10;\alpha}$ for the α values above:

Probability in the right tail	0.1	0.05	0.025	0.01
$f_{4,10;\alpha}$	2.61	3.48	4.47	5.99

These are obtained from A-1 in Appendix II and are the colored entries in the tables. The locations of the above *f* values along the horizontal axis and the corresponding probabilities are illustrated in Figure 12-5.

FIGURE 12-5

Probability density curve for *F* distribution with 4,10 df. Area to the right of 2.61 is 0.10, etc.

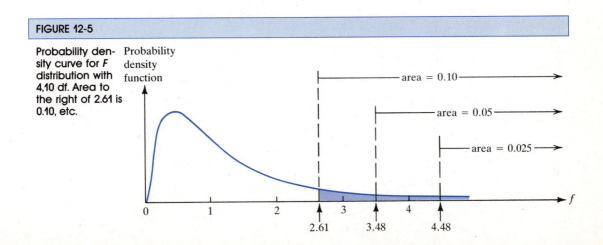

Returning to the testing aspect, we have agreed that if the variance ratio is substantially *larger than* 1, it will cast doubts on the validity of H_0. Now it can be shown that the statistic giving the variance ratio on page 441 has an F distribution with $(k - 1)$ and $k(n - 1)$ degrees of freedom. Consequently, *the critical region of the test consists of values in the right tail of the F distribution with $(k - 1)$ and $k(n - 1)$ degrees of freedom.* If the computed value of the ratio exceeds the table value of the F distribution for a given α, we shall reject the null hypothesis at that level of significance and conclude that at least two of the population means are different from each other.

In our illustration, since $SSB = 18.667$ and $SSE = 38$, we get

$$f = \frac{18.667/2}{38/9} = 2.21.$$

Since $k = 3$ and $n = 4$ in our example, the corresponding F distribution has 2 (or $3 - 1$) and 9 (or $3(4 - 1)$) degrees of freedom. The table value at the 5 percent level of significance is $f_{2,9;0.05} = 4.26$. Since the computed value 2.21 does not exceed the table value 4.26, we do not reject the null hypothesis. There is no evidence indicating that the brands of gasoline are different.

THE ANOVA TABLE

We have obtained two estimates of σ^2. Actually, if the null hypothesis is true, it is possible to obtain one more estimate as

$$\frac{\sum_{i=1}^{k} \sum_{j=1}^{n} (x_{i,j} - \bar{\bar{x}})^2}{N - 1}$$

where N is the total number of observations. N is equal to kn for the data in Table 12-4 and equal to 3×4, or 12, for the data in Table 12-1. The numerator of this expression is called the **total sum of squares** and is usually denoted as SST. To compute it, we would subtract each observation from the grand mean $\bar{\bar{x}}$, square these differences, and then total them. The denominator $N - 1$ is the number of degrees of freedom associated with SST. Thus

$$SST = \sum_{i=1}^{k} \sum_{j=1}^{n} (x_{i,j} - \bar{\bar{x}})^2.$$

For the data in Table 12-1, since $\bar{\bar{x}} = 17.67$, we see that

$$
\begin{aligned}
SST = {} & (15 - 17.67)^2 + (19 - 17.67)^2 + (14 - 17.67)^2 + (16 - 17.67)^2 \\
& + (19 - 17.67)^2 + (17 - 17.67)^2 + (16 - 17.67)^2 + (20 - 17.67)^2 \\
& + (22 - 17.67)^2 + (17 - 17.67)^2 + (19 - 17.67)^2 + (18 - 17.67)^2 \\
= {} & 56.667.
\end{aligned}
$$

In summary, when we consider the three sums of squares that we have computed, we will see that $SST = 56.667$, $SSB = 18.667$, $SSE = 38$, and interestingly enough,

$$SST = SSB + SSE.$$

The partition of the total sum of squares in this way as a sum of the two sums of squares, SSB and SSE, always holds. In practice we would therefore usually compute SST and SSB and then obtain SSE by subtracting as $SSE = SST - SSB$. Also notice that *the degrees of freedom associated with SST is equal to the sum of the degrees of freedom associated with SSB and SSE*. This relation also is always true. These relations enable us to present the results of our procedure in a table called the analysis of variance table (abbreviated as the ANOVA table), as in Table 12-5. The last column of the ANOVA table gives the variance ratio.

TABLE 12-5

Analysis of variance table for the data of Table 12-1.

Source of variation	Degrees of freedom	Sum of squares	Mean sum of squares	Computed variance ratio
Between	2	18.667	9.334	2.21
Within (error)	9	38.0	4.222	
Total	11	56.667		

The statistical test procedure for comparing means of k populations is summarized as follows:

COMPARISON OF k POPULATION MEANS

1. H_0: $\mu_1 = \mu_2 = \cdots \mu_k$
2. H_A: Not all the population means are equal.
3. Test statistic: The variance ratio which has F distribution with $k - 1$ and $N - k$ degrees of freedom. To compute its value prepare the ANOVA table.
4. At the level of significance α, the decision rule is *Reject H_0 if the computed value of the test statistic is greater than $f_{k-1, N-k; \alpha}$*. (*Note*: The rejection region is in the right tail of the distribution.)

EXAMPLE 1   Analyze the data in Table 12-3, which is reproduced below:

	Brand 1	Brand 2	Brand 3
	15.2	18.5	19.6
	16.1	17.5	19.3
	16.8	18.2	18.4
	15.9	17.8	18.7
Mean mileage per gallon	16.0	18.0	19.0

SOLUTION

1. H_0: $\mu_1 = \mu_2 = \mu_3$
2. H_A: Not all the brand means are equal.
3. The test criterion is provided by the variance ratio, which we shall compute next. Since the brand means are the same here as for the data in Table 12-1, SSB will be the same as before. We have

$$SSB = 18.667$$

$$\begin{aligned} SST = {} & (15.2 - 17.67)^2 + (16.1 - 17.67)^2 + (16.8 - 17.67)^2 \\ & + (15.9 - 17.67)^2 + (18.5 - 17.67)^2 + (17.5 - 17.67)^2 \\ & + (18.2 - 17.67)^2 + (17.8 - 17.67)^2 + (19.6 - 17.67)^2 \\ & + (19.3 - 17.67)^2 + (18.4 - 17.67)^2 + (18.7 - 17.67)^2 \\ = {} & 21.447. \end{aligned}$$

Therefore,

$$\begin{aligned} SSE &= 21.447 - 18.667 \\ &= 2.78. \end{aligned}$$

Table 12-6 is the ANOVA table summarizing these results and other computations.

TABLE 12-6				
ANOVA table for the data of Table 12-3.				
Source of variation	Degrees of freedom	Sum of squares	Mean sum of squares	Computed variance ratio
Between	2	18.667	9.334	30.11
Within	9	2.780	0.310	
Total	11	21.447		

4. Using $\alpha = 0.05$, the decision rule is *Reject H_0 if the computed value of the variance ratio is greater than 4.26.* (Here 4.26 is the table value of the F distribution with 2 and 9 degrees of freedom for $\alpha = 0.05$.)

5. The computed value is 30.11, as can be seen from the last column of the ANOVA table. Since it is greater than 4.26, we reject H_0 and conclude that the brand means are different. In our earlier discussion we had suspected that this was the case. ■

SECTION 12-1   EXERCISES

1. Suppose 5 populations are being compared for their means and samples of size 7 are drawn from each population. How many df does each of the following have?

   (a) *SSB*   (b) *SSE*   (c) *SST*

2. In an analysis of variance problem having 5 categories when samples of size 8 from each category were picked, the within sum of squares was found to be 85.75. Find an estimate of σ^2.

3. In Exercise 2 suppose the between sum of squares was given as 13.40. Assuming that the population means of the 5 categories are equal, obtain another estimate of σ^2.

4. The daily dissolved oxygen concentration (in milligrams per liter) of three streams was observed on 6 days for each stream. The observations are given in the following table:

Stream 1	Stream 2	Stream 3
4.02	2.88	2.76
3.98	3.06	3.50
3.85	3.70	2.82
4.32	2.98	3.22
4.16	3.88	3.15
3.92	3.28	3.80

   It may be assumed that the dissolved oxygen concentrations of the streams are normally distributed with the same variance, and are statistically independent. At the 5 percent level, are the mean concentrations of the streams significantly different?

5. When samples of reinforcing bars supplied by four companies were tested for their yield strength, the following data were obtained.

Company A	Company B	Company C	Company D
22	18	25	24
19	28	20	18
15	26	25	25
18	24	24	19
24	20	26	21
19	21	18	22

   Assume that the yield strength of the bars is normally distributed for each company with the same variance σ^2. At the 5 percent level of significance, are the mean yield strengths different for the companies?

6. The following data were obtained when five brands of sanders were tested for their trouble-free service (in hours):

Brand A	Brand B	Brand C	Brand D	Brand E
680	900	520	835	1020
735	580	680	450	630
540	600	720	700	700
865	780	925	730	550

If all the assumptions necessary for applying the ANOVA method hold, at the 5 percent level, are the five brands different?

12-2 ONE-WAY CLASSIFICATION— ARBITRARY SAMPLE SIZES

The ANOVA technique for one-way classification is not limited to equal sample sizes. We assumed equal sample sizes of n in order to simplify matters in our previous discussion. With slight modifications in the formulas, we can apply the technique to any number of samples and any number of observations within each sample. In the following discussion we shall assume that there are k samples 1, 2, . . . , k with $n_1, n_2, . . . , n_k$ observations, respectively. A general representation of the response collected is given in Table 12-7 where $x_{i,\bullet}$ is the total for Sample i.

TABLE 12-7

A general representation of the response with unequal sample sizes.

	Sample 1	Sample 2 . . . Sample k	
	$x_{1,1}$	$x_{2,1}$ $\cdots$	$x_{k,1}$
	$x_{1,2}$	$x_{2,2}$ $\cdots$	$x_{k,2}$
	$\vdots$	$\vdots$	$\vdots$
	x_{1,n_1}	x_{2,n_2} $\cdots$	x_{k,n_k}
Sample size	n_1	n_2 $\cdots$	n_k
Sample total	$x_{1,\bullet}$	$x_{2,\bullet}$ $\cdots$	$x_{k,\bullet}$
Sample mean	$\bar{x}_1$	$\bar{x}_2$ $\cdots$	$\bar{x}_k$

The total number of observations N is given by

$$N = n_1 + n_2 + \cdots + n_k$$

and the grand total, the sum of the entire set of observations, is given by $x_{1,\bullet} + x_{2,\bullet} + \cdots + x_{k,\bullet}$.

The computations involved in calculating the sums of squares in an ANOVA problem can get quite tedious. It is helpful to follow the shortcut routine given below:

STEP 1. Obtain what is called the **correction term** (abbreviated **C.T.**) as

$$C.T. = \frac{(\text{grand total})^2}{N}$$

STEP 2. $SSB = \dfrac{x_{1,\bullet}^2}{n_1} + \dfrac{x_{2,\bullet}^2}{n_2} + \cdots + \dfrac{x_{k,\bullet}^2}{n_k} - C.T.$

where $x_{i,\bullet}$ is the total of the ith sample.

STEP 3. $SST = \displaystyle\sum_{i=1}^{k} \sum_{j=1}^{n_i} x_{i,j}^2 - C.T.$

That is, square each of the observations and find the sum of these squares. From the total thus obtained, subtract the correction term.

STEP 4. SSE is found by subtracting SSB from SST.

The results can be presented now in an ANOVA table, Table 12-8.

TABLE 12-8

Analysis of variance table when the sample sizes are arbitrary.

Source of variation	Degrees of freedom	Sum of squares	Mean sum of squares	Computed ratio
Between	$k - 1$	$\displaystyle\sum_{i=1}^{k} \frac{x_{i,\bullet}^2}{n_i} - C.T.$	$SSB/(k - 1)$	$\dfrac{SSB/(k - 1)}{SSE/(N - k)}$
Within (error)	$N - k$	$SST - SSB$	$SSE/(N - k)$	
Total	$N - 1$	$\displaystyle\sum_{i=1}^{k} \sum_{j=1}^{n_i} x_{i,j}^2 - C.T.$		

We shall illustrate the procedure using the following data:

	Sample 1	Sample 2	Sample 3	Sample 4
	3	6	11	3
	7	8	7	10
	8	12	16	
	15		9	
			15	
Sample size	4	3	5	2
Sample total	33	26	58	13

Here $n_1 = 4$, $n_2 = 3$, $n_3 = 5$, $n_4 = 2$. Hence $N = 4 + 3 + 5 + 2 = 14$. Also, the grand total is $33 + 26 + 58 + 13 = 130$. We now carry out the following steps:

1. $C.T. = \dfrac{(130)^2}{14}$

    $= 1207.143$

2. $SSB = \dfrac{33^2}{4} + \dfrac{26^2}{3} + \dfrac{58^2}{5} + \dfrac{13^2}{2} - 1207.143$

    $= 1254.883 - 1207.143$

    $= 47.740$

3. $SST = 3^2 + 7^2 + 8^2 + 15^2 + 6^2 + 8^2 + 12^2 + 11^2 + 7^2 + 16^2 + 9^2$
    $\qquad + 15^2 + 3^2 + 10^2 - 1207.143$

    $= 1432.0 - 1207.143$

    $= 224.857$

4. $SSE = 224.857 - 47.740$

    $= 177.117$

The computed quantities are now presented in the following ANOVA table.

Source of variation	Degrees of freedom	Sum of squares	Mean sum of squares	Computed ratio
Between	3	47.740	15.913	0.898
Within	10	177.117	17.712	
Total	13	224.857		

EXAMPLE 1

In order to test whether Terry's productivity is the same on the five weekdays of a week, unknown to Terry, her employer kept records on 18 randomly picked days, as the data in Table 12-9 on page 451 shows. (Productivity is measured in terms of market value, in dollars, of the items produced by Terry.) Carry out the test using $\alpha = 0.05$.

SOLUTION

We shall assume that Terry's productivity on any weekday is normally distributed with the same variance and that her productivity on one day does not influence her productivity on another day. Our approach now consists of the following steps:

1. H_0: $\mu_1 = \mu_2 = \mu_3 = \mu_4 = \mu_5$
2. H_A: At least two of the means are different. That is, the mean productivity on at least two days is different.
3. *Computations:*

    $C.T. = \dfrac{(2552)^2}{18} = 361{,}817$

TABLE 12-9					
Productivity on 15 randomly picked days classified according to the day of the week.					
	Monday	*Tuesday*	*Wednesday*	*Thursday*	*Friday*
	143	162	160	138	110
	128	136	132	168	130
	110	144	180	120	135
		158	160		
			138		
Sample size	3	4	5	3	3
Sample total	381	600	770	426	375
Sample mean	127	150	154	142	125

$$SST = 143^2 + 128^2 + \cdots + 130^2 + 135^2 - 361{,}817$$
$$= 368{,}334 - 361{,}817$$
$$= 6517$$
$$SSB = \frac{381^2}{3} + \frac{600^2}{4} + \frac{770^2}{5} + \frac{426^2}{3} + \frac{375^2}{3} - 361{,}817$$
$$= 364{,}334 - 361{,}817$$
$$= 2517$$
$$SSE = 6517 - 2517$$
$$= 4000$$

We get the ANOVA table presented in Table 12-10.

TABLE 12-10				
ANOVA table for data in Table 12-9.				
Source of variation	*Degrees of freedom*	*Sum of squares*	*Mean sum of squares*	*Computed ratio*
Between	4	2517	629.3	2.05
Within	13	4000	307.7	
Total	17	6517		

4. The F distribution has 4 and 13 degrees of freedom. From the table of F distribution, $f_{4,13;0.05} = 3.18$. The decision rule is *Reject H_0 if the computed value of the variance ratio exceeds 3.18.*

5. From the ANOVA table, the computed value is 2.05, and so we do not reject H_0. The evidence does not substantiate that the mean productivity of Terry is different on five weekdays. ■

EXAMPLE 2   Table 12-11 gives the hourly wages (in dollars) of 15 randomly picked workers classified according to their occupations. At the 1 percent level of significance, are the mean wages different for the four occupations?

SOLUTION   We take for granted the basic assumptions, namely, that the hourly wages for each occupation are normally distributed with the same variance and the hourly wages of one occupation are not influenced by those of another occupation.

TABLE 12-11				
Hourly wages of workers classified according to their occupations.				
	Plumber	*Electrician*	*Painter*	*Carpenter*
	17.6	18.5	16.8	17.4
	18.3	18.7	16.7	16.5
	17.6	17.7	16.6	16.8
		18.3	16.4	
		17.8		
Sample size	3	5	4	3
Sample total	53.5	91.0	66.5	50.7
Sample mean	17.83	18.20	16.63	16.90

1. H_0: $\mu_1 = \mu_2 = \mu_3 = \mu_4$
2. H_A: At least two of the population means are different.
3. *Computations:*

The grand total $= 53.5 + 91.0 + 66.5 + 50.7$
$$= 261.7$$
$$C.T. = \frac{(261.7)^2}{15}$$
$$= 4{,}565.793$$
$$SST = 17.6^2 + 18.3^2 + \cdots + 16.5^2 + 16.8^2 - 4{,}565.793$$
$$= 4{,}574.270 - 4{,}565.793$$
$$= 8.477$$
$$SSB = \frac{53.5^2}{3} + \frac{91.0^2}{5} + \frac{66.5^2}{4} + \frac{50.7^2}{3} - 4{,}565.793$$
$$= 4{,}572.675 - 4{,}565.793$$
$$= 6.882$$
$$SSE = 8.477 - 6.882$$
$$= 1.595$$

These and other results are summarized in Table 12-12.

TABLE 12-12				
ANOVA table for data in Table 12-11.				
Source of variation	Degrees of freedom	Sum of squares	Mean sum of squares	Computed ratio
Between	3	6.882	2.294	15.82
Within	11	1.595	0.145	
Total	14	8.477		

4. The F distribution has 3 and 11 degrees of freedom. Since $f_{3,11;0.01} = 6.22$, the decision rule is *Reject H_0 if the computed value of the variance ratio is greater than 6.22.*
5. Since the computed value is 15.82, we reject H_0 at the 1 percent level of significance. There is rather strong evidence that mean wages differ significantly according to occupations. ▬▬

SECTION 12-2   EXERCISES

1. Complete the analysis of variance table given below:

Source of variation	Degrees of freedom	Sum of squares	Mean sum of squares	Computed ratio
Between	7			
Within		187.2		
Total	25	397.9		

2. In an analysis of variance problem having 5 categories with 40 observations altogether, the within-samples sum of squares was found to be 147. Obtain an estimate of σ^2.

In each of the following exercises, assume that the sampled populations are normally distributed with the same variance and that the samples are independent.

3. A golfer wants to compare a new brand of golf ball that has come on the market with his old brand. The figures below give the distances (in feet) of six drives with the old brand and seven drives with the new brand:

Old brand	New brand
138	165
162	126
140	195
175	203
190	180
151	150
	185

At the 5 percent level, is there significant difference in the mean distance on a drive for the two brands? Carry out the test using

(a) the method for comparison of two means developed in Section 9-4 (Chapter 9)
(b) the ANOVA technique.

4. A farm was divided into 18 plots, and three concentrations of a fertilizer were allotted to them at random with each concentration tried on 6 plots. The following figures give the yield of grain (in bushels) for each plot:

Concentration 1	Concentration 2	Concentration 3
15	32	28
24	21	16
18	16	25
28	35	28
22	26	22
20	18	29

At the 5 percent level of significance, do the data provide evidence that the three concentrations are significantly different in their effect on the yield of grain?

5. The figures below give the tar content (in milligrams) in a pack of cigarettes for four brands, when five packs of each brand were tested:

Brand A	Brand B	Brand C	Brand D
340	358	335	358
323	320	318	345
319	340	330	330
330	360	320	350
335	338	338	340

Is there significant difference in the mean tar content in a pack for the four brands? Use $\alpha = 0.1$.

6. The figures below give the lengths (in inches) of trout caught in three lakes, Blue Lake, Clear Lake, Fresh Lake:

Blue Lake	Clear Lake	Fresh Lake
11.9	10.6	11.2
10.8	10.4	9.6
9.8	10.3	11.2
10.6	10.2	10.5
11.2		10.3
9.7		

Is there a significant difference in the true mean lengths of trout found in the three lakes? Use $\alpha = 0.025$.

7. A laboratory experiment was conducted to compare three brands of tires. The following figures give the number of miles when 5 tires of each brand were tested:

Brand A	Brand B	Brand C
32,000	34,000	37,000
35,000	27,000	29,000
41,000	36,000	41,000
31,000	30,000	37,000
36,000	27,000	38,000

Do the data lend support to the contention that there is a significant difference between the three brands? Use $\alpha = 0.025$.

8. Three horses, Prima Dona, Princess Jane, and Flying Princess, were timed running along a course. The figures below give times (in minutes) in five runs along the course:

Prima Dona	Princess Jane	Flying Princess
5.9	5.6	6.2
6.3	6.3	6.3
5.8	5.9	6.4
6.4	6.2	6.1
6.1	5.9	6.2

At the 5 percent level, is there a significant difference between the true mean time along the course for the three horses?

9. In order to compare three different types of insulations, a building contractor built 15 rooms, 5 with each type of insulation. The figures below refer to the drop in temperature in each room during a four-hour period on a night:

Insulation A	Insulation B	Insulation C
16	10	15
19	12	16
18	20	20
10	12	16
17	14	13

Do the data indicate a significant difference between the three types of insulations? Use a suitable α level.

10. The manager of a bank decided to compare the speed on the job of four tellers working in the bank. The following data give the amount of time (in minutes) that they spent serving their customers, picked at random:

Teller 1	Teller 2	Teller 3	Teller 4
0.8	8.7	9.7	0.7
1.6	4.2	2.0	3.2
8.8	0.2	5.9	4.2
7.7	9.0	2.3	6.7
3.3		5.8	7.5
			1.4

Is there evidence indicating that there is a significant difference between the true mean serving times for the four tellers? Use $\alpha = 0.025$.

11. An agriculture experimental station is interested in comparing three corn hybrids. Hybrids A and B are each planted on six plots and Hybrid C is planted on five plots. The yield per acre (in bushels) is recorded:

Hybrid A	Hybrid B	Hybrid C
86	83	79
94	75	75
99	62	53
61	86	
70	77	55
89	79	72

At the 10 percent level of significance, is the true mean yield per acre different for the three hybrids?

12. A pharmacology lab conducted an experiment to compare four pain relieving drugs. Twenty-four subjects were used with six allotted at random to drug A, seven to drug B, five to drug C, and six to drug D. The figures given below give the number of hours of pain relief provided subsequent to administering a drug:

Drug A	Drug B	Drug C	Drug D
8	9	2	2
7	3	3	9
3	1	5	8
8	8	9	2
1	1	6	7
4	9		5
	8		

At the 5 percent level of significance, are the drugs different in the true mean number of hours of relief provided?

KEY TERMS AND EXPRESSIONS

Analysis of variance (ANOVA)
One-way classification
experimental error
grand mean
within-samples sum of squares (SSE)
error sum of squares (SSE)
between means sum of squares (SSB)
total sum of squares (SST)
variance ratio
F distribution
correction term (C.T.)

KEY FORMULAS

correction term

$$C.T. = (\text{grand total})^2/N, \text{ where } N \text{ is the total number of observations}$$

between means sum of squares

$$SSB = \frac{x_{1,\bullet}^2}{n_1} + \frac{x_{2,\bullet}^2}{n_2} + \cdots + \frac{x_{k,\bullet}^2}{n_k} - C.T.$$

total sum of squares

$$SST = \sum_{i=1}^{k} \sum_{j=1}^{n_i} x_{i,j}^2 - C.T.$$

within-samples sum of squares

$$SSE = SST - SSB.$$

CHAPTER 12   TEST

1. What assumptions are made in the analysis of variance for a one-way classification problem?

2. State whether the following statement is true or false:

   the mean between means sum of squares + the mean within samples sum of squares

   = the mean total sum of squares

3. In an analysis of variance problem with 4 categories and 30 observations altogether, the within sum of squares was obtained as 32.24. Find an estimate of σ^2.

4. In a laboratory experiment where a certain strain of fruit flies were classified according to their genotypes, AA, Aa, and aa, some flies were picked at random from each genotype and the total wing length (in microunits) of each fly was measured. The data are presented in the table below:

AA	Aa	aa
10	8	10
17	16	9
19	13	16
14	10	9
	12	12

   Do the data provide sufficient evidence to suggest that the true mean wing lengths for the three genotypes are different? Use a suitable level of α.

13

NONPARAMETRIC METHODS

INTRODUCTION

Except in Chapter 10, where we tested independence of two attributes, the statistical test procedures with which we have been concerned have involved tests regarding population parameters where we had to make assumptions about some aspects of the distribution of the sampled population, for example, that it was normally distributed. In short, we have carried out tests of hypotheses about parameters of known population types. Such tests are commonly known as classical, standard, or **parametric tests.**

We can, of course, envision situations where the assumptions required in a classical test are not justified. For instance, in comparing means of two populations using the *t* test we assumed that the populations were normally distributed and had equal variances. What if these assumptions are not justified? Statistical tests that are not based on assumptions about the distribution or the parameters of the sampled population have been devised and they are referred to as **nonparametric** or **distribution-free tests.**

Nonparametric techniques have several advantages over classical tests. Computationally they are simpler, they are intuitively easier to understand, and, of course, there is no need to make assumptions about the distribution of the population under investigation. Also, many nonparametric tests can be applied to situations where classical tests cannot be applied because of the nature of the response variable. This is especially the case when the data cannot be measured quantitatively except by assigning ranks, as is the case in judging a contest.

The picture is not completely rosy. There are also some disadvantages in using nonparametric methods. These methods very often ignore the actual numerical value of an observation and take into account only its relative standing (ranking) in relation to other observations in the sample. As a result, acquired information is wasted. Furthermore, if indeed the assumptions under which a classical test can be applied hold true, then a nonparametric test is less efficient. This is understandable, because a nonparametric test assumes very little about the population distribution and hence cannot be expected to be as proficient as tests that make use of known properties of the population. For a given level of significance α, a nonparametric test tends to have a higher probability of accepting H_0 *erroneously* as compared with the classical test (when the assumptions of the latter test are met).

In this chapter we shall present some of the commonly applied nonparametric methods. These are (1) the *two-sample sign test;* (2) the rank-sum test—also called the Mann-Whitney *U* test; (3) the runs test, a test of randomness; and (4) Spearman's rank-correlation method. ■

13-1 THE TWO-SAMPLE SIGN TEST FOR MATCHED PAIRS

The **two-sample sign test** is one of the simplest nonparametric tests. It was originally developed to test for the median of a population with continuous distribution. However, it can also be used in treating data consisting of paired observations when the two samples are not independent and the differences of the matched pairs are not normally distributed. Recall that if the differences are normally distributed or approximately so, then we would test for the equality of means using the paired *t*-test developed in Chapter 9. In this section we shall discuss the sign test in the context of comparison of two populations using paired data. The test is appropriately named *sign test,* because rather than using numerical values from the two samples as data, it uses *plus* signs $+$ and *minus* signs $-$ as the information on which to base the test. Consider the following situations:

(a) To test whether a particular diet has any effect on reducing weight, the weight of each individual involved is obtained before and after the diet. Certainly we feel that the weights of each individual before and after the diet are related. But; at the same time, we also feel that weight changes for different individuals are independent.

(b) Ten judges grade two brands of wine, grading each brand on a scale of 1 to 5 points. The response of each judge as regards the two brands is certainly not independent. However, we do feel justified in assuming that the responses are independent from judge to judge.

We replace each pair of observations by a $+$ sign if the first component of the pair is greater than the second, and by a $-$ sign if the first component is smaller than the second. Under the sign test, no ties are allowed. Those pairs for which ties do occur are omitted from the sample. The effective sample size *n* is then given as the original number of pairs reduced by the number of ties. To see the general approach, let us consider the following illustration.

Suppose an airline company wants to compare two models of planes, one manufactured by an English company and another by a French company. It is decided to fly 15 planes of each model. Fifteen pilots of proven ability are each assigned to fly an English and a French model and are asked to rate them on a scale from 1 to 10. The results of the tests are given in Table 13-1.

The null hypothesis we wish to test is that the two types of planes are identical, that is, the population distributions of the two models are the same. There are various alternative hypotheses against which we could test the null hypothesis, such as, the two models are not identical, the French model is superior, or the English model is superior.

TABLE 13-1			
Scores given to English and French models			
1	7	4	+
2	3	5	−
3	5	7	−
4	6	3	+
5	8	6	+
6	4	4	← Tie
7	6	8	−
8	7	2	+
9	5	6	−
10	4	7	−
11	7	9	−
12	8	7	+
13	4	6	−
14	3	5	−
15	7	9	−

Suppose we wish to test

H_0: The two plane models are identical.

against

H_A: The French model is superior.

at the 5 percent level of significance. As a first step we obtain column 4 in Table 13-1 giving the sign of the difference for each pair, the score of a pilot for an English plane minus his or her score for a French plane. There is one tie. Hence we take the effective sample size n as $15 - 1$, or 14. There are 5 plus signs and 9 minus signs.

If there are too many plus signs our inclination will be to judge in favor of the English model. By the same token, too few plus signs (that is, a preponderance of minus signs) will lead us to lean in favor of the French model. Thus an appropriate test statistic for carrying out the test is

X = number of plus signs.

If the two models are not different, then a pilot is just as likely to decide in favor of an English model as in favor of the French model. That is, the probability of getting a + sign for a pair is the same as that of getting a − sign. Let p be the probability of getting a + sign, that is, p is the probability of a pilot saying that the English model is superior. The null hypothesis that the two models are identical in performance is then stated equivalently as

$$H_0: p = \frac{1}{2}.$$

If the French model is superior, then the probability of deciding in favor of the English model is less than that of deciding against it. In other words, the probability of getting a + sign for a pair is less than $\frac{1}{2}$. Thus the alternative hypothesis, the French model is superior, can be reformulated as

$$H_A: p < \frac{1}{2}.$$

In view of the alternative hypothesis, we would prefer to reject H_0 for small values of X. Thus the critical region consists of the left-tail of the distribution of X. Now, assuming that the responses of the pilots are independent, the distribution of X is binomial with n trials (the effective sample size after excluding the ties) and probability of success p. Under the null hypothesis, X has a binomial distribution with probability of success $\frac{1}{2}$. This distribution can be obtained from Appendix Table A-2 for appropriate n (14 in our example) and $p = \frac{1}{2}$. We reproduce the distribution below:

Successes x	$P(X = x)$		Successes x	$P(X = x)$
0	0.000		8	0.183
1	0.001		9	0.122
2	0.006		10	0.061
3	0.022		11	0.022
4	0.061		12	0.006
5	0.122		13	0.001
6	0.183		14	0.000
7	0.209			

$\begin{pmatrix} \text{Total} \\ \text{probability} \end{pmatrix} = 0.029$ (for $x = 0,1,2,3$)

$\begin{pmatrix} \text{Total} \\ \text{probability} \end{pmatrix} = 0.09$ (for $x = 0,1,2,3,4$)

From the distribution, $P(X \leq 4) = 0.09$. This probability is greater than 0.05 and so 4 is too large to be the critical value. On the other hand, $P(X \leq 3) = 0.029$ which is in between the commonly used significance levels of 0.05 and 0.01. Thus we agree to take the value 3 as the critical value and arrive at the following decision rule:

Reject H_0 if the observed number of + signs is less than or equal to 3.

In our data, there are 5 plus signs. Since this number is greater than 3 we do not reject H_0. There is no evidence to show that the French model is superior.

An alternative method: Since we are not in a position to provide a critical value corresponding to the exact level of significance α of 0.05, a better way to carry out the test is by providing the P-value. From our data, the observed number of plus signs x is 5, and

$$P(X \leq 5) = 0.001 + 0.006 + 0.022 + 0.061 + 0.122 = 0.212.$$

The P-value is 0.212 which is higher than $\alpha = 0.05$. Since the P-value is greater than the level of significance, we cannot reject H_0 at the 5 percent level.

EXAMPLE 1   In Table 13-2 are given the volumes of quarterly sales (in thousands of dollars) of 17 salespersons before and after a three-month course in marketing. Is the course beneficial in promoting sales?

TABLE 13-2			
Sales before and after a course in marketing (in thousands of dollars)			
Salesman	Sales after the course	Sales before the course	Sign of difference
1	6	3	+
2	9	8	+
3	12	7	+
4	8	10	−
5	8	6	+
6	7	7	←Tie
7	6	12	−
8	8	7	+
9	12	6	+
10	10	8	+
11	6	9	−
12	11	7	+
13	10	10	←Tie
14	10	7	+
15	9	4	+
16	8	5	+
17	11	7	+

SOLUTION   The null hypothesis H_0 and the alternative hypothesis H_A can be stated as follows:

H_0: The course in marketing does not affect sales.
H_A: The course improves sales.

The signs assigned to the differences in sales (sales after minus sales before) are given in column 4 of Table 13-2. There are two ties. Hence the effective sample size is $17 - 2$, or 15. There are 12 plus and 3 minus signs.

If p represents the probability of increased sales for a salesman after the training, then we are really interested in testing the null hypothesis

$$H_0: p = \frac{1}{2}$$

against the alternative hypothesis

$$H_A: p > \frac{1}{2}.$$

We will opt in favor of the training program and reject the null hypothesis if there is a preponderance of + signs. In other words, the critical region consists of values in the right tail of the distribution of X, the number of + signs. Assuming independence, the distribution of X is binomial with 15 trials and, under the null hypothesis, $p = \frac{1}{2}$. We reproduce its distribution below from Table A-2 corresponding to $n = 15$ (effective sample size) and $p = \frac{1}{2}$.

Successes	P(X = x)	Successes	P(X = x)
0	0.000	8	0.196
1	0.000	9	0.153
2	0.003	10	0.092
3	0.014	11	0.042
4	0.042	12	0.014
5	0.092	13	0.003
6	0.153	14	0.000	Total probability = 0.017
7	0.196	15	0.000

To test the hypothesis we shall first compute the P-value. Since the critical region consists of large values of X and since the observed number of plus signs is 12, the P-value is equal to

$$P(X \geqslant 12) = 0.014 + 0.003 = 0.017.$$

Since the P-value is less than 0.05 we reject the null hypothesis at the 5 percent level of significance. Thus at the 5 percent level of significance we conclude that the course is beneficial in promoting sales.   ▬

If n is large, even as low as 15, then a normal distribution with mean $\frac{n}{2}$ (or np with $p = \frac{1}{2}$) and variance $\frac{n}{4}$ (or $np(1 - p)$ with $p = \frac{1}{2}$) provides a reasonably good approximation for the distribution of X under H_0. Consequently, the test can be carried out using the statistic

$$\frac{X - \dfrac{n}{2}}{\sqrt{\dfrac{n}{4}}} = \frac{2X - n}{\sqrt{n}}.$$

The critical region is determined as usual on the basis of the level of significance and the nature of the alternative hypothesis.

Let us apply this approach in Example 1 where $n = 15$, which is large enough for a normal approximation. Since $n = 15$ and $x = 12$, the computed value of the test statistic is

$$\frac{2x - n}{\sqrt{n}} = \frac{2(12) - 15}{\sqrt{15}} = 2.324.$$

Since the computed value is greater than 1.645 ($z_{0.05}$), we reject the null hypothesis at the 5 percent level of significance. We arrive at the same conclusion as before.

In summary, use the binomial tables for n less than 15 and use the normal approximation for n greater than or equal to 15.

SECTION 13-1 EXERCISES

1. In a controlled experiment, a home economist tested a diet on 15 men. The following figures give their weights before and after the diet:

Weight before	138	158	160	172	160	210	185	245
Weight after	143	149	159	180	160	192	170	205

Weight before	182	172	132	250	180	129	176
Weight after	180	163	137	247	176	133	167

Use the sign test to test the null hypothesis that the diet is not effective in reducing a man's weight against the alternative hypothesis that it is effective in reducing weight. Use the 5 percent level of significance.

2. A student can take a certain course in statistics either as a lecture course or as a personalized system of instruction course (PSI), where the student sets his or her own pace. Fifteen instructors who have taught both the lecture course and the PSI course were asked to give their evaluation of the effectiveness of the two methods by assigning scores from 0 to 5. The following figures give the instructors' ratings:

Instructor	PSI	Lecture
A	4	1
B	2	5
C	4	3
D	4	2
E	3	4
F	3	1
G	0	4
H	2	4

Instructor	PSI	Lecture
I	4	0
J	3	3
K	1	0
L	4	2
M	5	3
N	5	0
O	4	1

Using the sign test, at the 5 percent level of significance, test the null hypothesis that there is no preference for one type of course over the other.

3. Under a special assistance program, 50 students were provided special assistance in their studies. At the end of the program, 36 showed a marked improvement, 4 did not show any change, and 10 showed a deterioration. Using the sign test, at the 2 percent level of significance, test the null hypothesis that the program is not beneficial.

4. Two presidential candidates, candidate A and candidate B, were rated by eighteen randomly picked voters on a scale of 0 to 5, as follows:

Candidate A	0	1	5	3	5	4	1	4	3	4	1	4	0	5	4	3	4	5
Candidate B	4	4	3	0	0	2	3	2	2	0	2	1	0	2	2	1	2	1

Use the sign test to test the hypothesis that there is no significant preference for one candidate over the other.

5. To study the effect of temperature on skiing, 16 skiers were clocked on a downhill course on two different days, once when the temperature was subzero and once when it was considerably warmer. The figures are given below:

Skier	Time (subzero) (in minutes)	Time (warm) (in minutes)
A	8.20	8.35
B	8.01	8.10
C	9.98	10.01
D	8.85	8.90
E	7.88	7.91
F	8.26	8.20
G	9.30	9.10
H	8.50	8.38
I	9.66	9.69
J	8.76	8.70
K	7.98	8.78
L	8.98	9.21
M	9.22	9.86
N	8.62	8.60
O	9.00	9.00
P	8.44	8.48

At the 5 percent level of significance, test the null hypothesis that temperature has no effect on skiing skill against the alternative hypothesis that it does have an effect.

6. The figures below give the systolic blood pressure of fourteen subjects before and after yogic meditation. At the 5 percent level of significance, does meditation help lower systolic blood pressure?

Subject	Before	After		Subject	Before	After
A	138	139		H	150	144
B	124	124		I	160	152
C	175	172		J	145	147
D	158	155		K	156	158
E	153	147		L	150	148
F	160	157		M	148	146
G	148	140		N	166	155

7. The following data represent two-hour blood glucose levels (in milligrams per milliliter) of 17 patients before and after a treatment:

Patient number	Blood glucose level Before	Blood glucose level After		Patient number	Blood glucose level Before	Blood glucose level After
1	174	168		9	194	180
2	157	159		10	144	135
3	135	130		11	108	110
4	102	105		12	224	198
5	144	140		13	192	180
6	132	134		14	187	189
7	131	122		15	137	139
8	112	112		16	181	173
				17	159	148

At the 5 percent level of significance, is it true that the treatment results in reducing blood glucose level?

8. The following figures give uric acid levels (in milligrams per 100 milliliters) of twelve subjects before and after a special diet for one week. Do the data indicate that the diet reduces the uric acid level of a subject?

Subject	Uric Acid Before	Uric Acid After	Subject	Uric Acid Before	Uric Acid After
A	5.2	5.2	G	5.9	5.5
B	6.3	6.2	H	6.0	6.1
C	6.4	6.3	I	6.1	6.1
D	5.5	5.6	J	5.7	5.8
E	5.8	5.4	K	5.9	5.6
F	5.7	5.6	L	6.3	6.4

9. The productivity of twelve workers was observed during one week when they were under strict supervision and during another week when they were not. The following scores provide a measure of their productivity:

Worker	With supervision	With no supervision
A	81	72
B	59	52
C	78	83
D	85	88
E	60	58
F	50	30
G	77	82
H	79	83
I	53	40
J	68	66
K	89	76
L	67	63

Do the data indicate that strict supervision affects productivity? Use $\alpha = 0.05$.

13-2 THE RANK-SUM TEST

The *rank-sum test* which we develop in this section is a procedure for comparing two populations when independent samples are drawn from them. It will be recalled that for comparing the means of two populations by using the t-test, we assumed that the populations were normally distributed, both of them having the same variance. In many instances, it may happen that the two populations under investigation have distributions which, even at best, are not approximately normal thereby invalidating the use of the standard t-test. Two nonparametric tests based on "rank-sums" were proposed in the 1940s for comparing two populations when there are indications that conditions necessary for applying the classical t-test are not met. One test, proposed by F. Wilcoxon in 1945, is called the **Wilcoxon rank-sum test** and the other, proposed by Mann and Whitney in 1947, is commonly referred to as the **Mann-Whitney U test.** Since these two tests are equivalent and lead to identical conclusions we shall develop only the Mann-Whitney U test in what follows.

To show how this test is applied, let us consider the data in Table 13-3, which refer to the charitable contributions (in dollars) made during a year by two samples of households picked at random from two communities that are in the same economic bracket.

We are interested in testing whether the two communities are similar with regard to helping charitable causes. We might consider comparing the means by using the t-test, as was done in Chapter 9. But a quick glance will point out that the data in Community B show considerably more variability than the data in Community A, thus making the assumption of equality of variances in the populations highly questionable. Besides, it seems that even the assumption of normality in the populations is open to question.

TABLE 13-3	
Contributions to charities (in dollars)	
Community A	*Community B*
120	42
155	190
88	320
420	680
360	82
60	130
650	500
82	32
	160
	82
	890
	1020

Under the rank-sum test, the data in the two samples are combined and then the observations are arranged in order of their size as though the entire data represent one sample. That is, they are arranged by rank. The smallest observation is assigned the rank 1, the next highest the rank 2, and so on, until all observations are ranked. This ranking is shown in Table 13-4. Whenever there are tied observations, we take the mean rank for these observations and assign this value to each of the tied observations. For instance, in our data the fourth, fifth, and sixth observations are

TABLE 13-4

The contributors in the two communities combined and ranked according to their contributions

Contributions (in dollars)	Community	Rank	
32	B	1	
42	B	2	
60	A	3	
82	B	5	⎫ Tied observations
82	B	5	⎬ all assigned the
82	A	5	⎭ rank 5.
88	A	7	
120	A	8	
130	B	9	
155	A	10	
160	B	11	
190	B	12	
320	B	13	
360	A	14	
420	A	15	
500	B	16	
650	A	17	
680	B	18	
890	B	19 ·	
1020	B	20	

all 82. Hence we assign the rank $\dfrac{4 + 5 + 6}{3} = 5$ to the three observations. As a result of the ranking, the households in Community A occupy ranks 3, 5, 7, 8, 10, 14, 15, and 17, while those in Community B occupy ranks 1, 2, 5, 5, 9, 11, 12, 13, 16, 18, 19, and 20.

The **rank sum** for each community is now found. The sum of the ranks occupied by the households in Community A is

$$3 + 5 + 7 + 8 + 10 + 14 + 15 + 17 = 79$$

and that in Community B is

$$1 + 2 + 5 + 5 + 9 + 11 + 12 + 13 + 16 + 18 + 19 + 20 = 131.$$

The null hypothesis under test is

H_0: The two communities have identical distributions of contributions to charities.

against the alternative hypothesis

H_A: The distributions are different.

We now formulate a test procedure to carry out the test.

Suppose m represents the size of one of the samples and R_1 the sum of the ranks assigned to its members. If n is the size of the other sample, then the test is based on the statistic

$$U = mn + \frac{m(m + 1)}{2} - R_1.$$

If the null hypothesis that the two populations are identical is valid, it can be shown that the distribution of U has mean μ_U and standard deviation σ_U given as follows:

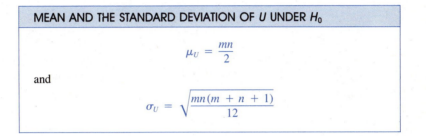

MEAN AND THE STANDARD DEVIATION OF U UNDER H_0

$$\mu_U = \frac{mn}{2}$$

and

$$\sigma_U = \sqrt{\frac{mn(m + n + 1)}{12}}$$

In our case, since $m = 8$, $n = 12$, and $R_1 = 79$, we get

$$u = (8)(12) + \frac{8(8 + 1)}{2} - 79 = 53$$

$$\mu_U = \frac{(8)(12)}{2} = 48$$

and

$$\sigma_U = \sqrt{\frac{(8)(12)(8 + 12 + 1)}{12}} = 12.96.$$

If, additionally, both m and n are greater than 8, the distribution of U is approximately normal. Changing to the z scale, we can then use the test statistic

$$\frac{U - \dfrac{mn}{2}}{\sqrt{\dfrac{mn(m + n + 1)}{12}}}.$$

Suppose we agree to carry out the test at the 5 percent level. In our example, the alternative hypothesis is two-sided. Hence, using a two-tailed test, we will reject the null hypothesis if the computed value is less than $-z_{\alpha/2}$ or greater than $z_{\alpha/2}$. Since $u = 53$, $\mu_U = 48$, and $\sigma_U = 12.96$, we get

$$\frac{u - \dfrac{mn}{2}}{\sqrt{\dfrac{mn(m + n + 1)}{12}}} = \frac{53 - 48}{12.96} = 0.39.$$

We do not reject the null hypothesis at the 5 percent level of significance. In other words, there is no evidence to indicate that the communities are dissimilar with regard to their contributions to charity.

EXAMPLE 1   A class of 18 students was divided into two groups, Group I of 8 students and Group II of 10 students. Group I was given special training using audiovisual facilities, whereas in Group II no such facilities were used. A common test was given to the two groups and their scores are as recorded in Table 13-5. Would it be reasonable to say that the two instructional methods produce different results? (Use $\alpha = 0.05$.)

TABLE 13-5	
Scores of two groups of students	
Group I	*Group II*
95	49
66	39
42	80
82	65
71	30
68	62
52	53
86	49
	55
	60

TABLE 13-6		
Students of the two groups combined and ranked according to their scores		
Score	*Group*	*Rank*
30	II	1
39	II	2
42	I	3
49	II	4.5 ⎤ Tied
49	II	4.5 ⎦ observations
52	I	6
53	II	7
55	II	8
60	II	9
62	II	10
65	II	11
66	I	12
68	I	13
71	I	14
80	II	15
82	I	16
86	I	17
95	I	18

SOLUTION   The null hypothesis is

H_0: The two methods produce identical results.

We shall test this against the alternative hypothesis that the two methods do not produce the same results at the 5 percent level of significance. The combined data of the two samples are arranged according to magnitude in Table 13-6. R_1, the sum of the ranks occupied by Group I, is

$$3 + 6 + 12 + 13 + 14 + 16 + 17 + 18 = 99.$$

Now $m = 8$ and $n = 10$. Therefore,

$$u = (8)(10) + \frac{(8)(9)}{2} - 99 = 17$$

$$\mu_U = \frac{(8)(10)}{2} = 40$$

and

$$\sigma_U = \sqrt{\frac{(8)(10)(8 + 10 + 1)}{12}} = 11.25.$$

Hence

$$\frac{u - \mu_U}{\sigma_U} = \frac{17 - 40}{11.25} = -2.04.$$

At the 5 percent level of significance, we reject the null hypothesis, since the computed value is less than -1.96. We conclude that there is evidence that the two instructional methods produce different results and, as a matter of fact, that students will benefit by the use of audiovisual facilities. ▄▄▄

SECTION 13-2 EXERCISES

1. Ten voters were picked at random from those who voted in favor of a certain proposition, and twelve from those who voted against it. The following figures give their ages:

In favor	28	33	27	31	29	25	58	30	25	41	
Against	31	43	45	37	40	41	48	35	33	39	42	36

Using the rank-sum test, test the null hypothesis that the age distributions of those voting in favor and those voting against the proposition were identical. Use a 5 percent level of significance.

2. The following figures relate to the number of hours needed by two groups of workers to learn some skills:

Group I	7.2	6.6	7.7	6.9	7.4	8.0	6.5	8.1	7.5	7.1
Group II	7.4	7.9	6.2	8.2	7.8	8.6	7.6	7.1	7.0	6.7

Use the rank-sum test to test the null hypothesis that the two samples come from identical populations. Use a 0.02 level of significance.

3. Twelve students who took a course under a lecture session program and fifteen students who took it under a self-pace program were given a common final test. The following figures give the scores of the students:

Lecture	46	89	85	61	87	35	57	87	76	79	49	92		
Self-pace	92	41	36	72	34	67	99	72	83	82	76	40	84	67	24

Use the rank-sum test to test the null hypothesis that the two methods of instruction do not produce significantly different results.

4. In the first grade in a school, the teacher divided the class into two groups, Group A, consisting of 10 students, and Group B, consisting of 14 students. Students in each group were asked to memorize, in a one-hour period, spellings of forty words from a given list. Sudents in Group A, however, were secretly promised an incentive that any student in their group would get as many pieces of candy as the number of words spelled correctly. No similar promise was made to students in Group B. The following figures refer to the number of words spelled correctly:

Group A	27	22	31	16	39	36	25	24	32	19				
Group B	28	11	32	35	6	24	14	16	39	19	26	20	10	40

Using the rank-sum test, at the 0.05 level of significance, test the null hypothesis that the incentive is ineffective.

5. Horses of two breeds, Breed A and Breed B, were entered in a race. They finished in the following order:

Standings	1	2	3	4	5	6	7	8	9	10
Breed	A	B	B	A	A	A	A	B	A	A

Standings	11	12	13	14	15	16	17	18	19
Breed	A	B	A	B	B	A	B	B	B

Using the rank-sum test, test the null hypothesis that there is no difference between the two breeds of racing horses. Use the 0.05 level of significance.

13-3 THE RUNS TEST

In order to get a good cross section of the population, we require that the sample drawn be a random sample. We have relied heavily on this premise and have often initiated our discussion by saying, "Suppose we have a random sample . . . ." Intuitively, the concept of drawing a random sample seems rather easy, and as a result, we have not given much consideration to it. We shall now address the important question of testing the randomness of a sample.

In recent years, several tests for randomness have been developed. The one presented here is based on the order in which the observations are obtained and is called a **runs test.** We begin with the definition of a run.

If there is a sequence of symbols of two kinds, then a **run** is defined as a maximum-length unbroken subsequence of identical symbols.

For example, consider the following sequence of letters d and r where we have drawn slashes to separate each block of consecutive identical letters:

$$d\,d\,d\,d\,/\,r\,r\,r\,/\,d\,/\,r\,r\,/\,d\,d\,d\,/\,r\,r$$

There are 6 runs, with 3 runs of r's and 3 of d's.

If in a sample drawn from a population containing letters d and r we obtained any of the following sequences

$d\,d\,d\,.\,.\,.\,d$, or

$r\,r\,r\,.\,.\,.\,r$, or

$d\,d\,.\,.\,.\,d\,r\,r\,.\,.\,.\,r$, or

$d\,r\,d\,r\,.\,.\,.\,d\,r\,d\,r$

we would immediately question their randomness. Thus too few runs or too many of them would clearly be indicative of a lack of randomness. We shall now provide a test when the case is not as clear-cut as above.

As a specific example, suppose 25 people are interviewed as to whether they are Democrats (recorded d) or Republicans (recorded r). Their responses (in the order in which they were interviewed) are as follows:

$$d\,d\,d\,/\,r\,r\,/\,d\,/\,r\,r\,/\,d\,d\,d\,/\,r\,r\,r\,r\,/\,d\,d\,d\,/\,r\,/\,d\,d\,d\,/\,r\,/\,d\,d$$

Counting the different blocks, we see that there are 11 runs. Also, there are 15 letters d and 10 letters r. We want to test the null hypothesis

H_0: The sequence is random

against the alternative hypothesis

H_A: The sequence is not random

using, for example, $\alpha = 0.05$.

The test of hypothesis regarding randomness is based on the distribution of runs. Suppose R denotes the total number of runs, m the number of symbols of one kind, and n the number of the other kind. It can be shown that the mean μ_R and the standard deviation σ_R of R are given as follows:

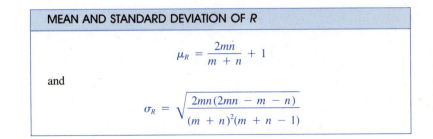

MEAN AND STANDARD DEVIATION OF R

$$\mu_R = \frac{2mn}{m+n} + 1$$

and

$$\sigma_R = \sqrt{\frac{2mn(2mn - m - n)}{(m+n)^2(m+n-1)}}$$

In our illustration, $m = 15$ and $n = 10$. (Since the formulas for μ_R and σ_R are symmetric in m and n, it would not matter if we took $m = 10$ and $n = 15$.) Hence

$$\mu_R = \frac{2(15)(10)}{15 + 10} + 1 = 13$$

and

$$\sigma_R = \sqrt{\frac{2(15)(10)[2(15)(10) - 15 - 10]}{(15 + 10)^2(15 + 10 - 1)}}$$
$$= \sqrt{5.5}$$
$$= 2.345$$

Now if m and n are large (both greater than 10), the distribution of R can be approximated by a normal distribution with mean μ_R and standard deviation σ_R defined above. The decision criterion is then provided by computing the value of

$$\frac{R - \mu_R}{\sigma_R} = \left[R - \left(\frac{2mn}{m + n} + 1 \right) \right] \Bigg/ \sqrt{\frac{2mn(2mn - m - n)}{(m + n)^2(m + n - 1)}}$$

where the distribution of the statistic is, approximately, the standard normal distribution.

In our example, $R = 11$, $\mu_R = 13$, and $\sigma_R = 2.345$. Hence the computed value of $(R - \mu_R)/\sigma_R$ is

$$\frac{11 - 13}{2.345} = -0.85$$

The alternative hypothesis is that there is a lack of randomness. Hence with a two-tailed test, we will reject the null hypothesis if the computed value is less than $-z_{\alpha/2}$ or greater than $z_{\alpha/2}$. Taking $\alpha = 0.05$, since $z_{0.025} = 1.96$, we see that there is no evidence showing a lack of randomness.

EXAMPLE 1   Lewis believes that his statistics instructor's lectures have a pattern to their coherence. He records whether the lecture is understandable on 25 consecutive class days, recording g if he understands the lecture and b if he does not. The following sequence gives the results.

$$g\,g\,g\,/\,b\,b\,/\,g\,/\,b\,/\,g\,g\,g\,/\,b\,/\,g\,/\,b\,b\,/\,g\,g\,/\,b\,/\,g\,g\,g\,/\,b\,b\,b\,/\,g\,/\,b$$

Using $\alpha = 0.05$, test for randomness.

SOLUTION   We have

H_0: There is randomness.

H_A: There is lack of randomness.

Separating runs by slashes, we see that there are 14 runs. There are 14 letters g and 11 letters b. Thus $m = 14$, $n = 11$, and $R = 14$. As a result,

$$\mu_R = \frac{2(14)(11)}{14 + 11} + 1$$

$$= 13.32$$

and

$$\sigma_R = \sqrt{\frac{2(14)(11)[2(14)(11) - 14 - 11]}{(14 + 11)^2(14 + 11 - 1)}}$$

$$= \sqrt{5.811}$$

$$= 2.41.$$

Therefore, the computed value of $(R - \mu_R)/\sigma_R$ is

$$\frac{14 - 13.32}{2.41} = 0.282$$

There is no evidence to show that the coherence of the lectures has any pattern, since the computed value is between -1.96 and 1.96.

The runs test can also be applied to check randomness of a sequence of numerical data. This can be accomplished by converting the original data into a sequence of symbols by writing the letter a, say, if a value is above the median and the letter b if it is below the median. If a value is equal to the median, it is deleted from consideration. This done, we proceed as before to check the randomness of the sequence of letters a and b.

EXAMPLE 2   Suppose the annual rainfall (in inches) over 24 consecutive years is given as follows:

38, 29, 56, 41, 43, 37, 57, 46, 62, 38, 54, 59,
60, 36, 62, 43, 57, 40, 46, 35, 62, 46, 52, 55

Test for randomness at the 5 percent level of significance.

SOLUTION   We state H_0 and H_A as

H_0: There is randomness in the sequence.
H_A: There is no randomness.

Arranging the data in order of magnitude as

29, 35, 36, 37, 38, 38, 40, 41, 43, 43, 46, 46,
46, 52, 54, 55, 56, 57, 57, 59, 60, 62, 62, 62

it can be seen that the median is 46. In the original data (taken in that order), if the value is above the median, we replace it by the letter a, and if it is below the median, we replace it by the letter b. The three observations equal to 46 are deleted, since 46 is the median. The following sequence of letters is obtained as a result:

$$b\ b\ /\ a\ /\ b\ b\ b\ /\ a\ a\ /\ b\ /\ a\ a\ a\ /\ b\ /\ a\ /\ b\ /\ a\ /\ b\ b\ /\ a\ a\ a$$

There are 10 b's, 11 a's, and 12 runs; that is, $m = 10$, $n = 11$, and $R = 12$. Hence

$$\mu_R = \frac{2(10)(11)}{10 + 11} + 1$$

$$= 11.48$$

$$\sigma_R = \sqrt{\frac{2(10)(11)[2(10)(11) - 10 - 11]}{(10 + 11)^2(10 + 11 - 1)}}$$

$$= \sqrt{4.96}$$

$$= 2.23$$

and the computed value of $(R - \mu_R)/\sigma_R$ is

$$\frac{12 - 11.48}{2.23} = 0.23.$$

Since the computed value is so close to zero, there is hardly any evidence to question randomness. ■

SECTION 13-3 EXERCISES

1. A true-false test consisting of 40 questions had the following sequence of answers (recorded T for true and F for false):

   F T T F T F T F F T T T F T F F T F T F
   T T F T F T F F T F T F T T T F F T F F

   Test the hypothesis that the questions are arranged such that the sequence of answers $(T$–$F)$ is random. Use $\alpha = 0.05$.

2. A customer who regularly visits a particular restaurant orders either a bowl of soup or a tossed salad. The waitress recorded the orders (B for bowl of soup and T for tossed salad) over thirty consecutive days:

   T T B T T T T B B T T T B T B
   B T T T B T T B T B T T T B B

   At the 5 percent level of significance, test the null hypothesis that there is randomness.

3. At the ticket counter in a movie house, the following standing order in a queue was observed (recorded M for male, F for female):

   M   F   M   M   F   M   M   M   F   F   M   F   F
   M   M   F   F   F   M   F   M   F   M   M   M

   At the 5 percent level, test whether the members of the two sexes were standing in random order.

4. In a quality-control inspection, forty successive tubes produced by a machine are inspected as to whether they are defective (d) or nondefective (g). The following data were obtained:

   g   g   g   g   g   d   d   g   g   g   g   g   g   g   g   g   d   g   d   g   g
   g   g   d   g   g   g   d   d   g   g   g   d   g   g   g   g   g   g   d   d   g

   Test the hypothesis that there was no particular trend in which the defective tubes occurred. Use $\alpha = 0.025$.

5. A copper wire coated with enamel was inspected, and the following figures give the number of defects on 30 consecutive pieces, each of length 1000 yards:

   9   1   11   8   9   3   12   9   14   8
   10   12   15   7   15   10   13   6   10   8
   12   9   15   6   14   16   12   4   10   13

   At the 5 percent level of significance, test for randomness of the number of defects.

6. A volunteer helping in a hospital spent the following number of hours on 30 consecutive days:

   4   6   5   4   6   2   1   5   4   6   3   4   3   2   7
   4   6   4   5   1   3   5   4   3   6   8   3   2   6   5

   Test for randomness of the number of hours spent, at the 5 percent level of significance.

7. Recording D if there was a drought during a year and N if the precipitation was normal, the following arrangement indicates the nature of precipitation during 50 consecutive years from 1936 to 1985:

   N   N   D   N   D   N   N   D   N   D   N   N   N
   D   N   N   D   N   D   N   D   N   N   N   D   N
   N   D   D   N   D   N   N   D   N   N   D   N   D
   N   D   N   N   D   N   D   N   N   N   D

   Use the 5 percent level of significance to test whether the occurrence of drought years and normal years may be regarded as random.

8. The following sentence, "*Then, his voice thick with emotion, he echoed one of his most familiar campaign themes,*" was obtained from a composition. At the 5 percent level of significance, test whether consonants and vowels occur at random.

9. Pick a column from the random numbers table (Appendix II, Table A-8) and thus obtain 50 successive digits one below the other. Test for randomness of the digits using $\alpha = 0.05$.

10. The following figures represent the number of migratory birds observed in a certain marshland over 25 consecutive years:

248   510   410   380   310   540   395   480   360
330   501   489   380   398   440   370   488   362
620   405   378   260   440   524   320

At the 5 percent level of significance, test for randomness of the number of birds.

13-4 RANK CORRELATION

The reader is already familiar with the concept of coefficient of correlation. It will be recalled that to carry out tests regarding the population coefficient of correlation, one of the assumptions is that the two variables involved have a joint normal distribution. When this assumption is not met, we can often use the rank correlation coefficient, a statistic developed by C. Spearman in 1904, to test if there is an association between two variables. This nonparametric method can be used to test an association between any two numerical variables x and y measured on each item by ranking the x values and y values in order of their magnitude (see Example 3 later). However, it is most useful to analyze the relationship between two variables that cannot be expressed by exact measurement but from which ranked data can be obtained. The ranking procedure might be based on some qualitative factor such as taste, appearance, or some other criterion. For example, suppose two judges are asked to judge a beauty contest in which seven contestants have entered. Each judge ranks the contestants from 1 to 7. Consider the rankings as given in Table 13-7.

TABLE 13-7	
Rankings in a beauty contest by two judges	
Judge 1 *x*	*Judge 2* *y*
2	4
7	6
1	1
3	3
5	5
6	2
4	7

The **rank correlation coefficient,** which we shall denote as r_{rank}, is so named because it is a correlation coefficient between two variables where, it turns out, the variables are expressed as ranks. We could compute it by employing the formula in Section 11-1. However, a quicker method is to use the following formula:

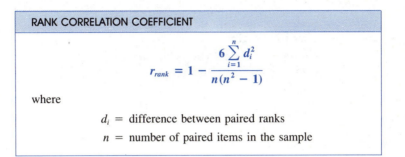

RANK CORRELATION COEFFICIENT

$$r_{rank} = 1 - \frac{6 \sum\limits_{i=1}^{n} d_i^2}{n(n^2 - 1)}$$

where

$$d_i = \text{difference between paired ranks}$$
$$n = \text{number of paired items in the sample}$$

The computations for finding r_{rank} for the data in Table 13-7 are presented in Table 13-8. Since $n = 7$ and $\sum\limits_{i=1}^{7} d_i^2 = 30$, substituting in the formula, we get

$$r_{rank} = 1 - \frac{6(30)}{7(7^2 - 1)}$$
$$= 0.464.$$

TABLE 13-8			
Computations for finding r_{rank} for Table 13-7			
Judge 1 *x*	*Judge 2* *y*	*d* *x − y*	*d²*
2	4	−2	4
7	6	1	1
1	1	0	0
3	3	0	0
5	5	0	0
6	2	4	16
4	7	−3	9
Total			30

TABLE 13-9		
Perfect agreement in the rankings of the two judges		
Judge 1 *x*	*Judge 2* *y*	*d* *x − y*
2	2	0
7	7	0
1	1	0
3	3	0
5	5	0
6	6	0
4	4	0

Next, suppose there is a perfect agreement between the rankings of the two judges, as shown in Table 13-9. In this case $\sum\limits_{i=1}^{7} d_i^2 = 0$ and we get

$$r_{rank} = 1 - \frac{6(0)}{7(7^2 - 1)} = 1$$

At the other extreme, assume that the ranks assigned by the judges are exactly inverse, as in Table 13-10. An individual who gets the highest score from one judge

TABLE 13-10

Assignments of ranks by the two judges are inverse

Judge 1 x	Judge 2 y	$x - y$	d^2
2	6	−4	16
7	1	6	36
1	7	−6	36
3	5	−2	4
5	3	2	4
6	2	4	16
4	4	0	0
Total			112

gets the lowest score from the other, and so on. Substituting in the formula,

$$r_{rank} = 1 - \frac{6(112)}{7(7^2 - 1)}$$

$$= -1$$

In summary, the rank correlation coefficient, as would be expected, ranges between −1 and 1, attaining the value of 1 when there is a perfect agreement and a value of −1 when there is an inverse agreement. Thus

$$-1 \leq r_{rank} \leq 1$$

EXAMPLE 1   Ten applicants applying for a professor's position at a university are ranked by the chairman of the department and the dean. The results are shown in Table 13-11 (along with other computations). Find the rank correlation coefficient.

SOLUTION   From Table 13-11, $\sum_{i=1}^{10} d_i^2 = 96$. Since $n = 10$, we get

$$r_{rank} = 1 - \frac{6(96)}{10(10^2 - 1)}$$

$$= 0.42. \quad \blacksquare$$

TABLE 13-11				
Rankings of applicants by the chairman and the dean, along with other computations				
Applicant	*Ranking by the chairman* x	*Ranking by the dean* y	d	d^2
A	4	9	-5	25
B	3	5	-2	4
C	6	10	-4	16
D	7	6	1	1
E	10	8	2	4
F	5	1	4	16
G	1	3	-2	4
H	8	7	1	1
I	9	4	5	25
J	2	2	0	0
Total				96

If the sample size is moderately large (even as low as 10), it can be shown that the sampling distribution of r_{rank} is approximately normal with a mean equal to the population rank correlation coefficient, ρ_{rank}, and the standard deviation $\sigma_{r_{rank}}$ given as follows:

STANDARD DEVIATION OF r_{rank}

$$\sigma_{r_{rank}} = \frac{1}{\sqrt{n-1}} .$$

Thus the test statistic employed is

$$\frac{r_{rank} - \rho_{rank}}{\sigma_{r_{rank}}}$$

and its distribution is approximately standard normal. Upon simplification, this test statistic reduces to

$$\sqrt{n-1}(r_{rank} - \rho_{rank}).$$

The procedure for testing a hypothesis is given in the following example.

EXAMPLE 2   Using the data given in Example 1, test the hypothesis that there is no relation between the rankings given 10 applicants by the chairman of the department and the dean.

SOLUTION   We want to test the null hypothesis that the rankings of the dean and the chairman are not related, that is, $\rho_{rank} = 0$. Actually, the statement is that the rankings are randomly matched. The alternative hypothesis is that $\rho_{rank} \neq 0$. Thus we have

H_0: Rankings of the dean and the chairman are not related.

H_A: Rankings are related.

We therefore compute

$$\sqrt{n-1}(r_{rank} - 0) = \sqrt{n-1}(r_{rank}).$$

In our example, $r_{rank} = 0.42$ and $\sqrt{n-1} = \sqrt{9} = 3$. Hence

$$\sqrt{n-1}(r_{rank}) = 3(0.42) = 1.26.$$

Since the computed value is between -1.96 and 1.96, we do not reject the null hypothesis at the 5 percent level. There is no evidence to indicate that the rankings of the chairman and the dean are related. ▪

EXAMPLE 3   Two wire services, International and Global, ranked 12 football teams as shown in Table 13-12. Test the hypothesis that $\rho_{rank} = 0$, at the 5 percent level of significance.

SOLUTION   The necessary computations are shown in Table 13-12. Since $n = 12$ and $\sum\limits_{i=1}^{8} d_i^2 = 28$, we get

$$r_{rank} = 1 - \frac{6(28)}{12(12^2 - 1)}$$

$$= 0.9.$$

TABLE 13-12

Rankings of twelve football teams by two wire services, and other computations

Team	Global x	International y	d	d^2
A	7	8	-1	1
B	1	3	-2	4
C	3	2	1	1
D	8	7	1	1
E	6	5	1	1
F	5	4	1	1
G	4	6	-2	4
H	2	1	1	1
I	10	9	1	1
J	9	12	-3	9
K	11	11	0	0
L	12	10	2	4
Total				28

We want to test

$$H_0: \rho_{rank} = 0 \quad \text{against} \quad H_A: \rho_{rank} \neq 0.$$

We therefore compute

$$\sqrt{n-1}(r_{rank}) = \sqrt{12-1}(0.9) = 2.98.$$

Since the computed value is greater than 1.96, we reject the null hypothesis at the 5 percent level and conclude that the rankings of the two wire services are related. ▬

In the next example we shall compute Spearman's rank correlation when the data consist of actual numerical values x and y.

EXAMPLE 4   A physical education department has obtained the following data on heights and weights of 12 athletes, as given in Table 13-13. Find the rank correlation and test whether the height and weight of an athlete are related.

TABLE 13-13

Heights and weights of 12 athletes

Athlete	Height (in inches) x	Weight (in pounds) y
A	68	152
B	64	162
C	72	182
D	67	197
E	71	149
F	73	224
G	70	184
H	69	159
I	66	167
J	74	230
K	76	218
L	65	146

SOLUTION   As a first step, we replace the x values by their ranks assigned in order of increasing magnitude, the smallest value receiving a rank of 1 and the largest a rank of 12. The same is done with the y values. The resultant rankings and pairings are shown in Table 13-14. Since $n = 12$, we get

$$r_{rank} = 1 - \frac{6(106)}{12(12^2 - 1)}$$

$$= 0.63.$$

TABLE 13-14

Heights and weights replaced by ranks

Athlete	Ranks assigned to heights	Ranks assigned to weights	d	d^2
A	5	3	2	4
B	1	5	−4	16
C	9	7	2	4
D	4	9	−5	25
E	8	2	6	36
F	10	11	−1	1
G	7	8	−1	1
H	6	4	2	4
I	3	6	−3	9
J	11	12	−1	1
K	12	10	2	4
L	2	1	1	1
Total				106

The null hypothesis is that the height and weight of an athlete are uncorrelated in the population. Now

$$\sqrt{n-1}(r_{rank}) = \sqrt{11}(0.63) = 2.09.$$

At the 5 percent level, the value is significant. We reject the null hypothesis, concluding that height and weight are correlated. ■

SECTION 13-4 EXERCISES

1. In a 1000-meter race in which 11 mothers and a teenage daughter of each participated, the following standings were observed among mothers and daughters. Find the rank correlation and test the null hypothesis that it is zero in the population. Use $\alpha = 0.05$.

Family name	Mother's rank	Daughter's rank
Tucker	2	4
Moore	5	7
Jackson	7	6
Howard	3	5
Alvarez	1	1
Fletcher	8	10
Harper	4	3
Chen	6	2
Price	11	8
Salerno	10	9
Mason	9	11

2. The following table gives the standings of 12 tennis pros at the end of six tournaments in the United States and at the end of as many tournaments in Europe:

Player	Standing in Europe	Standing in U.S.
Rod	3	1
Jimmy	2	3
Ken	1	9
Raul	4	5
Ille	5	2
Manuel	7	8
Vijay	10	6
Tony	12	7
John	6	10
Sandy	9	11
Mark	11	12
Arthur	8	4

Find the rank correlation coefficient. Using $\alpha = 0.05$, test the null hypothesis that standings in Europe and the United States are not related.

3. Ten vice-presidents in a company were ranked according to their affability and the level of competence. Find the rank correlation coefficient and test the null hypothesis that affability and competence are not related. Use $\alpha = 0.025$.

Name	Affability rank	Competence rank
Maroney	5	3
Jorgenson	8	1
Kennard	2	2
Ozawa	1	7
Meyer	6	4
Keating	3	10
Purcell	7	5
Ramirez	4	6
Simmons	10	8
Chen	9	9

4. The scores of 11 students on the midterm exam and the final exam were as follows:

Student	Midterm score	Final exam score
Rita	82	94
Tom	81	93
Hank	68	74
Mary	78	81
Pat	92	96
John	76	67
Jeff	54	53
Ron	86	89
Elaine	90	92
Jan	62	45
Dennis	52	61

Compute the rank correlation coefficient and test the hypothesis that it is zero in the population. Use $\alpha = 0.1$.

5. In the table below are listed 14 comparable cars according to their Environmental Protection Agency (EPA) rating based on gasoline mileage and their versatility:

Car make	Mileage	Versatility
A	20.0	8
B	23.0	5
C	28.0	7
D	27.0	3
E	24.0	10
F	25.5	9
G	30.0	4
H	22.5	6
I	26.0	1
J	31.0	2
K	29.0	14
L	26.5	11
M	28.5	13
N	23.5	12

Compute the rank correlation coefficient, and at the 5 percent level, test the null hypothesis that it is zero in the population against the alternate hypothesis that it is not zero.

6. The following figures give the IQ's of father and son for twelve families. Calculate the rank-correlation coefficient. At the 5 percent level of significance, is there an association between father's IQ and son's IQ?

Family	Father	Son
Jones	110	130
McKie	140	120
Lutnetsky	138	122
Wong	166	150
Turnbull	120	140
Howard	128	116
Patil	142	156
Salazar	118	94
Tatum	95	125
Adler	150	133
Bentsen	108	100
Conklin	133	110

KEY TERMS AND EXPRESSIONS

parametric tests   rank
nonparametric tests   rank sum
distribution-free tests   runs test
two-sample sign test   run
Wilcoxon rank-sum test   rank correlation coefficient
Mann-Whitney U test

KEY FORMULAS

Sign test $X = $ number of plus signs

Under H_0, X has binomial distribution with $p = \frac{1}{2}$. If $n \geq 15$, use the test statistic $\dfrac{2X - n}{\sqrt{n}}$ whose distribution is approximately standard normal.

Mann-Whitney U test $$U = mn + \frac{m(m + 1)}{2} - R_1$$

where m is the size of one of the samples, R_1 is the sum of the ranks assigned to its members, and n is the size of the other sample.

If m and n are both greater than 8, then the distribution of the following statistic is approximately standard normal:

$$\frac{U - \dfrac{mn}{2}}{\sqrt{\dfrac{mn(m + n + 1)}{12}}}$$

The runs test   If R is the number of runs, m is the number of symbols of one kind, n the number of the other kind, and if m and n are both greater than 10, then the distribution of

$$\left[R - \left(\frac{2mn}{m + n} + 1 \right) \right] \Big/ \sqrt{\frac{2mn(2mn - m - n)}{(m + n)^2(m + n - 1)}}$$

is approximately standard normal.

Rank correlation $$r_{rank} = 1 - \frac{6 \sum_{i=1}^{n} d_i^2}{n(n^2 - 1)}$$

where

$d_i = $ difference between paired ranks
$n = $ number of paired items.

If n is greater than 10, the statistic $\sqrt{n - 1}(r_{rank} - \rho_{rank})$ has a distribution which is approximately standard normal.

CHAPTER 13 TEST

1. Discuss the advantages and disadvantages of nonparametric tests over parametric tests.

2. When would you use the rank-sum test in preference to the classical t test?

3. What would you conclude about the randomness of a sequence in the following instances?

   (a) There were too many runs.   (b) There were too few runs.

4. Suppose we have 20 letters m and 15 letters n and they are arranged in a random sequence. If R represents the number of runs, find the mean number of runs and the variance of the number. What is the approximate distribution of R?

5. A construction company wants to compare two mixes used in manufacturing structural beams. The figures below represent the strengths of 22 experimental beams (in pounds per square inch):

Mix 1	Mix 2
996	1135
1035	860
1002	990
986	1200
895	870
1085	976
1110	1009
985	1182
1030	886
960	1090
	1230
	940

   At the 5 percent level of significance, does one mix give stronger beams than the other? Use the rank-sum test.

6. A laboratory experiment was conducted to test whether a diuretic agent was effective in inducing more urine in dogs. The figures below give a one-day collection of urine (in cubic centimeters) from 13 dogs of a certain breed before and after the diuretic agent was administered:

Dog number	Urine before	Urine after
1	980	1200
2	1600	1480
3	850	850
4	1700	1500
5	960	1040
6	1010	980
7	860	1010
8	1850	1900
9	980	1060
10	1030	990
11	1400	1400
12	750	880
13	1130	1040

   Use the sign test at the 5 percent level of significance to test the null hypothesis that the diuretic agent is not effective.

7. The written part of a driver's test consists of 30 questions, each with two choices. The following data give the correct choice (sequentially) for each of the questions (recorded 1 if the first is correct, 2 if the second is correct):

1 2 1 2 2 2 1 2 2 1 2 1 2 1 2
1 2 2 2 1 2 1 2 1 2 2 1 1 2 1

At the 5 percent level of significance, test the null hypothesis that the choices are arranged at random.

8. The following table gives the rankings of 10 players based on their ability in tennis and Ping-Pong. Calculate the rank correlation coefficient for the data. At $\alpha = 0.01$, test the null hypothesis of no correlation.

Player	Rank in tennis	Rank in Ping-Pong
1	4	7
2	2	4
3	3	2
4	1	8
5	8	3
6	9	5
7	7	6
8	6	9
9	5	1
10	10	10

SOLUTIONS TO ODD-NUMBERED EXERCISES*

CHAPTER 1

Section 1-1

1. (a) quantitative   (b) qualitative   (c) qualitative   (j) qualitative

    (d) quantitative   (e) quantitative   (f) qualitative

    (g) quantitative   (h) qualitative   (i) quantitative

3. (a) continuous   (b) discrete   (c) continuous   (j) discrete

    (d) discrete   (e) discrete   (f) continuous

    (g) continuous   (h) continuous   (i) continuous

5. The totality of all the tires conceivably manufactured by the plant. Infinite, if the plant is in operation forever; otherwise finite.

7. (a) Governor, Senator.   (b) President.

9. (a) 64.85   (b) 8.31   (c) 56.54   (d) 13

Section 1-2

1.

Weight	Frequency
15.6	6
15.8	4
15.9	5
16.0	6
16.2	9

3.

Class	Frequency
5.00–9.99	7
10.00–14.99	8
15.00–19.99	7
20.00–24.99	3
25.00–29.99	2
30.00–34.99	1

7. (a) 10   (f) 23   (In parts (b), (c), (d), and (e) no answers possible.)

9. (a) 8, 13, 18, 23, 28, 33

    (b) 5.5, 10.5, 15.5, 20.5, 25.5, 30.5, 35.5

    (c) 5

11. (a) 14.05, 18.05, 24.05, 30.05, 36.05, 44.05

    (b) 12.05, 16.05, 20.05, 28.05, 32.05, 40.05, 48.05

13. (a) During most months the earnings were less than 1900 dollars; rare months when the earnings were much higher.

    (b) 999.5, 1199.5, 1399.5, 1699.5, 2099.5, 2499.5, 3000

    (c) 899.5, 1099.5, 1299.5, 1499.5, 1899.5, 2299.5, 2699.5, 3300.5.

    (d) The longest class is 2700–3300 with length 601.

    (e) 1300–1499.

15. (a) 387.5, 412.5, 437.5, 462.5, 487.5, 512.5, 537.5

    (b) 25

    (c) 388–412, 413–437, 438–462, 463–487, 488–512, 513–537.

Section 1-3

1.

Relative frequency	Central angle (in degrees)
0.274	98.7
0.503	181.2
0.223	80.1

3.

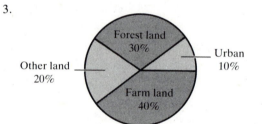

5.

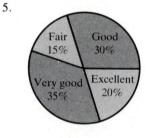

7.

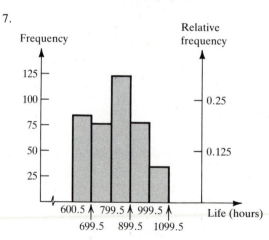

11. (a) 49   (b) 46   (c) 94   (d) 48

13.

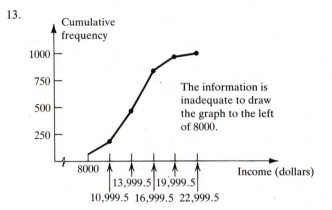

The information is inadequate to draw the graph to the left of 8000.

15. (a)

Speed	Cumulative frequency
less than 47.5	0
less than 50.5	12
less than 53.5	44
less than 56.5	94
less than 59.5	179
less than 62.5	194
less than 65.5	200

(b) Frequency

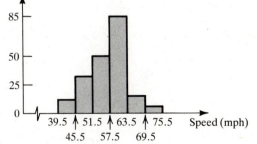

(c)

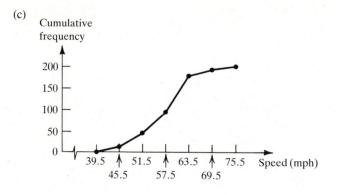

Chapter 1 Test

2. (a) quantitative; discrete   (b) qualitative

   (c) quantitative; discrete   (d) quantitative; continuous

   (e) quantitative; continuous   (f) quantitative; discrete

   (g) quantitative; discrete   (h) quantitative; continuous

   (i) quantitative; discrete   (j) quantitative; discrete

   (k) quantitative; discrete   (l) qualitative

   (m) qualitative

3.

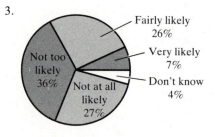

4. (a)

Class	Frequency
2.30–2.38	3
2.39–2.47	8
2.48–2.56	3
2.57–2.65	7
2.66–2.74	8
2.75–2.83	1

(b)

Class mark	Boundary lower	Boundary upper
2.34	2.295	2.385
2.43	2.385	2.475
2.52	2.475	2.565
2.61	2.565	2.655
2.70	2.655	2.745
2.79	2.745	2.835

(c)

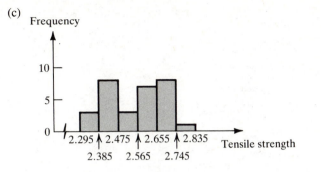

(d) Cumulative frequency distribution:

Tensile strength	Cumulative frequency
less than 2.295	0
less than 2.385	3
less than 2.475	11
less than 2.565	14
less than 2.655	21
less than 2.745	29
less than 2.835	30

(e)

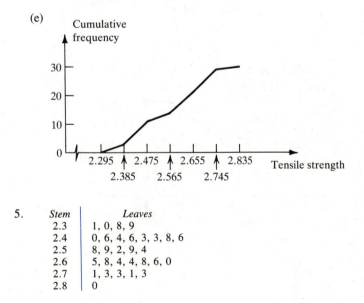

5.

Stem	Leaves
2.3	1, 0, 8, 9
2.4	0, 6, 4, 6, 3, 3, 8, 6
2.5	8, 9, 2, 9, 4
2.6	5, 8, 4, 4, 8, 6, 0
2.7	1, 3, 3, 1, 3
2.8	0

---

*In some instances the author has chosen to give answers to even-numbered questions.

6. (a)

Amount (in dollars)	Cumulative frequency
less than -0.5	0
less than 199.5	120
less than 399.5	205
less than 599.5	279
less than 799.5	348
less than 999.5	354

(b)

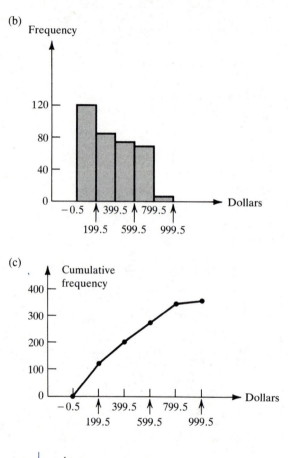

(c)

7.

stem	leaves
7.	3, 6, 5
8.	2, 9, 8, 8, 4
9.	8, 6
10.	6, 8
11.	5, 6, 9
12.	3, 2, 0, 5
13.	0
14.	2, 5, 4, 8, 9

8. (a)

Amount of magnesium	Cumulative frequency
less than 6.05	0
less than 7.05	6
less than 8.05	14
less than 9.05	28
less than 10.05	44
less than 11.05	54
less than 12.05	60

(b)

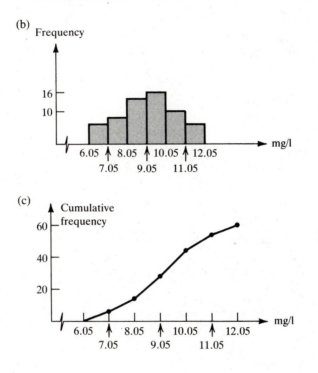

(c)

CHAPTER 2

Section 2-1

1. (a) $3 + 3^2 + 3^3 + 3^4 + 3^5 + 3^6$

(b) $\frac{2}{6} + \frac{2}{7} + \frac{2}{8} + \frac{2}{9}$

(c) $(2 + 1^2) + (2^2 + 2^2) + (2^3 + 3^2) + (2^4 + 4^2)$

(d) $1(0) + 2(1) + 3(2) + 4(3) + 5(4)$

(e) $(1^2 + 2(1) - 1) + (2^2 + 2(2) - 1) + (3^2 + 2(3) - 1) + (4^2 + 2(4) - 1)$

(f) $\frac{1}{4} + \frac{2}{5} + \frac{3}{6} + \frac{4}{7} + \frac{5}{8}$

3. -7

4. (a) 11   (c) 13   (e) 64

5. (a) $\displaystyle\sum_{i=1}^{n} 3x_i + \sum_{i=1}^{n} 2y_i$

    (b) $\displaystyle\sum_{i=1}^{n} 2x_i^2 + \sum_{i=1}^{n} 2y_i^2$

    (c) $\displaystyle\sum_{i=1}^{n} 2(x_i + y_i)^2$

    (d) $\displaystyle\sum_{i=1}^{n} x_i^2 y_i$

7. (a) 42   (b) 320   (c) 1764   (d) 2684

9. (a) 226   (b) 254

11. 100

Section 2-2

1. 67

3. 166.25

5. (a) False   (b) True   (c) True

7. $853,333.33

9. Yes

11. 48

13. (a) 27   (b) 50 + 27   (c) 10,100 + 27

15. 90 inches

17. 15.94

19. 135 miles

21. 2, 3, 8, 10, 15; and −6, 4, 8, 12, 23.

23. 0.5

25. (a) 3.5   (b) 3.29

27. 14

29. no mode

31. −4.0 and −6.8; bimodal

33. no mode

35. 3.5

Section 2-3

1. no; consider −7, −7, 0, 0, 0, 0, 4, 4, 3, 3, which has 0 as the mean, median, mode but is not symmetric.

3. the median

5. mean = 20; median = 20

7. (a) 27.5   (b) 14   (c) 12; mean least satisfactory

9. median; 51.

Section 2-4

1. 43, 56, 69, 75, 82, 87; range = 44

3. 10,900

5. $98.25

7. (a) 4.75

   (b) 1.25, 1.75, 2.75, 5.25, 2.75, 0.25, 0.75, 1.25

   (c) 2.0

9. 0.92

11. 2

13. All values are identical.

15. 3

17. standard deviation 10.1; variance 101.67

19. standard deviation 6.0; variance 36.5

21. all values are identical.

23. mean = 2.49; standard deviation = 0.102

25. 0.0000145

27. mean = 150; variance = 11,111.11

29. mean = 78; standard deviation = 12; variance = 144

31. variance = 0.049; standard deviation = 0.22.

33. Store A.

35. (a) mean = 1.2752; standard deviation = 0.005.

   (b) (1.2652, 1.2852)

   (c) 9   (d) at least 7   (e) Yes.

37. (a) (30.88, 51.52)   (b) 0.826   (c) 49

Section 2-5

1. $\bar{x} = 153.5$; $s^2 = 60.6$

3. $\bar{x} = 15.94$; $s^2 = 0.0486$

5. $s^2 = 15,425.501$; $s = 124.2$

7. (a) 104.7   (b) 629.551

9. (a) 8.15   (b) 2.486

Chapter 2 Test

2. The arithmetic mean

4. (a) 337.994   (b) 138.24

5. 8

6. (a) 2.2   (b) 2   (c) 2

7. (a) 10.95   (b) 8.0–11.9; 9.95

8. (a) 8.75   (b) 36.603

9. (a) Bob (20.0%); Judith (17.2%); Raphael (22.4%)
    George (16.9%); Laura (15.1%); Roy (15.5%)

    (b) 17.85

11. s^2 (ohms)^2; s (ohms).

12. mean deviation = 3.46; standard deviation = 4.30

13. variance = 4.42; standard deviation = 2.1

14. variance = 228,102.56; standard deviation = 477.6

15.
75C	71C	88B	99A	80C
77C	76C	70C	74C	68C
49F	61D	87B	75C	88B
72C	74C	69C	77C	59D

16. 15; yes, there are actually 18.

17. (a) 4.62   (b) 4.55   (c) 1.095

18. $\bar{x} = 2001$, $s^2 = 40.67$, $s = 6.4$.

CHAPTER 3

Section 3-2

1. S = {sss, ssd, sds, dss, dds, dsd, sdd, ddd}

3. (a) G_1G_2, G_1G_3, G_2D_3, G_4D_1, D_1D_2, D_2D_3

    (b) G_1D_1, G_1D_2, G_1D_3, G_3D_2, G_3D_3, G_4D_1

5. (c) ;   (d)

7. (a) 5/26   (b) 21/26   (c) 5/13   (d) 1/26   (e) 7/13

9. (a) 1/400   (b) 3/400   (c) 1/10

11. Using initials, (a) {{H,C}, {H,S}, {H,E}, {C,S}, {C,E}, {E,S}}

    (b) {{H,S}, {C,S}, {E,S}}

13. (a) 1/26   (b) 5/26   (c) 21/26   (d) 5/13   (e) 3/13

    (f) 3/26   (g) 1/13

15. 0.3

17. (a) 0.4   (b) 0.4   (c) 0.432   (d) 0.784

Section 3-3

1. 0.65

3. 0.8

5. 0.08

7. 0.35

9. A and B are mutually exclusive.

11. $0 \leq \alpha \leq 1$; the drug is not good when in fact it is not good.

13. No; would get $P(A \text{ or } B) = 1.1$, greater than 1.

15. 8/9

17. (a) 0.3   (b) 0.1

19. (a) 1/3   (b) 2/9

21. (a) 0.85   (b) 0.15

23. (a) 0.95   (b) 0.55   (c) 0.95   (d) 0.55

25. 0.1

Section 3-4

1. (a) 20   (b) 6720   (c) 720   (d) 720
   (e) 24   (f) 380   (g) 28   (h) 15
   (i) 364   (j) 1   (k) 1   (l) 125

3. 30

5. 5184; {(1,2,3,4,H,T), (1,1,3,2,T,T), (3,3,6,6,T,H)}

7. 60

9. $\binom{10}{2}$

11. (a) 210   (b) 90

13. 8008

15. 970,200

17. (a) 1/3   (b) 1/6

19. 25/234

21. 151,200/2,324,784

23. (a) 10/21   (b) 5/42   (c) 37/42   (d) 25/42

25. 51/56

27. (a) 1/39   (b) 38/39   (c) 196/455   (d) 32/455   (e) 8/1365

Section 3-5

1. (a) 1/3   (b) 1/3   (c) 5/8

3. (a) 0.92   (b) 0.96   (c) 0.84   (d) 0.78

5. 6 percent

7. 0.06

9. (a) 0.12   (b) 0.63   (c) 0.75

11. (a) 0.2   (b) 0.12   (c) 0.56   (d) 0.28

13. (a) 0.3   (b) 0.86

15. (a) no   (b) yes

17. (a) 0.42   (b) 0.28

19. (a) 0.0025   (b) 0.9025   (c) 0.0975

21. (a) 0.16   (b) 0.3   (c) 0.36   (d) 0.24   (e) 0.7

23. (a) 0.048   (b) 0.952

25. (a) 0.25   (b) 0.06   (c) 0.56

27. 0.000024

29. (a) 0.28   (b) 0.18   (c) 0.42   (d) 0.54

31. 0.72

33. 0.98

Chapter 3 Test

3. 60

4. (a) 0.26   (b) 0.18   (c) 0.24   (d) 0.15

5. (a) 0.8   (b) 0.74   (c) 0.52   (d) 0.76

6. 0.45

7. 0

8. (a) neither   (b) mutually exclusive   (c) independent

9. 0.675

10. (a) 0.86   (b) 0.14

11. (a) 1/120   (b) 1/20   (c) 1/4   (d) 1/6

12. (a) 0.005   (b) 0.095   (c) 0.14   (d) 0.855

13. 1/3

14. (a) 0.855   (b) 0.995   (c) 0.005

15. $28{,}800/\binom{41}{5}$

16. 0.748

CHAPTER 4

Section 4-1

1. The random variables assume the values (a) 1, 2, 3, 4, 5, 6
   (b) 0, 1, 2, 3, 4, 5   (c) 4, −3

3. continuous

5. discrete

7. continuous

9. continuous

11. discrete

13. 0, 1, 2, 3, 4, 5, 6; discrete

15. discrete

Section 4-2

1. Values assumed 0, 1, 2, and 3 with respective probabilities 0.05, 0.45, 0.45, 0.05

3. (a) There are 2 Republicans on the committee; $\binom{15}{4}\binom{10}{2}/\binom{25}{6}$

 (b) There are either 2, or 3, or 4 Republicans on the committee;

$$\left[\binom{15}{4}\binom{10}{2} + \binom{15}{3}\binom{10}{3} + \binom{15}{2}\binom{10}{4}\right]/\binom{25}{6}$$

 (c) $4 \leqslant X \leqslant 6$; $\left[\binom{15}{2}\binom{10}{4} + \binom{15}{1}\binom{10}{5} + \binom{15}{0}\binom{10}{6}\right]/\binom{25}{6}$

5. (a) 2.1 (b) 1.39

7. $-20¢$

13. (a) 0.4 (b) 0.4 (c) 0.55

15. $P(X = 2) = 0.2; P(X = 4) = 0.5$

16. (a) Profit, $18.00; probability 0.5

 (b)

Profit	−32	−7	18	43	68
Probability	0.1	0.2	0.5	0.1	0.1

 (c) $15.50

17. (a) $21.00 (b) $24.00 (most profitable to bake 2 cakes)

19. (a) 0.75 (b) 0.8874

Section 4-3

9. (a)

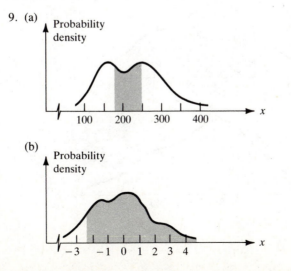

 (b)

(c)
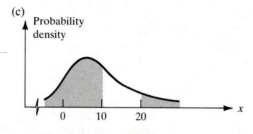

11. (a) 0.2   (b) 0.4   (c) 0.11
    (a) 1800   (b) 1600   (c) 1400

Chapter 4 Test

2.

x	0	1	2	3
$p(x)$	$\dfrac{\binom{4}{0}\binom{5}{3}}{\binom{9}{3}}$	$\dfrac{\binom{4}{1}\binom{5}{2}}{\binom{9}{3}}$	$\dfrac{\binom{4}{2}\binom{5}{1}}{\binom{9}{3}}$	$\dfrac{\binom{4}{3}\binom{5}{0}}{\binom{9}{3}}$

3. (a) (i)   0.1   (ii) 0.1   (iii) 0.4
       (iv) 0   (v) 0.9   (vi) 0.3

   (b) \$121.60   (c) 11,325.44 (dollars)^2.

4. 30¢

5. \$2.10

6. \$61

7. (a)

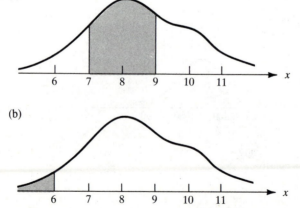

(b)

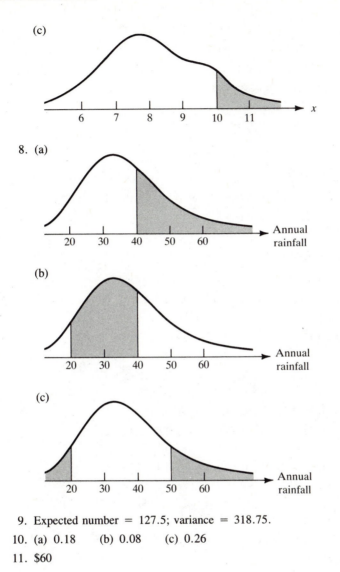

(c)

8. (a)

(b)

(c)

9. Expected number $= 127.5$; variance $= 318.75$.

10. (a) 0.18   (b) 0.08   (c) 0.26

11. $60

CHAPTER 5

Section 5-1

1. $(0.4)^3(0.6)^4$

3. $1/64$

5. (a) 0.230   (b) 0.233   (c) 0.158   (d) 0.273

7. 0.251

9. (a) 0   (b) 0.046   (c) 0.168

11. (a) 0.002   (b) 0.177

13. mean $= 10$; variance $= 9$

15. (a) 0.207   (b) 8.4   (c) 3.36

16. (a) 0.033   (b) 0.8   (c) 0.8485

17. 6.6

18. (a) 0.346   (b) 0.526   (c) 0.13   (d) 0.346

19. (a) 34.6   (b) 52.6   (c) 13   (d) 34.6

21. (a) 0.037   (b) 0.094   (c) 0   (d) 0.041   (e) 0.344

23. (a) $(0.1)^8$ (b) 0.005

Section 5-2

1. (a) 0.3849   (c) 0.8849   (e) 0.1525
   (g) 0.8306   (i) 0.2450   (k) 0.7517

2. (a) 0.48   (c) -1.26 (e) 2.16
   (g) 0.71   (i) -2.26 (k) -0.42

3. (a) 1.33   (c) -1.97 (e) 0.32

4. (a) 0.0122   (c) 0.0268   (e) 0.0818

5. (a) 0.3345   (b) 0.7734   (c) 0.7066
   (d) 0.3085   (e) 0.6915   (f) 0.0122

7. (a) 0.4772   (b) 0.9332   (c) 0.927
   (d) 0.9332   (e) 0.1525   (f) 0.1525

9. (a) 0.0228   (b) 0.1525   (c) 0.1359

11. 1.906 inches

13. 1.632

15. (a) 0.6554   (b) 655

17. (a) 0.8413   (b) 5384

18. 0.383

19. (a) $\binom{6}{2}(0.383)^2(0.617)^4$ (b) $(0.383)^6$

21. (a) 0.0139   (b) 0.3462

23. $(0.0475)^2$

Section 5-3

1. (a) 140   (b) 42   (c) 0.0638   (d) 0.7887

3. (a) 0.0638   (b) 0.4681   (c) 0.9292

5. 0.0643

7. (a) 0.0344   (b) 0.0436
9. 0.0023
11. (a) 0.0472   (b) 0.102
13. 0.1314
15. 0.0022
17. 0.0068

Chapter 5 Test

3. 0.395
4. 0.346
5. (a) 0.273   (b) 0.261   (c) 0.478
6. (a) 0.201   (b) 0.878   (c) 0.323
7. 0.0562
8. 0.0526
9. Approximately 4 percent.
10. (a) (i) 0.0139   (ii) 0.8413
    (b) (i) 7   (ii) 421
11. 20.47 percent.
12. 62.93 percent.

CHAPTER 6

Section 6-1

1. (a) 112.8   (b) 112.8
2. (a) 5.603   (b) 4.823
3. mean = 60; standard deviation = 1.598
5. (a) 401.48   (b) 437.5
7. (a) 8.5   (b) 6.917
8. (a) mean = 8.5; variance = 1.729
   (b) mean = 8.5; variance = 0.692
9. (a) mean = 2.3; variance = 21.81
   (b) mean = 2.3; variance = 0.2181
   (c) approximately normal with mean 2.3, variance 0.2181
   (d) (i) 0.8185   (ii) 0.0228   (iii) 0.0228
11. 0.0228
13. (a) 0.0062   (b) 0.0668   (c) 0.9544
15. 0.7971

17. 0.0062

19. 0.8413

21. (a) 0.1   (b) 0.0009

    (c) approximately normal with mean 0.1, variance 0.0009

    (d) 0.0228

23. (a) 0.0011   (b) 0.0071

25. 0.3085

27. (a) 0.0146   (b) 0.6043

Section 6-2

1. 195, 134, 238, 015, 242, 116, 042, 178

Chapter 6 Test

1. False

2. Many physical phenomena can be explained through this distribution. The central limit theorem accords it a place of prominence.

3. (a) True   (b) False   (c) True   (d) False   (e) True   (f) False

4. (a) 1.6   (b) 32/29

5. (a) False   (b) True

6. 0.0207

7. 0.16

8. (a) 0.003   (b) 0.9505

9. (a) normal with mean 35,000, standard deviation 1000; exact.

    (b) 0.0062

10. (a) 0.0823   (b) 0.7842

CHAPTER 7

Section 7-1

1. (a) consistency   (b) efficiency; estimator 1

    (c) sufficiency   (d) unbiasedness

    (e) unbiasedness   (f) unbiasedness

3. Parameter

4. (a) x/n (b) $\bar{x}$ (c) p (d) p (e) μ

5. (a) estimate of mean = 90; estimate of variance = 625; estimate of standard deviation = 25

    (b) estimate of mean = 18.5; estimate of variance = 1.764; estimate of standard deviation = 1.328

7. $3375.00

11. 0.205

13. 0.68

Section 7-2

1. (a) 0.04   (b) 0.08   (c) 0.16   (d) 0.03

2. (a) 1.645   (c) -1.73

3. (13.371, 17.029)

5. (147.598, 157.002)

7. (29.161, 33.239)

9. (a) (22.355, 25.645)   (b) (22.04, 25.96)

   (c) (21.67, 26.33)

11. (a) 1.76   (b) 13.32

13. (b)

15. 49

Section 7-3

1. (a) 1.734   (b) 2.16   (c) -3.143 (d) -2.262

2. (a) 2.447   (c) -2.624 (e) 2.921

3. (12.663, 17.737)

5. (143.132, 161.468)

7. (21.718, 25.882)

9. (4.805, 6.195)

11. (15.66, 21.54)

13. (148.03, 149.77)

15. (224.517, 235.483)

Section 7-4

1. (a) 14.067   (c) 6.571   (e) 9.542

2. (a) 11.07   (c) 18.475

3. $\alpha/2 = 0.05; 1 - (\alpha/2) = 0.95$

   (a) 7.962   (b) 26.296   (c) 2.733   (d) 2.167

4. (a) 26.119   (c) 1.646

5. (0.618, 6.416)

7. (8.301, 22.824)

9. (0.167, 0.530)

11. (a) (9.53, 12.47)   (b) (2.00, 14.07)

Section 7-5

1. (0.184, 0.416)
3. (0.545, 0.695)
5. (0.707, 0.893)
7. (0.311, 0.489)
9. 3,393
11. Not acceptable; p is supposed to be between 0 and 1.

Chapter 7 Test

1. (a) 1.62   (b) 0.05886   (c) 0.243
2. (a) -2.67 (b) 12.67   (c) 3.559
3. (a) 3.52   (b) 0.347   (c) 0.589
4. (a) 3.76   (b) 4.8695   (c) 2.207
5. 0.573
6. 0.375
7. (9.916, 17.684); normal distribution
8. (0.888, 4.386); normal distribution
9. $(-4.951, -1.849)$
10. (0.255, 0.420)
11. (6.228, 8.972); normal distribution
12. (a) 11.07   (b) 2.04   (c) 9.97   (d) 0.48
13. (47.835, 50.165)
14. (741.34, 778.66)
15. (a) (5.088, 5.512)   (b) (0.152, 0.502)
16. Point estimate 0.5; interval estimate at 95 percent confidence level (0.345, 0.655).
17. (153.249, 172.151)
18. (a) (1.289, 1.611)   (b) (0.016, 0.154)

CHAPTER 8

Section 8-1

1. (a) 3.05   (b) 0.5
3. -0.66
5. $(-1.16, 2.76)$
7. $(-1.923, 0.293)$
9. (a) $(-0.489, 3.689)$ (b) $(-4.381, 1.181)$
11. (2.01, 2.80)
13. (23.686, 26.114)

15. $(-18.886, -11.114)$

17. $(-4601.08, -3398.92)$ (confidence interval for $\mu_E - \mu_W$)

Section 8-2

1. 1.2758

3. 0.0238

5. 0.5685

7. (a) 1.7378   (b) 20   (c) (0.417, 2.984)

9. $(-211.106, 31.106)$ (confidence interval for $\mu_A - \mu_B$)

11. $(-2.663, -0.087)$

Section 8-3

1. $(-0.189, 0.119)$

3. $(-0.099, 0.118)$

5. $(0.019, 0.181)$ Point estimate 0.1; 95 percent confidence interval for $\mu_S - \mu_G$

7. $(-0.20, 0.05)$

9. $(-0.05, 0.13)$

11. $(-0.112, 0.072)$

Chapter 8 Test

1. $(-0.225, 4.825)$

2. $(-5.898, -3.702)$

3. $(-0.237, 0.126)$

4. $(-0.112, 0.120)$

5. $(2.367, 4.033)$

6. $(-0.639, -0.028)$

7. $(-2.833, -0.807)$

8. $(-152.09, 1030.83)$

9. $(-0.042, 0.087)$

10. $(-1.309, 5.309)$

CHAPTER 9

Section 9-1

1. (a) The serum is indeed effective and the decision is to not market it; the serum is not effective and the decision is to market it.

(b) The serum is effective and the decision is to not market it.

(c) The serum is not effective and the decision is to market it.

3. H_0: The serum is not effective.
   H_A: The serum is effective.

5. Type I error: concluding that the business venture will flourish when in fact it does not.

   Type II error: concluding that the business venture will not flourish when in fact it does flourish.

   The more serious error is to conclude that the business venture will flourish when in fact it does not.

7. (a) F   (b) T   (c) F   (d) T   (e) T   (f) F   (g) T

9. The actual state of affairs: The mean strengths of the two types of steel wires are different.

   Decision taken: Accept that the mean strengths are the same.

11. The conclusion is that the mean IQ of the region is higher than that for the rest of the nation. There is a 5 percent chance of arriving at this conclusion when in fact this is not the case.

Section 9-2

1.
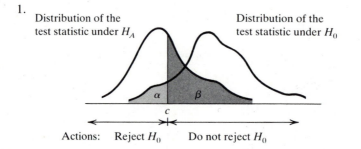

Distribution of the test statistic under H_A

Distribution of the test statistic under H_0

Actions: Reject H_0 Do not reject H_0

3. Reject H_0 if the observed chip is one of the chips 2, 4, 6, 8, 10; $\alpha = \beta = 0$.

5. Reject H_0 if the observed chip is one of the chips 6, 7, 8, 9. Here $\beta = 1/5$.

7. No. As the level of significance becomes smaller, the critical region becomes smaller.

9. Computed value $= \dfrac{\bar{x} - \mu_0}{\sigma/\sqrt{n}} = \dfrac{97 - 100}{3/\sqrt{16}} = -4$; $z_{0.05} = 1.645$; reject H_0.

11. (a) $\dfrac{18.6 - 20}{4/\sqrt{9}} = -1.05$ (b) 0.1469

13. (a) 0.685   (b) 0.05

14. (a) H_0: The stockbroker is guessing (that is, $p = 0.5$)

    H_A: The stockbroker is not guessing (that is, $p = 0.8$)

    (b) Type I error: decision that the stockbroker is not guessing when she is guessing.

    Type II error: decision that the stockbroker is guessing when she is not guessing.

15. $\alpha = 0.073$; $\beta = 0.206$.

Section 9-3

1. Test statistic: $\dfrac{\overline{X} - 10.2}{1.3/\sqrt{16}}$

   Decision rule: reject H_0 if the computed value is greater than 1.645.

3. Test statistic: $\dfrac{\overline{X} - 32.8}{6.8/\sqrt{14}}$

   Decision rule: reject H_0 if the computed value is less than -2.33.

5. Test statistic: $\dfrac{\overline{X} + 3.3}{4.8/\sqrt{20}}$

   Decision rule: reject H_0 if the computed value is less than -2.33 or greater than 2.33.

7. Computed value $= 1.83$

   (a) Do not reject H_0 since $z_{0.025} = 1.96$

   (b) Reject H_0 since $z_{0.05} = 1.645$.

9. $H_0: \mu = 50$; $H_A: \mu > 50$; computed value $= 2.875$; reject H_0; the company is correct in its claim.

11. $H_0: \mu = 5$; $H_A: \mu > 5$; computed value $= 2$; reject H_0. The contention of the management is justified.

13. Test statistic: $\dfrac{\overline{X} - 110}{4.32/\sqrt{25}}$

    Decision rule: reject H_0 if the computed value is less than -2.064.

15. Test statistic: $\dfrac{\overline{X} - 17.8}{2.8/\sqrt{20}}$

    Decision rule: reject H_0 if the computed value is greater than 2.093.

17. Test statistic: $\dfrac{\overline{X} - 11.6}{3.2/\sqrt{12}}$

    Decision rule: reject H_0 if the computed value is less than -1.796 or greater than 1.796.

19. $H_0: \mu = 13$; $H_A: \mu < 13$; computed value $= -1.845$; do not reject H_0 since $-t_{9,0.025} = -2.262$.

21. $H_0: \mu = 36{,}000$; $H_A: \mu > 36{,}000$; computed value $= 1.42$; $t_{15,0.05} = 1.753$; do not reject H_0.

23. Test statistic: $\dfrac{\overline{X} - 9.8}{1.6/\sqrt{46}}$

    Decision rule: reject H_0 if the computed value is less than -1.645.

25. Test statistic: $\dfrac{\overline{X} + 3}{2.4/\sqrt{80}}$

    Decision rule: reject H_0 if the computed value is greater than 1.28

27. $H_0: \mu = 750$; $H_A: \mu > 750$; computed value $= 1.67$; do not reject H_0.

29. Test statistic: $\dfrac{15S^2}{8.2}$

Decision rule: reject H_0 if the computed value is less than 6.262 or greater than 27.488.

31. Test statistic: $\dfrac{13S^2}{6.8^2}$

Decision rule: reject H_0 if the computed value is greater than 22.362.

33. Test statistic: $\dfrac{7S^2}{0.78}$

Decision rule: reject H_0 if the computed value is less than 1.239.

35. Computed value $= 3.5$

    (a) Do not reject H_0 since $\chi^2_{10,0.975} = 3.247$ and $\chi^2_{10,0.025} = 20.483$.

    (b) Reject H_0 since $\chi^2_{10,0.95} = 3.940$

37. H_0: $\sigma^2 = 2.4$; H_A: $\sigma^2 \neq 2.4$; computed value 15.83; do not reject H_0 since $\chi^2_{9,0.975} = 2.7$ and $\chi^2_{9,0.025} = 19.023$.

39. Test statistic: $\dfrac{(X/60) - 0.4}{\sqrt{\dfrac{(0.4)(0.6)}{60}}}$

Decision rule: reject H_0 if the computed value is less than -1.96 or greater than 1.96.

41. Test statistic: $\dfrac{(X/36) - 0.7}{\sqrt{\dfrac{(0.7)(0.3)}{36}}}$

Decision rule: reject H_0 if the computed value is greater than 1.645.

43. Test statistic: $\dfrac{(X/120) - 0.8}{\sqrt{\dfrac{(0.8)(0.2)}{120}}}$

Decision rule: reject H_0 if the computed value is less than -2.33.

45. H_0: $p = 0.8$; H_A: $p > 0.8$; computed value $= 2$; reject H_0.

47. H_0: $p = 0.5$; H_A: $p > 0.5$; computed value $= 2.4$; reject H_0; evidence shows that the claim is correct.

49. H_0: $p = 0.4$; H_A: $p < 0.4$; computed value $= -1.633$; do not reject H_0; P-value $= 0.0516$

Section 9-4

(In the following, μ_1 refers to the mean of the population mentioned first in a problem and μ_2 to the mean of the population mentioned next; the same hold true for p_1 and p_2).

1. Computed value $= 0.716$

    (a) Do not reject H_0 since $z_{0.025} = 1.96$

    (b) Do not reject H_0 since $z_{0.05} = 1.645$.

3. Computed value $= -2.9$

    (a) Reject H_0 since $z_{0.05} = 1.645$

    (b) Reject H_0 since $-z_{0.1} = -1.28$.

5. H_0: $\mu_1 = \mu_2$; H_A: $\mu_1 \neq \mu_2$; computed value $= -4.80$; reject H_0.

7. H_0: $\mu_1 = \mu_2$; H_A: $\mu_1 \neq \mu_2$; computed value $= -3.845$; reject H_0.

9. H_0: $\mu_1 = \mu_2$; H_A: $\mu_1 > \mu_2$; computed value $= 1.131$; do not reject H_0 since $t_{23,0.1} = 1.319$.

11. H_0: $\mu_1 = \mu_2$; H_A: $\mu_1 > \mu_2$; computed value $= 1.591$; reject H_0 since $t_{20,0.1} = 1.325$.

13. H_0: $\mu_1 = \mu_2$; H_A: $\mu_1 < \mu_2$; computed value $= -2.326$; reject H_0 since $-t_{12,0.05} = -1.782$; the reaction time is increased significantly by alcoholic stimulant.

15. H_0: $\mu_1 = \mu_2$; H_A: $\mu_1 \neq \mu_2$; computed value $= -2.056$; do not reject H_0 since $t_{20,0.025} = 2.086$.

17. Computed value $= -2.967$

   (a) Reject H_0 since $t_{14,0.025} = 2.145$

   (b) Reject H_0 since $-t_{14,0.05} = -1.761$

19. H_0: $\mu_D = 0$; H_A: $\mu_D < 0$; $n = 9$; $\bar{d} = -6.78$; $s_d = 13.526$; computed value $= -1.504$; do not reject H_0 since $t_{8,0.05} = 1.86$.

21. $-2.609 < \mu_D < 0.859$

23. $-2.390 < \mu_D < 0.036$

25. H_0: $p_D = p_R$; H_A: $p_D < p_R$; computed value $= -1.127$; $z_{0.01} = 2.33$; do not reject H_0.

27. H_0: $p_1 = p_2$; H_A: $p_1 < p_2$; computed value $= -1.273$; do not reject H_0 that the probability of getting cold with vaccine is the same as that without the vaccine; the vaccine is not effective.

29. H_0: $p_{student} = p_{graduate}$; H_A: $p_{student} > p_{graduate}$; computed value 2.37; $z_{0.05} = 1.645$; reject H_0; college students tend to be more in favor of environmental controls than college graduates.

Chapter 9 Test

2. On the basis of the alternative hypothesis we decide whether the test is right-tailed, or left-tailed, or two-tailed.

3. Type I error: To assert that Vitamin C is not effective in controlling colds when in fact it is.

   Type II error: To assert that Vitamin C is effective in controlling colds when in fact it is not.

4. (a) F   (b) F   (c) T   (d) T

5. (a) normal with mean 0.4, variance $= \dfrac{(0.4)(0.6)}{100}$, or 0.0024

   (b) normal with mean 0.3, variance $= \dfrac{(0.3)(0.7)}{100}$, or 0.0021

6. (a) H_0: $\mu = 16$; H_A: $\mu \neq 16$; computed value $= 0.625$; $t_{9,0.025} = 2.262$; do not reject H_0.

   (b) H_0: $\sigma^2 = 0.16$; H_A: $\sigma^2 \neq 0.16$; computed value $= 14.375$; $\chi^2_{9,0.025} = 19.023$; $\chi^2_{9,0.975} = 2.700$; do not reject H_0.

7. $H_0: p = 0.8$; $H_A: p \neq 0.8$; computed value $= -1.37$; $z_{0.025} = 1.96$; do not reject H_0.

8. $H_0: \mu_1 = \mu_2$; $H_A: \mu_1 \neq \mu_2$; $s_p = 133.954$; computed value $= -1.574$; $t_{16,0.025} = 2.12$; do not reject H_0.

9. $H_0: p_1 = p_2$; $H_A: p_1 \neq p_2$; computed value $= -1.39$; $z_{0.05} = 1.645$; do not reject H_0.

10. $H_0: \sigma = 0.1$; $H_A: \sigma < 0.1$; computed value $= 26.173$; $\chi^2_{29,0.95} = 17.708$; do not reject H_0.

11. $H_0: \mu_1 = \mu_2$; $H_A: \mu_1 \neq \mu_2$; $s_p = 1.519$; computed value $= 1.53$; $t_{25,0.025} = 2.06$; do not reject H_0.

CHAPTER 10

Section 10-1

1. (a) Computed χ^2 value $= 0.775$; since $\chi^2_{1,0.05} = 3.841$, do not reject H_0 that the segregation ratio is 3:1.

   (b) Computed value of $\dfrac{\dfrac{x}{n} - 0.75}{\sqrt{\dfrac{(0.75)(0.25)}{172}}}$ with $x = 134$, $n = 172$ is 0.88; do not reject H_0 (same conclusion in (a) and (b)).

3. Computed value $= 9.586$; $\chi^2_{5,0.05} = 11.07$; do not reject H_0.

5. Computed value $= 6.635$; $\chi^2_{3,0.05} = 7.815$; do not reject the hypothesis that the segregation ratio is 9:3:3:1.

7. (a) 1/16, 4/16, 6/16, 4/16, 1/16   (b) 10, 40, 60, 40, 10

   (c) Computed value $= 5.842$; $\chi^2_{4,0.05} = 9.488$; do not reject H_0.

9. Computed value $= 11.4$; $\chi^2_{4,0.05} = 9.488$; reject H_0.

Section 10-2

1. Computed value $= 17.017$; $\chi^2_{1,0.025} = 5.024$; reject H_0.

3. Computed value $= 16.837$; $\chi^2_{2,0.01} = 9.21$; reject H_0

5. Computed value $= 35.1$; $\chi^2_{9,0.05} = 16.919$; reject H_0

7. Computed value $= 12.746$; $\chi^2_{4,0.05} = 9.488$; reject H_0 that weight and blood pressure are not related.

Section 10-3

1. (a) Computed value $= 1.924$; $\chi^2_{1,0.05} = 3.841$; do not reject H_0; there is no significant difference in proportions.

   (b) Computed value $= 1.387$; $z_{0.025} = 1.96$; do not reject H_0.

3. Computed value $= 1.443$; $\chi^2_{2,0.05} = 5.991$; do not reject H_0; there is no significant difference in the proportions of successful launches from the three pads.

5. Computed value $= 5.093$; $\chi^2_{3,0.025} = 9.348$; do not reject H_0.

7. Computed value $= 6.945$; $\chi^2_{2,0.05} = 5.991$; reject H_0; the evidence supports Professor Kieval's assertion.

9. Computed value $= 11.356$; $\chi^2_{4,0.05} = 9.488$; reject H_0.

11. Computed value $= 6.298$; $\chi^2_{2,0.01} = 9.21$; do not reject H_0; the true proportions of patients getting relief are not significantly different.

Chapter 10 Test

1. H_0: The data is in conformity with a hypothesized distribution. That is, probability that an item belongs to category c_1 is p_1, that it belongs to category c_2 is p_2, and so on.

   H_A: The data is not in conformity.

2. H_0: The two attributes on the basis of which the observations are classified are independent.

   H_A: The two attributes are not independent.

3. Computed value 6.1; $\chi^2_{3,0.05} = 7.815$; do not reject H_0 that numbers were picked at random.

4. Computed value 29.62; $\chi^2_{2,0.05} = 5.991$; reject H_0 that eye color and hair color are not related.

5. Computed value 3.5; P-value greater than 0.05.

6. Computed value 11.93; $\chi^2_{4,0.05} = 9.488$; reject H_0; the attitude towards detente is not the same in the three cities.

CHAPTER 11

Section 11-1

1.

x	3	0	-2	-10.75	4.25
y	-4.5	1.5	5.5	23	-7

3. (a) positive   (b) inverse   (c) positive

5. (a) -0.75 (b) -1.5

7. (a) 1   (b) -1

9. (a) 492; 1254; 34,628; 227,948; 88,420

   (b) 0.712

11. (a)

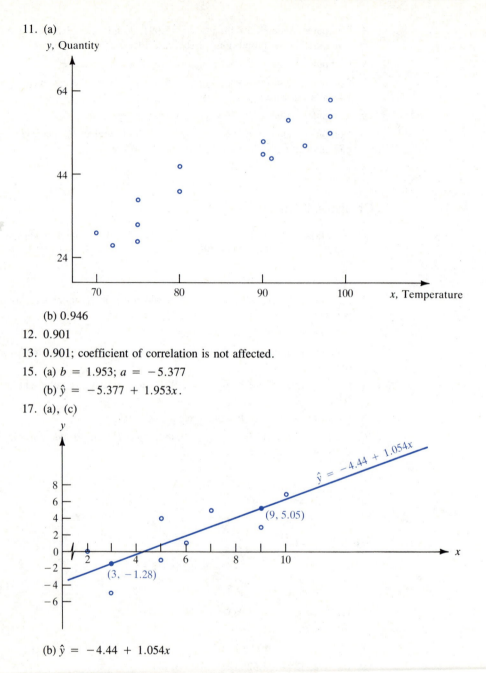

(b) 0.946

12. 0.901

13. 0.901; coefficient of correlation is not affected.

15. (a) $b = 1.953$; $a = -5.377$

(b) $\hat{y} = -5.377 + 1.953x$.

17. (a), (c)

(b) $\hat{y} = -4.44 + 1.054x$

19. (a) 59.38   (b) 71.68

20. (a) $\hat{y} = 14.33 + 0.826x$

    (b) $\hat{y} = 111.56 - 0.584x$

(c)

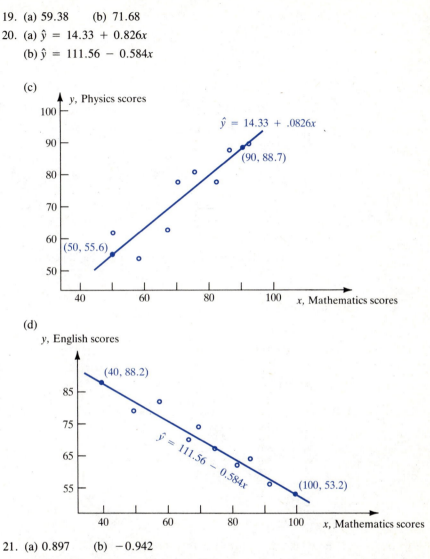

(d)

21. (a) 0.897   (b) -0.942

    (c) The signs of the correlation coefficients are the same as the signs of the corresponding regression coefficients.

23. $\hat{y} = -45.47 + 1.058x$; 44,460 gallons.

25. (a), (b)

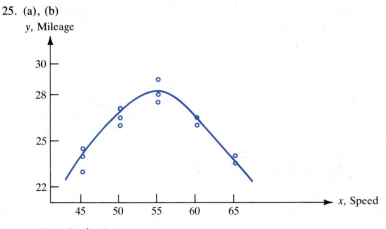

y, Mileage

(c) 55 miles/gallon

26. -0.053; there is a curvilinear relation (not a linear relation).

Section 11-2

1. (a) 8; 194; 32.5; 5300; 146.75; 880

   (b) $a = 0.3212$; $b = 0.1543$; $\hat{y} = 0.3212 + 0.1543x$

   (c)

x	*y*	$\hat{y}$
12	2.0	2.1728
16	3.0	2.7900
16	2.5	2.7900
24	4.0	4.0244
24	4.5	4.0244
30	5.0	4.9502
36	6.0	5.8760
36	5.5	5.8760

   (d) 0.0907

   $$\sum_{i=1}^{8} (y_i - \hat{y}_i)^2 = 0.5441$$

3. 3.322

5. (a) 0.000152   (b) (0.1303, 0.1783)

7. (a) 11   (b) $s_a^2 = 0.3483$; $s_b^2 = 0.11$

9. (a) $b = 0.0229$; $a = 0.4294$; $s_e^2 = 0.4281$; $s_b = 0.0135$

   $$\frac{b - B_0}{s_b} = 1.695; \quad t_{6,0.025} = 2.447; \text{ do not reject } H_0$$

   (b) $b = 0.0387$; $a = -0.1792$; $s_e^2 = 0.04287$; $s_b = 0.0049$

   $$\frac{b - B_0}{s_b} = 7.986; \quad t_{5,0.025} = 2.571; \text{ reject } H_0$$

11. (a)

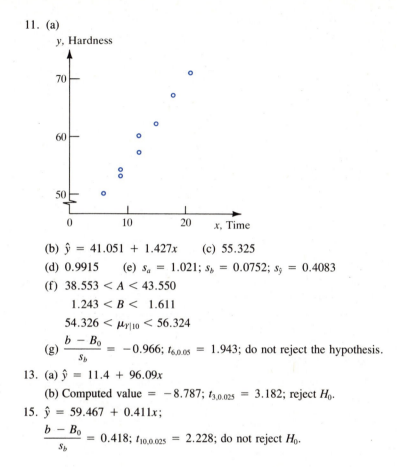

y, Hardness

(b) $\hat{y} = 41.051 + 1.427x$ (c) 55.325

(d) 0.9915   (e) $s_a = 1.021$; $s_b = 0.0752$; $s_{\hat{y}} = 0.4083$

(f) $38.553 < A < 43.550$

     $1.243 < B < 1.611$

     $54.326 < \mu_{Y|10} < 56.324$

(g) $\dfrac{b - B_0}{s_b} = -0.966$; $t_{6,0.05} = 1.943$; do not reject the hypothesis.

13. (a) $\hat{y} = 11.4 + 96.09x$

     (b) Computed value $= -8.787$; $t_{3,0.025} = 3.182$; reject H_0.

15. $\hat{y} = 59.467 + 0.411x$;

     $\dfrac{b - B_0}{s_b} = 0.418$; $t_{10,0.025} = 2.228$; do not reject H_0.

Section 11-3

1. (a) 0.6654   (b) 63.528

   (c) 1040.9   (d) 812.27

   (e)

Component	Sum of squares	Explained/total
explained (regression)	228.63	0.22
unexplained	812.27	
Total	1040.90	

   (f) 0.22

3. 0.7

5. (a) 0.7128   (b) 0.5081

   (c) 50.81 percent of the total variation in annual wheat yield is explained by the linear regression relation.

7. Computed value of $r\sqrt{\dfrac{n-2}{1-r^2}} = -4.619$; $t_{64,0.025} = 2.0$; reject H_0.

9. Computed value $= -1.732$; $t_{9,0.025} = 2.262$; do not reject H_0.

11. Computed value $= \pm 5.881$; $t_{36,0.025} = 2.021$; reject H_0. '

13. (a)

Component	Sum of squares	Explained/total
explained (regression)	1673.613	0.5067
unexplained	1629.244	
Total	3302.857	

   (b) 0.712

   (c) Computed value $= 2.266$; $t_{5,0.025} = 2.571$; do not reject the hypothesis.

15. (a) 81 percent

   (b) ± 0.9; however, from our experience we suspect that the correlation is $+0.9$ rather than -0.9.

   (c) Computed value $= 16.52$; P-value less than 0.005.

Chapter 11 Test

2. For an increase of $1,000 in advertising, the sales will increase by $420,000.

4. (a)

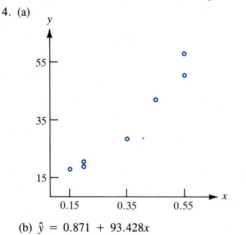

   (b) $\hat{y} = 0.871 + 93.428x$

   (c) 0.975

5. The regression coefficient b and the correlation coefficient r must have the same sign.

6. (a) $\hat{y} = 2.01 + 0.0797x$

   (b) 0.934

7. (a) 0.6561

   (b) Computed value $= 12.04$; $t_{76;0.025} = 2.0$; reject H_0.

   (c) 262 years.

8. $r = 0.943$; computed value $= 8.042$; $t_{8,0.025} = 2.306$; reject H_0.

9. estimate of B, 11.441; computed value $= -0.39$; $t_{6,0.025} = 2.447$; do not reject H_0.

10. (a) $\hat{y} = 468.709 + 6.01x$

    (b) (5.657, 6.359)

    (c) computed value $= 2.814$; $t_{6,0.025} = 2.447$; reject H_0 at the 5-percent level.

    (d) (2793.28, 2950.68)

    (e) (2639.91, 3104.06)

11. (a)

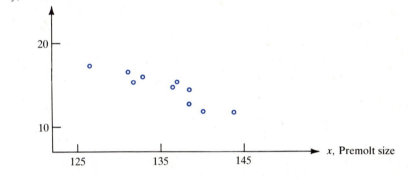

    (b) $\hat{y} = 59.4 - 0.331x$;   (c) -0.915

    (d) Computed value $= -6.42$; $t_{8,0.025} = 2.306$; reject H_0.

CHAPTER 12

Section 12-1

1. (a) 4   (b) 30   (c) 34

2. 2.45

3. 3.35

5.

Source	df	Sum of squares	Mean sum of squares	Computed ratio
between	3	47.125	15.71	1.483
within	20	211.833	10.59	
Total	23	258.958		

$f_{3,20;0.05} = 3.10$; do not reject H_0; the mean yield strengths are not different.

Section 12-2

1.
7	210.7	30.1	2.89
18	187.2	10.4	
25	397.9		

3. (b)

Source	df	Sum of squares	Mean sum of squares	Computed ratio
between	1	518.37	518.37	0.882
within	11	6463.32	587.58	
Total	12	6981.69		

There is no significant difference.

5.

Source	df	Sum of squares	Mean sum of squares	Computed ratio
between	3	1148.55	382.85	2.88
within	16	2130.00	133.125	
Total	19	3278.55		

$f_{3,16;0.1} = 2.46$; there is a significant difference.

7.

Source	df	Sum of squares	Mean sum of squares	Computed ratio
between	2	84,933,330	42,466,665	2.45
within	12	208,000,000	17,333,333	
Total	14	292,933,330		

$f_{2,12;0.025} = 5.10$; there is no significant difference between the brands.

9.

Source	df	Sum of squares	Mean sum of squares	Computed ratio
between	2	19.2	9.6	0.85
within	12	135.2	11.3	
Total	14	154.4		

$f_{2,12;0.05} = 3.89$; there is no significant difference.

11.

Source	df	Sum of squares	Mean sum of squares	Computed ratio
between	2	736.84	368.42	2.583
within	14	1997.63	142.69	
Total	16	2734.47		

$f_{2,14;0.1} = 2.73$; the mean yield is not significantly different.

Chapter 12 Test

2. False   3. 1.24

4.

Source	df	Sum of squares	Mean sum of squares	Computed ratio
between	2	35.9	17.95	1.68
within	11	117.6	10.69	
Total	13	153.5		

$f_{2,11;0.05} = 3.98$; do not reject H_0.

CHAPTER 13

Section 13-1

1. Do not reject H_0.
3. Reject H_0 concluding that the program is beneficial.
5. Do not reject H_0; temperature has no effect on skiing skills.
7. Do not reject H_0; the treatment is ineffective.
9. Do not reject H_0; supervision does not effect productivity.

Section 13-2

1. $m = 10; n = 12; R_1 = 76.5; u = 98.5; \mu_U = 60; \sigma_U = 15.166; \dfrac{u - \mu_U}{\sigma_U} = 2.54$; reject H_0; the age distributions are not identical.

3. $m = 12; n = 15; R_1 = 185; u = 73; \mu_U = 90; \sigma_U = 20.494; \dfrac{u - \mu_U}{\sigma_U} = -0.83$; do not reject H_0.

5. $m = 10; n = 9; R_1 = 82; u = 63; \mu_U = 45; \sigma_U = 12.247; \dfrac{u - \mu_U}{\sigma_U} = 1.47$; do not reject H_0.

Section 13-3

1. $m = 20; n = 20; R = 29; \mu_R = 21; \sigma_R = 3.121; \dfrac{R - \mu_R}{\sigma_R} = 2.563$; reject H_0 that the sequence of answers is random.

3. $m = 14; n = 11; R = 15; \mu_R = 13.32; \sigma_R = 2.41; \dfrac{R - \mu_R}{\sigma_R} = 0.70$; do not reject H_0 that there was randomness.

5. $m = 13; n = 13; R = 18; \mu_R = 14; \sigma_R = 2.498; \dfrac{R - \mu_R}{\sigma_R} = 1.60$; do not reject H_0 that the number of defects are randomly distributed along the length of the wire.

7. Reject the hypothesis of randomness.

Section 13-4

1. $r_{rank} = 0.782$; $\sqrt{n-1}(r_{rank}) = 2.47$; reject H_0.

3. $r_{rank} = 0.067$; $\sqrt{n-1}(r_{rank}) = 0.2$; do not reject H_0.

5. $r_{rank} = 0.077$; do not reject H_0.

Chapter 13 Test

3. (a), (b): lack of randomness.

4. mean $= 18.14$; variance $= 8.139$; the approximate distribution is normal.

5. The sum of the ranks assigned to readings in mix 1 is 113. $u = 62$; $\mu_U = 60$; $\sigma_U = 15.17$; computed value $= 0.13$; do not reject H_0.

6. $n = 11$; $x = 5$; do not reject H_0.

7. computed value $= \dfrac{23 - 15.73}{2.641} = 2.75$; $z_{0.025} = 1.96$; reject H_0.

8. $r_{rank} = 0.212$; computed value $= 0.636$; do not reject H_0.

APPENDIX I

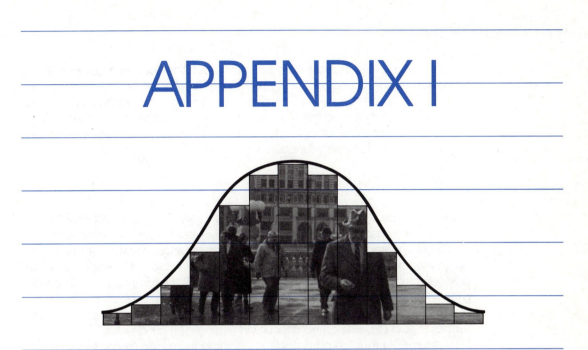

POISSON DISTRIBUTION

INTRODUCTION

The Poisson distribution provides a probabilistic model for a wide class of phenomena. Typical examples are the number of telephone calls during a given period of time, the number of particles emitted from a radioactive source in a given period of time, and the number of bacteria on a plate of a given size.

We say that a random variable X has a Poisson distribution with parameter λ (lambda) if its probability function is given by

$$p(x) = P(X = x) = \frac{e^{-\lambda}\lambda^x}{x!}$$

where x can assume any value 0, 1, 2, ... (the three dots mean *ad infinitum*).

Thus a Poisson random variable is a discrete random variable and any nonnegative integer is in its range. Here e is a universal constant (base of the natural logarithms) and has the value 2.7182 .... In Table A-1 in Appendix II, we give the probabilities $p(x)$ for certain values of λ. For example, if $\lambda = 3.2$, then $p(3) = 0.2226$, $p(5) = 0.1140$, and so on. Most scientific pocket calculators, available on the market, are equipped to find $e^{-\lambda}$ for any value of λ and the reader should experience no difficulty in finding $p(x)$ for any λ.

It turns out that the parameter λ is actually the mean (expected value) of the random variable. The variance of the random variable is also λ. Thus, the mean and the variance for a Poisson distribution coincide, both being equal to λ. ■

EXAMPLE 1

In an experiment conducted by Rutherford and Geiger, the mean number of α-particles emitted by a radioactive source during 7.5 seconds was found to be 3.9. Assuming that the number of α-particles emitted has a Poisson distribution, what is the probability that during 7.5 seconds the source will emit

(a) exactly 2 α-particles
(b) at most 2 α-particles
(c) at least 2 α-particles.

SOLUTION

The mean number of particles is 3.9. That is, $\lambda = 3.9$. Hence, referring to Table A-1 with $\lambda = 3.9$, we get

(a) $P(\text{exactly 2 } \alpha\text{-particles}) = p(2) = 0.1539$
(b) $P(\text{at most 2 } \alpha\text{-particles}) = p(0) + p(1) + p(2)$
$$= 0.0202 + 0.0789 + 0.1539$$
$$= 0.2530$$
(c) "At least 2 α-particles" means *not* 0 or 1 α-particle. Hence

$$P(\text{at least 2 } \alpha\text{-particles}) = 1 - P(0 \text{ or } 1 \text{ } \alpha\text{-particle})$$
$$= 1 - [p(0) + p(1)]$$
$$= 1 - (0.0202 + 0.0789)$$
$$= 0.9009. \quad ■$$

POISSON APPROXIMATION TO THE BINOMIAL PROBABILITIES

In Section 5-3, we saw that if p is not very near to 0 or 1, then for large n the binomial distribution can be approximated by a normal distribution. If p is near 0 or 1, the normal approximation is unsatisfactory. It can be shown, however, that if p is small and n is large, the Poisson distribution provides a satisfactory approximation to the binomial distribution. The statement of this result is as follows:

POISSON APPROXIMATION TO THE BINOMIAL

If n is large and p is small, the binomial probabilities $b(x; n, p)$ can be approximated by the Poisson probabilities with parameter np. Specifically

$$b(x; n, p) \approx \frac{e^{-np}(np)^x}{x!}, \quad x = 0, 1, 2 \dots .$$

If p is near 1, then of course, the probability of failure q will be small and, consequently, one can approximate the binomial probabilities of the number of failures by Poisson probabilities with parameter nq.

EXAMPLE 2   A computer has 1000 electronic components. The probability that a component will fail during one year of operation is 0.0022. Assuming the components function independently, find the approximate probability that during one year of operation

(a) 3 components fail
(b) between 1 and 3 components, inclusive, fail.

SOLUTION   (a) Let X represent the number of components failing. We are given that $n = 1000$ and $p = 0.0022$. The exact answer to the probability that 3 components fail is

$$b(3; \ 1000, \ 0.0022) \ \binom{1000}{3}(0.0022)^3(0.9978)^{997}.$$

Since p is small and n is large, we can use the Poisson approximation with parameter $np = 1000(0.0022) = 2.2$, and obtain an approximate answer as

$$b(3; \ 1000, \ 0.0022) \approx \frac{e^{-2.2}(2.2)^3}{3!} = 0.1966$$

referring to Table A-1 with $\lambda = 2.2$ and $x = 3$.

(b) Here we get

$$P(1 \leq X \leq 3) = b(1; 1000, 0.0022) + b(2; 1000, 0.0022)$$
$$+ \; b(3; 1000, 0.0022)$$
$$\approx \frac{e^{-2.2}(2.2)^1}{1!} + \frac{e^{-2.2}(2.2)^2}{2!} + \frac{e^{-2.2}(2.2)^3}{3!}$$
$$= 0.2438 + 0.2681 + 0.1966$$
$$= 0.7085. \quad \blacksquare$$

APPENDIX 1 EXERCISES

1. The number of elementary particles, recorded by a device in a space vehicle during a one-day flight, has a Poisson distribution with a mean of 3.5 particles. Find the probability that during a one-day flight there will be

   (a) no particle recorded
   (b) at least 4 particles recorded.

2. Suppose the number of telephone calls arriving in an office during a 15-minute period has a Poisson distribution with mean number of calls equal to 2.5. Find the probability that during a 15-minute period there will be

   (a) no calls
   (b) between 2 and 6 calls, inclusive.

3. Suppose a biological cell contains 400 genes. When treated radioactively, the probability that a gene will change into a mutant gene is 0.006 and is independent of the condition of the other genes. What is the approximate probability that there are at most 4 mutant genes after the treatment?

4. A book has 350 pages. The probability that a page has misprints (one or more) is 0.004. Find the approximate probability that the book has no page with misprints.

5. A salesman of heavy appliances has determined that the probability that a customer inquiring about a product will actually buy it is 0.01. What is the approximate probability that when 120 people make inquiries (independently), two or less will buy?

APPENDIX II

TABLES

TABLE A-1

Poisson Probabilities*

x	λ 0.1	0.2	0.3	0.4	0.5	0.6	0.7	0.8	0.9	1.0
0	0.9048	0.8187	0.7408	0.6703	0.6065	0.5488	0.4966	0.4493	0.4066	0.3679
1	0.0905	0.1637	0.2222	0.2681	0.3033	0.3293	0.3476	0.3595	0.3659	0.3679
2	0.0045	0.0164	0.0333	0.0536	0.0758	0.0988	0.1216	0.1438	0.1647	0.1839
3	0.0002	0.0011	0.0033	0.0072	0.0126	0.0198	0.0284	0.0383	0.0494	0.0613
4		0.0001	0.0002	0.0007	0.0016	0.0030	0.0050	0.0077	0.0111	0.0153
5				0.0001	0.0002	0.0003	0.0007	0.0012	0.0020	0.0031
6							0.0001	0.0002	0.0003	0.0005
7										0.0001

x	λ 1.1	1.2	1.3	1.4	1.5	1.6	1.7	1.8	1.9	2.0
0	0.3329	0.3012	0.2725	0.2466	0.2231	0.2019	0.1827	0.1653	0.1496	0.1353
1	0.3662	0.3614	0.3543	0.3452	0.3347	0.3230	0.3106	0.2975	0.2841	0.2707
2	0.2014	0.2169	0.2303	0.2417	0.2510	0.2584	0.2640	0.2678	0.2700	0.2707
3	0.0738	0.0867	0.0998	0.1127	0.1255	0.1378	0.1496	0.1607	0.1710	0.1804
4	0.0203	0.0260	0.0324	0.0395	0.0471	0.0551	0.0636	0.0723	0.0812	0.0902
5	0.0045	0.0062	0.0084	0.0111	0.0141	0.0176	0.0215	0.0260	0.0309	0.0361
6	0.0008	0.0012	0.0018	0.0026	0.0035	0.0047	0.0061	0.0078	0.0098	0.0120
7	0.0001	0.0002	0.0003	0.0005	0.0008	0.0011	0.0015	0.0020	0.0027	0.0034
8			0.0001	0.0001	0.0001	0.0002	0.0003	0.0005	0.0006	0.0009
9							0.0001	0.0001	0.0001	0.0002

x	λ 2.1	2.2	2.3	2.4	2.5	2.6	2.7	2.8	2.9	3.0
0	0.1225	0.1108	0.1003	0.0907	0.0821	0.0743	0.0672	0.0608	0.0550	0.0498
1	0.2572	0.2438	0.2306	0.2176	0.2052	0.1931	0.1815	0.1703	0.1596	0.1494
2	0.2700	0.2681	0.2652	0.2613	0.2565	0.2510	0.2450	0.2384	0.2314	0.2240
3	0.1890	0.1966	0.2033	0.2090	0.2138	0.2176	0.2205	0.2225	0.2237	0.2240
4	0.0992	0.1082	0.1169	0.1254	0.1336	0.1414	0.1488	0.1557	0.1622	0.1680
5	0.0416	0.0476	0.0538	0.0602	0.0668	0.0735	0.0804	0.0872	0.0940	0.1008
6	0.0146	0.0174	0.0206	0.0241	0.0278	0.0319	0.0362	0.0407	0.0455	0.0504
7	0.0044	0.0055	0.0068	0.0083	0.0099	0.0118	0.0139	0.0163	0.0188	0.0216
8	0.0011	0.0015	0.0019	0.0025	0.0031	0.0038	0.0046	0.0057	0.0068	0.0081
9	0.0003	0.0004	0.0005	0.0007	0.0009	0.0011	0.0014	0.0018	0.0022	0.0027
10	0.0001	0.0001	0.0001	0.0002	0.0002	0.0003	0.0004	0.0005	0.0006	0.0008
11						0.0001	0.0001	0.0001	0.0002	0.0002
12										0.0001

x	λ 3.1	3.2	3.3	3.4	3.5	3.6	3.7	3.8	3.9	4.0
0	0.0450	0.0408	0.0369	0.0334	0.0302	0.0273	0.0247	0.0224	0.0202	0.0183
1	0.1397	0.1304	0.1217	0.1135	0.1057	0.0983	0.0915	0.0850	0.0789	0.0733
2	0.2165	0.2087	0.2008	0.1929	0.1850	0.1771	0.1692	0.1615	0.1539	0.1465
3	0.2236	0.2226	0.2209	0.2186	0.2158	0.2125	0.2087	0.2046	0.2001	0.1954
4	0.1734	0.1781	0.1823	0.1858	0.1887	0.1912	0.1931	0.1944	0.1951	0.1954

TABLE A-1 (continued)

					λ					
x	3.1	3.2	3.3	3.4	3.5	3.6	3.7	3.8	3.9	4.0
5	0.1075	0.1140	0.1203	0.1264	0.1322	0.1377	0.1429	0.1477	0.1522	0.1563
6	0.0555	0.0607	0.0662	0.0716	0.0771	0.0826	0.0881	0.0936	0.0989	0.1042
7	0.0246	0.0278	0.0312	0.0348	0.0385	0.0425	0.0466	0.0508	0.0551	0.0595
8	0.0095	0.0111	0.0128	0.0148	0.0169	0.0191	0.0215	0.0241	0.0269	0.0298
9	0.0033	0.0040	0.0047	0.0056	0.0066	0.0076	0.0089	0.0102	0.0116	0.0132
10	0.0010	0.0013	0.0016	0.0018	0.0023	0.0028	0.0033	0.0039	0.0045	0.0053
11	0.0003	0.0004	0.0005	0.0006	0.0007	0.0009	0.0011	0.0013	0.0016	0.0019
12	0.0001	0.0001	0.0001	0.0002	0.0002	0.0003	0.0003	0.0004	0.0005	0.0006
13					0.0001	0.0001	0.0001	0.0001	0.0002	0.0002
14										0.0001

*The entries in the table give the Poisson probabilities $p(x) = \dfrac{e^{-\lambda}\lambda^x}{x!}$ for certain values of λ. For example, for $\lambda = 3.2$, $x = 3$, $p(3) = 0.2226$.

TABLE A-2

Binomial probabilities, denoted $b(x; n, p)$

							p					
n	x	0.05	0.1	0.2	0.3	0.4	0.5	0.6	0.7	0.8	0.9	0.95
2	0	0.902	0.810	0.640	0.490	0.360	0.250	0.160	0.090	0.040	0.010	0.002
	1	0.095	0.180	0.320	0.420	0.480	0.500	0.480	0.420	0.320	0.180	0.095
	2	0.002	0.010	0.040	0.090	0.160	0.250	0.360	0.490	0.640	0.810	0.902
3	0	0.857	0.729	0.512	0.343	0.216	0.125	0.064	0.027	0.008	0.001	
	1	0.135	0.243	0.384	0.441	0.432	0.375	0.288	0.189	0.096	0.027	0.007
	2	0.007	0.027	0.096	0.189	0.288	0.375	0.432	0.441	0.384	0.243	0.135
	3		0.001	0.008	0.027	0.064	0.125	0.216	0.343	0.512	0.729	0.857
4	0	0.815	0.656	0.410	0.240	0.130	0.062	0.026	0.008	0.002		
	1	0.171	0.292	0.410	0.412	0.346	0.250	0.154	0.076	0.026	0.004	
	2	0.014	0.049	0.154	0.265	0.346	0.375	0.346	0.265	0.154	0.049	0.014
	3		0.004	0.026	0.076	0.154	0.250	0.346	0.412	0.410	0.292	0.171
	4			0.0002	0.008	0.026	0.062	0.130	0.240	0.410	0.656	0.815
5	0	0.774	0.590	0.328	0.168	0.078	0.031	0.010	0.002			
	1	0.204	0.328	0.410	0.360	0.259	0.156	0.077	0.028	0.006		
	2	0.021	0.073	0.205	0.309	0.346	0.312	0.230	0.132	0.051	0.008	0.001
	3	0.001	0.008	0.051	0.132	0.230	0.312	0.346	0.309	0.205	0.073	0.021
	4			0.006	0.028	0.077	0.156	0.259	0.360	0.410	0.328	0.204
	5				0.002	0.010	0.031	0.078	0.168	0.328	0.590	0.774
6	0	0.735	0.531	0.262	0.118	0.047	0.016	0.004	0.001			
	1	0.232	0.354	0.393	0.303	0.187	0.094	0.037	0.010	0.002		
	2	0.031	0.098	0.246	0.324	0.311	0.234	0.138	0.060	0.015	0.001	
	3	0.002	0.015	0.082	0.185	0.276	0.312	0.276	0.185	0.082	0.015	0.002
	4		0.001	0.015	0.060	0.138	0.234	0.311	0.324	0.246	0.098	0.031
	5			0.002	0.010	0.037	0.094	0.187	0.303	0.393	0.354	0.232
	6				0.001	0.004	0.016	0.047	0.118	0.262	0.531	0.735

TABLE A-2 (continued)

							p					
n	x	0.05	0.1	0.2	0.3	0.4	0.5	0.6	0.7	0.8	0.9	0.95
7	0	0.698	0.478	0.210	0.082	0.028	0.008	0.002				
	1	0.257	0.372	0.367	0.247	0.131	0.055	0.017	0.004			
	2	0.041	0.124	0.275	0.318	0.261	0.164	0.077	0.025	0.004		
	3	0.004	0.023	0.115	0.227	0.290	0.273	0.194	0.097	0.029	0.003	
	4		0.003	0.029	0.097	0.194	0.273	0.290	0.227	0.115	0.023	0.004
	5			0.004	0.025	0.077	0.164	0.261	0.318	0.275	0.124	0.041
	6				0.004	0.017	0.055	0.131	0.247	0.367	0.372	0.257
	7					0.002	0.008	0.028	0.082	0.210	0.478	0.698
8	0	0.663	0.430	0.168	0.058	0.017	0.004	0.001				
	1	0.279	0.383	9.336	0.198	0.090	0.031	0.008	0.001			
	2	0.051	0.149	0.294	0.296	0.209	0.109	0.041	0.010	0.001		
	3	0.005	0.033	0.147	0.254	0.279	0.219	0.124	0.047	0.009		
	4		0.005	0.046	0.136	0.232	0.273	0.232	0.136	0.046	0.005	
	5			0.009	0.047	0.124	0.219	0.279	0.254	0.147	0.033	0.005
	6			0.001	0.010	0.041	0.109	0.209	0.296	0.294	0.149	0.051
	7				0.001	0.008	0.031	0.090	0.198	0.336	0.383	0.279
	8					0.001	0.004	0.017	0.058	0.168	0.430	0.663
9	0	0.630	0.387	0.134	0.040	0.010	0.02					
	1	0.299	0.387	0.302	0.156	0.060	0.018	0.004				
	2	0.063	0.172	0.302	0.267	0.161	0.070	0.021	0.004			
	3	0.008	0.045	0.176	0.267	0.251	0.164	0.074	0.021	0.003		
	4	0.001	0.007	0.066	0.172	0.251	0.246	0.167	0.074	0.017	0.001	
	5		0.001	0.017	0.074	0.167	0.246	0.251	0.172	0.066	0.007	0.001
	6			0.003	0.021	0.074	0.164	0.251	0.267	0.176	0.045	0.008
	7				0.004	0.021	0.070	0.161	0.267	0.302	0.172	0.063
	8					0.004	0.018	0.060	0.156	0.302	0.387	0.299
	9						0.002	0.010	0.040	0.134	0.387	0.630
10	0	0.599	0.349	0.107	0.028	0.006	0.001					
	1	0.315	0.387	0.268	0.121	0.040	0.010	0.002				
	2	0.075	0.194	0.302	0.233	0.121	0.044	0.011	0.001			
	3	0.010	0.057	0.201	0.267	0.215	0.117	0.042	0.009	0.001		
	4	0.001	0.011	0.088	0.200	0.251	0.205	0.111	0.037	0.006		
	5		0.001	0.026	0.103	0.201	0.246	0.201	0.103	0.026	0.001	
	6			0.006	0.037	0.111	0.205	0.251	0.200	0.088	0.011	0.001
	7			0.001	0.009	0.042	0.117	0.215	0.267	0.201	0.057	0.010
	8				0.001	0.011	0.044	0.121	0.233	0.302	0.194	0.075
	9					0.002	0.010	0.040	0.121	0.268	0.387	0.315
	10						0.001	0.006	0.028	0.107	0.349	0.599

TABLE A-2 (continued)

							p					
n	x	0.05	0.1	0.2	0.3	0.4	0.5	0.6	0.7	0.8	0.9	0.95
11	0	0.569	0.314	0.086	0.020	0.004						
	1	0.329	0.384	0.236	0.093	0.027	0.005	0.001				
	2	0.087	0.213	0.295	0.200	0.089	0.027	0.005	0.001			
	3	0.014	0.071	0.221	0.257	0.177	0.081	0.023	0.004			
	4	0.001	0.016	0.111	0.220	0.236	0.161	0.070	0.017	0.002		
	5		0.002	0.039	0.132	0.221	0.226	0.147	0.057	0.010		
	6			0.010	0.057	0.147	0.226	0.221	0.132	0.039	0.002	
	7			0.002	0.017	0.070	0.161	0.236	0.220	0.111	0.016	0.001
	8				0.004	0.023	0.081	0.177	0.257	0.221	0.071	0.014
	9				0.001	0.005	0.027	0.089	0.200	0.295	0.213	0.087
	10					0.001	0.005	0.027	0.093	0.236	0.384	0.329
	11							0.004	.0.020	0.086	0.314	0.569
12	0	0.540	0.282	0.069	0.014	0.002						
	1	0.341	0.377	0.206	0.071	0.017	0.003					
	2	0.099	0.230	0.283	0.168	0.064	0.016	0.002				
	3	0.017	0.085	0.236	0.240	0.142	0.054	0.012	0.001			
	4	0.002	0.021	0.133	0.231	0.213	0.121	0.042	0.008	0.001		
	5		0.004	0.053	0.158	0.227	0.193	0.101	0.029	0.003		
	6			0.016	0.079	0.177	0.226	0.177	0.079	0.016		
	7			0.003	0.029	0.101	0.193	0.227	0.158	0.053	0.004	
	8			0.001	0.008	0.042	0.121	0.213	0.231	0.133	0.021	0.002
	9				0.001	0.012	0.054	0.142	0.240	0.236	0.085	0.017
	10					0.002	0.016	0.064	0.168	0.283	0.230	0.099
	11						0.003	0.017	0.071	0.206	0.377	0.341
	12							0.002	0.014	0.069	0.282	0.540
13	0	0.513	0.254	0.055	0.010	0.001						
	1	0.351	0.367	0.179	0.054	0.011	0.002					
	2	0.011	0.245	0.268	0.139	0.045	0.010	0.001				
	3	0.021	0.100	0.246	0.218	0.111	0.035	0.006	0.001			
	4	0.003	0.028	0.154	0.234	0.184	0.087	0.024	0.003			
	5		0.006	0.069	0.180	0.221	0.157	0.066	0.014	0.001		
	6		0.001	0.023	0.103	0.197	0.209	0.131	0.044	0.006		
	7			0.006	0.044	0.131	0.209	0.197	0.103	0.023	0.001	
	8			0.001	0.014	0.066	0.157	0.221	0.180	0.069	0.006	
	9				0.003	0.024	0.087	0.184	0.234	0.154	0.028	0.003
	10				0.001	0.006	0.035	0.111	0.218	0.246	0.100	0.021
	11					0.001	0.010	0.045	0.139	0.268	0.245	0.111
	12						0.002	0.011	0.054	0.179	0.367	0.351
	13							0.001	0.010	0.055	0.254	0.513

TABLE A-2 (continued)

							p					
n	x	0.05	0.1	0.2	0.3	0.4	0.5	0.6	0.7	0.8	0.9	0.95
14	0	0.488	0.229	0.044	0.007	0.001						
	1	0.359	0.356	0.154	0.041	0.007	0.001					
	2	0.123	0.257	0.250	0.113	0.032	0.006	0.001				
	3	0.026	0.114	0.250	0.194	0.085	0.022	0.003				
	4	0.004	0.035	0.172	0.229	0.155	0.061	0.014	0.001			
	5		0.008	0.086	0.196	0.207	0.122	0.041	0.007			
	6		0.001	0.032	0.126	0.207	0.183	0.092	0.023	0.002		
	7			0.009	0.062	0.157	0.209	0.157	0.062	0.009		
	8			0.002	0.023	0.092	0.183	0.207	0.126	0.032	0.001	
	9				0.007	0.041	0.122	0.207	0.196	0.086	0.008	
	10				0.001	0.014	0.061	0.155	0.229	0.172	0.035	0.004
	11					0.003	0.022	0.085	0.194	0.250	0.114	0.026
	12					0.001	0.006	0.032	0.113	0.250	0.257	0.123
	13						0.001	0.007	0.041	0.154	0.356	0.359
	14							0.001	0.007	0.044	0.229	0.488
15	0	0.463	0.206	0.035	0.005							
	1	0.366	0.343	0.132	0.031	0.005						
	2	0.135	0.267	0.231	0.092	0.022	0.003					
	3	0.031	0.129	0.250	0.170	0.063	0.014	0.002				
	4	0.005	0.043	0.188	0.219	0.127	0.042	0.007	0.001			
	5	0.001	0.010	0.103	0.206	0.186	0.092	0.024	0.003			
	6		0.002	0.043	0.147	0.207	0.153	0.061	0.012	0.001		
	7			0.014	0.081	0.177	0.196	0.118	0.035	0.003		
	8			0.003	0.035	0.118	0.196	0.177	0.081	0.014		
	9			0.001	0.012	0.061	0.153	0.207	0.147	0.043	0.002	
	10				0.003	0.024	0.092	0.186	0.206	0.103	0.010	0.001
	11				0.001	0.007	0.042	0.127	0.219	0.188	0.043	0.005
	12					0.002	0.014	0.063	0.170	0.250	0.129	0.031
	13						0.003	0.022	0.092	0.231	0.267	0.135
	14							0.005	0.031	0.132	0.343	0.366
	15								0.005	0.035	0.206	0.463

TABLE A-3

The standard normal distribution (area from 0 to z, z positive)

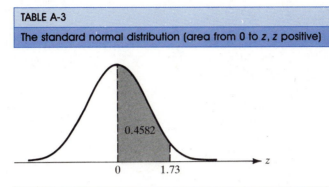

z	0.00	0.01	0.02	0.03	0.04	0.05	0.06	0.07	0.08	0.09
0.0	0.0000	0.0040	0.0080	0.0120	0.0160	0.0199	0.0239	0.0279	0.0319	0.0359
0.1	0.0398	0.0438	0.0478	0.0517	0.0557	0.0596	0.0636	0.0675	0.0714	0.0753
0.2	0.0793	0.0832	0.0871	0.0910	0.0948	0.0987	0.1026	0.1064	0.1103	0.1141
0.3	0.1179	0.1217	0.1255	0.1293	0.1331	0.1368	0.1406	0.1443	0.1480	0.1517
0.4	0.1554	0.1591	0.1628	0.1664	0.1700	0.1736	0.1772	0.1808	0.1844	0.1879
0.5	0.1915	0.1950	0.1985	0.2019	0.2054	0.2088	0.2123	0.2157	0.2190	0.2224
0.6	0.2257	0.2291	0.2324	0.2357	0.2389	0.2422	0.2454	0.2486	0.2517	0.2549
0.7	0.2580	0.2611	0.2642	0.2673	0.2704	0.2734	0.2764	0.2794	0.2823	0.2852
0.8	0.2881	0.2910	0.2939	0.2967	0.2995	0.3023	0.3051	0.3078	0.3106	0.3133
0.9	0.3159	9.3186	0.3212	0.3238	0.3264	0.3289	0.3315	0.3340	0.3365	0.3389
1.0	0.3413	0.3438	0.3461	0.3485	0.3508	0.3531	0.3554	0.3577	0.3599	0.3621
1.1	0.3643	0.3665	0.3686	0.3708	0.3729	0.3749	0.3770	0.3790	0.3810	0.3830
1.2	0.3849	0.3869	0.3888	0.3907	0.3925	0.3944	0.3962	0.3980	0.3997	0.4015
1.3	0.4032	0.4049	0.4066	0.4082	0.4099	0.4115	0.4131	0.4147	0.4162	0.4177
1.4	0.4192	0.4207	0.4222	0.4236	0.4251	0.4265	0.4279	0.4292	0.4306	0.4319
1.5	0.4332	0.4345	0.4357	0.4370	0.4382	0.4394	0.4406	0.4418	0.4429	0.4441
1.6	0.4452	0.4463	0.4474	0.4484	0.4495	0.4505	0.4515	0.4525	0.4535	0.4545
1.7	0.4554	0.4564	0.4573	0.4582	0.4591	0.4599	0.4608	0.4616	0.4625	0.4633
1.8	0.4641	0.4649	0.4656	0.4664	0.4671	0.4678	0.4686	0.4692	0.4699	0.4706
1.9	0.4713	0.4719	0.4726	0.4732	0.4738	0.4774	0.4750	0.4756	0.4761	0.4767
2.0	0.4772	0.4778	0.4783	0.4788	0.4793	0.4798	0.4803	0.4808	0.4812	0.4817
2.1	0.4821	0.4826	0.4830	0.4834	0.4838	0.4842	0.4846	0.4850	0.4854	0.4857
2.2	0.4861	0.4864	0.4868	0.4871	0.4875	0.4878	0.4881	0.4884	0.4887	0.4890
2.3	0.4893	0.4896	0.4898	0.4901	0.4904	0.4906	0.4909	0.4911	0.4913	0.4916
2.4	0.4918	0.4920	0.4922	0.4925	0.4927	0.4929	0.4931	0.4932	0.4934	0.4936
2.5	0.4938	0.4940	0.4941	0.4943	0.4945	0.4946	0.4948	0.4949	0.4951	0.4952
2.6	0.4953	0.4955	0.4956	0.4957	0.4959	0.4960	0.4961	0.4962	0.4963	0.4964
2.7	0.4965	0.4966	0.4967	0.4968	0.4969	0.4970	0.4971	0.4972	0.4973	0.4974
2.8	0.4974	0.4975	0.4976	0.4977	0.4977	0.4978	0.4979	0.4979	0.4980	0.4981
2.9	0.4981	0.4982	0.4982	0.4983	0.4984	0.4984	0.4985	0.4985	0.4986	0.4986
3.0	0.4987	0.4987	0.4987	0.4988	0.4988	0.4989	0.4989	0.4989	0.4990	0.4990

TABLE A-4

The standard normal distribution (areas in the right tail)

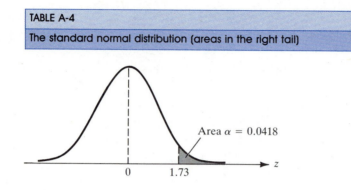

Area $\alpha = 0.0418$

z	0.00	0.01	0.02	0.03	0.04	0.05	0.06	0.07	0.08	0.09
0.0	0.5000	0.4960	0.4920	0.4880	0.4840	0.4801	0.4761	0.4721	0.4681	0.4641
0.1	0.4602	0.4562	0.4522	0.4483	0.4443	0.4404	0.4364	0.4325	0.4286	0.4247
0.2	0.4207	0.4168	0.4129	0.4090	0.4052	0.4013	0.3974	0.3936	0.3897	0.3859
0.3	0.3821	0.3783	0.3745	0.3707	0.3669	0.3632	0.3594	0.3557	0.3520	0.3483
0.4	0.3446	0.3409	0.3372	0.3336	0.3300	0.3264	0.3228	0.3192	0.3156	0.3121
0.5	0.3085	0.3050	0.3015	0.2981	0.2946	0.2912	0.2877	0.2843	0.2810	0.2776
0.6	0.2743	0.2709	0.2676	0.2643	0.2611	0.2578	0.2546	0.2514	0.2483	0.2451
0.7	0.2420	0.2389	0.2358	0.2327	0.2296	0.2266	0.2236	0.2206	0.2177	0.2148
0.8	0.2119	0.2090	0.2061	0.2033	0.2005	0.1977	0.1949	0.1922	0.1894	0.1867
0.9	0.1841	0.1814	0.1788	0.1762	0.1736	0.1711	0.1685	0.1660	0.1635	0.1611
1.0	0.1587	0.1562	0.1539	0.1515	0.1492	0.1469	0.1446	0.1423	0.1401	0.1379
1.1	0.1357	0.1335	0.1314	0.1292	0.1271	0.1251	0.1230	0.1210	0.1190	0.1170
1.2	0.1151	0.1131	0.1112	0.1093	0.1075	0.1056	0.1038	0.1020	0.1003	0.0985
1.3	0.0968	0.0951	0.0934	0.0918	0.0901	0.0885	0.0869	0.0853	0.0838	0.0823
1.4	0.0808	0.0793	0.0778	0.0764	0.0749	0.0735	0.0721	0.0708	0.0694	0.0681
1.5	0.0668	0.0655	0.0643	0.0630	0.0618	0.0606	0.0594	0.0582	0.0571	0.0559
1.6	0.0548	0.0537	0.0526	0.0516	0.0505	0.0495	0.0485	0.0475	0.0465	0.0455
1.7	0.0446	0.0436	0.0427	0.0418	0.0409	0.0401	0.0392	0.0384	0.0375	0.0367
1.8	0.0359	0.0351	0.0344	0.0336	0.0329	0.0322	0.0314	0.0307	0.0301	0.0294
1.9	0.0287	0.0281	0.0274	0.0268	0.0262	0.0256	0.0250	0.0244	0.0239	0.0233
2.0	0.0228	0.0222	0.0217	0.0212	0.0207	0.0202	0.0197	0.0192	0.0188	0.0183
2.1	0.0179	0.0174	0.0170	0.0166	0.0162	0.0158	0.0154	0.0150	0.0146	0.0143
2.2	0.0139	0.0136	0.0132	0.0129	0.0125	0.0122	0.0119	0.0116	0.0113	0.0110
2.3	0.0107	0.0104	0.0102	0.0099	0.0096	0.0094	0.0091	0.0089	0.0087	0.0084
2.4	0.0082	0.0080	0.0078	0.0075	0.0073	0.0071	0.0069	0.0068	0.0066	0.0064
2.5	0.0062	0.0060	0.0059	0.0057	0.0055	0.0054	0.0052	0.0051	0.0049	0.0048
2.6	0.0047	0.0045	0.0044	0.0043	0.0041	0.0040	0.0039	0.0038	0.0037	0.0036
2.7	0.0035	0.0034	0.0033	0.0032	0.0031	0.0030	0.0029	0.0028	0.0027	0.0026
2.8	0.0026	0.0025	0.0024	0.0023	0.0023	0.0022	0.0021	0.0021	0.0020	0.0019
2.9	0.0019	0.0018	0.0018	0.0017	0.0016	0.0016	0.0015	0.0015	0.0014	0.0014
3.0	0.0013	0.0013	0.0013	0.0012	0.0012	0.0011	0.0011	0.0011	0.0010	0.0010

TABLE A-5

Values of *t* for given probability levels

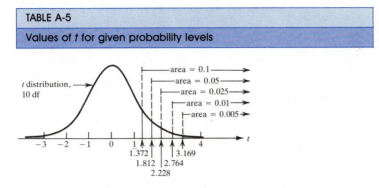

Degrees of Freedom ν	α, area in the right tail				
	0.1	0.05	0.025	0.01	0.005
	$t_{\nu,0.1}$	$t_{\nu,0.05}$	$t_{\nu,025}$	$t_{\nu,0.01}$	$t_{\nu,0.005}$
1	3.078	6.314	12.706	31.821	63.657
2	1.886	2.920	4.303	6.965	9.925
3	1.638	2.353	3.182	4.541	5.841
4	1.533	2.132	2.776	3.747	4.604
5	1.476	2.015	2.571	3.365	4.032
6	1.440	1.943	2.447	3.143	3.707
7	1.415	1.895	2.365	2.998	3.499
8	1.397	1.860	2.306	2.896	3.355
9	1.383	1.833	2.262	2.821	3.250
10	1.372	1.812	2.228	2.764	3.169
11	1.363	1.796	2.201	2.718	3.106
12	1.356	1.782	2.179	2.681	3.055
13	1.350	1.771	2.160	2.650	3.012
14	1.345	1.761	2.145	2.624	2.977
15	1.341	1.753	2.131	2.602	2.947
16	1.337	1.746	2.120	2.583	2.921
17	1.333	1.740	2.110	2.567	2.898
18	1.330	1.734	2.101	2.552	2.878
19	1.328	1.729	2.093	2.539	2.861
20	1.325	1.725	2.086	2.528	2.845
21	1.323	1.721	2.080	2.518	2.831
22	1.321	1.717	2.074	2.508	2.819
23	1.319	1.714	2.069	2.500	2.807
24	1.318	1.711	2.064	2.492	2.797
25	1.316	1.708	2.060	2.485	2.787

TABLE A-5 (continued)

Degrees of Freedom ν	α, area in the right tail				
	0.1	0.05	0.025	0.01	0.005
	$t_{\nu,0.1}$	$t_{\nu,0.05}$	$t_{\nu,025}$	$t_{\nu,0.01}$	$t_{\nu,0.005}$
26	1.315	1.706	2.056	2.479	2.779
27	1.314	1.703	2.052	2.473	2.771
28	1.313	1.701	2.048	2.467	2.763
29	1.311	1.699	2.045	2.462	2.756
30	1.310	1.697	2.042	2.457	2.750
40	1.303	1.684	2.021	2.423	2.704
60	1.296	1.671	2.000	2.390	2.660
120	1.290	1.661	1.984	2.358	2.626
∞	1.282	1.645	1.960	2.326	2.576

Table A-4 is taken from Table III of Fisher and Yates: *Statistical Tables for Biological, Agricultural and Medical Research,* published by Longman Group Ltd., London (previously published by Oliver & Boyd, Edinburgh), and by permission of the authors and publishers.

TABLE A-6

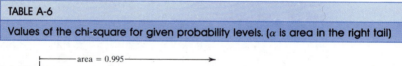

Values of the chi-square for given probability levels. (α is area in the right tail)

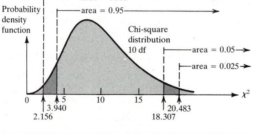

Degrees of freedom ν	α, area in the right tail							
	0.995	0.99	0.975	0.95	0.05	0.025	0.01	0.005
	$\chi^2_{\nu,0.995}$	$\chi^2_{\nu,0.99}$	$\chi^2_{\nu,0.975}$	$\chi^2_{\nu,0.95}$	$\chi^2_{\nu,0.05}$	$\chi^2_{\nu,0.025}$	$\chi^2_{\nu,0.01}$	$\chi^2_{\nu,0.005}$
1	0.0^4393	0.0^3157	0.0^3982	0.0^2393	3.841	5.024	6.635	7.879
2	0.0100	0.0201	0.0506	0.103	5.991	7.378	9.210	10.597
3	0.0717	0.115	0.216	0.352	7.815	9.348	11.345	12.838
4	0.207	0.297	0.484	0.711	9.488	11.143	13.277	14.860
5	0.412	0.554	0.831	1.145	11.070	12.832	15.086	16.750

TABLE A-6 (continued)

Degrees of freedom ν	$\chi^2_{\nu,0.995}$	$\chi^2_{\nu,0.99}$	$\chi^2_{\nu,0.975}$	$\chi^2_{\nu,0.95}$	$\chi^2_{\nu,0.05}$	$\chi^2_{\nu,0.025}$	$\chi^2_{\nu,0.01}$	$\chi^2_{\nu,0.005}$
	0.995	0.99	0.975	0.95	0.05	0.025	0.01	0.005
6	0.676	0.872	1.237	1.635	12.592	14.449	16.812	18.548
7	0.989	1.239	1.690	2.167	14.067	16.013	18.475	20.278
8	1.344	1.646	2.180	2.733	15.507	17.535	20.090	21.955
9	1.735	2.088	2.700	3.325	16.919	19.023	21.666	23.589
10	2.156	2.558	3.247	3.940	18.307	20.483	23.209	25.188
11	2.603	3.053	3.816	4.575	19.675	21.920	24.725	26.757
12	3.074	3.571	4.404	5.226	21.026	23.337	26.217	28.300
13	3.565	4.107	5.009	5.892	22.362	24.736	27.688	29.819
14	4.075	4.660	5.629	6.571	23.685	26.119	29.141	31.319
15	4.601	5.229	6.262	7.261	24.996	27.488	30.578	32.801
16	5.142	5.812	6.908	7.962	26.296	28.845	32.000	34.267
17	5.697	6.408	7.564	8.672	27.587	30.191	33.409	35.718
18	6.265	7.015	8.231	9.390	28.869	31.526	34.805	37.156
19	6.844	7.633	8.907	10.117	30.144	32.852	36.191	38.582
20	7.434	8.260	9.591	10.851	31.410	34.170	37.566	39.997
21	8.034	8.897	10.283	11.591	32.671	35.479	38.932	41.401
22	8.643	9.542	10.982	12.338	33.924	36.781	40.289	42.796
23	9.260	10.196	11.689	13.091	35.172	38.076	41.638	44.181
24	9.886	10.856	12.401	13.848	36.415	39.364	42.980	45.558
25	10.520	11.524	13.120	14.611	37.652	40.646	44.314	46.928
26	11.160	12.198	13.844	15.379	38.885	41.923	45.652	48.290
27	11.808	12.879	14.573	16.151	40.113	43.194	46.963	49.645
28	12.461	13.565	15.308	16.928	41.337	44.461	48.278	50.993
29	13.121	14.256	16.047	17.708	42.557	45.722	49.588	52.336
30	13.787	14.953	16.791	18.493	43.773	46.979	50.892	53.672

Abridged from Table 8 of *Biometrika Tables for Statisticians*, Vol 1, by permission of Biometrika Trustees.

TABLE A-7

Percentage points of the F distribution.

(a) Upper 10 percent points

Upper 10 percent points

v_2 \ v_1	1	2	3	4	5	6	7	8	9	10	12	15	20	24	30	40	60	120	∞
1	39.86	49.50	53.59	55.83	57.24	58.20	58.91	59.44	59.86	60.19	60.71	61.22	61.74	62.00	62.26	62.53	62.79	63.06	63.33
2	8.53	9.00	9.16	9.24	9.29	9.33	9.35	9.37	8.38	9.39	9.41	9.42	9.44	9.45	9.46	9.47	9.47	9.48	9.49
3	5.54	5.46	5.39	5.34	5.31	5.28	5.27	5.25	5.24	5.23	5.22	5.20	5.18	5.18	5.17	5.16	5.15	5.14	5.13
4	4.54	4.32	4.19	4.11	4.05	4.01	3.98	3.95	3.94	3.92	3.90	3.87	3.84	3.83	3.82	3.80	3.79	3.78	3.76
5	4.06	3.78	3.62	3.52	3.45	3.40	3.37	3.34	3.32	3.30	3.27	3.24	3.21	3.19	3.17	3.16	3.14	3.12	3.10
6	3.78	3.46	3.29	3.18	3.11	3.05	3.01	2.98	2.96	2.94	2.90	2.87	2.84	2.82	2.80	2.78	2.76	2.74	2.72
7	3.59	3.26	3.07	2.96	2.88	2.83	2.78	2.75	2.72	2.70	2.67	2.63	2.59	2.58	2.56	2.54	2.51	2.49	2.47
8	3.46	3.11	2.92	2.81	2.73	2.67	2.62	2.59	2.56	2.54	2.50	2.46	2.42	2.40	2.38	2.36	2.34	2.32	2.29
9	3.36	3.01	2.81	2.69	2.61	2.55	2.51	2.47	2.44	2.42	2.38	2.34	2.30	2.28	2.25	2.23	2.21	2.18	2.16
10	3.29	2.92	2.73	2.61	2.52	2.46	2.41	2.38	2.35	2.32	2.28	2.24	2.20	2.18	2.16	2.13	2.11	2.08	2.06
11	3.23	2.86	2.66	2.54	2.45	2.39	2.34	2.30	2.27	2.25	2.21	2.17	2.12	2.10	2.08	2.05	2.03	2.00	1.97
12	3.18	2.81	2.61	2.48	2.39	2.33	2.28	2.24	2.21	2.19	2.15	2.10	2.06	2.04	2.01	1.99	1.96	1.93	1.90
13	3.14	2.76	2.56	2.43	2.35	2.28	2.23	2.20	2.16	2.14	2.10	2.05	2.01	1.98	1.96	1.93	1.90	1.88	1.85
14	3.10	2.73	2.52	2.39	2.31	2.24	2.19	2.15	2.12	2.10	2.05	2.01	1.96	1.94	1.91	1.89	1.86	1.83	1.80
15	3.07	2.70	2.49	2.36	2.27	2.21	2.16	2.12	2.09	2.06	2.02	1.97	1.92	1.90	1.87	1.85	1.82	1.79	1.76
16	3.05	2.67	2.46	2.33	2.24	2.18	2.13	2.09	2.06	2.03	1.99	1.94	1.89	1.87	1.84	1.81	1.78	1.75	1.72
17	3.03	2.64	2.44	2.31	2.22	2.15	2.10	2.06	2.03	2.00	1.96	1.91	1.86	1.84	1.81	1.78	1.75	1.72	1.69
18	3.01	2.62	2.42	2.29	2.20	2.13	2.08	2.04	2.00	1.98	1.93	1.89	1.84	1.81	1.78	1.75	1.72	1.69	1.66
19	2.99	2.61	2.40	2.27	2.18	2.11	2.06	2.02	1.98	1.96	1.91	1.86	1.81	1.79	1.76	1.73	1.70	1.67	1.63

Degrees of freedom of the numerator

Degrees of freedom of the denominator

TABLE A-7 (continued)

Upper 10 percent points

v_2 \ v_1	1	2	3	4	5	6	7	8	9	10	12	15	20	24	30	40	60	120	∞
									Degrees of freedom of the numerator										
20	2.97	2.59	2.38	2.25	2.16	2.09	2.04	2.00	1.96	1.94	1.89	1.84	1.79	1.77	1.74	1.71	1.68	1.64	1.61
21	2.96	2.57	2.36	2.23	2.14	2.08	2.02	1.98	1.95	1.92	1.87	1.83	1.78	1.75	1.72	1.69	1.66	1.62	1.59
22	2.95	2.56	2.35	2.22	2.13	2.06	2.01	1.97	1.93	1.90	1.86	1.81	1.76	1.73	1.70	1.67	1.64	1.60	1.57
23	2.94	2.55	2.34	2.21	2.11	2.05	1.99	1.95	1.92	1.89	1.84	1.80	1.74	1.72	1.69	1.66	1.62	1.59	1.55
24	2.93	2.54	2.33	2.19	2.10	2.04	1.98	1.94	1.91	1.88	1.83	1.78	1.73	1.70	1.67	1.64	1.61	1.57	1.53
25	2.92	2.53	2.32	2.18	2.09	2.02	1.97	1.93	1.89	1.87	1.82	1.77	1.72	1.69	1.66	1.63	1.59	1.56	1.52
26	2.91	2.52	2.31	2.17	2.08	2.01	1.96	1.92	1.88	1.86	1.81	1.76	1.71	1.68	1.65	1.61	1.58	1.54	1.50
27	2.90	2.51	2.30	2.17	2.07	2.00	1.95	1.91	1.87	1.85	1.80	1.75	1.70	1.67	1.64	1.60	1.57	1.53	1.49
28	2.89	2.50	2.29	2.16	2.06	2.00	1.94	1.90	1.87	1.84	1.79	1.74	1.69	1.66	1.63	1.59	1.56	1.52	1.48
29	2.89	2.50	2.28	2.15	2.06	1.99	1.93	1.89	1.86	1.83	1.78	1.73	1.68	1.65	1.62	1.58	1.55	1.51	1.47
30	2.88	2.49	2.28	2.14	2.05	1.98	1.93	1.88	1.85	1.82	1.77	1.72	1.67	1.64	1.61	1.57	1.54	1.50	1.46
40	2.84	2.44	2.23	2.09	2.00	1.93	1.87	1.83	1.79	1.76	1.71	1.66	1.61	1.57	1.54	1.51	1.47	1.42	1.38
60	2.79	2.39	2.18	2.04	1.95	1.87	1.82	1.77	1.74	1.71	1.66	1.60	1.54	1.51	1.48	1.44	1.40	1.35	1.29
120	2.75	2.35	2.13	1.99	1.90	1.82	1.77	1.72	1.68	1.65	1.60	1.55	1.48	1.45	1.41	1.37	1.32	1.26	1.19
∞	2.71	2.30	2.08	1.94	1.85	1.77	1.72	1.67	1.63	1.60	1.55	1.49	1.42	1.38	1.34	1.30	1.24	1.17	1.00

TABLE A-7 (continued)

Upper 5 percent points

Probability density function

(b) Upper 5 percent points

										Degrees of freedom of the numerator									
v_2 \ v_1	1	2	3	4	5	6	7	8	9	10	12	15	20	24	30	40	60	120	∞
1	161.4	199.5	215.7	224.6	230.2	234.0	236.8	238.9	240.5	241.9	243.9	245.9	248.0	249.1	250.1	251.1	252.2	253.3	254.3
2	18.51	19.00	19.16	19.25	19.30	19.33	19.35	19.37	19.38	19.40	19.41	19.43	19.45	19.45	19.46	19.47	19.48	19.49	19.50
3	10.13	9.55	9.28	9.12	9.01	8.94	8.89	8.85	8.81	8.79	8.74	8.70	8.66	8.64	8.62	8.59	8.57	8.55	8.53
4	7.71	6.94	6.59	6.39	6.26	6.16	6.09	6.04	6.00	5.96	5.91	5.86	5.80	5.77	5.75	5.72	5.69	5.66	5.63
5	6.61	5.79	5.41	5.19	5.05	4.95	4.88	4.82	4.77	4.74	4.68	4.62	4.56	4.53	4.50	4.46	4.43	4.40	4.36
6	5.99	5.14	4.76	4.53	4.39	4.28	4.21	4.15	4.10	4.06	4.00	3.94	3.87	3.84	3.81	3.77	3.74	3.70	3.67
7	5.59	4.74	4.35	4.12	3.97	3.87	3.79	3.73	3.68	3.64	3.57	3.51	3.44	3.41	3.38	3.34	3.30	3.27	3.23
8	5.32	4.46	4.07	3.84	3.69	3.58	3.50	3.44	3.39	3.35	3.28	3.22	3.15	3.12	3.08	3.04	3.01	2.97	2.93
9	5.12	4.26	3.86	3.63	3.48	3.37	3.29	3.23	3.18	3.14	3.07	3.01	2.94	2.90	2.86	2.83	2.79	2.75	2.71
10	4.96	4.10	3.71	3.48	3.33	3.22	3.14	3.07	3.02	2.98	2.91	2.85	2.77	2.74	2.70	2.66	2.62	2.58	2.54
11	4.84	3.98	3.59	3.36	3.20	3.09	3.01	2.95	2.90	2.85	2.79	2.72	2.65	2.61	2.57	2.53	2.49	2.45	2.40
12	4.75	3.89	3.49	3.26	3.11	3.00	2.91	2.85	2.80	2.75	2.69	2.62	2.54	2.51	2.47	2.43	2.38	2.34	2.30
13	4.67	3.81	3.41	3.18	3.03	2.92	2.83	2.77	2.71	2.67	2.60	2.53	2.46	2.42	2.38	2.34	2.30	2.25	2.21
14	4.60	3.74	3.34	3.11	2.96	2.85	2.76	2.70	2.65	2.60	2.53	2.46	2.39	2.35	2.31	2.27	2.22	2.18	2.13
15	4.54	3.68	3.29	3.06	2.90	2.79	2.71	2.64	2.59	2.54	2.48	2.40	2.33	2.29	2.25	2.20	2.16	2.11	2.07
16	4.49	3.63	3.24	3.01	2.85	2.74	2.66	2.59	2.54	2.49	2.42	2.35	2.28	2.24	2.19	2.15	2.11	2.06	2.01
17	4.45	3.59	3.20	2.96	2.81	2.70	2.61	2.55	2.49	2.45	2.38	2.31	2.23	2.19	2.15	2.10	2.06	2.01	1.96
18	4.41	3.55	3.16	2.93	2.77	2.66	2.58	2.51	2.46	2.41	2.34	2.27	2.19	2.15	2.11	2.06	2.02	1.97	1.92
19	4.38	3.52	3.13	2.90	2.74	2.63	2.54	2.48	2.42	2.38	2.31	2.23	2.16	2.11	2.07	2.03	1.98	1.93	1.88

Degrees of freedom of the denominator

TABLE A-7 (continued)

Upper 5 percent points

Degrees of freedom of the numerator

v_2 \ v_1	1	2	3	4	5	6	7	8	9	10	12	15	20	24	30	40	60	120	∞
20	4.35	3.49	3.10	2.87	2.71	2.60	2.51	2.45	2.39	2.35	2.28	2.20	2.12	2.08	2.04	1.99	1.95	1.90	1.84
21	4.32	3.47	3.07	2.84	2.68	2.57	2.49	2.42	2.37	2.32	2.25	2.18	2.10	2.05	2.01	1.96	1.92	1.87	1.81
22	4.30	3.44	3.05	2.82	2.66	2.55	2.46	2.40	2.34	2.30	2.23	2.15	2.07	2.03	1.98	1.94	1.89	1.84	1.78
23	4.28	3.42	3.03	2.80	2.64	2.53	2.44	2.37	2.32	2.27	2.20	2.13	2.05	2.01	1.96	1.91	1.86	1.81	1.76
24	4.26	3.40	3.01	2.78	2.62	2.51	2.42	2.36	2.30	2.25	2.18	2.11	2.03	1.98	1.94	1.89	1.84	1.79	1.73
25	4.24	3.99	2.99	2.76	2.60	2.49	2.40	2.34	2.28	2.24	2.16	2.09	2.01	1.96	1.92	1.87	1.82	1.77	1.71
26	4.23	3.37	2.98	2.74	2.59	2.47	2.39	2.32	2.27	2.22	2.15	2.07	1.99	1.95	1.90	1.85	1.80	1.75	1.69
27	4.21	3.35	2.96	2.73	2.57	2.46	2.37	2.31	2.25	2.20	2.13	2.06	1.97	1.93	1.88	1.84	1.79	1.73	1.67
28	4.20	3.34	2.95	2.71	2.56	2.45	2.36	2.29	2.24	2.19	2.12	2.04	1.96	1.91	1.87	1.82	1.77	1.71	1.65
29	4.18	3.33	2.93	2.70	2.55	2.43	2.35	2.28	2.22	2.18	2.10	2.03	1.94	1.90	1.85	1.81	1.75	1.70	1.64
30	4.17	3.32	2.92	2.69	2.53	2.42	2.33	2.27	2.21	2.16	2.09	2.01	1.93	1.89	1.84	1.79	1.74	1.68	1.62
40	4.08	3.23	2.84	2.61	2.45	2.34	2.25	2.18	2.12	2.08	2.00	1.92	1.84	1.79	1.74	1.69	1.64	1.58	1.51
60	4.00	3.15	2.76	2.53	2.37	2.25	2.17	2.10	2.04	1.99	1.92	1.84	1.75	1.70	1.65	1.59	1.53	1.47	1.39
120	3.92	3.07	2.68	2.45	2.29	2.17	2.09	2.02	1.96	1.91	1.83	1.75	1.66	1.61	1.55	1.50	1.43	1.35	1.25
∞	3.84	3.00	2.60	2.37	2.21	2.10	2.01	1.94	1.88	1.83	1.75	1.67	1.57	1.52	1.46	1.39	1.32	1.22	1.00

TABLE A-7 (continued)

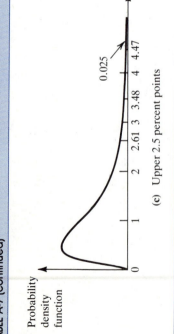

Probability density function

(e) Upper 2.5 percent points

Upper 2.5 percent points

v_2 \ v_1	1	2	3	4	5	6	7	8	9	10	12	15	20	24	30	40	60	120	∞
1	647.8	799.5	864.2	899.6	921.8	937.1	948.2	956.7	963.3	968.6	976.7	984.9	993.1	997.2	1001.	1006.	1010.	1014.	1018.
2	38.51	39.00	39.17	39.25	39.30	39.33	39.36	39.37	39.39	39.40	39.41	39.43	39.45	39.46	39.46	39.47	39.48	39.49	39.50
3	17.44	16.04	15.44	15.10	14.88	14.73	14.62	14.54	14.47	14.42	14.34	14.25	14.17	14.12	14.08	13.99	13.99	13.95	13.90
4	12.22	10.65	9.98	9.60	9.36	9.20	9.07	8.98	8.90	8.84	8.75	8.66	8.56	8.51	8.46	8.41	8.36	8.31	8.26
5	10.01	8.43	7.76	7.39	7.15	6.98	6.85	6.76	6.68	6.62	6.52	6.43	6.33	6.28	6.23	6.18	6.12	6.07	6.02
6	8.81	7.26	6.60	6.23	5.99	5.82	5.70	5.60	5.52	5.46	5.37	5.27	5.17	5.12	5.07	5.01	4.96	4.90	4.85
7	8.07	6.54	5.89	5.52	5.29	5.12	4.99	4.90	4.82	4.76	4.67	4.57	4.47	4.42	4.36	4.31	4.25	4.20	4.14
8	7.57	6.06	5.42	5.05	4.82	4.65	4.53	4.43	4.36	4.30	4.20	4.10	4.00	3.95	3.89	3.84	3.78	3.73	3.67
9	7.21	5.71	5.08	4.72	4.48	4.32	4.20	4.10	4.03	3.96	3.87	3.77	3.67	3.61	3.56	3.51	3.45	3.39	3.33
10	6.94	5.46	4.83	4.47	4.24	4.07	3.95	3.85	3.78	3.72	3.62	3.52	3.42	3.37	3.31	3.26	3.20	3.14	3.08
11	6.72	5.26	4.63	4.28	4.04	3.88	3.76	3.66	3.59	3.53	3.43	3.33	3.23	3.17	3.12	3.06	3.00	2.94	2.88
12	6.55	5.10	4.47	4.12	3.89	3.73	3.61	3.51	3.44	3.37	3.28	3.18	3.07	3.02	2.96	2.91	2.85	2.79	2.72
13	6.41	4.97	4.35	4.00	3.77	3.60	3.48	3.39	3.31	3.25	3.15	3.05	2.95	2.89	2.84	2.78	2.72	2.66	2.60
14	6.30	4.86	4.24	3.89	3.66	3.50	3.38	3.29	3.21	3.15	3.05	2.95	2.84	2.79	2.73	2.67	2.61	2.55	2.49
15	6.20	4.77	4.15	3.80	3.58	3.41	3.29	3.20	3.12	3.06	2.96	2.86	2.76	2.70	2.64	2.59	2.52	2.46	2.40
16	6.12	4.69	4.08	3.73	3.50	3.34	3.22	3.12	3.05	2.99	2.89	2.79	2.68	2.63	2.57	2.51	2.45	2.38	2.32
17	6.04	4.62	4.01	3.66	3.44	3.28	3.16	3.06	2.98	2.92	2.82	2.72	2.62	2.56	2.50	2.44	2.38	2.32	2.25
18	5.98	4.56	3.95	3.61	3.38	3.22	3.10	3.01	2.93	2.87	2.77	2.67	2.56	2.50	2.44	2.38	2.32	2.26	2.19
19	5.92	4.51	3.90	3.56	3.33	3.17	3.05	2.96	2.88	2.82	2.72	2.62	2.51	2.45	2.39	2.33	2.27	2.20	2.13

Degrees of freedom of the numerator

Degrees of freedom of the denominator

TABLE A-7 (continued)

Upper 2.5 percent points

v_2	Degrees of freedom of the numerator																		
	1	2	3	4	5	6	7	8	9	10	12	15	20	24	30	40	60	120	∞
20	5.87	4.46	3.86	3.51	3.29	3.13	3.01	2.91	2.84	2.77	2.68	2.57	2.46	2.41	2.35	2.29	2.22	2.16	2.09
21	5.83	4.42	3.82	3.48	3.25	3.09	2.97	2.87	2.80	2.73	2.64	2.53	2.42	2.37	2.31	2.25	2.18	2.11	2.04
22	5.79	4.38	3.78	3.44	3.22	3.05	2.93	2.84	2.76	2.70	2.60	2.50	2.39	2.33	2.27	2.21	2.14	2.08	2.00
23	5.75	4.35	3.75	3.41	3.18	3.02	2.90	2.81	2.73	2.67	2.57	2.47	2.36	2.30	2.24	2.18	2.11	2.04	1.97
24	5.72	4.32	3.72	3.38	3.15	2.99	2.87	2.78	2.70	2.64	2.54	2.44	2.33	2.27	2.21	2.15	2.08	2.01	1.94
25	5.69	4.29	3.69	3.35	3.13	2.97	2.85	2.75	2.68	2.61	2.51	2.41	2.30	2.24	2.18	2.12	2.05	1.98	1.91
26	5.66	4.27	3.67	3.33	3.10	2.94	2.82	2.73	2.65	2.59	2.49	2.39	2.28	2.22	2.16	2.09	2.03	1.95	1.88
27	5.63	4.24	3.65	3.31	3.08	2.92	2.80	2.71	2.63	2.57	2.47	2.36	2.25	2.19	2.13	2.07	2.00	1.93	1.85
28	5.61	4.22	3.63	3.29	3.06	2.90	2.78	2.69	2.61	2.55	2.45	2.34	2.23	2.17	2.11	2.05	1.98	1.91	1.83
29	5.59	4.20	3.61	3.27	3.04	2.88	2.76	2.67	2.59	2.53	2.43	2.32	2.21	2.15	2.09	2.03	1.96	1.89	1.81
30	5.57	4.18	3.59	3.25	3.03	2.87	2.75	2.65	2.57	2.51	2.41	2.31	2.20	2.14	2.07	2.01	1.94	1.87	1.79
40	5.42	4.05	3.46	3.13	2.90	2.74	2.62	2.53	2.45	2.39	2.29	2.18	2.07	2.01	1.94	1.88	1.80	1.72	1.64
60	5.29	3.93	3.34	3.01	2.79	2.63	2.51	2.41	2.33	2.27	2.17	2.06	1.94	1.88	1.82	1.74	1.67	1.58	1.48
120	5.15	3.80	3.23	2.89	2.67	2.52	2.39	2.30	2.22	2.16	2.05	1.94	1.82	1.76	1.69	1.61	1.53	1.43	1.31
∞	5.02	3.69	3.12	2.79	2.57	2.41	2.29	2.19	2.11	2.05	1.94	1.83	1.71	1.64	1.57	1.48	1.39	1.27	1.00

TABLE A-7 (continued)

Probability density function

0.01

0   1   2   2.61   3   3.48   4   4.47   5.99 f

(d) Upper 1 percent points

Upper 1 percent points

v_2 \\ v_1	1	2	3	4	5	6	7	8	9	10	12	15	20	24	30	40	60	120	∞
1	4052	4999.5	5403	5625	5764	5859	5928	5981	6022	6056	6106	6157	6209	6235	6261	6287	6313	6339	6366
2	98.50	99.00	99.17	99.25	99.30	99.33	99.36	99.37	99.39	99.40	99.42	99.43	99.45	99.46	99.47	99.47	99.48	99.49	99.50
3	34.12	30.82	29.46	28.71	28.24	27.91	27.67	27.49	27.35	27.23	27.05	26.87	26.69	26.60	26.50	26.41	26.32	26.22	26.13
4	21.20	18.00	16.69	15.98	15.52	15.21	14.98	14.80	14.66	14.55	14.37	14.20	14.02	13.93	13.84	13.75	13.65	13.56	13.46
5	16.26	13.27	12.06	11.39	10.97	10.67	10.46	10.29	10.16	10.05	9.89	9.72	9.55	9.47	9.38	9.29	9.20	9.11	9.02
6	13.75	10.92	9.78	9.15	8.75	8.47	8.26	8.10	7.98	7.87	7.72	7.56	7.40	7.31	7.23	7.14	7.06	6.97	6.88
7	12.25	9.55	8.45	7.85	7.46	7.19	6.99	6.84	6.72	6.62	6.47	6.31	6.16	6.07	5.99	5.91	5.82	5.74	5.65
8	11.26	8.65	7.59	7.01	6.63	6.37	6.18	6.03	5.91	5.81	5.67	5.52	5.36	5.28	5.20	5.12	5.03	4.95	4.86
9	10.56	8.02	6.99	6.42	6.06	5.80	5.61	5.47	5.35	5.26	5.11	4.96	4.81	4.73	4.65	4.57	4.48	4.40	4.31
10	10.04	7.56	6.55	5.99	5.64	5.39	5.20	5.06	4.94	4.85	4.71	4.56	4.41	4.33	4.25	4.17	4.08	4.00	3.91
11	9.65	7.21	6.22	5.67	5.32	5.07	4.89	4.74	4.63	4.54	4.40	4.25	4.10	4.02	3.94	3.86	3.78	3.69	3.60
12	9.33	6.93	5.95	5.41	5.06	4.82	4.64	4.50	4.39	4.30	4.16	4.01	3.86	3.78	3.70	3.62	3.54	3.45	3.36
13	9.07	6.70	5.74	5.21	4.86	4.62	4.41	4.30	4.19	4.10	3.96	3.82	3.66	3.59	3.51	3.43	3.34	3.25	3.17
14	8.86	6.51	5.56	5.04	4.69	4.46	4.28	4.14	4.03	3.94	3.80	3.66	3.51	3.43	3.35	3.27	3.18	3.09	3.00
15	8.68	6.36	5.42	4.89	4.56	4.32	4.14	4.00	3.89	3.80	3.67	3.52	3.37	3.29	3.21	3.13	3.05	2.96	2.87
16	8.53	6.23	5.29	4.77	4.44	4.20	4.03	3.89	3.78	3.69	3.55	3.41	3.26	3.18	3.10	3.02	2.93	2.84	2.75
17	8.40	6.11	5.18	4.67	4.34	4.10	3.93	3.79	3.68	3.59	3.46	3.31	3.16	3.08	3.00	2.92	2.83	2.75	2.65
18	8.29	6.01	5.09	4.58	4.25	4.01	3.84	3.71	3.60	3.51	3.37	3.23	3.08	3.00	2.92	2.84	2.75	2.66	2.57
19	8.18	5.93	5.01	4.50	4.17	3.94	3.77	3.63	3.52	3.43	3.30	3.15	3.00	2.92	2.84	2.76	2.67	2.58	2.49

Degrees of freedom of the numerator

Degrees of freedom of the numerator

TABLE A-7 (continued)

Upper 1 percent points

v_2	\multicolumn{19}{c}{Degrees of freedom of the numerator}

v_2 \ v_1	1	2	3	4	5	6	7	8	9	10	12	15	20	24	30	40	60	120	∞
20	8.10	5.85	4.94	4.43	4.10	3.87	3.70	3.56	3.46	3.37	3.23	3.09	2.94	2.86	2.78	2.69	2.61	2.52	2.42
21	8.02	5.78	4.87	4.37	4.04	3.81	3.64	3.51	3.40	3.31	3.17	3.03	2.88	2.80	2.72	2.64	2.55	2.46	2.36
22	7.95	5.72	4.82	4.31	3.99	3.76	3.59	3.45	3.35	3.26	3.12	2.98	2.83	2.75	2.67	2.58	2.50	2.40	2.31
23	7.88	5.66	4.76	4.26	3.94	3.71	3.54	3.41	3.30	3.21	3.07	2.93	2.78	2.70	2.62	2.54	2.45	2.35	2.26
24	7.82	5.61	4.72	4.22	3.90	3.67	3.50	3.36	3.26	3.17	3.03	2.89	2.74	2.66	2.58	2.49	2.40	2.31	2.21
25	7.77	5.57	4.68	4.18	3.85	3.63	3.46	3.32	3.22	3.13	2.99	2.85	2.70	2.62	2.54	2.45	2.36	2.27	2.17
26	7.72	5.53	4.64	4.14	3.82	3.59	3.42	3.29	3.18	3.09	2.96	2.81	2.66	2.58	2.50	2.42	2.33	2.23	2.13
27	7.68	5.49	4.60	4.11	3.78	3.56	3.39	3.26	3.15	3.06	2.93	2.78	2.63	2.55	2.47	2.38	2.29	2.20	2.10
28	7.64	5.45	4.57	4.07	3.75	3.53	3.36	3.23	3.12	3.03	2.90	2.75	2.60	2.52	2.44	2.35	2.26	2.17	2.06
29	7.60	5.42	4.54	4.04	3.73	3.50	3.33	3.20	3.09	3.00	2.87	2.73	2.57	2.49	2.41	2.33	2.23	2.14	2.03
30	7.56	5.39	4.51	4.02	3.70	3.47	3.30	3.17	3.07	2.98	2.84	2.70	2.55	2.47	2.39	2.30	2.21	2.11	2.01
40	7.31	5.18	4.31	3.83	3.51	3.29	3.12	2.99	2.89	2.80	2.66	2.52	2.37	2.29	2.20	2.11	2.02	1.92	1.80
60	7.08	4.98	4.13	3.65	3.34	3.12	2.95	2.82	2.72	2.63	2.50	2.35	2.20	2.12	2.03	1.94	1.84	1.73	1.60
120	6.85	4.79	3.95	3.48	3.17	2.96	2.79	2.66	2.56	2.47	2.34	2.19	2.03	1.95	1.86	1.76	1.66	1.53	1.38
∞	6.63	4.61	3.78	3.32	3.02	2.80	2.64	2.51	2.41	2.32	2.18	2.04	1.88	1.79	1.70	1.59	1.47	1.32	1.00

This table is reproduced from Table 18 of the *Biometrika Tables for Statisticians*, Vol. 1, with the kind permission of the Biometrika Trustees.

TABLE A-8

Random digits.

12159	66144	05091	13446	45653	13684	66024	91410	51351	22772
30156	90519	95785	47544	66735	35754	11088	67310	19720	08379
59069	01722	53338	41942	65118	71236	01932	70343	25812	62275
54107	58081	82470	59407	13475	95872	16268	78436	39251	64247
99681	81295	06315	28212	45029	57701	96327	85436	33614	29070
27252	37875	53679	01889	35714	63534	63791	76342	47717	73684
93259	74585	11863	78985	03881	46567	93696	93521	54970	37607
84068	43759	75814	32261	12728	09636	22336	75629	01017	45503
68582	97054	28251	63787	57285	18854	35006	16343	51867	67979
60646	11298	19680	10087	66391	70853	24423	73007	74958	29020
97437	52922	80739	59178	50628	61017	51652	40915	94696	67843
58009	20681	98823	50979	01237	70152	13711	73916	87902	84759
77211	70110	93803	60135	22881	13423	30999	07104	27400	25414
54256	84591	65302	99257	92970	28924	36632	54044	91798	78018
36493	69330	94069	39544	14050	03476	25804	49350	92525	87941
87569	22661	55970	52623	35419	76660	42394	63210	62626	00581
22896	62237	39635	63725	10463	87944	92075	90914	30599	35671
02697	33230	64527	97210	41359	79399	13941	88378	68503	33609
20080	15652	37216	00679	02088	34138	13953	68939	05630	27653
20550	95151	60557	57449	77115	87372	02574	07851	22428	39189
72771	11672	67492	42904	64647	94354	45994	42538	54885	15983
38472	43379	76295	69406	96510	16529	83500	28590	49787	29822
24511	56510	72654	13277	45031	42235	96502	25567	23653	36707
01054	06674	58283	82831	97048	42983	06471	12350	49990	04809
94437	94907	95274	26487	60496	78222	43032	04276	70800	17378
97842	69095	25982	03484	25173	05982	14624	31653	17170	92785
53047	13486	69712	33567	82313	87631	03197	02438	12374	40329
40770	47013	63306	48154	80970	87976	04939	21233	20572	31013
52733	66251	69661	58387	72096	21355	51659	19003	75556	33095
41749	46502	18378	83141	63920	85516	75743	66317	45428	45940
10271	85184	46468	38860	24039	80949	51211	35411	40470	16070
98791	48848	68129	51024	53044	55039	71290	26484	70682	56255
30196	09295	47685	56768	29285	06272	98789	47188	35063	24158
99373	64343	92433	06388	65713	35386	43370	19254	55014	98621
27768	27552	42156	23239	46823	91077	06306	17756	84459	92513
67791	35910	56921	51976	78475	15336	92544	82601	17996	72268
64018	44004	08136	56129	77024	82650	18163	29158	33935	94262
79715	33859	10835	94936	02857	87486	70613	41909	80667	52176
20190	40737	82688	07099	65255	52767	65930	45861	32575	93731
82421	01208	49762	66360	00231	87540	88302	62686	38456	25872

TABLE A-8 (continued)

00083	81269	35320	72064	10472	92080	80447	15259	62654	70882
56558	09762	20813	48719	35530	96437	96343	21212	32567	34305
41183	20460	08608	75273	43401	25888	73405	35639	92114	48006
39977	10603	35052	53751	64219	36235	84687	42091	42587	16996
29310	84031	03052	51356	44747	19678	14619	03600	08066	93899
47360	03571	95657	85065	80919	14890	97623	57375	77855	15735
48481	98262	50414	41929	05977	78903	47602	52154	47901	84523
48097	56362	16342	75261	27751	28715	21871	37943	17850	90999
20648	30751	96515	51581	43877	94494	80164	02115	09738	51938
60704	10107	59220	64220	23944	34684	83696	82344	19020	84834
25906	55812	91669	89718	23250	04815	17631	71270	15212	79339
71700	66325	76311	17034	21766	07705	48877	42317	04105	42030
03016	85673	47567	12916	68175	33340	96422	80878	24803	47406
51869	68055	17763	50758	06197	20462	50212	61152	09784	36499
32800	33619	44111	31961	55915	32361	73344	41828	77001	73221
19033	15965	24010	82435	61754	71963	47249	65025	72642	51584
68532	20248	51858	92816	90176	78898	15669	78373	87683	53987
32624	64167	43811	47066	86723	91338	21399	57289	64512	76225
06896	13417	33832	99204	01869	22261	02591	65699	28016	36314
15643	72081	43395	27506	45992	63462	69685	59743	10789	83838
02655	14695	35834	28596	82124	43435	86245	03798	83691	57642
28169	30715	75768	32547	28968	76950	63945	47760	46249	40666
44827	78135	42579	65416	44144	07656	03949	26107	62280	40485
35245	53285	43338	22138	34026	00753	81526	47610	59944	02252
97720	31595	84314	78326	15899	54686	23842	38401	79888	64313
96119	74549	17179	37592	80940	53257	49241	92202	44984	98852
88216	11170	60618	35715	76336	12700	91283	71581	47406	13439
29428	72252	96633	65837	28356	10246	52263	38764	78422	90000
57020	50087	62210	76160	99240	14258	99000	77387	61801	93913
46518	21876	20660	60898	35424	96365	35586	84191	23140	98133
96912	98556	81929	73875	99719	04683	52514	50332	94262	68185
18365	85411	16407	85145	40903	22339	59446	25583	59193	88923
79863	03951	18832	89268	66313	68361	28818	07181	83525	42084
95044	01570	85590	69782	95322	42212	37103	47333	55931	76976
14954	64265	08799	65488	79740	70215	37568	39848	56313	86346
49227	10283	49155	20201	52220	60997	78868	90255	59693	42359
86673	39383	86702	45646	00020	31072	25273	43667	49225	35578
30442	30784	12122	64652	87058	00246	66809	90213	85167	47616
00585	86972	72633	57337	76142	75035	93392	83142	56102	07470
58299	63144	61881	92204	68931	39479	06376	69292	74701	94930

TABLE A-8 (continued)

59842	85732	60448	74306	56895	97138	09514	14733	32763	41907
63578	50517	45400	05435	19713	23480	77878	47612	30154	05205
04638	41365	37885	36303	86611	03942	52933	48006	78687	39614
92139	74190	82902	28458	75624	24817	60091	18320	00991	95534
50496	12184	93603	23585	21626	30149	29708	72907	26905	03942
79485	31771	86195	01580	95535	29550	85860	55704	93395	68575
24104	67713	33278	45818	04408	50762	96800	59926	87225	94079
88542	84112	39205	49448	93668	11231	32445	69237	78294	65571
39954	27322	28657	46837	80829	15769	51005	91973	59555	10699
51764	20132	26744	59451	24426	93136	97392	76192	03512	76835
29085	80646	94619	09620	07512	66945	87580	48101	05973	76955
63166	14260	28167	61932	19123	43138	18608	73115	40175	05768
01103	68710	40632	59878	91195	72530	91708	50102	86958	60608
18031	90380	49055	55267	31208	81041	83200	25310	68162	91569
82124	16104	14324	10556	65863	02737	43070	00293	78355	92542
83443	15963	24036	15700	70067	00306	02331	53632	10004	72661
13126	67011	23482	61148	29214	11958	27569	44972	38537	69399
31844	91106	05210	30675	95843	20081	65857	72739	63990	10627
28138	68226	65633	33619	74139	57886	74823	94054	45760	40250
26292	36728	39389	84023	85120	06707	89701	12968	53511	49679
66677	57037	75632	09737	19172	18524	19920	79653	67386	48610
33568	61248	93534	00096	67637	21500	85447	63971	08600	55290
70551	71480	92572	27863	00367	77983	49226	69089	89523	57676
10832	50993	90483	66691	87680	10618	14705	32507	77765	74713
28856	35125	59697	29078	29165	50225	78650	33304	17317	33497
79872	66425	01553	55736	67129	66077	76136	36858	92126	34143
19480	53710	86437	71272	14237	17374	34525	70864	37316	15849
75368	39736	15713	84441	40463	97703	44277	24260	09284	98421
76223	93540	59992	91093	77951	04668	77613	10655	14759	94607
38993	82617	85490	56227	46984	89975	12727	34354	34498	58699
38475	68290	66451	15683	89570	24799	62775	61476	85338	94374
45712	72135	99597	22317	14929	83304	58805	45952	54166	06511
47613	55353	52248	01038	34471	41854	73294	94662	09926	44544
55512	45546	00352	89543	63076	62056	09210	36946	32114	50815
65697	03207	57833	27002	95131	05194	91931	33872	17684	77560
70741	03589	54335	50715	62424	20945	49530	88690	52815	34342
33004	17068	51494	53038	38255	18158	63751	44279	91954	78787
04122	77272	80245	51166	76277	13895	18823	05736	20492	22357
32115	01951	67968	51498	82075	96771	96527	52954	11649	49712
87189	31344	97705	26424	79585	15829	92914	64283	44940	01806
09546	76235	00582	65658	55783	53592	46602	57663	14667	71212
21765	29910	50763	77505	57949	85202	39491	46077	69583	84572
16934	59911	22790	93603	13805	42794	27915	91231	51858	48399
18776	22383	52940	98907	79405	73795	03552	04272	06388	51853
41488	90158	39267	61639	88854	58318	82398	94890	73580	87792

TABLE A-8 (continued)

84867	74055	63182	09192	47756	13791	73837	41443	83747	81256
46779	15579	97278	02677	36922	39826	00932	07150	51089	34142
76145	23408	14038	73888	29423	60738	55806	31998	12821	68379
05483	94743	88796	40173	24085	27398	40523	08570	12091	47730
69255	72428	33431	06186	08386	90139	36214	56712	59704	44895
24212	55567	11831	74160	55868	41657	21755	55426	28591	78206
71350	01169	62429	67192	89912	02402	32078	23227	99553	81219
29020	47291	75514	86384	70145	10491	98777	02533	77676	70458
97530	44651	49644	65718	58660	32793	57059	73484	84470	02112
75596	45727	02374	38086	85321	88521	18992	12940	30189	84026
32235	70425	47161	99015	78308	68746	18932	81745	13714	09601
88590	81787	29137	06616	23506	93413	93199	79503	82394	54764
23395	96060	57098	48890	06131	59647	10580	57042	20211	86994
95790	90688	64878	37752	90466	22768	01843	99499	35842	32184
10910	18277	52633	37227	95731	56639	71329	17690	56293	94392
75288	48328	88973	77899	26111	15014	30216	22236	72901	49626
73024	59219	80867	44331	80990	30296	47530	96640	79462	50845
26586	63772	11278	08598	96406	87996	90243	70530	98040	66436
29077	82702	01762	12078	60659	21436	50670	64601	66717	57684
40107	12987	36466	81268	29108	66157	27443	82875	04289	93856
17788	08023	14266	81079	96293	59991	34847	51054	26524	37035
86032	60360	84837	00912	26527	63684	60704	85236	33467	01641
29125	40001	99700	78836	79937	55608	26941	92279	29113	00396
63463	64969	37360	81225	15569	37200	93398	94244	27103	11242
30349	16191	49713	15246	15640	77334	90879	57999	48409	33489
88293	45401	12350	19040	81561	01155	85253	49479	66144	36486
74400	78899	06127	24365	88646	03944	87215	27085	16372	53548
87891	01263	68595	82315	46193	33306	66011	32972	92802	15708
03275	22982	83272	43570	29817	17323	45466	20498	08228	69682
95126	94417	09943	03316	64978	79651	97371	17634	62956	17714
40601	87085	51394	58140	80641	11547	57397	79825	62665	78796
78981	34943	92315	98737	85007	17558	77808	03537	46872	63504
54784	95754	48786	94402	62005	16589	08267	61878	69562	52089
24596	32468	84259	39563	77353	36180	34350	53331	18863	25424
45777	74383	58012	32923	44106	52121	30948	98812	39109	30748
94305	58672	25440	61911	16594	96747	73515	02587	72237	03004
74423	24195	01270	62733	08841	92999	82907	48049	05309	81854
87243	87076	04321	45628	31593	19783	24674	39222	37639	97573
95279	96674	37550	92395	17821	47986	56822	50499	37384	59050
10669	10427	09299	84045	39184	37102	94060	37288	30669	51740
13983	63696	08604	96115	34877	93673	16767	37258	32137	88081
25520	13548	78676	39472	41957	58478	45877	84216	79001	11734
81754	89975	07744	82864	63632	81103	66748	28539	81244	71407
99249	06587	10989	74629	17264	09220	16293	21958	99275	33075
25692	49122	53820	61407	66479	50095	43462	66380	52481	17731

TABLE A-8 (continued)

71062	25537	03787	00508	70346	72869	66697	66426	92430	03241
98835	64436	73086	73216	37983	82543	88681	78385	26327	23654
31631	10493	96276	59088	17480	13347	60948	92191	33074	18170
48222	82827	64181	51377	39636	28938	83204	89441	06125	58756
97700	56405	25858	42033	11729	91962	62762	16535	61465	59891
02050	26216	39216	73287	56887	26522	20127	64522	58368	00710
75973	65876	36777	52742	08272	03969	14125	68998	93575	95479
01373	65250	08317	65902	05034	12660	92903	11015	52893	11155
83866	83199	85406	39590	17954	62519	87775	56581	74495	67094
12597	05140	36748	15697	43206	03194	47216	45552	19465	09413
33387	64696	72102	44081	67644	91011	54737	95279	96971	81654
76084	35282	10085	53580	28924	87971	56631	72301	11510	75086
73769	66032	11119	88801	30169	73856	09346	25188	54381	24477
040307	32351	86338	86420	10259	55707	52711	02482	41349	78143
29933	72064	94327	94859	18890	84470	58420	55774	74052	88976
83787	23267	80357	20523	58215	19706	45552	81944	90820	48073
00437	77777	57832	01058	25654	59456	36924	17398	72197	19795
87816	27435	01573	42907	98043	21332	21732	42079	91177	01928
82756	67233	82534	40832	48525	53269	26225	89933	32494	84807
70426	94290	17064	68483	64842	47695	13623	55646	29305	51719
42711	61143	40516	12203	14367	95095	44703	64297	13381	40965
76510	88343	65246	28697	10606	83368	24310	90199	84181	33045
57261	17829	07486	29959	71893	99581	39680	13235	11465	41203
99713	22576	71336	09523	09491	18354	69516	29568	16788	95639
51808	19367	76265	97323	96197	72764	60674	51051	76627	51398
00180	75937	95135	68397	27720	23011	58415	18328	39360	64450
24387	70467	99776	54982	87625	30810	34591	20667	09254	04537
44516	44771	71969	11540	51710	40042	19607	72235	29191	98230
23436	83139	43405	03055	14175	05963	91920	06619	99717	46565
30311	13461	76003	22280	12694	95205	12666	86677	85569	76320

From *A Million Random Digits with 100,000 Normal Deviates*, 1955, with the kind permission of the Rand Corporation.

INDEX